NITROGEN FIXATION WITH NON-LEGUMES

Developments in Plant and Soil Sciences

VOLUME 35

Nitrogen Fixation
with Non-Legumes

The Fourth International Symposium on 'Nitrogen Fixation
with Non-Legumes', Rio de Janeiro, 23–28 August 1987

Edited by

F. A. SKINNER,
Harpenden, Hertfordshire, UK

R. M. BODDEY
EMBRAPA, Rio de Janeiro, Brazil

and

I. FENDRIK
University of Hannover, Hannover, FRG

Sponsored by
Empresa Brasileira de Pesquisa Agropecuaria (EMBRAPA)
The Government of the Federal Republic of Germany (BMFT)
Brazilian National Research Council (CNPq)
EMBRAPA-Inter-American Institute for Cooperation in Agriculture (IICA) agreement financed by the
Inter-American Development Bank (BID)
Third World Academy of Sciences
Latin-American Academy of Sciences
Brazilian Academy of Sciences

Kluwer Academic Publishers
DORDRECHT / BOSTON / LONDON

Library of Congress Cataloging in Publication Data

International Symposium on Nitrogen Fixation with Non-legumes (4th :
   1987 : Rio de Janeiro, Brazil)
    Nitrogen fixation with non-legumes : proceedings of the 4th
International Symposium on Nitrogen Fixation with Non-legumes, Rio
de Janeiro, Brazil, 23-28 August 1987 / edited by F.A. Skinner, R.M.
Boddey, I. Fendrik.
        p.   cm. -- (Developments in plant and soil sciences)
    "Partly first published in Plant and soil, vol. 110."

    1. Nitrogen--Fixation--Congresses.   I. Skinner, F. A. (Frederick
Arthur), 1919-   . II. Boddey, R. M. III. Fendrik, I. IV. Title.
V. Series.
QR89.7I59   1987
589.9'0133--dc19   88-7999

ISBN-13: 978-94-010-6888-8   e-ISBN-13: 978-94-009-0889-5
DOI: 10.1007/978-94-009-0889-5

Proceedings of the Fourth International Symposium on
Nitrogen Fixation with Non-legumes, Rio de Janeiro,
23–28 August 1987.

The Proceedings were edited by F.A. Skinner, R.M. Boddey
and I. Fendrik on behalf of EMBRAPA, Brazil and the
Institut für Biophysik, University of Hannover, FRG.

Published by Kluwer Academic Publishers,
P.O. Box 17, 3300 AA Dordrecht, The Netherlands.

Kluwer Academic Publishers incorporates
the publishing programmes of
Martinus Nijhoff, Dr W. Junk, D. Reidel, and MTP Press.

Sold and distributed in the U.S.A. and Canada
by Kluwer Academic Publishers,
101 Philip Drive, Norwell, MA 02061, U.S.A.

In all other countries, sold and distributed
by Kluwer Academic Publishers Group,
P.O. Box 322, 3300 AH Dordrecht, The Netherlands.

Contents

*Chapters indicated with an asterisk were first published in *Plant and Soil*, Volume 110 (1988).

Preface

The Fourth International Symposium on Nitrogen Fixation with Non-Legumes, held from August 23 to 29, 1987 in Rio de Janeiro, was organized as a joint effort of the National Program for Research in Soil Biology of EMBRAPA, Brazil and the Biophysics Department of the University of Hannover, FRG.

Since the first International Symposium on Nitrogen Fixation with Non-Legumes organized by Peter Vose in Piracicaba, Brazil, in 1979, considerable progress has been made in clarifying the mechanisms involved in plant-diazotroph interactions other than those associated with legumes. It is gratifying that the 1987 Symposium returned to Brazil now that several important new breakthroughs have been made. *Frankia* strains have been isolated in pure culture and reinoculated into their hosts and plant-microbe specificities demonstrated. Although *Parasponia* is still the only non-legume able to nodulate with rhizobia, similarities in the nodule structure with primitive tree legumes indicate an exciting new link between host-plant families, and this is beginning to clarify evolutionary steps leading to the highly sophisticated *Rhizobium*-legume symbiosis. Comparisons between the various systems were chosen as the highlights of the present Symposium and were so expertly introduced in Janet Sprent's opening lecture.

Much can be learned from the symbiotic systems and this should help us to understand the mechanisms involved in the less well-defined plant-diazotroph associations. Plant growth-promoting effects other than N_2-fixation have been demonstrated in several systems. Besides azospirilla several new root-infecting diazotrophs are now known, and contributions of N_2-fixation in economically important amounts have been firmly established for sugar cane and certain forage grasses.

During the Symposium a wealth of information was presented in 48 papers and 35 posters, and was actively discussed by the 120 participants from 22 different countries. The present volume brings together most of the orally presented papers.

Our special thanks go to all members of the Organizing Committee, both German and Brazilian, who were able to overcome difficulties due to distance and are fully responsible for the success of the meeting, and also to Dr. Fred Skinner who is taking from us the most difficult task of editing the proceedings.

The Symposium would not have been possible without generous financial support from EMBRAPA, the German Federal Ministry of Science and Technology, the Brazilian Research Council, (CNPq), the Inter-American Institutes for Cooperation in Agriculture, financed by the Inter-American Bank for Development (BID), the Third World Academy of Sciences, and the Latin-American Academy of Sciences.

Johanna Döbereiner and E.-G. Niemann
Chairmen of the Organizing Committee

Session 1

Mechanisms of infection of plants
by N_2-fixing microorganisms

F. A. Skinner et al. (Eds.), Nitrogen fixation with non-legumes, 3–11.

Mechanisms of infection of plants by nitrogen fixing organisms

J. I. SPRENT and S. M. de FARIA
Department of Biological Sciences, University of Dundee, Dundee, DD1 4HN, UK

Key words: cell wall, *Frankia*, infection, legume, non-legume, *Rhizobium*

Abstract

Heterotrophic nitrogen-fixing microorganisms can enter plants via wounds, root hairs or intact epidermises. All at some stage need the ability to digest primary cell walls and/or middle lamellas. None appears to digest secondary walls. The ability of any organism to infect a particular plant reflects (a) the enzymes produced by the microorganism (and possibly, as part of its reaction, the plant); (b) the exact nature of the primary wall; (c) the distribution of secondary walls. Plants may respond to infection by hypersensitive and other reactions which could be triggered by production of cell wall fragments. Infection threads of secondary wall material may be essential for root hair infection and where cell boundaries are crossed. Entry into host cells other than by infection threads involves a delicate balance between endophyte and host. This may only be achieved in one or a few cells, which may then divide repeatedly to produce a symbiotic structure.

Introduction

Three modes of entry of nitrogen fixing organisms into roots (occasionally stems) are known (Dart, 1977): (a) through wounds particularly where lateral or adventitious roots occur; (b) through root hairs; (c) between undamaged epidermal cells. Excluding photosynthetic species, which are not considered here, there are also three major groups of heterotrophic nitrogen fixing organism which invade roots: (a) rhizobia (used here in a general sense to include both *Rhizobium* and *Bradyrhizobium*); (b) *Frankia*; (c) other bacteria, including *Azospirillum* and certain enterobacteria. Table 1 indicates the known relationships between the three modes of entry and the three types of organism. Subsequent movement of endophytes may be intercellular, presumably following digestion of middle lamellas, via intercellular spaces, or by crossing cells (when bacteria are confined by host cell wall material, forming infection threads).

This paper discusses these modes of entry and movement between tissues and examines some factors of plant and endophyte origin which may limit the establishment of symbioses between particular organisms. An understanding of these limitations is an essential prerequisite for developing novel symbioses.

Pre-infection processes

Before entry, bacteria must colonize the root surface. This may involve at least four stages: (a) movement towards the root (chemotaxis, electrotaxis); (b) recognition; (c) adhesion; (d) provision, by the plant, of nutrients for the endophyte. If it cannot readily fix nitrogen *ex planta* (*i.e.* most rhizobia and *Frankia*) suitable sources of combined N and C must be available. The C:N ratio of plant exudates varies widely; for example it is generally lower in legumes than grasses. To support rhizobia

Table 1. Types of nitrogen fixing organism and possible modes of entry into plant tissues

Organism	Method of entry		
	Wound	Hair	Epidermis
Rhizobia	+	+	+
Frankia	?	+	+
Others, *e.g.* *Azospirillum*	?	?	?

Table 2. Some sources of variation within structure of young roots, for details see Esau (1965) and Cutter (1969) and for point 3, Justin and Armstrong (1987)

1. Histological origin of epidermal cells
2. Presence and structure of exodermis
3. Arrangement of cortical cells
4. Pattern of root hair production
5. Vascular anatomy

prior to infection may require a considerable investment of N and C on the part of the host (Sprent and Raven, 1985). In published studies on *Frankia* infection, surface colonization appears less extensive and thus, possibly, less expensive. Organisms such as *Azospirillum* which readily fix nitrogen in pure culture could have an advantage when exudates have a high C:N ratio, *e.g.* on cereal roots (see discussion in Hubbell and Gaskins, 1984).

In addition to using plant exudates, rhizosphere microorganisms may digest some host cell wall material. The ability of potential endophytes to do this is closely linked to infection, and will be discussed later.

Plant anatomical features relevant to infection

There are major differences in root structure between monocotyledons and dicotyledons, and also many variants within these types (Table 2). Each of the factors listed is relevant to bacterial entry into and passage through roots. Points 1–3 reflect major variations in cell wall structure and presence of intercellular spaces. Point 4 is clearly relevant to root hair infection: occurrence and structure of root hairs varies with species, some having none (see discussion in Sprent *et al.*, 1987). Vascular anatomy may be relevant to the production of nodule meristems, via hormone balance (*e.g.* Syono *et al.*, 1976).

Many of these features are modified by environment. For example, whether soil air is water-vapour saturated or not may affect wall structure and ion uptake properties in some species (Clarkson *et al.*, 1987). Production and size of root hairs is greatly affected by soil physical and chemical factors (Sprent and McInroy, 1984; Barber and Silverbush, 1984). Biotic factors may also be involved, for example, both *Azospirillum* (Yahalom *et al.*, 1987) and *Klebsiella* (Haahtela *et al.*, 1986)

may stimulate root hair production; pseudomonads may 'help' in the infection of *Alnus* by *Frankia* (Knowlton and Dawson, 1983).

Modes of infection

Wounds

If a major function of the surface of an organ such as a root is protection, then wounding by breaching that protection could make infection easier. *Agrobacterium*, which is closely related to *Rhizobium*, is the classic obligate wound-infecting parasite. Not only do wounds provide entry points, they expose recognition sites of middle lamellar origin and possibly also enable direct transfer of bacterial plasmid DNA to the host at plasmodesmata (Huang, 1986). The two best-studied cases of wound infection by nitrogen fixing organisms are the legumes, *Arachis hypogaea* (Chandler, 1978) and *Stylosanthes* spp. (Chandler *et al.*, 1982), although the phenomenon is probably common in some legume tribes (Sprent *et al.*, 1988). For these and other reasons, it has been suggested that legume nodules may have arisen from a wound tumour, with rhizobia evolving from an ancestor of *Agrobacterium* (Sprent and Raven, 1985; see also Trinick, 1982).

Arachis and *Stylosanthes* nodules never show structures resembling infection threads. In the former, rhizobia proceed between cells, presumably by digesting walls, and eventually penetrate host cells in an undefined matrix which may form from partially broken cell wall fragments. This rather amorphous mass of dividing rhizobia and surrounding wall material is sometimes called 'zooglea'. Bacteria are released from it into host-derived membrane-bound packets in cells which then divide repeatedly to form a nodule with uniformly infected central tissue. The matrix of the zoogleae may be electron-dense and with some similarities in appearance to that illustrated around intercellular *Frankia* infections (Miller and Baker, 1985), intercellular invasion by rhizobia of *Parasponia* (Lancelle and Torrey, 1984) and *Mimosa scabrella* (Faria *et al.*, 1988). We do not know whether this matrix is of host and/or endophyte origin. If it is a product of host cell wall breakdown, the necessary enzymes could be of host and/or endophyte origin (see Huang, 1986). Fragments of cell walls produced in this way may effect a hyper-sensitive or regulatory reaction on the part of the plant (see McNeil *et al.*, 1984). Since cell-wall structure varies between species and cell type, a range of host responses to endophytes might be expected, accounting possibly for a major observed difference in the passage of rhizobia through roots of *Arachis*

and *Stylosanthes*. In the latter, outer root cells penetrated by rhizobia die – possibly a hypersen-sitive reaction. It is only when the infection reaches deep into the root cortex that cells are not killed; the subsequent development of the nodule is similar to that of *Arachis*. In both genera, hairs are produced near the lateral root junctions: in *Arachis*, this is the only part of the root which normally

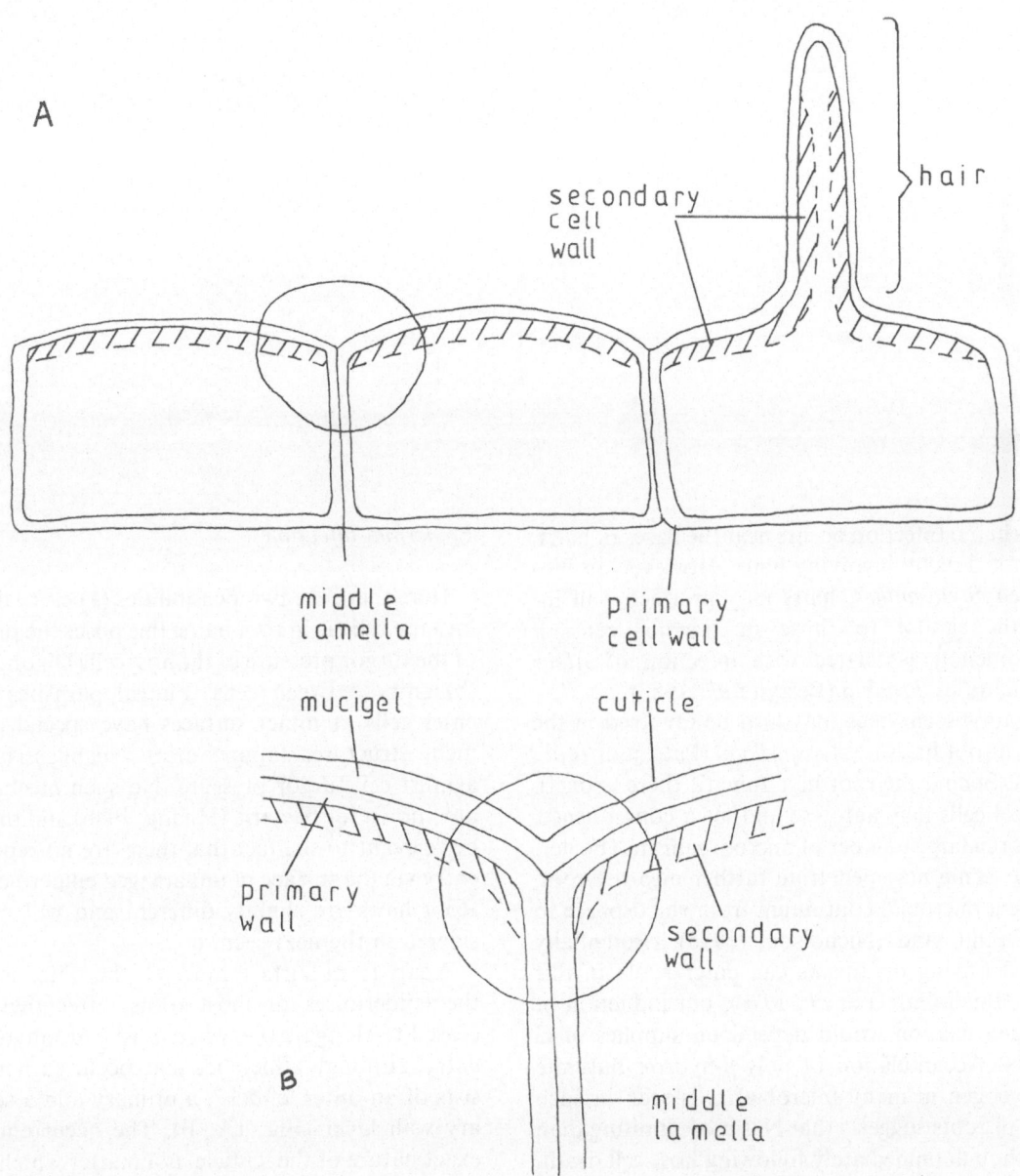

Fig. 1. **A**. Diagram to show features of a typical root epidermis. Note however that there is wide variation, both in the nature of cell walls and the production of hairs. **B**. Enlargement of the circled portion of A. **C**. (overleaf) Enlargement of the circled portion of B to illustrate pathway of bacterial invasion.

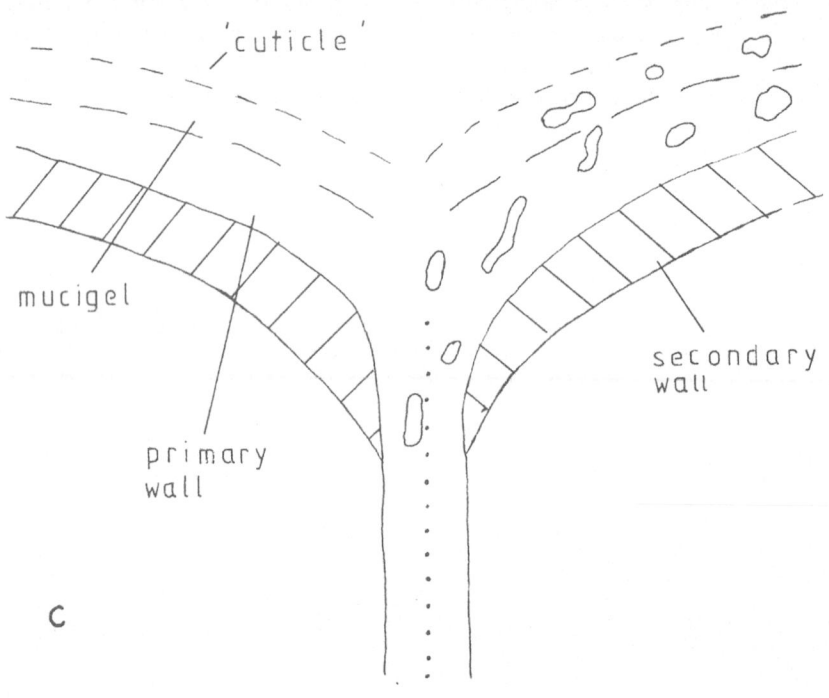

C

bears hairs. Infection occurs near the bases of hairs, but not directly involving hairs. However, in inoculated *Stylosanthes*, hairs may produce wall ingrowths similar to those in transfer cells, a phenomenon associated with infection of *Alnus* root hairs by *Frankia* (Berry *et al.*, 1986).

Pectolytic enzymes may also be involved in the lysis of root hairs by *Azospirillum* (Patriquin *et al.*, 1983). Behind the root hair zone (if there is one!), cortical cells may autolyse and, as a consequence, support a large number of microorganisms (Foster, 1986). Some may penetrate further into the root, giving a microbial continuum from rhizosphere to stele (Old and Nicholson, 1978). Potentially nitrogen-fixing organisms can enter roots in this way (Umali-Garcia *et al.*, 1978), but induction of nitrogen fixation would depend on supplies of C and N. Accumulation of poly-β-hydroxybutyrate or glycogen in many microbial cells seen in such parts of roots suggests that N is more limiting than C, although immediately following host cell death, cytoplasm rich in N may be available.

Epidermal infection

Here there are two possibilities (a) entry directly into the cell, as in root hairs: this poses the problem of the turgor pressure of the host cell (Dixon, 1969); (b) entry between cells. Fungal parasites which enter cells at intact surfaces have special attachment structures (appressoria) which assist entry against cell turgor pressure. No such mechanisms are known for bacteria (Huang, 1986) and this may be relevant to the fact that there are no reports of entry via the surface of undamaged epidermal cells. Root hairs are slightly different and will be considered in the next section.

Apart from surface mucigel, the outer walls of the epidermises of most roots, after they have ceased to elongate (*i.e.* when most endophytes gain entry, although adherence may occur earlier) consists of an outer 'cuticle', a primary and a secondary wall layer (Fig. 1A, B). The occurrence and exact nature of the 'cuticle' is a matter which is not universally agreed; although not always seen in axenic systems, it appears to be common in soils (Foster, 1986), and is easily abraded by soil particles. Once breached, the 'cuticle' admits entry of soil bacteria to the primary wall, which has a fairly loose construction and appears to be relatively easily broken down by rhizosphere bacteria (Foster, 1981). The latter, however, never penetrate the secondary wall. Comparison of micrographs of Foster (1981) with those of Bender *et al.* (1988) for rhizobia colonizing *Parasponia andersonii*; Miller and Baker (1986) for *Frankia* colonizing *Elaeagnus angustifolius* and Faria *et al.*, 1988) for rhizobia colonizing *Mimosa scabrella*, show inoculated endophytes in a similar position to other rhizosphere organisms. In no system is the secondary wall breached. In *Elaeagnus* and *Mimosa*, the endophyte penetrates the radial walls, presumably by digesting the middle lamella, then proceeds between cells and through intercellular spaces as in wound infections (Fig. 1C). In *Parasponia*, the colonizing bacteria stimulate cell division in the outer cortex. In due course these newly-formed cells rupture the epidermis and infection follows — making this effectively a wound infection. Occasionally, in *Mimosa*, bacteria appear to enter epidermal cells, but always from the radial or tangential wall where there is no secondary layer. Turgor of neighbouring cells could avoid pressure problems. Contents of invaded epidermal cells are indistinct, perhaps indicating incipient death. This could precede cell collapse in a fashion similar to that observed in *Stylosanthes*, where epidermal hairs can also be invaded from the inner wall (Chandler *et al.*, 1982). In *Parasponia rigida* (Lancelle and Torrey, 1984), large proliferations of intercellular rhizobia can lead to host cell damage and death. This may also occur in some woody legume nodules (Faria *et al.*, 1987).

Thus there are many common features in the infection processes of some legumes and non-legumes: individual plant-endophyte pairs may be expected to show one from a range of responses.

Formation of infection threads

In *Parasponia*, as nodules develop, intercellular passage of rhizobia is followed by entry into cells and formation of infection threads. These infection threads are discrete units (as in legumes) whose walls, made by the host plant, change in nature as cells differentiate into the nitrogen fixing form (Smith *et al.*, 1986). A similar change from intercellular passage of rhizobia to infection threads has been found in all Caesalpinioid legumes studied, together with four Papilionoid genera (Faria *et al.*, 1987). In all these cases the nitrogen fixing rhizobia are not released from the infection thread: we regard this as a primitive state. In advanced legumes, bacteria are released into membrane-bound structures. This membrane is of plant origin, but may incorporate elements from the rhizobial surface (Bradley *et al.*, 1986): it may serve to control transport of materials to and from the nitrogen fixing endophyte. However, this sophisticated system is clearly not an essential part of an active symbiosis.

In many actinorhizas, the endophyte hyphae are encapsulated by material believed to be, at least in part, of host cell origin (*e.g.* Berry *et al.*, 1986). This material is less clearly defined than the wall of the infection thread of *Parasponia* and some legumes. However, now that more actinorhizas have been studied it is clear that there are many grades of hyphal penetration (and concomitant encapsulation) of host cells (compare, for *e.g.* Newcomb and Pankhurst, 1982a, b).

The function of infection threads

The thread wall is generally agreed to be a plant product, produced as a response to invading bacteria. It fuses with cell walls as it crosses from cell to cell: for this reason and because of staining reactions, in legumes it tends to be thought of as having a rather similar structure to the primary wall of nodule cells. Recent work by Higashi *et al.* (1987), suggests that this is not the case. These workers used the cell-wall degrading enzyme mixture 'driselase' and found that all walls of the infected region of *Astragalus indica* were digested except those of the infection threads. Their beautiful micrographs show clearly that at the host cell boundary the infection threads are funnel shaped. This is rather like a fungal appressorium, attaching the infection thread firmly to the host cell wall and

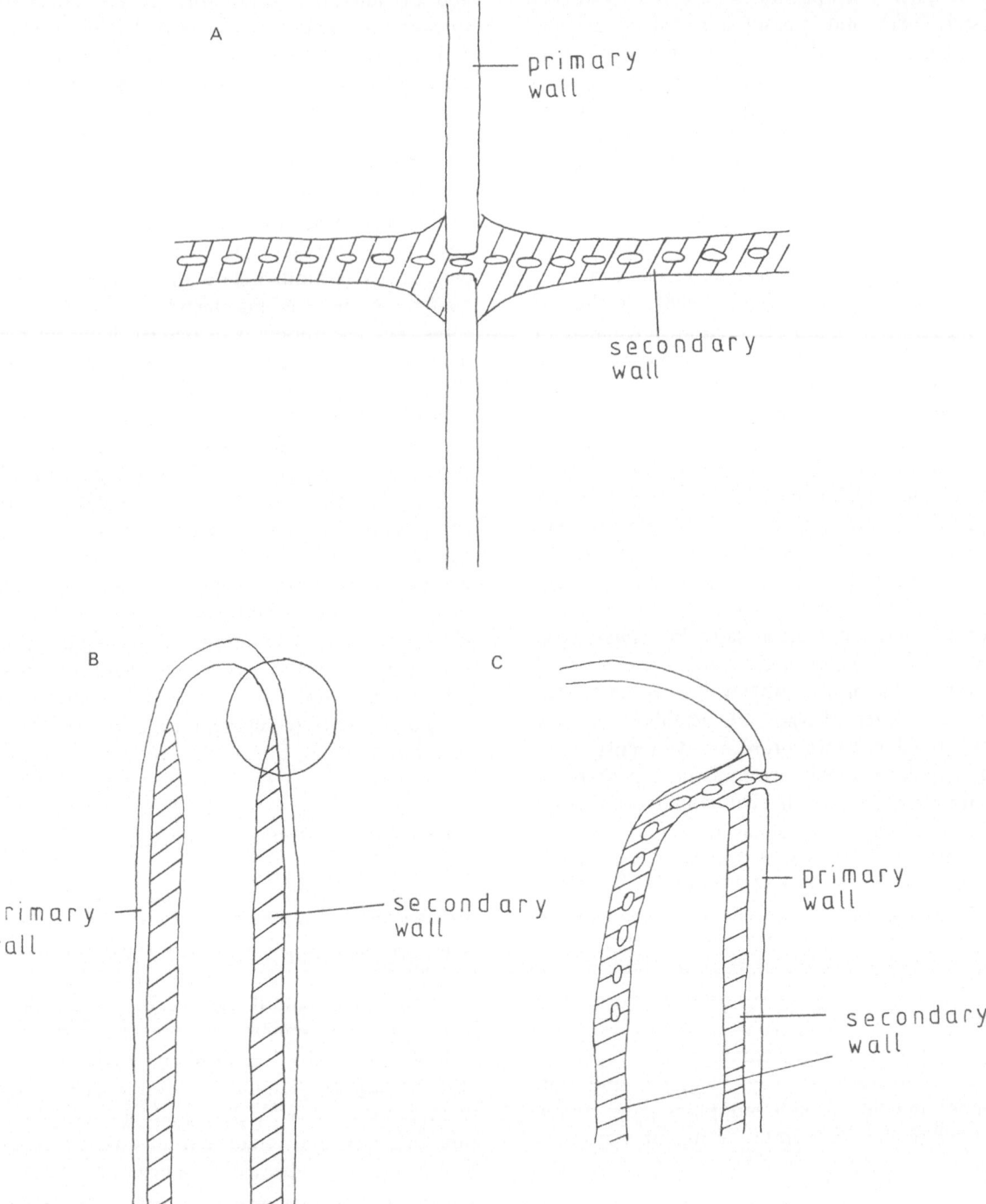

Fig. 2. **A**. Diagram to show how bacteria may cross cell boundaries by digesting the primary wall. The infection thread has secondary wall material. **B**. Tip of root hair to show possible relation between primary and secondary wall material at point of rhizobial infection. The hair is drawn uncurled for simplicity. **C**. Enlargement of circled portion of B to illustrate how infection threads may form following invasion of primary wall. Compare with A.

allowing rhizobia to digest a hole in the latter so that they can infect the next cell. A firm attachment of this type would prevent any effects of back pressure from the cell being invaded (Fig. 2A). It is presumably also necessary that the thread wall is structurally distinct from the host wall, otherwise it would also be digested by bacterial enzymes. The last argument could also be applied to infection threads in root hairs.

Root hair infections

Although these occur in both legumes and non-legumes, there appear to be some major differences between these groups as well as differences in detail among groups. In legumes, root hair infections are confined to growing or potentially growing parts of cell walls. These may be points where root hairs are about to emerge (*e.g.* soybean; Turgeon and Bauer, 1985) or near the tip of an elongating root hair (Callaham and Torrey, 1981), whose growth may be redirected by the bacteria (Batenberg *et al.*, 1986). Such zones may be characterized by having only primary cell wall material (Fig. 2B, C) and locally differentiated plasmamembranes (Volkmann, 1984). In infectible legume hairs the rhizobia appear to be able to invade only the primary wall. Infection thread walls are continuous with the secondary wall although whether this is formed only after breaching of the primary wall, as seems to be the case in clover (Callaham and Torrey, 1981) or is present but invaginates at this time (as Turgeon and Bauer, 1985 suggest) remains to be seen. Either way, rhizobia appear not to be able to digest the secondary wall and we may perhaps infer that the enzyme resistance of the infection thread network, observed by Higashi *et al.* (1987) is because of this. It is certainly consistent with proposals made above for intercellular infections.

However, not all uninfected root hairs have the same structure. In *Alnus rubra*, hairs do not have a secondary wall unless they are infected with *Frankia* (Berry *et al.*, 1986). Infection leads to extensive deposition of several layers of cell wall material, accompanied by formation of wall ingrowths in the region of the infection site. These ingrowths are continuous with the rather irregular hyphal-encapsulating layer.

General discussion

Most information on infection processes is from rhizobia or *Frankia*. Although numerous other bacteria associate with and may penetrate roots, there is no clear evidence as to the processes of infection. Most of the questions posed by Hubbell and Gaskins (1984) with respect to *Azospirillum* remain unresolved. Undoubtedly, it and other bacteria can have major effects on root morphology and functioning (Okon and Kapulnik, 1986), which alone could be important. The suggestion that *A. brasilense* can increase the number of epidermal cells which can form infectible root hairs in three legume species (Yahalom *et al.*, 1987), taken with other microbial interactions such as the 'helper' role of some pseudomonads in *Frankia* infection of *Alnus* (Knowlton and Dawson, 1983) show that in natural conditions infection processes can be modified.

Recent work, discussed here, suggests that many of the widely accepted dogmas for 'normal' symbioses, for example hair infection and, in the case of legumes, the necessity for bacteria to be released from infection threads before they differentiate into active nitrogen fixing forms are not universal. Particularly striking is the similarity of intercellular infection processes whether involving rhizobia in legumes or *Parasponia*, or *Frankia* and non-legumes.

In his review of nodule morphogenesis and differentiation, Newcomb (1981) pointed out that only the 'tip of the iceberg' had been exposed. It now appears that the iceberg may have many tips and that a comparative study of a range of systems, together with an intensive study of a few of them, may help us to understand and hence exploit the possibilities for increasing the range of nitrogen fixing systems.

Acknowledgements

We should like to thank J A Raven, J M Sutherland and numerous others for helpful discussions. Financial support from EMBRAPA (supporting S.M.de F.), the British Council-CNPq (for supporting the cooperative programme between Brazil (J Döbereiner and colleagues) and Britain (J I Sprent and colleagues) has been essential in enabling this work to be carried out. S M McInroy, H Bennett and G Hay have provided excellent technical assistance. G Bender kindly sent a preprint of his *Parasponia* paper.

References

Barber S A and Silberbush M 1984 Plant root morphology and nutrient uptake. *In* Roots, Nutrient and Water Influx, and Plant Growth. Eds. S A Barber and D R Bouldin. pp. 65–87. ASA Special Publication No. 49. Soil Science Society of America, Madison.

Batenburg F H D Van, Jonker R and Kijne J W 1986 *Rhizobium* induces marked root hair curling by redirection of tip growth: A computer simulation. Physiol. Plant. 66, 476–480.

Bender G L, Nayudu M, Goydych W and Rolfe B G 1988 Early infection events in the nodulation of the non-legume *Parasponia andersonii* by *Bradyrhizobium*. Plant Science *In press*

Berry A M, McIntyre L and McCully M E 1986 Fine structure of root hair infection leading to nodulation in the *Frankia-Alnus* symbiosis. Can. J. Bot. 64, 292–305.

Bradley D J, Butcher G W, Galfre G, Wood E A and Brewin N J 1986 Physical association between the peribacteroid membrane and lipopolysaccharide from the bacteroid outer membrane in *Rhizobium*-infected pea root nodule cells. J. Cell Sci. 85, 47–61.

Callaham D A and Torrey J G 1981 The structural basis for infection of root hairs in *Trifolium repens* by *Rhizobium*. Can. J. Bot. 59, 1647–1664.

Chandler M R 1978 Some observations on infection of *Arachis hypogaea* L. by *Rhizobium*. J. Expt. Bot. 29, 749–755.

Chandler M R, Date R A and Roughley R J 1982 Infection and root-nodule development in *Stylosanthes* species by *Rhizobium* J. Expt. Bot. 33, 47–57.

Clarkson D T, Robards A W, Stephens J E and Stark M 1987 Suberin lamellae in the hypodermis of maize (*Zea mays*) roots: Development and factors affecting the permeability of hypodermal layers. Plant, Cell Environ. 10, 83–93.

Cutter E G 1969 Plant Anatomy: Experiment and Interpretation. Part I. Cells and Tissues. Edward Arnold, London.

Dart P 1977 Infection and development of leguminous nodules. *In* A Treatise on Dinitrogen Fixation; Section III; Biology. Eds. R W F Hardy and W S Silver. pp. 367–472. Wiley Interscience, New York.

Dixon R O D 1969 Rhizobia (with particular reference to relationships with host plants). Ann. Rev. Microbiol. 23, 137–158.

Esau K 1965 Plant Anatomy, 2nd ed. Wiley, New York.

Faria S M, McInroy S G and Sprent J I 1987 The occurrence of infected cells, with persistent infection threads, in legume root nodules. Can. J. Bot. 65, 553–558.

Faria S M, Hay G T and Sprent J I 1988 Entry of rhizobia into roots of *Mimosa scabrella* Bentham occurs between epidermal cells. J. Gen. Microbiol. 134, *In press*.

Foster R C 1981 The ultrastructure and histochemistry of the rhizosphere.New Phytol. 89, 263–273.

Foster R C 1986 The ultrastructure of the rhizoplane and rhizosphere. Annu. Rev. Phytopath. 24, 211–234.

Haahtela K, Laakso T and Korhonen T K 1986 Associative nitrogen fixation by *Klebsiella* spp.: Adhesion sites and inoculation effects on grass roots. Appl. Environ. Microbiol. 52, 1974–1079.

Higashi S, Kushiyama K and Abe M 1987 Electron microscopic observations of infection threads in driselase treated nodules of *Astragalus sinicus*. Can. J. Microbiol. 32, 947–952.

Huang J-S 1986 Ultrastructure of bacterial penetration in plants. Annu. Rev. Phytopathol. 24, 141–157.

Hubbell D H and Gaskins M H 1984 Associative N_2 fixation with *Azospirillum*. *In* Biological Nitrogen Fixation — Ecology, Technology and Physiology. Ed. M Alexander. pp. 201–224. Plenum Press, New York.

Justin S H F W and Armstrong W 1987 The anatomical characteristics of roots and plant response to soil flooding. New Phytol. 106, 465–495.

Knowlton S and Dawson J O 1983 Effects of *Pseudomonas cepacia* and cultural factors on the nodulation of *Alnus rubra* roots by *Frankia*. Can. J. Bot. 61, 2877–2882.

Lancelle S A and Torrey J G 1984 Early development of *Rhizobium*-induced root nodules of *Parasponia rigida*. I. Infection and early nodule initiation. Protoplasma 123, 26–37.

McNeil M, Darvill A G, Fry S C and Albersheim P 1984 Structure and function of the primary cell walls of plants. Annu. Rev. Biochem. 53, 625–663.

Miller I M and Baker D D 1985 The initiation, development and structure of root nodules in *Elaeagnus angustifolia* L. (Eleagnaceae). Protoplasma 128, 107–119.

Miller I M and Baker D D 1986 Nodulation of actinorhizal plants by *Frankia* strains capable of both root hair infection and intercellular penetrations. Protoplasma 131, 82–91.

Newcomb W 1981 Nodule morphogenesis and differentiation. Int. Rev. Cytol. Supplement 13, 247–298.

Newcomb W and Pankhurst C E 1982a Fine structure of actinorhizal root nodules of *Coriaria arborea*. N.Z. J. Bot. 20, 93–103.

Newcomb W and Pankhurst C E 1982b Ultrastructure of actinorhizal root nodules of *Discaria toumatou* Raoul (Rhamnaceae). J. Bot. 20, 105–113.

Okon Y and Kapulnik Y 1986 Development and function of *Azospirillum*-inoculated roots. Plant and Soil 90, 3–16.

Old K M and Nicholson T H 1978 The root cortex as part of a microbial continuum. *In* Microbial Ecology. Eds. M W Loutit and J A R Miles. pp. 291–294. Springer-Verlag, Berlin and Heidelberg.

Patriquin D G, Döbereiner J and Jain D K 1983 Sites and processes of association between diazotrophs and grasses. Can. J. Microbiol. 29, 900–915.

Smith C A, Skvirsky R C and Hirsch A M 1986 Histochemical evidence for the presence of a suberin-like compound in *Rhizobium*-like nodules of the nonlegume *Parasponia rigida*. Can. J. Bot. 64, 1474–1483.

Sprent J I and McInroy S G 1984 Effects of salinity on growth and nodulation of *Arachis hypogaea*. *In* Advances in Nitrogen Fixation Research. Eds. C Veeger and W E Newton. p. 546. Martinus Nijhoff/Dr W Junk, Pudoc.

Sprent J I and Raven J A 1985 Evolution of nitrogen-fixing symbioses. Proc. Roy. Soc. Edinburgh 85B, 215–237.

Sprent J I, Sutherland J M and Faria, S. M. de 1987 Some aspects of the biology of nitrogen-fixing organisms. Phil. Trans. R. Soc. London B., 317, 119–121.

Sprent J I, Sutherland J M and Faria S M de 1988 Structure and function of nodules from woody legumes. *In* Biology of Legumes. Eds. C H Stirton and J L Zarucchi. Missouri Botanic Gardens *In press.*

Syono K, Newcomb W and Torrey J G 1976 Cytokinin production in relation to the development of pea root nodules. Can. J. Bot. 54, 2155–2162.

Trinick M J 1982 Biology. *In* Nitrogen Fixation, Volume 2: *Rhizobium.* Ed. W J Broughton. pp. 76–146. Clarendon Press, Oxford.

Turgeon B G and Bauer W D 1985 Ultrastructure of infection-thread development during the infection of soybean by *Rhizobium japonicum.* Planta 163, 328–349.

Umali-Garcia M, Hubbell D H and Gaskins M H 1978 Process of infection of *Panicum maximum* by *Spirillum lipoferum.* Ecol. Bull. (Stockholm) 26, 373–379.

Volkmann D 1984 The plasma membrane of growing root hairs is composed of zones of local differentiation. Planta 163, 392–403.

Yahalom E, Okon Y and Dovrat A 1987 *Azospirillum* effects on susceptibility to *Rhizobium* nodulation and on nitrogen fixation of several forage legumes. Can. J. Microbiol. 33, 510–514.

Session 2

Frankia and *Parasponia* symbiosis

F. A. Skinner et al. (Eds.), Nitrogen fixation with non-legumes, 15–24.
© 1989 by Kluwer Academic Publishers.

Effective nodulation of *Coriaria myrtifolia* L. induced by *Frankia* strains from *Alnus glutinosa* L. (Gaertn.)

C. RODRIGUEZ-BARRUECO, F. SEVILLANO and T. M. NEBREDA
Unidad de Biologia Vegetal, IRNA-CSIC, Apartado 257, E-37071 Salamanca, Spain

Key words: actinorhizal association, *Frankia*, nitrogen-fixation, specificity

Abstract

The present contribution covers the cross-inoculation between two actinorhizae belonging to different genera and families, mainly *Alnus glutinosa* and *Coriaria myrtifolia*. *Frankia* strains isolated from *A. glutinosa* received from the Netherlands ($LDAgp_1r_1$, $LDAgn_1$) and from Scotland (UGL010708), induced a fully effective nodulation on *C. myrtifolia*. The same effect was caused by a nodule extract from *A. glutinosa*. The reverse, a crushed-nodule inoculum from *C. myrtifolia* nodulated all the *A. glutinosa* seedlings, though nodules formed were less effective than those induced by the other inocula. Re-isolation of those *Frankia* strains from the nodules formed on *A. glutinosa* was readily obtained, whereas attempts to re-isolate them from the nodules formed on *C. myrtifolia* failed, suggesting that isolation procedures different to those employed should be tried.

Introduction

From the literature on cross-inoculation among actinorhizal plants we can try to elucidate the degree of compatibility between host plants and endophytes, in order to formulate possible specificity groups. From the reports available it appears difficult, as yet, to define those groups, mainly because information is still lacking on all the cross-inoculations possible among the increasing number of host species being studied. Although information on some crosses has been hindered by the lack of pure cultures of the respective nodule endophytes, progress on the extent to which specificity exists was made in the past through the use of a crushed-nodule suspension as inoculum. Some crosses were made repeatedly by several workers, the results being uniform both as regards infectivity and effectivity. Any contradictory results were interpreted in terms of the methods employed to prepare the inoculum, its final concentration, plant nutritional status or on a too drastic nodule surface sterilizaton. Also the intervention of environmental factors was suggested. The partial success in the symbiotic performance of some crosses was also attributed to an actual adaptation of the microbe to a host with a particular geographic distribution. At present, the availability of pure *Frankia* isolates permits more rigorous cross-inoculation tests.

The present experiment covers partially the cross-inoculation between two actinorhizal species belonging to different genera and families, *i.e. Alnus glutinosa* and *Coriaria myrtifolia*. The latter is one of the 13 *Coriaria* species known to bear actinorhizae (Cañizo and Rodríguez-Barrueco, 1978). The two species studied here are known to be dissimilar in growth habits and distribution. The different growth habitats of the two hosts, with a restricted occurrence of *C. myrtifolia* to northern Mediterranean coastal Spain and southern France (Bond and Montserrat, 1958), and the more widespread distribution of alder throughout humid areas of Spain and of Europe generally, where *Coriaria* is not present, adds a point of interest to the trial concerning a probable geographic adaptation of *Frankia* to a particular host. Also, it is well known that the morphology and location of *Frankia* in the nodules of *A. glutinosa* differ greatly from those of *C. myrtifolia.* Thus, vesicle clusters of *Frankia*, typically prominent in alder, are not present in *Coriaria* nodules, where the microbe forms, at the most, club-shaped terminal hyphae. Besides, the vesicles in alder are commonly disposed peripherally within the infected cell (Becking *et al.*, 1964) whereas in *Coriaria*, as well as in *Datisca*, the wider terminal hyphae are directed inwards towards the vacuole. This, and the diverse biochemical characteristics of the two kinds of nodules, might indicate, in principle, the existence of a strict specificity of each of the endophytes towards their respective host plants. Confirming whether this should be true or otherwise is the aim of this contribution. It should be pointed out that while *Frankia* isolates from alder nodules are readily available, only non-infective isolates from *Coriaria* have been reported in the literature, thus introducing further complexity to the trial (Chaudhary and Sajjad Mirza, 1987).

Materials and methods

Germination and growth of the plants

Seeds of *A. glutinosa* were collected in February 1986 from adult trees growing on the banks of the river Tormes, Salamanca. The seed of *C. myrtifolia* was collected in August 1985 from plants growing in Figueras, Costa Brava. To obtain fast and regular germination, the seeds were stored in a refrigerator at 4C for one month before sowing. The alder seed was surface sterilized by immersion for 20 min in a solution of Vim Clorex (Lever Ibérica, S.A., Madrid) with continuous shaking, followed by successive washings with sterile distilled water. To improve germination, *Coriaria* seed was immersed in conc. sulphuric acid for five min, and then washed with abundant sterile distilled water.

The two types of seeds were sown on Peralite (Perlite Española, S.A., Barcelona) in separate shallow plastic trays of horticultural type, previously swabbed with methylated spirit as a precaution against the presence of the nodule endophyte, followed by ample rinsing; the Peralite was moistened with a very dilute nitrogen-free Crone's solution containing 1/8th of the following amounts: KCl, 0.75 g; $CaSO_4.2H_2O$, 0.5 g; $MgSO_4.7H_2O$,

0.5 g; $Ca_3(PO_4)_2$, 0.25 g; $Fe_3(PO_4)_2$. $8H_2O$, 0.25 g. The salts were added to 1 litre of distilled water. One $ml.l^{-1}$ of the following minor element concentrate was added to each litre of Crone's medium: 0.62 g H_3BO_3; 0.40 g of each of Na_2SiO_3, $MnCl_2.4H_2O$ and $KMnO_4$; 0.055 g of each of $CuSO_4.5H_2O$, $ZnSO_4.7H_2O$, $Al_2(SO_4)_3$, $Ni(SO_4)$.- $6H_2O$, $CoCl_2$, and TiO_2; 0.035 g of each of $Li_2SO_4.H_2O$, $SnCl_2.2H_2O$, KI, BrI, and Na_2- $MoO_4.2H_2O$; and distilled water 1 litre. The Peralite used required the addition of an amount of solution equal to 3.3 times the weight of the Peralite for the latter to be moistened sufficiently. The weight of the filled and sown tray was noted, and restored daily to that weight by the addition of distilled water (Bond and Wheeler, 1980). The germination trays were finally put in a glasshouse at 18–28°C.

Seedlings of *A. glutinosa* and *C. myrtifolia* were transplanted to the above solution at 1/4 strength, contained in 12 vitrified earthenware 1-litre jars covered with plastic squares with holes for the plants, and allowing 6 pots for each host plant. The seedlings were transplanted one month after germination when they were at the 2-4 leaf-stage and allowed to grow for 10 days until inoculation.

Inoculation procedures

Pure strains of *Frankia* Agp_1r_1 and Agn_1, were kindly supplied by A.D.L. Akkermans (Dept. Microbiology, Wageningen Agric. Univ. The Netherlands), and *Frankia* UGL 1.1.8. now identified as UGL010708 (Hooker and Wheeler, 1987), was obtained from C.T. Wheeler (Bot. Dept. Univ. of Glasgow, UK), all of them isolated originally from *A. glutinosa* nodules. The strains were subcultured and grown in BuCT medium (C.T. Wheeler, personal communication) for 40 days at 28°C. The

Table 1. Yield of *Frankia* strains in BuCT medium and amount of inoculum applied to each plant. Each value is the mean of four replicates

Frankia strain	Initial *Frankia* ($\mu g\,ml^{-1}$)	Growth at 30 days ($\mu g\,ml^{-1}$)	*Frankia* added to each plant (μg)
$LDAgp_1r_1$	0.05	12.0	12.0
$LDAgn_1$	0.07	17.0	17.0
UGL010708	0.09	26.6	19.5

medium (at pH 6.8) contained (g.l^{-1}): CaCl$_2$, 0.1; MgSO$_4$.7H$_2$O, 0.67; Na propionate, 0.5; trace elements (Baker and Torrey, 1979), 1 ml; casamino acids, 1.0; Tween 80, 0.5. The cultures were then concentrated by centrifugation at 1600 g and washed three times in phosphate buffer. The final growth data are based on fresh weight and total organic carbon determinations by Beckman Mod. 915A (Table 1). The colonies were broken by repeated passage through a 0.8 mm needle until satisfactory breakdown of hyphae was achieved, thus obtaining a fairly uniform *Frankia* suspension to be used as inoculum. Crushed-nodule inocula were prepared from healthy nodule-lobes detached from *A. glutinosa* and *C. myrtifolia* plants grown on Peralite for 10–12 months in a glasshouse. Once cleaned, washed and blotted dry, the nodules were surface sterilized in 1% chloramine T (Diem *et al.*, 1982) and then crushed on sand in a mortar and the suspension diluted in distilled water to a 10% wt/vol concentration.

A fraction of the crushed-nodule suspension from *Coriaria* was homogenized in a Sorvall blender (Mod. Omni Mixer 17106) at full speed for 5 min. The suspension thus obtained was filtered through a 90 μm pore nylon cloth, and subsequently through a 0.22 μm Millipore filter. The filtrate was examined under an Optiphot Nikon microscope to check for absence of *Frankia*, and again tested in *C. myrtifolia* plants to check for lack of infectivity. This filtrate was used eventually as an additive to the *Frankia* inocula from *Alnus*, prepared as above, to help infection and further nodulation of *Coriaria* plants, at the rate of 1 ml/plant.

The inoculation was carried out by applying with a syringe 1 ml of Agp$_1$r$_1$, Agn$_1$, UGL010708, and of a suspension of crushed alder and *Coriaria* nodules respectively, to the top part of the root system of each of 5 alder plants in each of 5 pots. One more pot was left uninoculated as control. The same procedure was followed for *Coriaria*, employing in this instance three plants per pot.

Nitrogen-fixing potential

Three months after inoculation, the acetylene-reducing activity (ARA) of the nodules was measured after incubation of samples under 90% air–10% acetylene gas mixture, and data expressed as n mol ethylene.h^{-1}.g^{-1} fresh weight. The ethylene was measured in a Varian Mod. Vista 6000 gas chromatograph equipped with an alumina column. Also, plant dry weight and total nitrogen content at harvest were determined.

Re-isolation trials

Re-isolation of *Frankia* from the nodules induced both on *A. glutinosa* and *C. myrtifolia* by the various inocula tested was made following Benson's method (Benson, 1982). Detached nodules were washed in distilled water followed by surface sterilization with 30% Na hypochlorite. They were then washed with abundant sterile distilled water and homogenized in 15 ml of BuCT medium. The homogenate was filtered successively through 90 μm and 20 μm nylon mesh netting. Those fractions retained on the filters were washed with 30 ml of BuCT medium. That on the 20 μm mesh was collected and, when examined under a phase contrast microscope, showed the presence of abundant endophyte masses. This fraction was diluted in BuCT medium, and 0.8 ml were plated on a 1.5% agar layer and a further 0.8% agar layer placed on top. Plates were incubated at 28°C in the dark.

Results

The *Frankia* strains subcultured in BuCT medium showed normal whitish growth and produced, at 30 days incubation, the biomass reported in Table 1. At that time numerous sporangia and some round vesicles could be seen under the microscope on strain UGL010708 (Fig. 1). However, strains LDAgp$_1$r$_1$ and LDAgn$_1$ did not show either of those structures and only abundant hyphal masses were observed (Fig. 2). The strains had similar growth rates.

At 18 days after noculation all the alder plants had nodules irrespective of the inoculum source, except for the *Coriaria* inoculum which infected only 60% of the alder plants in that period, eventually reaching 100% ten days later (Table 2). However, no nodulation symptoms were detected on *Coriaria* plants when inoculated with heterologous

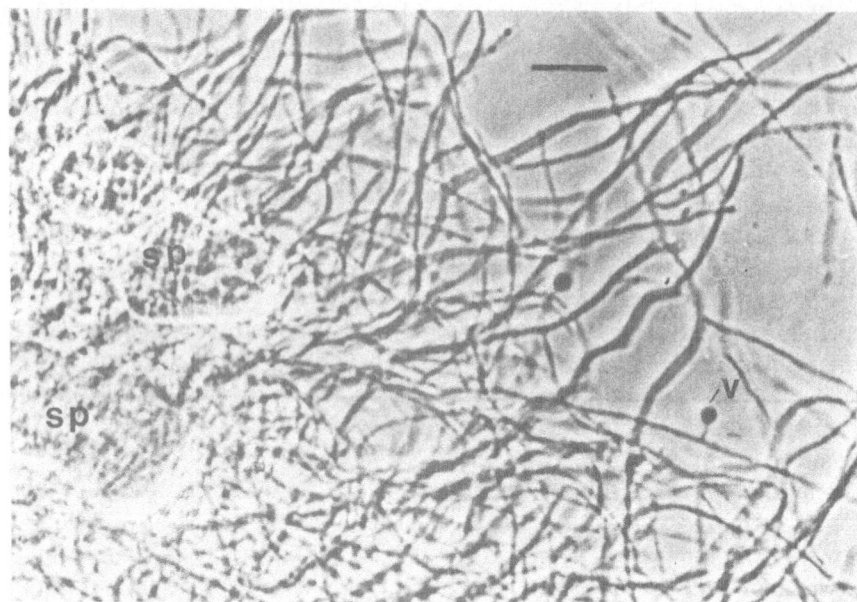

Fig. 1. Sporulation (sp) and vesicles (v) in free-living colonies of *Frankia* UGL010708 grown in BuCT liquid medium. Bar = 10 μm.

inocula, either as crushed alder nodules or with the *Frankia* strains from alder. In trying to seek an explanation for the latter results, a microscopical examination of the root systems of the inoculated but non-nodulated *Coriaria* plants showed fewer root hairs when pure *Frankia* strains from alder were applied, than when the inoculum was in the form of a crushed alder nodule suspension. This course of events led the authors to think that substances in the nodule might be responsible. Consequently, an addition to *Coriaria* non-nodulated plants of an endophyte-free 0.22 μm Millipore

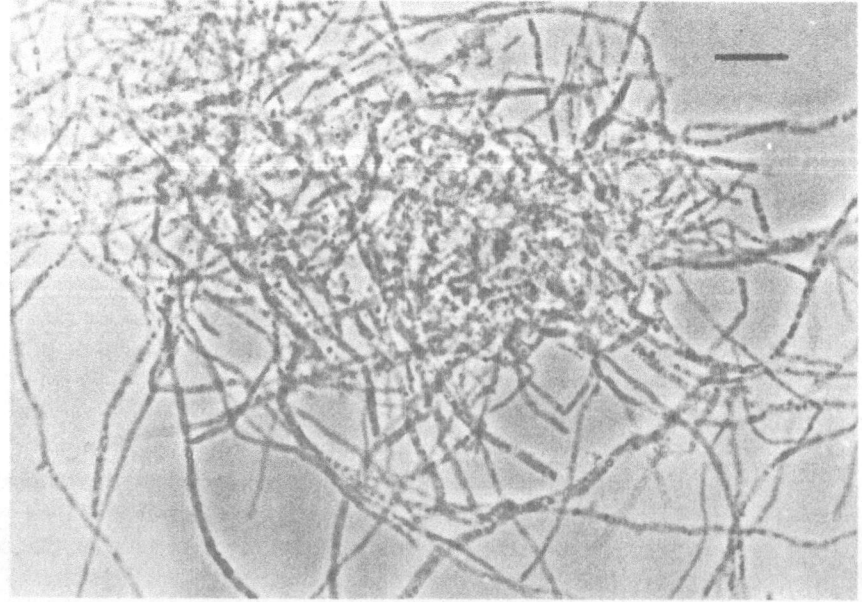

Fig. 2. Frankia spore-forming strain LDAgp$_1$r$_1$ grown in BuCT liquid medium. Note the lack of sporangia and vesicles. Bar = 10 μm.

Table 2. Nodulation of *A. glutinosa* and *C. myrtifolia* after treatment with the inocula shown. Data are ratios of nodulated to inoculated plants

Nodulation time, days	Host plant	Inocula				
		Crushed nodule extract		UGL010708	Agp$_1$r$_1$	Agn$_1$
		A. glutinosa	*C. myrtifolia*			
18	*C.m.*	0/3	1/3	0/3	0/3	0/3
18	*A.g.*	5/5	3/5	5/5	4/5	5/5
28	*C.m.*	0/3	3/3	0/3	0/3	0/3
28	*A.g.*	5/5	5/5	5/5	5/5	5/5
17*	*C.m.*	3/3	3/3	3/3	3/3	3/3

* Days taken by plants to nodulate after the second inoculation. This included the addition of a 0.22 μm nodule extract filtrate from *C. myrtifolia*.

filtrate crushed nodule suspension, this time from *Coriaria*, was made jointly with the application of the *Frankia* strains from alder. At 17 days after that second inoculation nodule initiation was visible to the naked eye in all the *Coriaria* plants inoculated with either of the inocula from alder. No nodulation was observed on *Coriaria* control plants when treated with filtrate only.

The plants were harvested five months after inoculation when they were all showing a vigorous growth indicative of active nitrogen fixation.

Nodules of *A. glutinosa* plants induced by pure *Frankia* strains were less numerous though bigger in size than corresponding nodules induced by the crushed nodule suspension from alder. The latter kind of nodules were less active in fixation than those induced by pure *Frankia* strains, except for strain LDAgn$_1$ which showed an activity comparable with that when a crushed nodule suspension from alder was used.

Plant growth and nitrogen data at harvest (Table 3) indicate strain UGL010708 as the most effective

Table 3. Plant growth and nitrogen data of *A. glutinosa* and *C. myrtifolia* plants nodulated with various inocula. Age of plants: five months. Figures in parentheses are minimum and maximum of five replicates

Inoculum	Host plant	Number of nodules	ARA. (nmols $C_2H_4 h^{-1} g^{-1}$ nod. f.w.)	Plant d.w. (g)	Plant (%N)	N fixed g nodule d.w. (mg)
LDAgp$_1$r$_1$	*A.g.*	85 (68–101)	308 (212–442)	3.1 (2.5–3.6)	2.9 (2.7–3.0)	537 (466–622)
	C.m.	34 (12–51)	336 (171–476)	0.4 (0.3–0.7)	1.5 (1.4–1.7)	343 (298–462)
LDAgn$_1$	*A.g.*	49 (44–53)	152 (132–171)	2.5 (2.5–2.7)	1.5 (1.4–1.6)	557 (554–587)
	C.m.	50 (48–51)	331 (209–493)	0.4 (0.2–0.6)	1.5 (1.3–1.7)	227 (133–336)
UGL 1.1.8 (UGL010708)	*A.g.*	108 (65–137)	269 (135–369)	4.9 (2.2–6.6)	2.2 (2.0–2.5)	873 (853–932)
	C.m.	84 (52–107)	493 (415–537)	1.6 (1.3–2.6)	1.4 (1.3–1.6)	519 (392–728)
A.g. crushed nodules	*A.g.*	120 (72–188)	153 (92–262)	1.7 (1.1–2.7)	2.0 (1.9–2.4)	333 (271–433)
	C.m.	49 (38–51)	590 (410–771)	0.5 (0.3–0.6)	1.8 (1.8–1.9)	320 (307–346)
C.m. crushed nodules	*A.g.*	61 (49–74)	189 (134–219)	0.6 (0.5–0.8)	1.7 (1.4–2.1)	177 (149–232)
	C.m.	99 (63–123)	200 (161–232)	2.4 (1.5–2.9)	1.9 (1.8–1.9)	463 (356–519)

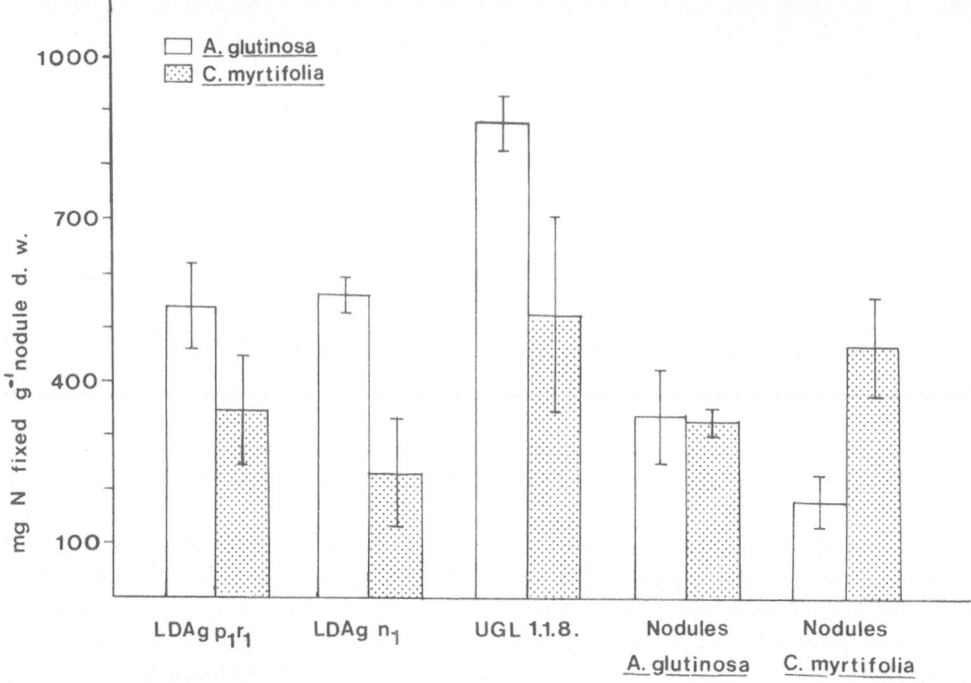

Fig. 3. Nitrogen fixed per g nodule dry weight by 5 month-old *C. myrtifolia* (dotted bars) and *A. glutinosa* (open bars) plants nodulated by different inocula. Strain designation UGL 1.1.8. = UGL010708.

inoculum applied, as reflected by the dry weight accumulated by plants of alder and *Coriaria* respectively. Nitrogen fixed per gram of nodule dry weight for all the combinations can be seen in Fig. 3.

Frankia re-isolates were readily obtained from nodules of *A. glutinosa* inoculated with the various *Frankia* strains, and the alder crushed nodule suspension. In contrast, attempts to reisolate those same *Frankia* strains from nodules of *C. myrtifolia*, including those induced by a homologous crude nodule extract, failed.

Discussion

Behaviour of Frankia *strains in culture*

Although *Frankia* strain UGL010708 in subculture was morphologically similar to other *Frankia* strains described by Callaham *et al.* (1978) and Burggraaf (1984), neither sporangia nor vesicles were observed in strains $LDAgp_1r_1$ and $LDAgn_1$. This entirely hyphal phase of growth of

$LDAgp_1 r_1$ and $LDAgn_1$, in the medium tried, does not necessarily preclude the possibility of vesicles or sporangia being formed in culture, a characteristic nor reported for strain $LDAgn_1$, although the growth conditions and composition of the medium were suitable for the formation of these structures of other strains (Tisa *et al.*, 1983; Hafeez *et al.*, 1984). In fact, no conclusive existence of non-spore forming (type N) and spore-forming (type P) strains (Normand and Lalonde, 1982), has been demonstrated (Torrey, 1987). Growth conditions appeared to be optimal for sporangia formation and occasional vesicle development in strain UGL010708. Biomass accumulated by this strain in BuCT medium was the largest for the three strains tested, a fact which agrees with the finding that the presence of nitrogen-fixing vesicles leads to more growth than that made by non-spore forming strains (Tjepkema, 1981).

Despite sporulation of strain UGL010708 in pure culture, an energy-demanding process, symbiotic performance within the nodules was high. In this present work, however, we did not check for the presence of sporangia within the alder nodules formed, so we were not able to correlate those structures with actual nitrogen-fixing rates in that host. It is known that type P strains do not consistently develop sporangia within the nodules. Conversely, type N *Frankia* strains will occasionally induce type P nodules on seedlings (Burggraaf, 1984).

From the above results with UGL010708, it appears that selection for the purpose of inoculation does not necessarily have to be directed towards type N strains as suggested previously (Torrey, 1987). Host plant environment provided will eventually affect strain morphology within the nodules (Lalonde, 1979; Newcomb *et al.*, 1980). That effect was observed in the present work, where strain LDAgp$_1$r$_1$ adopted within the *C. myrtifolia* nodules the typical morphology presented in a normal *C. myrtifolia* nodule (Fig. 4). Also, strain morphology in culture is greatly determined by growth conditions and composition of the medium (Tjepkema *et al.*, 1980, 1981).

Nodulation of host plants with different inocula

The two host plants from Spain were effectively nodulated by *Frankia*, originally isolated from alders grown as far apart as Scotland and The Netherlands. By way of comparison, two species of *Casuarina* (*C. equisetifolia* sp. *incana*, *C. cunninghamiana*) differed in growth responses to inoculation with *Frankia* from *Casuarina* grown in five different geographical locations (Reddell and Bowen, 1985). The established symbiosis varied from ineffective to highly effective in those *Casuarina-Frankia* source combinations tried. On the other hand, only two of the five *Frankia* strains were able to infect *C. cunninghamiana*. Furthermore, it has been reported that morphologically and biochemically distinct strains have been isolated from nodules of the same host species within the same geographic zone (Benson, 1982; Lechevalier *et al.*, 1983).

More nodules were formed by those *Frankia* strains known to sporulate, mainly LDAgp$_1$r$_1$ and UGL010708, the latter being most infective; this applied to both hosts. Effectivity (mg N fixed per g nodule dry weight) of nodules induced by UGL010708 was the highest on both hosts. A crushed nodule inoculum prepared from *A. glutinosa* gave a similar effective nodulation on the two

plant hosts, whereas a nodule suspension from *Coriaria myrtifolia* gave a nodulation of *A. glutinosa* which was only half as effective as that on its homologous host.

Although both *A. glutinosa* and *C. myrtifolia* nodules induced by a crushed nodule suspension from *Coriaria* showed a similar ARA (Table 3), the lower effectivity of the nodules formed on *A. glutinosa* indicates that nitrogen fixed by the latter was not contributing regularly to plant growth or, at least, to the extent that *Coriaria* nodules induced by the same inoculum did. This could be an indication of incompatibility between the *Coriaria* endophyte and *A. glutinosa* plants. The finding that more than one *Frankia* strain may inhabit the same nodule (Benson and Hanna, 1983), or the same plant (Simonet *et al.*, 1985) should be kept in mind, so that only suggestions about full compatibility can be made at this stage.

In comparing the nodulation frequency obtained for both types of inoculum, either *Frankia* strains or crushed nodules, no differences are found, contrary to what was reported previously for *Comptonia* isolates, shown to be more infective than corresponding nodule extracts (Callaham *et al.*, 1979).

Host growth variation within the same species, as observed in the present trial, should also be taken into account because the same *Frankia* strain may fix N at different rates on hosts within the same pot (Table 3). Large variations in plant growth rate and nodulation were also observed in *Casuarina equisetifolia* seedlings (Sougoufara *et al.*, 1986), and recommendations were made to use micropropagation techniques to select clones for their nitrogen-fixing potential.

Importance of crushed nodule filtrate in nodulation of Coriaria

In the present trial, the *Frankia* strains from *A. glutinosa* nodules were readily re-isolated but the same *Frankia* strains could not be re-isolated from the *Coriaria* nodules. Thus, it appears that the tested *Frankia* strains from alder can modify their requirements within the *Coriaria* nodules, making it necessary to search for growth media or isolation methods different from those employed. Meanwhile, an indirect method to isolate the *Coriaria*

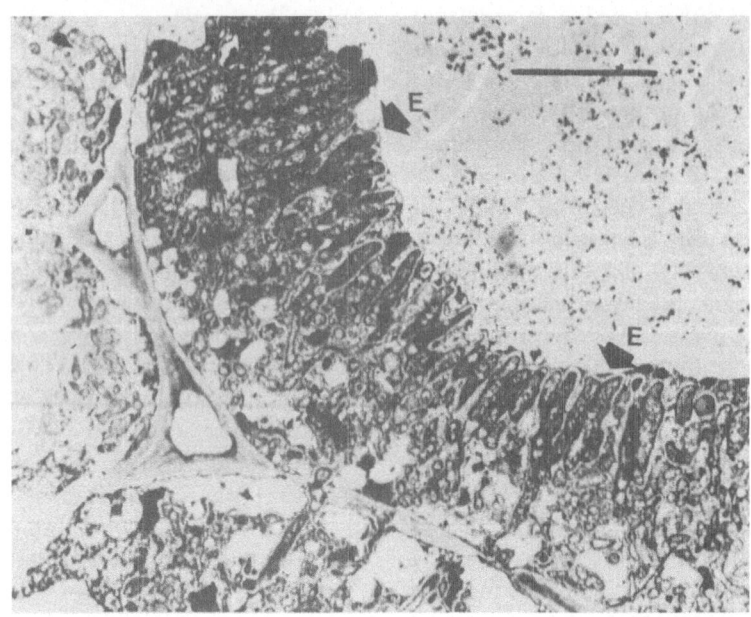

Fig. 4. Transmission electron micrograph of root nodule cell of *C. myrtifolia* grown in the greenhouse in Salamanca. The cell shows the endophyte (E) directing its widened terminal hyphae towards the centre. Bar = 10 μm.

endophyte might be through the previous nodulation of alder using a *Coriaria* crude nodule extract, followed by re-isolation of the strain from the nodules formed.

The favourable effect of an endophyte-free filtered nodule extract from *Coriaria* nodules on infection of all the *C. myrtifolia* plants by the three alder *Frankia* strains, could be due to the presence of wall-degrading enzymes or other types of substances of either plant or microbial origin. Figures 4 and 5 showing nodule cell-wall penetration by *Frankia* hyphae within a *C. myrtifolia* nodule suggest that an enzyme is involved in the digestion process. Despite the favourable effect of the nodule

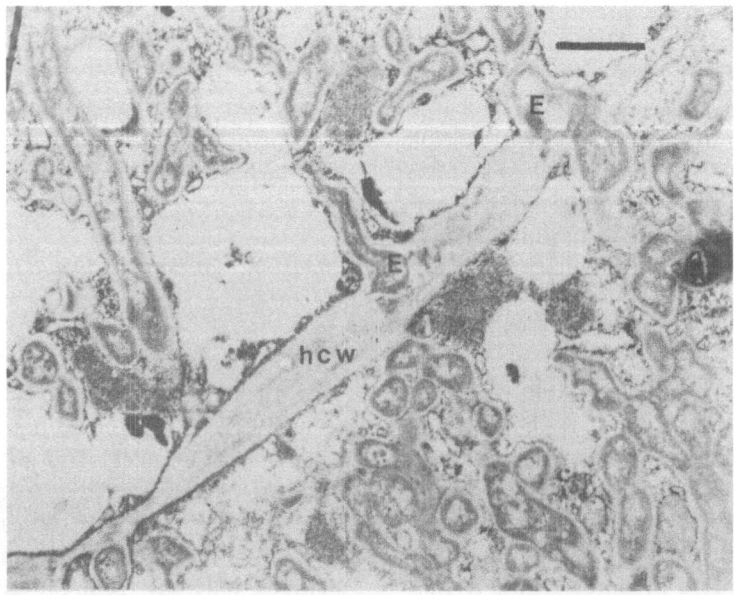

Fig. 5. Magnification of Fig. 4 showing host cell wall (hcs) penetration by the endophyte (E). Bar = 2 μm.

extract filtrate in the cross-inoculation experiment described, we are reluctant to believe that a wall-degrading agent was the only factor influencing the viability shown by the alder *Frankia* strains on *C. myrtifolia*.

Effectiveness of nodules from different inocula

The isolation of strain Cn3 from *Coriaria nepalensis*, able to sporulate in pure culture but unable to reinfect its own host, was reported recently (Chaudhary and Sajjad Mirza, 1987). Although this finding does not preclude the inclusion of strain Cn3 within the genus *Frankia*, it certainly cannot be taken as the true *C. nepalensis* endophyte as yet. Crushed-nodule suspensions of *Coriaria* failed repeatedly to infect homologous host plants (Kataoka, 1930; Allen *et al.*, 1966) until recently (Cañizo and Rodríguez-Barrueco, 1976). Other workers (J. D. Tjepkema, personal communication) have failed to nodulate *C. myrtifolia* (from Spain) and *C. arborea* (from New Zealand) with various inocula, mainly *Frankia* from *Myrica gale*, *Casuarina cunninghamiana* (CcI2) and crushed nodules from *Datisca*. A few ineffective nodules only on *Coriaria arborea* were obtained by employing crude extracts from *Coriaria* nodules. This state of things renders the results in this present contribution really surprising, and they may reflect how critical the growth conditions of both *Frankia* and plant can be at the time of securing nodulation on *Coriaria*. The finding that all the *C. myrtifolia* plants became effectively nodulated when inoculated with alder *Frankia* strains and with homologous crushed nodules respectively, in a reasonable period of time, conflicts with the reported results by other authors, suggesting that further trials should be made to identify and isolate the *Coriaria* endophyte, to fulfill Koch's postulates, and to define its taxonomic position.

Definition of cross-inoculation groups is controversial and extreme caution should be taken before a conclusion on the taxonomic status of an isolate is reached. There are data about strain CpI1 from *Comptonia* failing to nodulate *A. rubra* and other actinorhizal plants (Knowlton *et al.*, 1980). Also, even though *Frankia* isolated from *Colletia* (Rhamnales), failed to nodulate non-Rhamnales actinorhizals, one of the latter group, *Casuarina*

equisetifolia, was able to nodulate *H. rhamnoides*, *Elaeagnus angustifolia* and *Colletia spinosissima*, all belonging to the Rhamnales.

Acknowledgements

This research was supported by a grant from the Comisión Asesora de Investigación Científica y Técnica from Spain. We wish to thank J D Tjepkema for his preliminary results on inoculation of *Coriaria*, A D L Akkermans and C T Wheeler for the supply of *Frankia* strains, and E López for his technical assistance.

References

Allen J D, Silvester W B and Kalin N 1966 *Streptomyces* associated with root nodules of *Coriaria* in New Zealand. N.Z.J. Bot. 4, 57–65.
Baker D D and Torrey J G 1979 The isolation and cultivation of actinomycetous root nodule endophytes. *In* Symbiotic Nitrogen Fixation in the Management of Temperate Forests. Eds. J C Gordon, C T Wheeler and D A Perry. Oregon Sta. Univ. pp 38–56.
Becking J H, De Boer W E and Houwink A L 1964 Electron microscopy of the endophyte of *Alnus glutinosa*. Antonie v. Leeuwenhoek 30, 343–376.
Benson D R 1982 Isolation of *Frankia* strains from alder actinorhizal root nodules. Appl. Env. Microbiol. 44, 461–465.
Benson D R and Hanna D 1983 *Frankia* diversity in an alder stand as estimated by sodium dodecyl sulfate-polyacrylamide gel electrophoresis of whole-cell proteins. Can. J. Bot. 61, 2919–2923.
Bond G and Wheeler C T 1980 Non-legume nodule systems. *In* Methods for Evaluating Biological Nutrigen Fixation. Ed. F J Bergersen. John Wiley and Sons. pp 185–211.
Bond G and Montserrat P 1958 Root nodules of *Coriaria*. Nature (London) 182, 474–475.
Burggraaf A J P 1984 Isolation and characterization of *Frankia* strains from actinorhizal root nodules. Ph.D. Thesis. Univ. Leiden. NL. 139 pp.
Callaham D, Del Tredici P and Torrey J G 1978 Isolation and cultivation *in vitro* of the actinomycete causing root nodulation in *Comptonia*. Science 199, 899–902.
Callaham D, Newcomb W, Torrey J G and Peterson R L 1979 Root hair infection in actinomycete induced root nodule initiation in *Casuarina*, *Myrica* and *Comptonia*. Bot. Gaz. 140 (Suppl.), S1–S9.
Cañizo A and Rodríguez-Barrueco C 1976 The induction of root nodules on *Coriaria myrtifolia* L. plants growing in water culture. N.Z.J. Bot. 14, 271–274.
Cañizo A and Rodríguez-Barrueco C 1978 Nitrogen fixation by *Coriaria nepalensis* Wall. Rev. Ecol. Biol. Sol 15, 453–458.
Chaudhary A H and Sajjad Mirza M 1987 Isolation and charac-

terization of *Frankia* from nodules of actinorhizal plants of Pakistan. Physiol. Plant. 70, 255–258.

Diem H G, Gauthier D and Dommergues Y 1982 Isolement et culture *in vitro* d'une souche infective de *Frankia* isolée de nodules de *Casuarina*. C.R. Acad. Sci. Paris sér C, 295, 759–763.

Hafeez F, Akkermans A D L and Chaudhary A H 1984 Morphology, physiology and infectivity of two *Frankia* isolates An1 and An2 from root nodules of *Alnus nitida*. Plant and Soil 78, 45–59.

Hooker J E and Wheeler C T 1987 The effectivity of *Frankia* for nodulation and nitrogen fixation in *A. rubra* and *A. glutinosa*. Physiol. Plant. 70, 333–341.

Kataoka T 1930 On the significance of the root nodules of *Coriaria japonica* A. Gr. in the nitrogen nutrition of the plant. Jap. J. Bot. 5, 209–218.

Knowlton S, Berry A and Torrey J G 1980 Evidence that associated soil bacteria may influence root hair infection of actinorhizal plants by *Frankia*. Can. J. Microbiol. 26, 971–977.

Lalonde M 1979 Immunological and structural demonstration of nodulation of the European *Alnus glutinosa* host plant by an actinomycetal isolate from the North American *Comptonia peregrina* root nodule. Bot. Gaz. 140 (Suppl.), 535–543.

Lechevalier M P, Baker D D and Horrière F 1983 Physiology, chemistry, serology and infectivity of the two *Frankia* isolates from *A. incana* sp. *rugosa*. Can. J. Bot. 61, 2826–2833.

Newcomb W, Callaham D, Torrey J G and Peterson R L 1980 Morphogenesis and fine structure of actinomycetous endo-phyte on nitrogen-fixing root nodule of *Comptonia peregrina*. Bot. Gaz. (Suppl.), 522–534.

Normand P and Lalonde M 1982 Evaluation of *Frankia* strains isolated from provenances of two *Alnus* species. Can. J. Microbiol. 28, 1133–1142.

Reddell P and Bowen G D 1985 *Frankia* source affects growth, nodulation and nitrogen fixation in *Casuarina* species. New Phytol. 100, 115–122.

Simonet P, Normand P, Moiroud A and Lalonde M 1985 Restriction enzyme digestion patterns of *Frankia* plasmids. Plant and Soil 87, 49–60.

Sougoufara B, Duhoux E, Corbasson M and Dommergues Y 1986 Improvement of nitrogen fixation by *Casuarina equisetifolia* through clonal selection. *In* Proc. 18th World IUFRO Congress. Div. I. Ljubljana. Ed. R Hermann. Pl.12.00, S1.07.07, S2.02.00.

Tisa L, McBride M and Ensign J C 1983 Studies on growth and morphology of *Frankia* strains EAN1$_{pec}$, EuI1$_c$, CpI1 and ACN1^{AC}. Can. J. Bot. 61, 2768–2773.

Tjepkema J D, Ormerod W and Torrey J G 1980 Vesicle formation and acetylene-reducing activity in *Frankia* species CpI1 cultured in defined nutrient media. Nature (London) 287, 633–635.

Tjepkema J D, Ormerod W and Torrey J G 1981 Factors affecting vesicle formation and acetylene reduction (nitrogenase activity) in *Frankia* species CpI1. Can. J. Microbiol. 27, 815–823.

Torrey J G 1987 Endophyte sporulation in root nodules of actinorhizal plants. Physiol. Plant. 70, 279–288.

F. A. Skinner et al. (Eds.), Nitrogen fixation with non-legumes, 25–33.

Biology of the *Parasponia-Bradyrhizobium* symbiosis

M. J. TRINICK and P. A. HADOBAS
CSIRO, Division of Plant Industry, G.P.O. Box 1600, Canberra, ACT 2601, Australia

Key words: *Bradyrhizobium*, competition, haemoglobin, non-legume, *Parasponia*, symbiosis

Abstract

Parasponia remains the only non-legume known to nodulate with *Rhizobium/Bradyrhizobium*. It is a pioneer plant that is capable of rapid growth and fixing large quantities of nitrogen. In addition to its high agronomic potential, the symbiosis offers the scientist the unique opportunity of studying differences at the molecular level of both partners, and to investigate any possible extension of the symbiosis to other non-legumes of importance. Haemoglobin has been found in the nodule tissue of *Parasponia* and other nodulated non-legumes and the gene for it has been found and expressed in non-nodulating plants such as *Trema tomentosa* and *Celtis australis*. *Bradyrhizobium* strains isolated from species of *Parasponia* growing in Papua New Guinea form a group that are more specific in their host requirements than *Bradyrhizobium* strains from tropical legumes from the same area. They do not effectively nodulate (except CP283) tropical legumes, and *Parasponia* is not readily nodulated with *Rhizobium* and *Bradyrhizobium* strains from legumes. The effectiveness of the symbiosis is influenced by host species, the *Bradyrhizobium* strain and the environment. *Parasponia andersonii* forms a more effective symbiosis than the other species tested. In competition studies with strains from legumes, isolates from *Parasponia* always dominate in nodules on *Parasponia*.

Biology of the Parasponia-Bradyrhizobium symbiosis

Parasponia species, (family Ulmaceae, order Urticales), remain the only non-legumes known to nodulate with *Rhizobium* or *Bradyrhizobium*. Nodulation appears to be restricted to *Parasponia* (Akkermans *et al.*, 1978), since field-collected, closely related genera, including *Trema* spp., *Ulmus parvifolia*, *Celtis* spp. and *Gironniera subequalis*, or genera in the closely related family Urticaceae, do not have nodules. Both *Parasponia* spp. and *Trema* spp. have frequently been confused by botanists. Early reports of nodulation of *Trema* spp. in Malaya are now believed to be of *Parasponia* spp. *Parasponia* is easily distinguished from *Trema* by the imbricate perianth lobes of the male flowers and the intrapetiolar, connate stipules enclosing the terminal bud. *Parasponia* spp. are restricted to the Malay Archipelago while *Trema* spp. occur throughout the tropics and subtropics. In Australia their distribution extends to Northern Victoria (Soepadmo, 1977).

There have been a number of unconfirmed reports of similar nodulation in five species of the genera *Tribulus* and *Zyophyllum* in the Zygophyllaceae (Athar and Mamood, 1981). Although they isolated bacteria from the nodules on these plants, and were able to form effective associations with legumes, including *Vigna* spp., they were unable to test them on *Tribulus* spp., through lack of viable seed.

Parasponia forms a highly effective symbiosis with *Bradyrhizobium*. The root nodule bacteria from *Parasponia* form a distinct group. They differ from other *Bradyrhizobium* strains in cultural characteristics, host range, and particularly in forming ineffective associations with legume species. Under glasshouse conditions, with growth studies between 4 and 10 months, the symbiosis was found capable of fixing up to $850\,kg\,N\,ha^{-1}\,year^{-1}$ with a biomass production of 30t dry matter $ha^{-1}\,year^{-1}$ (Trinick, 1981). This compares very favourably with legumes. Soybeans in approximately five months can fix 313 kg N ha^{-1} with a dry matter yield of 15.3t. ha^{-1} (Herridge, 1982). *Leucaena* can yield more than 600 kg N ha^{-1} year^{-1} with the Hawaiian Giant (K8) (Vietmeyer and Cottom, 1977). The advantage, and hence the potential of *Parasponia* over legumes for grazing, is that this non-legume will nodulate under both acidic and alkaline conditions, appears less dependent on nutrients and will coppice readily.

Both *Rhizobium* and *Bradyrhizobium* are capable of nodulating a plant genus that is vastly different from the legumes. The genetic compatibility of the components of both *Parasponia* and root-nodule bacteria has resulted in this extension of the symbiosis. It is hoped that through the study of the DNA of both partners, it may be possible to extend the nitrogen-fixing symbiosis to important non-nodulating plants of agricultural, forestry and industrial importance. Early research in this direction has revealed unexpected findings such as the presence of the haemoglobin gene, not only in *Parasponia*, but also in *Trema* (Appleby *et al.*, 1986).

The group of *Bradyrhizobium* nodulating *Parasponia*

The classification of the root-nodule bacteria associated with legumes is unsatisfactory. Early studies with clovers, medics, peas, vetches, soybeans and, perhaps, some tropical species, conveniently grouped *Rhizobium* strains according to host range. For example many fast-growers were designated *Rhizobium trifolii*, *R. leguminosarum*, *R. phaseoli* or *R. meliloti* whilst some slow-growers were named *R. lupini* or *R. japonicum*, with *Rhizobium* spp. used to group the bulk of the isolates from tropical genera. This system has proved unsatisfactory as a result of many reported instances of bacteria nodulating plants in other plant groups, through both fast and slow-growing types, effectively nodulating the one host, and through the marked physiological differences occurring between the two bacterial types (Trinick, 1981). *Bradyrhizobium* was proposed (Jordan, 1982) to distinguish the slow-growers from the physiologically different fast-growing *Rhizobium*.

The root-nodule bacteria nodulating Parasponia

Rhizobium phaseoli, R. leguminosarum and *R. meliloti* do not form nodules on this host while *R. trifolii* strains induce ineffective nodulation with prolonged exposure of plants under conditions of severe nitrogen deficiency (Trinick and Galbraith, 1980; Becking, 1983; Trinick and Hadobas, 1986). Nodulation took 50 to 100 days to form and only one *Rhizobium* strain (NGR 66) of the 29 tested formed nodules on all replicated plants; one third of the strains nodulated 50% or more of the plants, whilst the others formed nodules on only 16–33% of plants. Attempts at isolation of the respective strains of *Rhizobium trifolii* showed that they were the only occupants of the nodules, sometimes present in large numbers. Their identity was confirmed using serological tests and a nodulation test on *Trifolium repens* (Trinick and Hadobas, 1986).

Fast-growing rhizobia from tropical legumes such as *Lablab purpureus*, (NGR 234), and the *Leucaena — Mimosa — Sesbania* group are able to form nodules on *Parasponia*. The association is usually ineffective but sometimes a trace of nitrogen is fixed and reflected in acetylene reduction assays, plant colour and plant weight.

Nodulation is delayed, (31 to 74 days), with *Bradyrhizobium* strains isolated from tropical legumes (*Crotalaria, Stylosanthes, Cajanus, Flemingia* and *Macroptilium*) and the association is frequently ineffective. When nitrogen fixation occurs, it is less effective than associations with *Bradyrhizobium* strains isolated from *Parasponia*.

Although *Parasponia* accepted infection by diverse strains and species of root-nodule bacteria, the developing plant-bacterial interaction showed varying degrees of incompatibility. Only *Bradyrhizobium* strains, isolated from *Parasponia* spp., were able to form highly effective associations.

The Bradyrhizobium *strains isolated from* Parasponia

Host range. Parasponia spp., growing in Papua New Guinea, are well nodulated and show rapid growth (Trinick, 1987). The 49 strains of *Bradyrhizobium,* isolated from *Parasponia* spp., fixed nitrogen in symbiosis with *P. andersonii, P. rugosa,* and *P. rigida*, with varying degrees of effectiveness. Only one (CP283) of 15 strains from *Parasponia* was able to nodulate effectively a wide host range of tropical legumes (Trinick and Galbraith, 1980), that were normally nodulated with slow-growing root-nodule bacteria. A further test with 34 strains on the promiscuous *Macroptilium atropurpureum* has shown that CP283 is exceptional and remains the only isolate to nodulate tropical legumes effectively. The isolates from *Parasponia* either failed to form nodules or formed nodules that fixed very little or no nitrogen. Legumes normally nodulated with fast-growing root nodule bacteria, and belonging to the pea, medic, and the *Leucaena — Mimosa — Sesbania* group, were not nodulated. However, *Trifolium* spp. (Trinick and Hadobas, 1986), and *Phaseolus vulgaris* (Trinick and Galbraith 1980), were ineffectively nodulated.

Time taken to nodulate legumes and Parasponia. All isolates (except CP283 which took longer than 60 days) were able to nodulate three *Parasponia* species in 9 to 12 days. However, *Bradyrhizobium* isolated from legumes, like CP283, took 31–74 days to form nodules on *Parasponia*. Thus *Parasponia* selects specific effective *Bradyrhizobium* strains from mixed soil populations that are able to induce early nodulation. All three species of *Parasponia* tested are more specific than most tropical legumes, all nodulating effectively with a group of *Bradyrhizobium* strains that are less successful on all tested legumes.

There was a correlation between the level of nitrogenase activity and the time required for nodulation to occur on *M. atropurpureum*, (siratro), when inoculated with isolates from *Parasponia*. Thus 27 ineffective strains, 11 with limited nitrogenase activity, 10 partially effective and one fully effective strain, (CP283), took 20–35, 14–38, 11–27, and 11–12 days, respectively, for nodules to appear. Plants inoculated with regular strains from legumes, were nodulated within 10 days. When re-isolates from nitrogen fixing nodules on *M. atropurpureum*, formed by *Parasponia* bradyrhizobia, were again tested on siratro, nodulation occurred in a shorter time and a greater percentage of the replicated plants were able to fix

itrogen. Similar studies with CP283, possibly a regular legume type, on *Parasponia* did not produce an improvement in time to nodulate, or improvement in effectiveness.

Cultural types of Bradyrhizobium *from* Parasponia. The *Bradyrhizobium* strains can be divided into five groups according to their growth characteristics on yeast extract — mannitol agar, (YMA). With extended incubation, (2–3 weeks), groups 1, 2, and 3 are capable of producing growth similar to *Rhizobium* strains from the *Leucaena — Mimosa* group. In 10 days at 27°C, Group 1, NGR 231 type, formed white, opaque colonies 4 mm in diameter and produced gum which was easily emulsified in water. The second most common type, (Group 2, CP299 type), formed similar colonies with a gum which was difficult to dissolve. Group 3, CP315 type, formed smaller colonies, 1 to 2 mm, greyish with readily soluble gum. Group 4, CP330 type, was rare and formed colonies less than 0.5 mm diameter that were white, opaque and without apparent gum production after 10 days growth on YMA. After three weeks growth, colonies were 2 mm diameter with a white centre surrounded by clear non-sticky gum. Group 5 is exceptional and represented only by CP283, which produces a misty growth with a colony size about 3 mm and has both MUC + and MUC − forms. The generation time in YMA broth of Groups 1, 2, 3 and 5 is 6–6.5 h and of Group 4, 8 h (Trinick and Appleby, 1984).

Competition for nodulation of Parasponia. *Bradyrhizobium* strains isolated from *Parasponia* in paired inocula with strains isolated from legumes, always prevented the legume bacteria from forming nodules on *Parasponia*. However, the legume isolate 8B4, like the exceptional CP283 from *Parasponia*, took 35 days to form nodules and was successful in forming dual occupancy with CP283 in 11% of the nodules, but failed to form nodules as the sole occupant. The more rapid nodulation of *Parasponia*, with its own isolates (all except CP283), appears to give those isolates a competitive advantage. Thus CP283, which also effectively nodulates legumes (Trinick and Galbraith, 1980), when in competition with other *Parasponia* isolates, is unable to compete for nodules. Some dual occupancy of nodules, (4%), was noted only when the number of CP283 was more than 100 times that of its competitor. The above experiments were conducted in plant tubes (25°C and 12/12 h, light/dark). Similar results were obtained when competition studies were duplicated using Leonard jars at glasshouse temperatures of 28/23°C (day/ night). The poor competitive nature of the *Parasponia* isolate CP283 was dramatically demonstrated when plants nodulated with CP283 were transferred to initially sterile pots in a glasshouse which had previously grown *Parasponia* fully effectively nodulated with CP299. Despite the initial additional heavy inoculation with CP283 at transplanting time, the dominant nodule occupant, at harvest time, was CP299. The higher effectiveness of CP299, and its rapid infection of *Parasponia* placed CP283 at a great competitive disadvantage.

When *Bradyrhizobium* isolates from *Parasponia* were in competition with isolates from legumes (32Hl, 8B4, CB756, NGR180, NGR169 and NGR120) for nodulation of the legume *M. atropurpureum*, the *Parasponia* isolates, except CP283, were non-competitive. The exceptional *Parasponia* isolate CP283 occupied all nodules in competition with NGR169 and most nodules with NGR120, was equally successful in nodule representation with 32Hl, 8B4 and NGR180, but was a poor competitor with CB756. CP283 is as effective on this legume, *M. atropurpureum*, as the other legume *Bradyrhizobium* strains used. These results further substantiate the greater kinship of CP283 with the cowpea type bradyrhizobia of tropical legumes, rather than with the *Parasponia* isolates (Trinick and Appleby, 1984).

Isolates from *Parasponia* and *Rhizobium trifolii* can co-inhabit *Parasponia* nodules even when both are presented to *Parasponia* in equal and high numbers. *Rhizobium trifolii* strain NGR66 is able to co-inhabit with many *Bradyrhizobium* strains in the majority of nodules formed as well as attaining relatively high nodule populations. Other *R. trifolii* strains were less successful and often had a total nodule population of less than one hundred. When the inoculant level of *Bradyrhizobium* is low, compared with *R. trifolii*, the success of the *Bradyrhizobium* is often reflected in the effectiveness of the association. In ineffective nodules, *R. trifolii* dominates with numbers in excess of 3×10^6 per nodule and with *Bradyrhizobium* less than 100 per

nodule. With increase in the level of effectiveness, the *Bradyrhizobium* component is raised and numbers are often higher than $10^5 \times R.$ *trifolii*. The *Bradyrhizobium* strains tested (CP283, NGR231, CP299 and CP314) were all able to have dual occupancy with *R. trifolii* strains (Trinick and Hadobas, 1986).

The Parasponia *group of* Bradyrhizobium. The *Bradyrhizobium* strains studied were isolated from *Parasponia* spp. growing in various parts of Papua New Guinea where tropical legumes and their symbionts were abundant. The data indicates that *Parasponia* has selected, from the heterogeneous soil population of *Bradyrhizobium*, a group of strains that are more specific in their host requirements. They, except CP283, are able to nodulate *Parasponia* rapidly and effectively, but not the tropical legumes tested; they appear culturally different to the majority of *Bradyrhizobium* strains isolated from Papua New Guinea soils; they have the competitive advantage, when mixed with other *Bradyrhizobium* for nodulation of *Parasponia*. The isolate CP283, from *Parasponia*, is the exception, and resembles *Bradyrhizobium* strains isolated from tropical legumes. Perhaps, this strain nodulated *Parasponia* in the presence of other, more competitive strains, by accident.

These strains of bradyrhizobia are better suited to nodulate the non-leguminous plant, *Parasponia*. A future study of the molecular biology of these strains, may reveal additional Nod genes that make their rapid infection and development of the symbiosis in the non-legume *Parasponia*, possible. These genetic factor (s) may be part of the additional genome required to extend the host range to other non-leguminous plants.

Factors affecting the effectiveness of the symbiosis in Parasponia

The bacterial symbiont. The root-nodule bacteria from *Parasponia* vary in their effectiveness on *P. andersonii*. Twenty six strains were tested and showed a range of effectiveness on plants grown in Leonard jars in a glasshouse maintained at 25–29/ 22°C, day/night, temperatures. The symbiotic response of eight of these strains is illustrated in Table 1. The most effective strains, (CP314, CP299,

Table 1. Effectiveness of eight strains of *Bradyrhizobium* isolated from *Parasponia* spp. on *P. andersonii*.

Bradyrhizobium strain	Plant top weight (g)	Nodule weight (g)
NGR231	11.90b[x]	2.970
CP283	8.65b	2.494
CP288	28.41c	5.283
CP277	31.67c	6.009
CP292	36.65c,d	4.747
CP299	43.85d	6.998
CP314	46.29d	7.302
CP289	44.03d	6.368
Uninoculated	0.18	

[x] Top weights with the same letter are not statistically different at the 5% level using Duncan's multiple range test.

CP289), fixed up to five times the quantity of nitrogen fixed by the least effective, (CP283). The delay in nodulation by CP283 contributes to its poor performance. Under conditions of higher radiation and long day length CP283 nodules *Parasponia* more rapidly and its relative effectiveness, compared with NGR231, becomes changed (Trinick, 1987).

The efficiency of the nodule tissue formed varied between *Bradyrhizobium* strains, with the most effective strains forming less nodule tissue per unit of plant top compared with the least effective strains. The ratio of plant top to nodule mass, for the highly effective symbiosis, was 6.5 (r = 0.82) compared with 3.6 (r = 0.89) for the least effective strains. The *Bradyrhizobium* strains tested were isolated from three species of *Parasponia* and their effectiveness on *P. andersonii* was not related to their host source (Trinick, 1987).

A comparison was made of the effectiveness of the symbiosis of *P. andersonii*, *P. rugosa* and *P. rigida* with seven strains of root-nodule bacteria known to have different effectiveness ratings with *P. andersonii*. There was a highly significant interaction between *Parasponia* species and *Bradyrhizobium* strains; strains most and least effective on *P. andersonii*, performed similarly on the other *Parasponia* species. The intermediate strains on *P. andersonii* sometimes performed better, and ranked best on *P. rigida* and *P. rugosa*. With the tested strains of *Bradyrhizobium*, *P. andersonii* was the most effective fixer of nitrogen and *P. rigida* the least.

Bradyrhizobium strains isolated from *Inocarpus* sp., *Centrosema pubescens*, *Crotalaria paulina*, *Ma-*

croptilium lathyroides, *Cajanus cajan*, *Stylosanthes gracilis* and *Flemingia congesta* were tested on *Parasponia*. Some strains were ineffective, others were less effective than the least effective *Parasponia* isolates, NGR231 and CP283.

The host plant. Early experiments with *Parasponia* seed collected in Bougainville, from a *Parasponia* thicket, produced a high level of variability between replicates (Trinick, 1980a). Often many plants failed to nodulate, despite the survival of *Bradyrhizobium* in high numbers, in the rhizosphere. Mature plants showed marked morphological variation. Seed collected from well-nodulated plants, highly effective in nitrogen fixation, exhibited much less variation in plant response to inoculation. Further plant selection for early nodulation with CP283, to reduce its excessive time to nodulate *P. andersonii*, has met with limited success.

Hybridization experiments between *P. andersonii*, *P. rugosa* and *P. rigida*, under glasshouse conditions, have indicated that they are incompatible. Their independent responses to *Bradyrhizobium* strains and differing capacities to fix atmospheric nitrogen, were probably developed early, and have remained characteristic for the species.

The environment. *Parasponia*, like legumes, is affected by the level of nitrate in the rooting medium, with stimulatory and inhibitory influences on the infection process and on nodule development as well as on the expression of nitrogenase activity. Levels of $7–14\,mg\,N\,l^{-1}$, stimulated nodule initiation of most *Bradyrhizobium* strains tested. Some strains, like CP283, preferred $7\,mg\,N\,l^{-1}$ and showed a significant reduction at higher levels, while others, *e.g.* NGR231, had an optimum response at $14\,mg\,N\,l^{-1}$: one strain, NGR120, ineffective on *Parasponia*, had increased nodule counts at $28\,mg\,N\,l^{-1}$. Some strains, however, like NGR231, were able to yield higher nodule counts with more than $140\,mg\,N\,l^{-1}$, compared with the counts obtained with no nitrogen in the rooting medium. The mass of nodules formed was generally increased with small quantities of nitrogen ($7–28\,mg\,N\,l^{-1}$). The time required for nodule initiation was reduced with a small addition of nitrogen, but was increased at higher levels. Total nitrogenase activity per plant was stimulated at $7 mg l^{-1}$, but higher levels of nitrate reduced nitrogenase activity (Trinick, 1987).

Parasponia grows best at temperatures above 25°C and below 32°C; growth rate is reduced rapidly below 25°C with little growth at 20°C. At temperatures constantly below 20°C, *Parasponia* stops growing and deteriorates rapidly. Optimum temperature for nitrogenase activity was between 25 and 30°C for two of the *Bradyrhizobium* strains used (NGR231 and CP283) whilst a wider temperature range of 25–35°C was associated with CP279, NGR45 and NGR170. Plants in agar-tube culture inoculated with NGR231 and CP283, grown at 25 and at 30°C, showed that the total amount of nitrogen fixed by CP283 at 30°C was significantly greater while no differences were recorded with NGR231. Many tropical legumes have similar optimum temperatures for nitrogen fixation (Trinick, 1982). At temperatures above 35°C, and below 20°C, nitrogenase activity decreases rapidly. At 15°C all acetylene-reducing activity is lost within 15 min and when the plants are returned to 25–30°C, recovery of nitrogen fixation is slow and only partial (25% of the pre-cold treatment) (Trinick, 1980b).

Parasponia is able to form dense stands on fresh volcanic ash, act as pioneers on lava streams and seems indifferent to soil fertility and soil types (Soepadmo, 1977). In laboratory studies (Trinick, unpublished) *Parasponia* is able to nodulate with the slow-growing bradyrhizobia under acid (pH 4.5) and alkaline (pH 8) conditions, but *Parasponia* prefers an acid to neutral root environment.

Nodulation and growth of *Parasponia* is improved with increased light intensity and day length, especially at day lengths longer than 12 h. In laboratory studies, increased shade levels reduced nodule weight, top dry weight, root development and total plant nitrogen. This is reflected in the field where *Parasponia* thrives, as a pioneer plant, in newly opened-up habitats receiving maximum radiation. *Parasponia* rapidly disappears with invasion by taller shade-producing species.

Extension of the Rhizobium/Bradyrhizobium symbiosis to non-legumes

The development of the effectively nitrogen-fixing legume/non-legume nodule, depends on 'the intrinsic properties of the partners, their mutual genetically controlled compatibility, and the influence exercised by the environment on each separately and on their interaction' (Vincent, 1980). Vincent (1980) describes the complexity of the symbiosis, lists and discusses an analysis of the symbiotic sequences from the point of view of the bacterium and host genome. The mutually successful and productive relationship between bacteria, actinomycetes, blue-green algae and other plants including legumes and non-legumes, further illustrates the diversity of different requirements of partners that permit effective symbiotic nitrogen-fixation. The discovery of *Parasponia* type infection (Trinick, 1979) in many genera of the Caesalpinioideae and in some genera of the Papilionoideae, but in none from the Mimosoideae, has suggested the possible primitive state of this type of symbiosis (Faria *et al.*, 1986; Faria *et al.*, 1987). In these legumes, the nodule structure does not resemble those on *Parasponia*. However, *Parasponia*, which is predominately monoecious, may be considered by some to be a reasonably advanced genus. The closely related genus, *Trema*, does not nodulate. One hundred and seven strains of root nodule bacteria, isolated from 30 tropical species, and temperate legumes, were unable to nodulate *T. tomentosa*, collected from Papua New Guinea, and various Australian States. This genus, although close taxonomically to *Parasponia*, is lacking in one or many of the 'intrinsic' properties essential for symbiotic nitrogen fixation.

Detailed studies by molecular biologists into the molecular basis of infection, host specificity, nodule development and nitrogen fixation, are beginning to define areas of the bacterial DNA responsible for various aspects of the process. For instance *Bradyrhizobium* strains nodulating *Parasponia* contain *nod* ABC genes that are present in other *Bradyrhizobium* and *Rhizobium* strains and they also contain an additional *nod* K gene (Scott, 1986; Marvel *et al.*, 1987). This work may eventually assist in extending the bacterial host range to other legumes, or even to agriculturally important, non-legumes.

A detailed knowledge of the structure and function of the various bacterial — host interactions is essential in any attempt to define essential attributes of the symbiosis. The *Parasponia* nodule, like other non-legume nodules, resembles a modified root, with a central stele surrounded by the infec-

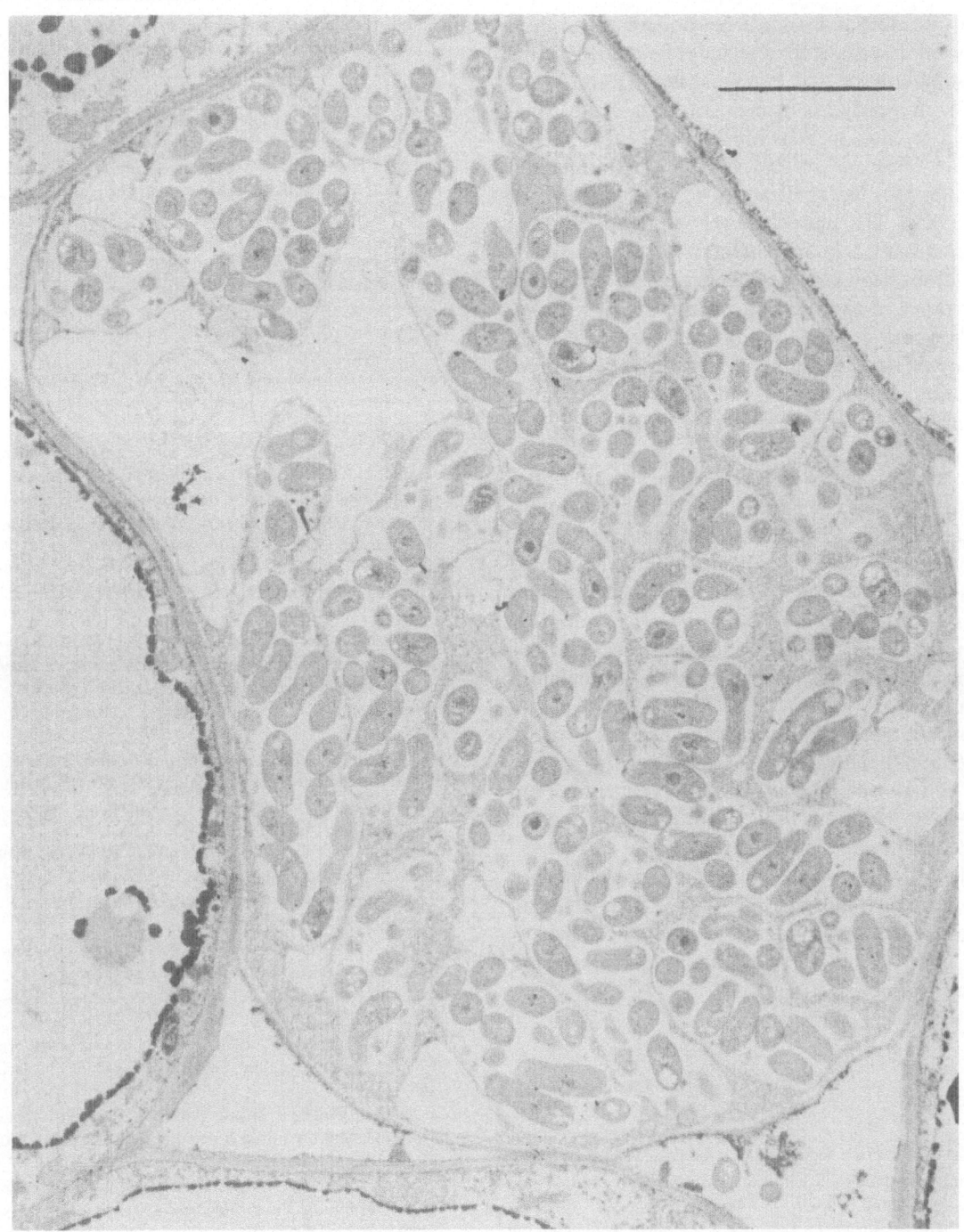

Fig. 1. Bradyrhizobium strain CP283 in infection threads in a nodule from *Parasponia andersonii*. Threads are seen in longitudinal and transverse section. Magnification bar = 5 μm. (Electron micrograph kindly supplied by Dr. D. J. Goodchild).

tion zone. Unlike legumes, the symbiont in *Parasponia* is not released into the host cytoplasm but is contained within infection threads (Fig. 1) (Trinick, 1979). Root infection, unlike that of most legumes is not via the root hair, but through breaks in the epidermis (Lancelle and Torrey, 1984). Infection in genera such as *Arachis* and *Stylosanthes* probably occurs through a similar pathway. The *Parasponia* type infection and nodule development has been reported in a number of legumes belonging to the Caesalpinioideae and Papilionoideae (Faria *et al.*, 1987). Recently, rhizobial mutants have been shown to enter through wounds and other breaks in the epidermis instead of the usual infection through root hairs (Hirsch *et al.*, 1985; Hrabek *et al.*, 1985). The understanding of these different systems may help in determining the factors or conditions required for a bacterium to extend its host range to another non-legume.

The presence of plant produced haemoglobin in *Parasponia* (Appleby *et al.*, 1983), in *Casuarina* (Fleming *et al.*, 1987) and other non-legumes (Tjepkema, 1983), has raised the possibility of the haemoglobin gene occurring in many non-nodulated plants. If this is true then the molecular biologist would not have to insert this gene into a future candidate selected for development of a symbiotic system. Recently, haemoglobin gene-related sequence in the non-nodulating close relative to *Parasponia, Trema tomentosa* was reported (Landsmann *et al.*, 1986). The expression of the haemoglobin gene has been shown in the non-nodulating *T. tomentosa, Celtis australis* and in aseptically grown *Parasponia* (Ulmaceae) (Bogusz *et al.*, 1988). These reports strengthen the suggestion by Landsmann *et al.*, (1986) that haemoglobin genes are ubiquitous in the plant kingdom and are expressed in non-symbiotic tissue.

The full study of the biology and molecular biology of all the range of symbiotic systems occurring in nature may lead to an unravelling of the genetically controlled factors governing the development of the symbiosis.

References

Akkermans A D L, Abdulkadir S and Trinick M J 1978 N_2-fixing root nodules in Ulmaceae: *Parasponia* or (and) *Trema* spp.? Plant and Soil 49, 145–149.

Appleby C A, Bogusz D, Dennis E S, Dudman W F, Fleming A I, Higgins J, Kortt A A, Landsmann J, Peacock W J, Tjepkema J D, Trinick M J, Wittenberg B A and Wittenberg J B 1986 The origin and survival of plant haemoglobin genes. *In* Proceedings of the 8th Australian Legume Nodulation Conference, held in association with The Australian Institute of Agricultural Science, AIAS Occasional Publication No 25, p77.

Appleby C A, Tjepkema J D and Trinick M J 1983 Haemog-

lobin in a non-legume plant, *Parasponia*: Possible genetic origin and function in nitrogen fixation. Science 220, 951–953.

Athar M and Mamood A 1981 Extension of *Rhizobium* host range to Zygophyllaceae. *In* Current Perspectives in Nitrogen Fixation. Eds. A H Gibson and W E Newton. p. 481. Aust. Acad. Science, Canberra.

Becking J H 1983 The *Parasponia parviflora — Rhizobium* symbiosis: Host specificity, growth and nitrogen fixation under various conditions. Plant and Soil 75, 309–342.

Bogusz D, Appleby C A, Landsmann J, Dennis E S, Trinick M J and Peacock J W 1988 Functioning haemoglobin genes in non-nodulating plants. Nature (London) 331, 178–180.

Faria S M de, McInroy S G and Sprent J I 1987 The occurrence of infected cells, with persistent infection threads, in legume root nodules. Can. J. Bot. 65, 553–558.

Faria S M de, Sutherland J M and Sprent J I 1986 A new type of infected cell in root nodules of *Andira* spp. (Leguminosae). Plant Sci. 45, 143–147.

Fleming A I, Wittenberg J B, Wittenberg B A, Dudman W F and Appleby C A 1987 The purification, characterisation and ligand binding kinetics of hemoglobins from root nodules of the non-leguminous *Casuarina glauca–Frankia* symbiosis. Biochim. Biophys. Acta (in press).

Herridge D F 1982 Assessment of nitrogen fixation. *In* Nitrogen Fixation in Legumes. Ed. J M Vincent. pp 123–136. Academic Press.

Hirsch A M, Drake D, Jacobs T W and Long S R 1985 Nodules are induced on alfalfa roots by *Agrobacterium tumefaciens* and *Rhizobium trifolii* containing small segments of the *Rhizobium meliloti* nodulation region. J. Bacteriol. 161, 223–230.

Hrabek E M, Truchet G L, Dazzo F B and Govers F 1985 Characterization of the anomalous infection and nodulation of subterranean clover roots by *Rhizobium leguminosarum* 1020. J. Gen. Microbiol. 131, 3287–3302.

Jordan J D 1982 Transfer of *Rhizobium japonicum* Buchanan 1980 to *Bradyrhizobium* gen. mov., a genus of slow-growing root nodule bacteria from leguminous plants. Inst. J. System. Bacteriol. 32, 136–139.

Lancelle S A and Torrey J G 1984 Early development of *Rhizobium*-induced root nodules of *Parasponia rigida* I. Infection and early nodule initiation. Protoplasma 123, 26–37.

Landsmann J, Dennis E S, Higgins T J V, Appleby C A, Kortt A A and Peacock W J 1986 Common evolutionary origin of legume and non-legume plant haemoglobins. Nature 324, 166–168.

Marvel D J, Torrey J G and Ausubel F M 1987 *Rhizobium* symbiotic genes required for nodulation of legume and non-legume hosts. Proc. Nat. Acad. Sci. USA 84, 1319–1323.

Scott K F 1986 Conserved nodulation genes from the non-legume symbiont *Bradyrhizobium* sp. (*Parasponia*). Nucleic Acid Res. 14, 2905–2919.

Soepadmo E 1977 Ulmaceae. *In* Flora Malesiana Ser. 1, vol. 8. Ed. C G G J van Steenis. pp 31–76. Noordhoff n.y. Jakarta.

Tjepkema J D 1983 Hemoglobins in the nitrogen-fixing root nodules of actinorhizal plants. Can. J. Bot. 61, 2924–2929.

Trinick M J 1979 Structure of nitrogen fixing root nodules formed on *Parasponia andersonii* Planch. Can. J. Microbiol. 25, 565–578.

Trinick M J 1980a Growth of *Parasponia* in agar tube culture and symbiotic effectiveness of isolates from *Parasponia* sp. New Phytol. 85, 37–45.

Trinick M J 1980b Effects of oxygen, temperature and other factors on the reduction of acetylene by root nodules formed by *Rhizobium* on *Parasponia andersonii* Planch. New Phytol. 86, 27–38.

Trinick M J 1981 The effective rhizobium symbiosis with the non-legume *Parasponia andersonii*. *In* Current Perspectives in Nitrogen Fixation. Eds. A H Gibson and W E Newton. p 480. Aust. Acad. Science, Canberra.

Trinick M J 1982 Biology. *In* Nitrogen Fixation, volume 2. Ed. W J Broughton. pp. 76–146. Oxford University Press, New York.

Trinick M J 1987 *Bradyrhizobium* of the non-legume, *Parasponia*. *In* Microbiology in Action, a tribute to James M. Vincecnt, held at the Sydney University July, 1986. Eds. I R Kennedy and W G Murrell. John Wiley and Sons, U.K. In press.

Trinick M J and Appleby C A 1984 The rhizobia of *Parasponia*. *In* The Seventh Australian Legume Nodulation Conference, held in association with the Australian Institute of Agricultural Science, AIAS Occasional Publication No 12. pp 87–88.

Trinick M J and Galbraith J 1980 The *Rhizobium* requirements of the non-legume *Parasponia* in relationship to the cross-inoculation group concept of legumes. New Phytol. 86, 17–26.

Trinick M J and Hadobas P A 1986 Nodulation of *Trifolium repens* with naturally modified *Rhizobium* from *Parasponia*. *In* The Eighth Australian Legume Nodulation Conference, held in association with the Australian Institute of Agricultural Science, AIAS Occasional Publication No 25, pp 87–88.

Vietmeyer N and Cottom B 1977 Leucaena — Promising Forage and Tree Crop for the Tropics. National Academy of Sciences, Washington, D.C.

Vincent J M 1980 Factors Controlling the Legume — *Rhizobium* Symbiosis. *In* Nitrogen Fixation, Volume 11. Eds. W E Newton and W H Orme-Johnson. pp. 103–129.

F. A. Skinner et al. (Eds.), Nitrogen fixation with non-legumes, 35–46.
© 1989 by Kluwer Academic Publishers.

Glasshouse evaluation of the growth of *Alnus rubra* and *Alnus glutinosa* on peat and acid brown earth soils when inoculated with four sources of *Frankia*

L. J. SHEPPARD[1], J. E. HOOKER[2], C. T. WHEELER[2] and R. I. SMITH[1]
[1]*Institute of Terrestrial Ecology, Bush Estate, Penicuik, Midlothian, UK and* [2]*Department of Botany, University of Glasgow, Glasgow G12 8QQ, UK*

Key words: acid brown earth, *Alnus glutinosa*, *Alnus rubra*, *Frankia*, nitrogen fixation, nodule, peat

Abstract

The effects of soil type (an acid peat and 2 acid brown earths) and *Frankia* source (3 spore-positive crushed nodule inocula and spore-negative crushed nodules containing the single *Frankia* ArI5) on nodulation, N content and growth of *Alnus glutinosa* and *A. rubra* were determined in a glasshouse pot experiment of two years duration. Plants on all soils required additional P for growth. Growth of both species was very poor on peat with *A. glutinosa* superior to *A. rubra*. The former species was also superior to *A. rubra* on an acid brown earth with low pH and low P content. Some plant-inoculum combinations were of notable effectivity on particular soils but soil type was the major source of variation in plant weight. Inoculation with crushed nodules containing *Frankia* ArI5 only gave poor infection of the host plant, suggesting that inoculation with locally-collected crushed nodules can be a preferred alternative to inoculation with *Frankia* isolates of untested effectivity. Evidence of adaptation of *Frankia* to particular soils was obtained. Thus, while the growth of all strains was stimulated by mineral soil extracts, inhibitory effects of peat extracts were more apparent with isolates from nodules from mineral soils than from peat, suggesting that survival of *Frankia* on peat may be improved by strain selection.

Introduction

Alnus species have been used widely for the afforestation of difficult and derelict land. Several species are tolerant of acid and of water-saturated soils and this has focussed interest on their potential for afforestation of acid forest soils and peatlands (Wheeler *et al.*, 1986; Zehetmayr, 1954). Species such as *Alnus glutinosa* (L.) Gaertn. or *A. rubra* Bong. generally grow well and show good nodulation on mineral soils. In contrast, growth often is poor on highly organic peats, although nodulation and growth may be improved by draining and by application of phosphorus or molybdenum (McVean, 1962, 1963; Zehetmayr, 1954). Increase in nodule number on such soils is likely to occur at or close to the initial sites of infection on the plant since peat generally is inhibitory to the survival and spread of *Frankia* (Arveby and Huss-

Danell, 1988; Hooker, 1987). By contrast, *Frankia* occurs and spreads widely in soils of lower organic content so that widespread nodulation of the root system may be observed even on acid soils (Hooker, 1987).

As in the legume-*Rhizobium* symbiosis, the *Frankia* strains which are indigenous to particular soils may not always give rise to the most effective nodules. Glasshouse tests have shown considerable variation in the effectivity in symbiotic nitrogen fixation of different *Frankia* strains, whether originating from soil inoculum (Wheeler *et al.*, 1981), from crushed nodule preparations (Reddell and Bowen, 1985) or as cultured isolates (Dillon and Baker, 1982; Hooker and Wheeler, 1987). Spore (+) *Frankia* strains—strains which produce spores within infected host plant cells (Torrey, 1987)—are common in some soils and often give rise to nodules which are less effective in nitrogen

Table 1. Physical and chemical properties of the irradiated soils

pH (0.01 M CaCl$_2$)	sand (%)	silt (%)	clay (%)	C (Tinsley) (%)	P (Truog) (%)	K (Extractable in M NH$_4$OAc pH7) (mg 100 g^{-1})	Ca (mg 100 g^{-1})	Mg (mg 100 g^{-1})	Total N (H$_2$SO$_4$ + H$_2$O$_2$) (%)	P (%)
Elibank (acid brown earth)										
4.1	60	29	11	7	2.4	18	29	7.4	0.44	0.10
Leadburn (acid brown earth)										
4.5	62	23	15	7	1.3	13	65	9.3	0.36	0.06
Leadburn (peat)										
2.9	–	–	–	49	–	7.4	44	56	1.01	0.02

fixation than spore (−) strains (Normand and Lalonde, 1982; Sellstedt *et al.*, 1986; van den Bosch and Torrey, 1984).

Effectivity of symbiotic nitrogen fixation by *Frankia* has usually been determined for plants grown on N-free media, such as Perlite or sand, and little information is available to show whether strains are adapted to particular soil types. The research reported here compared the symbiosis between different sources of *Frankia* and *A. rubra* or *A. glutinosa* grown in two nitrogen-deficient, acid brown earth soils and in acid peat, soils typical of those on which *Alnus* might be planted in Scotland. The utilization of crushed nodule preparations as inoculum permitted the inclusion of spore (+) *Frankia* in the comparison, since maintenance of this type of *Frankia* in culture has not been demonstrated unequivocally (Torrey, 1987). This technique also is of practical value for nurserymen, who are still more likely at present to use crushed nodules than cultured *Frankia* for seedling inoculation. The results of preliminary experiments to determine whether *Frankia* shows adaptation to inhibitory or stimulatory, water-soluble soil factors, which may influence growth and survival in soil, are also reported.

Materials and methods

Sources of inoculum

The characteristics of Scottish sites from which *A. glutinosa* (Milngavie) and *A. rubra* (Lennox) nodules were collected have been described previously (Hooker and Wheeler, 1987). Corvallis

and ArI5 nodules were North American sources of *Frankia*. Corvallis nodules were from water cultured *A. rubra*, maintained in the glasshouse for five years after inoculation of plants with field nodules from *A. rubra* (Wheeler *et al.*, 1981). Inoculum containing *Frankia* ArI5 was prepared from nodules harvested from *A. rubra* which had been grown in Perlite supplied with N-free nutrients for six months after inoculation with this *Frankia* strain. ArI5 gave rise to nodules which were always spore (−). At least some lobes of nodules formed after inoculation with the other *Frankia* sources were spore (+).

Sources, collection and properties of soil

Soils were collected from sites in Scotland, sieved to < 1 cm, mixed and stored in plastic bags. Peat from Leadburn came shredded from a local extraction site. Other soils were collected from the top 10 cm below the litter layer. Soils and peat from Elibank and Leadburn were sterilized by irradiation to 4 megarads 11 days after collection (Parker and Vincent, 1981). Other soils were stored at 4°C for up to 3 days, until used for preparation of aqueous extracts. Some chemical properties of the soils used for plant growth are described in Table 1. Soils used for preparation of aqueous extracts were an agricultural brown earth (Wolfson Hall), from underplanted *A. rubra* (Malcolm *et al.*, 1985) and acid peat soils from Leadburn (Table 1), Wauchope (Hooker and Wheeler, 1987) and Loch Ard (Ordnance Survey Reference NS 511922). The vegetation cover of the latter peat site included *Calluna vulgaris*, together with *Myrica gale* and *Molinia* spp.

Table 2. Origins of seed

Provenance	Species	Origin	Latitude	Altitude (m)
Loch Fyne	*Alnus glutinosa*	Scotland	56°9′	135
Timberlands	*Alnus rubra*	Vancouver Is, Canada	49°3′	168
Terrace	*Alnus rubra*	British Columbia, Canada	54°31′	–

Source and growth of plant material

Seed of two provenances of *Alnus rubra* and one of *A. glutinosa* (Table 2) were stratified at 2°C for 6 weeks then germinated at 18°C/6°C (day/night) in Perlite enriched with nitrogen free Crone's nutrients (Wheeler *et al.*, 1986). Seedlings were inoculated 32 days after sowing and each seed tray supplied with 4 mg N. After two months, two 2-3 cm seedlings were transplanted into each sterile 5 cm square pot containing irradiated soil. One month later the best seedling was repotted into a 12 cm, 1.25-l pot on a foil saucer and maintained in a cool, unlit glasshouse. At the beginning of the second growing season all the pots received 0.31 g P and 0.78 g K as K_2HPO_4. The plants were treated twice with gamma BHC (Murphy U.K. Ltd.,) and malathione to control an infestation of vine weevil (*Otiorrhyncus sulcatus*) and all weeds were removed.

Experimental design

A factorial design was used with three soil types (peat and Elibank and Leadburn brown earths), four *Frankia* sources (control—no *Frankia*, Lennox, Milngavie, Corvallis and ArI5), one provenance of *A. glutinosa* and two of *A. rubra*. The 45 treatment combinations (5 × 3 × 3) were replicated 20 times giving 900 plants in all. At the end of the first growing season following budset, first year heights were measured and the 20 plants for each treatment combination were allocated to the blocks (4 per block) on the basis of height, so that the tallest plants were in block 5. The blocks were split into the three soil types which for practical reasons were kept together, and plants randomly positioned subject to this constraint.

Harvesting and analysis

Plants were harvested 511 days after sowing. The number of nodules, and root, shoot and nodule dry weights were determined for each plant after drying in an oven at 80°C to constant weight. Shoot material was combined from the four replicates in each block and root and nodules from all plants within a single treatment. Concentrations of N, P, K, Ca and Mg were determined on the finely ground leaf and stem material following wet oxidation (Parkinson and Allen, 1975). The nitrogen content of roots plus nodules was determined by Kjeldahl analysis (Hooker and Wheeler, 1987). None of the treatments significantly influenced plant nutrient concentrations other than N content and hence only nitrogen data is presented in this paper.

Preparation of soil extracts

Soil was sieved to remove particles larger than 2 mm. Equal volumes of soil and distilled water were shaken on a rotary shaker at 20°C for 17 h. Larger soil particles were allowed to sediment after which the water was decanted and centrifuged at 10,000 g for 1 h. The supernatant liquid was filtered sequentially under vacuum through Whatmann 44 filter paper, Whatman GF/C glassfibre, Whatman 0.45 µm membrane filter and finally Whatman 0.2 µm membrane filter. Distilled water for preparation of media for control flasks was passed through the same sequence of filters.

BuCT medium (Malcolm *et al.*, 1985) containing mineral salts, casamino acids and Tween 80 was prepared in soil extracts or distilled water. All media were filter sterilized. Conical flasks (100 ml) containing 40 ml of each medium were each inoculated with 13 µg protein of *Frankia* UGL 013110,

UGL 013113 or with UGL 010701 as described previously (Hooker and Wheeler, 1987). Flasks were static during incubation in the dark at 28°C, before harvest of *Frankia* after 25 days (first two strains) or 42 days (last strain). The contents of flasks were collected by filtration under vacuum through 20 μm membrane filters, made up to volume and assayed for protein with Pierce protein assay reagent (Warner and Pierce, U.K. Ltd.) after sonication for 60 sec as described previously (Hooker and Wheeler, 1987).

Results

Interactions between Frankia, *host plant species,* A. rubra *provenance and soil type*

All surviving inoculated plants were nodulated at harvest, with the exception of plants grown in peat where only 70% were infected. The non-nodulated plants on peat were chlorotic and grew poorly and many died during the course of the experiment, presumably due to poor nutrition. Plant deaths also occurred randomly in Elibank and Leadburn

soils due to vine weevil infestation. These dead plants are included in the data analyses as missing values.

Plant nodulation

Some of the roots of the uninoculated control plants were found at harvest to bear nodules, indicative of a degree of cross-infection of plants during the 2-year growth period. Cross-infection probably occurred late in the two year growth period since most nodules were small and were dispersed towards the periphery of the root system, distant from the root crown where the initial infections were concentrated on inoculated plants. Mean dry weights of infected control plants were only different statistically from those of uninfected control plants for *A. rubra* (Timberlands) on Elibank soil, where there was a difference of 19%. In other instances, late nodulation had little effect on the growth of control plants.

Crushed nodule inoculum containing Ar15 was the least infective of the *Frankia* sources tested on both *A. glutinosa* and *A. rubra*, 18% of the plants

Table 3. Mean nodule numbers on *Alnus* inoculated with *Frankia* from different sources and grown in different soils

Soil	Inoculum	Host plant[1]		
		A. glutinosa	*A. rubra* (Terrace)	*A. rubra* (Timberlands)
Elibank	Control	1.3	0.8	1.9
	Ar15	1.1	3.8	8.5
	Corvallis	38.2	99.0	77.6
	Lennox	41.0	124.4	123.2
	Milngavie	10.3	125.0	81.8
Leadburn	Control	1.7	4.3	5.8
	Ar15	3.2	1.8	9.8
	Corvallis	71.9	115.9	90.3
	Lennox	109.8	143.4	147.2
	Milngavie	55.7	92.9	107.9
Peat	Control	0.7	0.0	0.0
	Ar15	0.2	0.1	0.2
	Corvallis	9.1	1.1	1.3
	Lennox	25.7	2.9	1.0
	Milngavie	8.5	5.1	1.8
		$P < 0.050$	$P < 0.010$	$P < 0.001$
LSD: all levels		27.40	36.01	46.01
same level		24.58	32.30	41.27
SED: all levels		13.98		
same level		12.54		

[1]. Locations in brackets are of *A. rubra* provenances

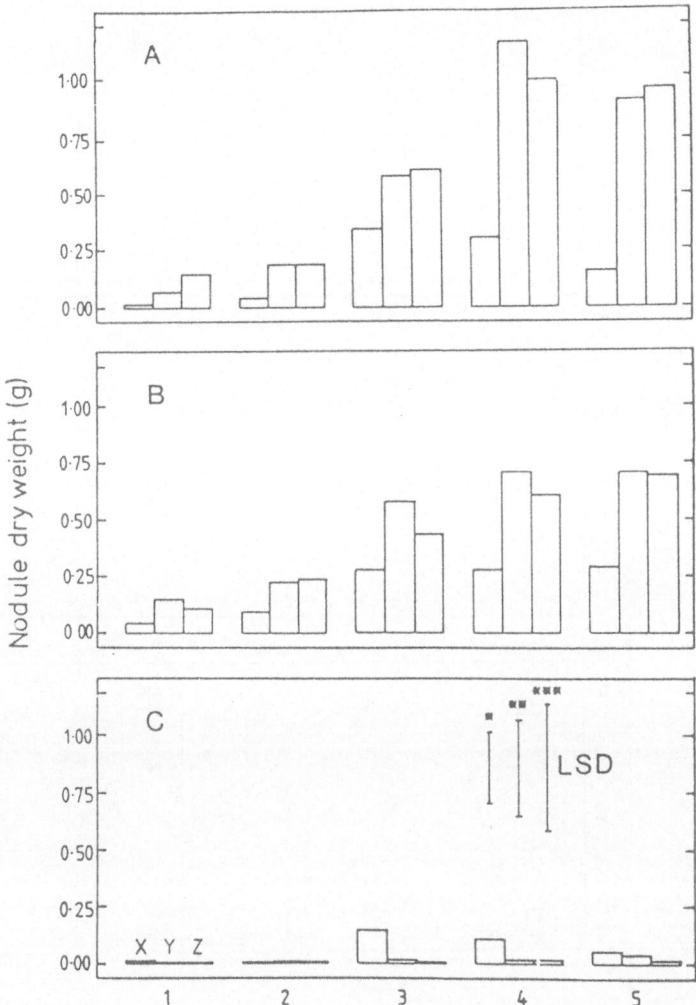

Fig. 1. Nodule dry weight of *Alnus* spp. growing in soils from A, Elibank; B, Leadburn and C, peat after inoculation with crushed nodules from 1, control—no inoculum; 2, ArI5; 3, Corvallis; 4, Lennox; 5, Milngavie. The letters X, Y, Z indicate the plant species; *A. glutinosa, A. rubra* (Terrace provenance) and *A. rubra* (Timberlands provenance) respectively.

being nodule-free at harvest. The number and weight of nodules per plant were generally not significantly greater than were formed by chance infection of control plants (Table 3 and Fig. 1). However, nodules always comprised a higher percentage of the dry weight of the root system in inoculated compared with control plants, the presence of nodules from the early stages of seedling growth retarding root development slightly.

Virtually all plants grown on the mineral soils and inoculated with crushed nodules containing Corvallis, Lennox or Milngavie *Frankia* nodulated,

and more than 75% produced densely nodulated root systems to which nodules contributed more than 5% of the dry weight. Within a species, varying the sources of inoculum gave rise to different nodulation responses, which were modified by the soil on which the plants were grown. *Alnus rubra* grown on peat was poorly nodulated irrespective of the source of inoculum. However, nodulation of *A. glutinosa* was improved with the Lennox or Corvallis *Frankia* sources.

On both Elibank and Leadburn soils, Corvallis inoculum generally gave rise to lower nodule num-

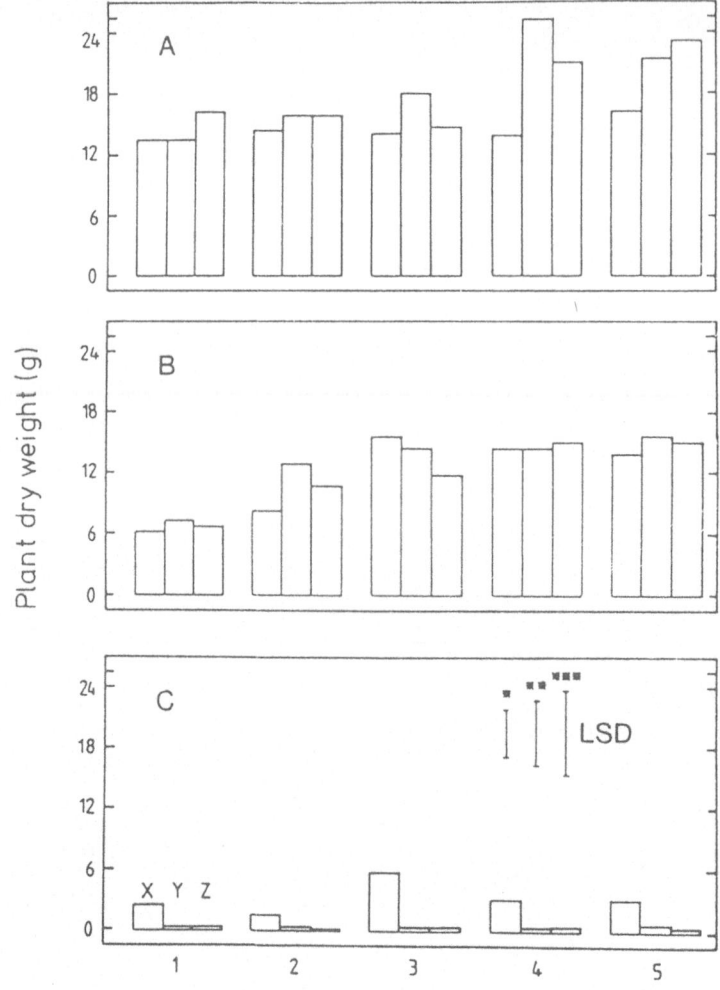

Fig. 2. Total dry weight of *Alnus* spp. growing in acid brown earth or peat soils after inoculation with crushed nodules. Designation of numbers and letters on the figure are as described in the legend to Figure 1.

bers and weights on *A. rubra* than Lennox or Milngavie inoculum, although differences between plants were smaller on the latter soil. Differences in the nodulation of *A. glutinosa* by these *Frankia* sources were also smaller on Leadburn than Elibank soil but with this species, poorest nodulation was with Milngavie inoculum on Elibank (Table 3 and Fig. 1). Analysis of Variance of the data showed that the main sources of variation were the soil and the inoculum. All three factors (host plant, inoculum and soil) showed significant interactions which affected the number and weight of nodules per plant.

Plant dry weight

Growth on peat of both species was poor, although the dry weight of *A. glutinosa* at harvest was substantially greater than that of *A. rubra*, with best growth by plants inoculated with Corvallis *Frankia*. The dry weight at harvest of *A. glutinosa* inoculated with this source of *Frankia* was about one third that of plants grown on the acid brown earth soils (Fig. 2).

Generally, there were few differences in dry weight accretion in the two *A. rubra* provenances. The dry weights of *A. rubra*, inoculated with

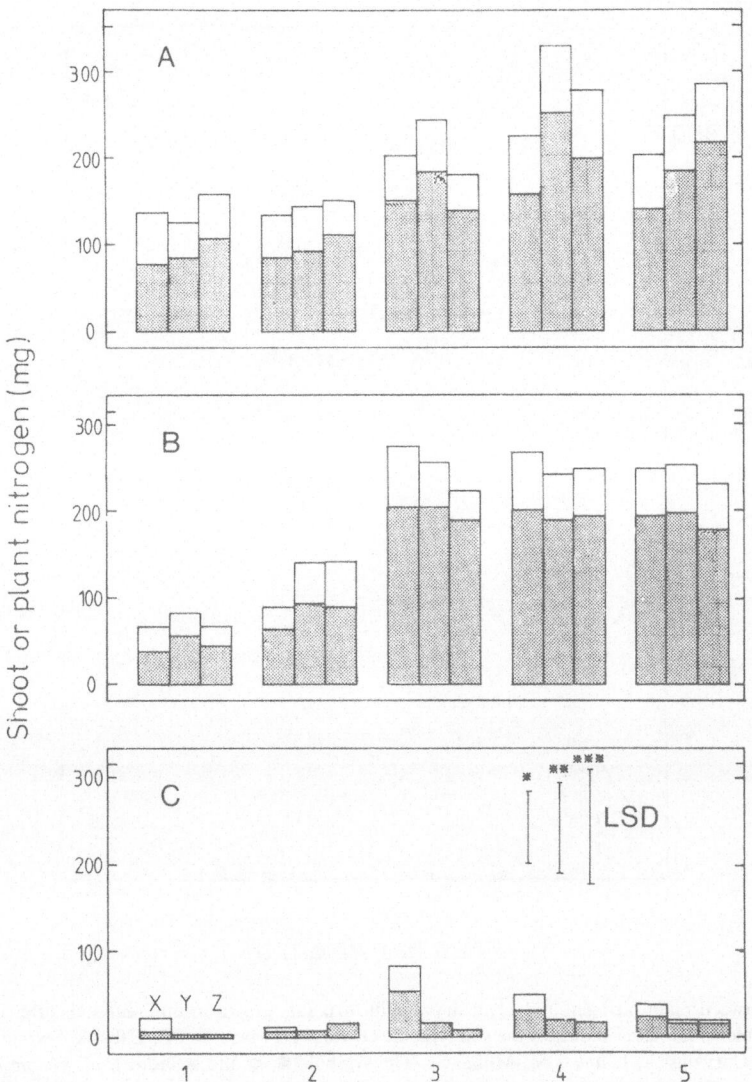

Fig. 3. Shoot (shaded area) or plant (non-shaded area) nitrogen of *Alnus* spp. growing in acid brown earth or peat soils after inoculation with crushed nodules. Designation of numbers and letters on the figure are as described in the legend to Figure 1.

Corvallis, Lennox or Milngavie *Frankia* and grown on Leadburn soil were not significantly different from controls. Best growth of *A. rubra* was with plants inoculated with Lennox *Frankia* and grown on Elibank soil. By contrast, dry weights of inoculated *A. glutinosa* on Elibank soil were increased significantly only with Milngavie inoculum, even though nodulation was also increased in plants inoculated with Corvallis and Lennox *Frankia*. On Leadburn soil, the dry weight of plants inoculated with these *Frankia* sources were increased about 1.5 times compared with controls.

Analysis of Variance of the data showed that soil was the major source of variation in plant dry weight. Inoculum had less effect. Soil, inoculum and host plant all showed significant interactions which affected plant dry weight at harvest.

Plant N content

The poor growth of plants on peat is reflected in their low N content, with *A. glutinosa* containing

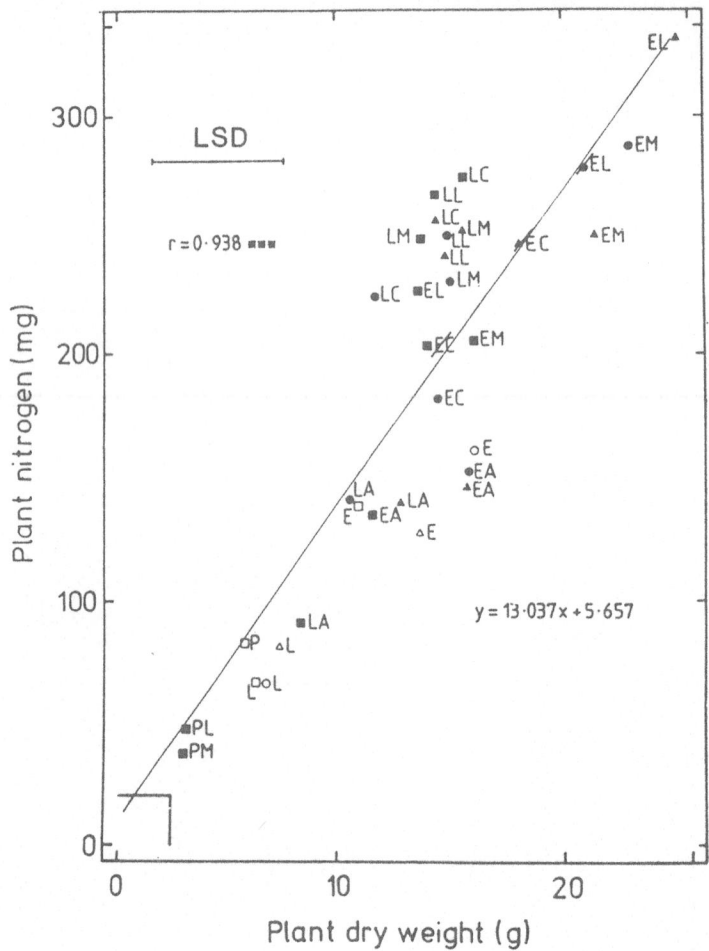

Fig. 4. Relationship between plant nitrogen and plant dry weight of plants grown in different soils after inoculation with different sources of crushed nodules. First letter indicates the soil type; L, Leadburn; P, peat and E, Elibank. Second letter indicates source of inoculum. A, ArI5; C, Corvallis; L, Lennox; M, Milngavie. The symbols ■, ● and ▲ indicate *A. glutinosa*, *A. rubra* (Timberlands provenance) and *A. rubra* (Terrace provenance) respectively. □, ○ and △ indicate corresponding non-inoculated controls. Data within box is for *A. rubra* grown in peat.

most N (Fig. 3). Inoculation with ArI5 had little effect on plant N on any of the soils but Corvallis, Lennox and Milngavie *Frankia* all gave significant increases with both *A. glutinosa* and *A. rubra*. Although the dry weight of control plants on Elibank was about twice that of plants on Leadburn soil, the % N content of controls was very similar (0.93 to 1.01% on the former and 0.99 to 1.10% on the latter soil). There was a highly significant, linear correlation between plant dry weight and N content for all associations together (Fig. 4), showing that N supply was a major factor determining growth on the Elibank and Leadburn soils. It is notable that the most effective of the *Frankia*-host

plant associations on Leadburn soil lie above the regression line and consequently have higher % N contents than the plants on Elibank, more of which lie on or below the regression line. Because plants developed greatest dry weight on Elibank soil, some factor(s) other than N supply must restrict growth on Leadburn soil.

Effect of soil extracts on Frankia *growth*

This experiment was carried out to determine whether compounds antagonistic to the growth of *Frankia* were present in peat soils. Aqueous ex-

Table 4. Effects of soil extracts on growth of *Frankia* in BuCT medium

Strain and source of extract	Protein per flask (μg + S.E.[1,2])	Final pH of culture medium
UGL 013113		
Control (no extract)	94.6 + 18.34	7.0
Loch Ard (peat)	73.5 + 7.74	6.8
Wauchope (peat)	862.9 + 16.45	7.7
Wolfson Hall (loam)	935.1 + 22.36	7.7
UGL 013110		
Control (no extract)	46.5 + 5.73	6.8
Loch Ard (peat)	7.9 + 1.04	6.8
Wauchope (peat)	41.4 + 5.16	6.9
Wolfson Hall (loam)	217.8 + 12.06	7.1
UGL 010701		
Control	459.8 + 22.89	7.7
Leadburn (peat)	423.6 + 15.60	7.7
Wauchope (peat)	516.4 + 15.68	7.7
Wolfson Hall (loam)	626.4 + 59.19	7.7

[1] Figures are means of duplicate measurements on the contents of 11 (UGL 013113 and UGL 013110) and 8 (UGL 010701) replicate flasks.

[2] Extracts in flasks with UGL 010701 were quarter strength of those used in flasks with UGL 013113 and UGL 013110, and were grown for 42 days.

tracts from three peats from Leadburn, Wauchope and Loch Ard were added to propionate medium in which were grown *Frankia* isolated from nodules from mineral soils (UGL 013110, UGL 010701) or from Wauchope peat (UGL 013113). Comparisons were with control flasks (propionate medium alone) or flasks which were supplemented with extract from a mineral soil on which alders nodulate well. From growth curves established previously (Hooker, 1987), cultures were harvested about the middle of the logarithmic growth phase so that differences between control and treatment flasks were not lost with the advent of the stationary phase.

The extracts elicited different growth responses with the various *Frankia* strains (Table 4). The peat and mineral soil extracts added to cultures of UGL 010701, isolated from *A. glutinosa* nodules from mineral soils, were more dilute than those supplied to the other two strains. However, the results showed a different growth response to the two peats. Leadburn peat extracts inhibited growth slightly, Wauchope peat had little effect and mineral soil extracts stimulated growth by more than one third. The growth of strain UGL 013110, isolated from *A. rubra* nodules, also from mineral soil, was increased almost 5 times by a more concentrated extract of mineral soil but again the response to peat extracts

varied according to source. Extracts of Wauchope peat affected growth little, as with strain UGL 010701, whereas Loch Ard peat extracts inhibited growth 6-fold. By contrast, growth of UGL 013113, isolated from nodules from Wauchope peat, was stimulated more than 9 times by extracts from this peat and from mineral soil and was inhibited only slightly by the Loch Ard peat extracts. Strains would not grow in peat or soil extracts alone.

These results show, therefore, that some mineral soils contain water soluble substances which stimulate the growth of *Frankia* but that some peats contain water soluble substances which can either stimulate, have no effect or inhibit *Frankia* growth, depending upon the strain.

Discussion

A notable feature of the experiments reported was that nodulation, the dry weight and the N content of alders inoculated with crushed nodules containing the spore($-$) *Frankia* ArI5 was consistently inferior to that of plants inoculated with spore ($+$) crushed nodule preparations. Lower infectivity of inoculum from spore ($-$) nodules compared with that from spore ($+$) nodules has been noted by others (Houwers and Akkermans, 1981) and could contribute to the poor performance in symbiosis of ArI5, noted here. It is likely that factor(s) attributable directly to the use of crushed nodule inoculum from ArI5 infected plants, rather than cultured *Frankia*, contribute to the poor infection since in a previous study, satisfactory nodulation was observed for alders inoculated with cultured ArI5 and grown in Perlite with pH close to neutral (Wheeler *et al.*, 1986). Inimical soil conditions, such as low pH, may also affect ArI5 more than the *Frankia* in the other crushed nodule inocula tested.

In general, on all soils nodulation of both *A. glutinosa* and *A. rubra* inoculated with Lennox *A. rubra* crushed nodules equalled, or was superior to that achieved with other sources of inoculum (Table 3 and Fig. 1). This result suggests that this inoculum is best suited overall for inoculation of the two species for growth on acid soils. It should be noted however that inoculation with *A. rubra* crushed nodules gave rise to a large number of small nodules on *A. glutinosa* whereas plants inoculated with *A. glutinosa* (Milngavie) crushed nodules formed a smaller number of large nodules. This is indicative of a degree of incompatibility between the partners in the heterologous inoculation.

Irrespective of the potential of a strain for nodulation and N fixation, it is clear that the properties of a particular soil can have an overriding influence on plant growth. Acid peat soil clearly interferes grossly with both nodulation and plant growth and will be considered further below. On the acid brown earth soils, greater nodule numbers on both *A. rubra* and *A. glutinosa* on Leadburn compared with Elibank probably reflected greater infection of plants on the less acid soil. However, at harvest both nodule and total plant dry weight was greater for plants from Elibank soil, even though the percentage N content of most inoculated plants was higher on Leadburn. Clearly, factor(s) other than pH and N supply must limit plant growth on Leadburn soil, the most likely factor from the soil analyses being the phosphorus content of the soil during the first year of growth — Elibank contained twice the P content of Leadburn soil. Differences in growth were less between *A. glutinosa* on the two soils than between *A. rubra*, illustrative of greater suitability of *A. glutinosa* for planting on the more difficult soil.

Growth of *Alnus* on acid peat in Britain can range from reasonably satisfactory to very poor with causes of failure ascribed variously to low pH, low nutrient availability and anaerobicity (Hooker, 1987; McVean 1962, 1963; Zehetmayr, 1954). The low pH of the Leadburn peat used in the present study will have been a major factor leading to the greatly reduced growth and death of *Alnus* on this soil. The root systems of plants were poorly developed, probably leading to water stress and nutrient shortage. Nodulation, growth and survival of *A. glutinosa* was substantially better than that of *A. rubra*. With the former species, choice of a suitable inoculum was more important for satisfactory N nutrition on peat than on the mineral soils. Thus, inoculation of *A. glutinosa* with Lennox crushed nodules gave the greatest number of nodules, but these remained relatively small. Corvallis inoculum gave plants with the greatest nodule weight, nitrogen content and total dry weight even though the nodule number per plant was little more than one third of that with Lennox inoculum. Inhibition of root development by heavy infection may have limited the ability of Lennox-inoculated *A. glutinosa* to absorb nutrients, thus restricting further plant growth. The ability of *A. glutinosa* to grow in very acid conditions has been noted by others (Troelstra *et al.*, 1986) and this species is clearly more suitable for planting on acid peats of the type studied here.

Frankia on peat soils generally has restricted occurrence and spreads relatively little (Arveby and Huss-Danell, 1988; Hooker, 1987). Because *Frankia* capable of nodulating *Alnus* occurs widely in mineral soils, even in acid soils which have been free of actinorhizal plants for many years (Huss-Danell and Frej, 1986; Rodriguez-Barrueco, 1968; Smolander and Sundman, 1987), experiments were carried out to ascertain whether peat contains toxic principles which may inhibit the growth of *Frankia*. The results obtained provided some indication of *Frankia* strain adaptation to particular peat soils. Thus, although the growth of all strains was stimulated by mineral soil extracts, inhibitory effects of peat extracts were more apparent with the *Frankia* isolates from nodules of plants growing on mineral soil.

The nature of the inhibitory or stimulatory compound(s) is unknown. Stimulatory or inhibitory effects of soil phenolics on *Frankia* growth have been reported (Perradin *et al.*, 1983; Vogel and Dawson, 1986). Humic acids will stimulate the growth of microorganisms from organic soils more than from a sandy soil (Visser, 1985) and will also stimulate *Frankia* growth (Wheeler, unpublished data). The phytotoxic and fungitoxic effects of short-chain fatty acids found in *Calluna* heathland soils have been reported (Jalal and Read, 1983). *Calluna vulgaris* was present only in the vegetation of peat from Loch Ard from which the more toxic extracts were obtained and conceivably could be the source of compounds inhibitory to *Frankia* growth. Further research is required to establish whether resistance to toxic principles is a useful property for the infectivity and survival of *Frankia* strains nodulating *Alnus* on peat.

The main findings of these experiments stress the need for testing *Frankia* in the soil of interest before using specific strains to inoculate alders for planting. Local soil effects on plant growth, infection and competition may be as, or more important than differences in effectivity of nitrogen fixation in determining the suitability of an *Alnus–Frankia* combination for a particular site. Multiple occurrence of *Frankia* strains in field nodules (Benson and Hanna, 1983; Hooker and Wheeler, 1987; Wheeler *et al.*, 1986) increases the possibility of infecting plants inoculated with field nodule preparations with strains adapted to particular soil conditions.

The superior performance of plants inoculated with preparations of field nodules compared with those containing ArI5 shows that nursery inoculation with locally-collected crushed nodules can be a satisfactory or indeed preferred alternative to inoculation with *Frankia* isolates where effectivity in particular conditions may not have been tested.

Acknowledgements

We would like to thank Dr M G R Cannell for helpful advice at various stages of the work. Mr D Cameron, Mrs M B Murray, Miss Anne Halliday and staff of ITE, Merlewood, are thanked for help with biomass and nutrient analysis. Thanks are also due to Mr. Stuart Allison, Glasgow, for contributing to assessments of effects of soil extracts on *Frankia* growth. J E Hooker and C T Wheeler were supported by a research grant from NERC.

References

Arveby A S and Huss-Danell K 1988 Presence and spreading of infective *Frankia* in peat and meadow soils in Sweden. Biol. Fertil. Soils 5, 1–6.

Benson D R and Hanna D 1983 *Frankia* diversity in an alder stand as estimated by sodium dodecyl sulphate-polyacrylamide gel electrophoresis of whole-cell proteins. Can. J. Bot. 61, 2919–2923.

Dillon J T and Baker D 1982 Variations in nitrogenase activity among pure cultured *Frankia* strains tested on actinorhizal plants as an indication of symbiotic compatibility. New Phytol. 92, 215–219.

Hooker J E 1987 Variation in *Frankia* strains isolated from *Alnus* root nodules. Ph.D. Thesis, University of Glasgow, p. 288.

Hooker J E and Wheeler C T 1987 The effectivity of *Frankia* for nodulation and nitrogen fixation in *Alnus rubra* and *Alnus glutinosa*. Physiol. Plant. 70, 333–341.

Houwers A and Akkermans A D L 1981 Influence of inoculation on yield of *Alnus glutinosa* in the Netherlands. Plant and Soil 61, 189–202.

Huss-Danell K and Frej A-K 1986 Distribution of *Frankia* in soils from forests and afforestation sites in northern Sweden. Plant and Soil 90, 407–418.

Jalal M A F and Read D J 1983 The organic acid composition of *Calluna* heathland soil with special reference to phyto- and fungitoxicity. I. Isolation and identification of organic acids. Plant and Soil 70, 257–272.

Malcolm D C, Hooker J E and Wheeler C T 1985 *Frankia* symbiosis as a source of nitrogen in forestry: A case study of symbiotic nitrogen fixation in a mixed *Alnus-Picea* plantation in Scotland. Proc. Roy. Soc. Edin. 85B, 263–282.

McVean D N 1962 The establishment of alder on peatland and its possible role in afforestation. Irish Forestry 19, 81–84.

McVean D N 1963 Scots Pine, Alder and Bog Myrtle. Report of the Nature Conservancy, London, 78–79.

Normand P and Lalonde M 1982 Evaluation of *Frankia* strains isolated from provenances of two *Alnus* species. Can. J. Microb. 28, 1133–1142.

Parker J A and Allen S E 1975 A wet oxidation procedure suitable for the determination of N and mineral nutrients in biological material. Commun. Soil Sci. and Plant Anal. 6, 1–11.

Perradin Y, Mottet M J and Lalonde M 1983 Influence of phenolics on *in vitro* growth of *Frankia* strains. Can. J. Bot. 6, 2807–2814.

Reddell P and Bowen G D 1985 Do single nodules of Casuarinaceae contain more than one *Frankia* strain? Plant and Soil 88, 275–279.

Rodriguez-Barrueco C 1968 The occurrence of the root-nodule endophyte of *Alnus glutinosa* and *Myrica gale* in soils. J. Gen. Microbiol. 52, 189–194.

Sellstedt A, Huss-Danell K and Ahlqvist A-S 1986 Nitrogen fixation and biomass production in symbioses between *Alnus incana* and *Frankia* strains with different hydrogen metabolism. Physiol. Plant 66, 99–107.

Smolander A and Sundman V 1987 *Frankia* in acid soils of forests devoid of actinorhizal plants. Physiol. Plant. 70, 297–303.

Torrey J G 1987 Endophyte sporulation in actinorhizal nodules. Physiol. Plant 70, 279–288.

Troelstra S R, Van Dijk C and Blacquière T 1986 Growth, ionic balance, proton excretion and nitrate reductase activity in *Alnus* and *Hippophae* supplied with different sources of nitrogen. Plant and Soil 91, 381–384.

Van den Bosch K A and Torrey J G 1984 Consequences of sporangial development for nodule function in root nodules of *Comptonia peregrina* and *Myrica gale*. Pl. Physiol. 76, 556–560.

Visser S A 1985 Physiological action of humic substances on microbial cells. Soil Biol. Biochem. 17, 457–462.

Vogel C S and Dawson J O 1986 *In vitro* growth of five *Frankia* isolates in the presence of four phenolic acids and juglone. Soil Biol. Biochem. 18, 227–231.

Wheeler C T, McLaughlin M E and Steele P 1981 A comparison of symbiotic nitrogen fixation in Scotland in *Alnus glutinosa* and *Alnus rubra*. Plant and Soil 61, 169–188.

Wheeler C T, Hooker J E, Crowe A and Berrie A M M 1986 The improvement and utilisation in forestry of nitrogen fixation by actinorhizal plants, with special reference to *Alnus* in Scotland. Plant and Soil 90, 393–406.

Zehetmayr J W L 1954 Afforestation of upland heaths. Forestry Commission Bulletin 32, 102–103.

F. A. Skinner et al. (Eds.), Nitrogen fixation with non-legumes, 47–53.
© 1989 by Kluwer Academic Publishers.

Ultrastructure of *Frankia* isolated from three Chilean shrubs (Rhamnaceae)

L. LONGERI and MIREYA ABARZUA
Departamento de Agronomia, Universidad de Concepción, Casilla 537, Chillán, Chile

Key words: actinorhizal nodules, Chile, *Discaria serratifolia*, *Discaria trinervis*, *Frankia*, nitrogen
fixation, *Retanilla ephedra*, symbiosis, ultrastructure

Abstract

Ultrastructure of *Frankia* strains DtI1, DsI1 and ReI1, isolated from *Discaria trinervis*, *D. serratifolia* and *Retanilla ephedra*, was determined by transmission and scanning electron microscopy, both in the actinorhizal nodules and in the free state. The three strains showed similar structural features. Septated hypha, of 0.6 to 0.9 μm in diameter, produced sporangia and vesicles *in vitro*, but only vesicles in actinorhizae. Vesicles were spherical, of 2.4 to 3.8 μm in diameter and septated. Comparative studies of the vesicles produced *in vitro* and *in vivo* show that the bilaminar layer surrounding the vesicle void area is continuous to the hyphal cell wall, suggesting that this bilaminar structure is the vesicle cell wall.

Introduction

Studies on the nitrogen-fixing symbiosis of non-legume actinorhizal plants have attracted many researchers due to their importance in forests and other natural ecosystems.

Before the first isolation and cultivation, by Callaham *et al.* (1978), of *Frankia* in the free living state from the root nodules of *Comptonia peregrina*, research on the micromorphology of the microsymbiont was considerably hampered. Since 1978, many other successful isolations have been performed from actinorhizal nodules of plants from host genera such as *Alnus* (Baker *et al.*, 1979; Burggraaf *et al.*, 1981; Benson, 1982; Baker and O'Keefe, 1984), *Casuarina* (Baker and O'Keefe, 1984), *Ceanothus* (Lechevalier and Ruan, 1984), *Colletia* (Gauthier *et al.*, 1984), *Cowania* (Baker and O'Keefe, 1984), *Elaeagnus* (Baker *et al.*, 1979), *Hippophae* (Shipton and Burggraaf, 1982), *Myrica* (Burggraaf *et al.*, 1981; Shipton and Burggraaf, 1982), *Purshia* (Baker and O'Keefe, 1984) and *Shepherdia* (Normand and Lalonde, 1986).

In the present paper, we report the *in vivo* and *in vitro* ultrastructure of *Frankia* causing root nodulation in three Chilean shrubs, *Discaria trinervis* (Gill, ex H. et A.) Reiche, *D. serratifolia* (Vent.) B. et H. ex Masters, and *Retanilla ephedra* (Vent.) Brongn., that are common to the central valley and the Coast Range in the VIII Region, Chile.

Materials and methods

Root nodules

Actinorhizal root nodules from *D. trinervis* and *D. serratifolia* were collected from native plants growing in the area near Concepción, Chile. *Retanilla ephedra* nodules were excised from greenhouse-grown plants.

Frankia *strains*

From the collected nodules, strains DtI1 from *D. trinervis*, DsI1 from *D. serratifolia* and ReI1 from *R. ephedra*, were isolated by the authors, either by the filter exclusion technique (Benson, 1982) or the double-layer agar technique (Diem and Dommergues, 1983) from excised nodule tips sterilized with sodium hypochlorite, and maintained by transfer in solid and liquid FMC medium (Benson, 1982). For vesicle induction, nitrogen-free FMC medium was used.

Transmission electron microscopy

Root nodule lobes were cut and fixed in 2% glutaraldehyde in 0.025 M potassium phosphate buffer, pH 6.8, for 16 h at room temperature. The fixed specimens were then rinsed in buffer, post-fixed in 1% buffered osmium tetroxide for 2 h at 4°C, rinsed again in buffer, dehydrated in an ethanol series (30, 50, 70, 80, 90 and 100%) for 20 min in each concentration and embedded in Epon-Araldite.

Frankia strains were cultured on the surface of FMC agar and after incubation, small agar cubes, each containing one colony, were cut and fixed in glutaraldehyde for 2 h at room temperature. Rinsing, post-fixation, dehydration and embedding were done as previously described.

Ultra-thin sections (40 nm) were cut with glass knives, picked up on uncoated grids, stained with uranyl acetate for 45 min and with lead citrate for 35 min, and examined with a Philips EM 200 microscope operating at 80 kV.

Scanning electron microscopy

Nodule lobes were fixed in 2% glutaraldehyde in 0.025 M potassium phosphate buffer, pH 6.8, for 3 h at 4°C, rinsed in buffer, dehydrated through increasing concentrations of ethanol and critical-point dried. The dried lobes were mounted on stubs, fractured to expose the interior of the nodule and sputter coated with gold. Solid culture medium bearing surface colonies of *Frankia* was cut in small pieces that were fixed, dehydrated and critical-point dried. *Frankia* colonies were harvested from liquid cultures by filtration through 0.45 μm, 12 mm diameter Millipore membranes which were then processed for SEM. The samples were examined with an ETEC-Autoscan microscope.

Results and discussion

The ultrastructure of *Frankia* in the actinorhizal nodules and in culture was common to the three host species studied. The isolated strains are morphologically similar to described *Frankia* strains (Lalonde, 1978, 1979; Baker *et al.*, 1980; Horriere *et al.*, 1983), with minimal differences in hypha,

vesicle and spore dimensions. Colonies of the three strains were dense mats of hyphae.

In the cytoplasm of the infected actinorhizal host cells, septate hypha of 0.6–0.9 μm in diameter and spherical vesicles 2–3 μm in diameter were observed (Figs. 1, 2 and 3). Vesicles and sporangia were produced by all the isolated strains (Figs. 4 and 5); vesicles but no spores were detected in TEM and SEM preparations of nodules. The vesicles, produced at the tips of hyphae, are septate and similar to those described in *Alnus* (Gardner, 1965; Lalonde, 1979), *Elaeagnus* (Baker *et al.*, 1980) and *Discaria* (Newcomb and Pankhurst, 1982). The absence of spores in the actinorhizal nodules may suggest that Sp− *Frankia* strains, that sporulate *in vitro* but not *in vivo*, naturally induce nodulation in *D. serratifolia*, *D. trinervis* and *R. ephedra*. It has been found that nodules infected by Sp− strains are more efficient than Sp+ nodules (Normand and Lalonde, 1982).

In TEM observations, hyphae showed the same structure when grown endophytically or *in vitro* (Figs. 1 and 6). The cell wall of the hypha is two-layered, only the inner layer and the plasma membrane participating in septation (Fig. 6). The cell wall, about 0.06 μm thick, is similar in structure and thickness to the wall described in *Comptonia* isolates (Lalonde, 1978, 1979). The cytoplasm contains zones of low electron density, probably of genetic material, and numerous ribosomes.

The vesicle structure was also similar in actinorhizae or in culture. Externally, an electron-dense laminar envelope encircles the vesicles; the vesicle cytoplasm is surrounded by a thin membrane which is continuous with the vesicular septa, and a wide void area, electron translucent, occurs between the external envelope and the cytoplasm. The external envelope seems to be continuous with the hyphal cell wall and has a similar thickness (Figs. 2 and 4).

Sporangia are formed by the swelling of the ends of hyphae, followed first by transverse septation (Fig. 7) and then by longitudinal septa, forming compartments that will originate spores. In a transverse cut (Fig. 8), a mature sporangium shows many orderly-arranged spherical or ovoid spores, 1.5–1.7 μm in size, that mark the sporangium wall (Fig. 5). The spore cytoplasm contains numerous ribosomes and lipid inclusions. The ovate sporangia were 3–14 μm long, according to the stage of

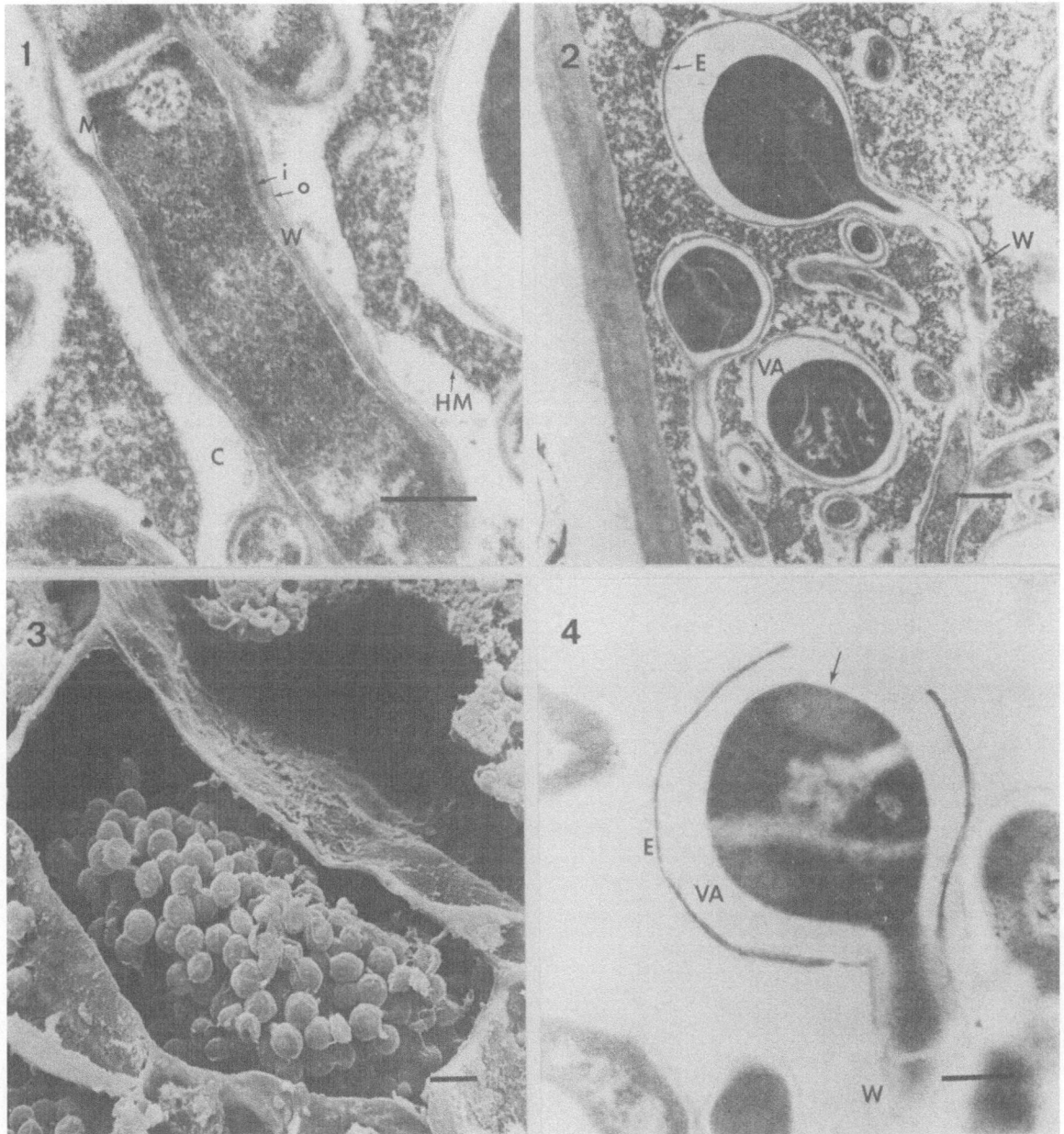

Fig. 1. Transmission electron microscopy (TEM) of hypha of the endophyte of *D. serratifolia* showing the outer (o) and inner (i) layers of the cell wall (W), the plasma membrane (M), the capsule (C) and the host plasma membrane (HM). Bar = 0.5 µm.

Fig. 2. TEM of hyphae and vesicles of *D. serratifolia* endophyte. Note the continuity between the hyphal cell wall (W), the electron-dense vesicle envelope (E) and the vesicle void area (VA). Bar = 1 µm.

Fig. 3. Scanning electron microscopy (SEM) of a nodular cell of *R. ephedra* containing a cluster of vesicles. Bar = 5 µm.

Fig. 4. TEM of vesicles produced by *Frankia* strain DsI1. Note the continuity between the electron-dense envelope surrounding the vesicle (E) and the hypha cell wall (W), the void area (VA) and the single-layer envelope (arrow) over the plasma membrane. Bar = 0.5 µm.

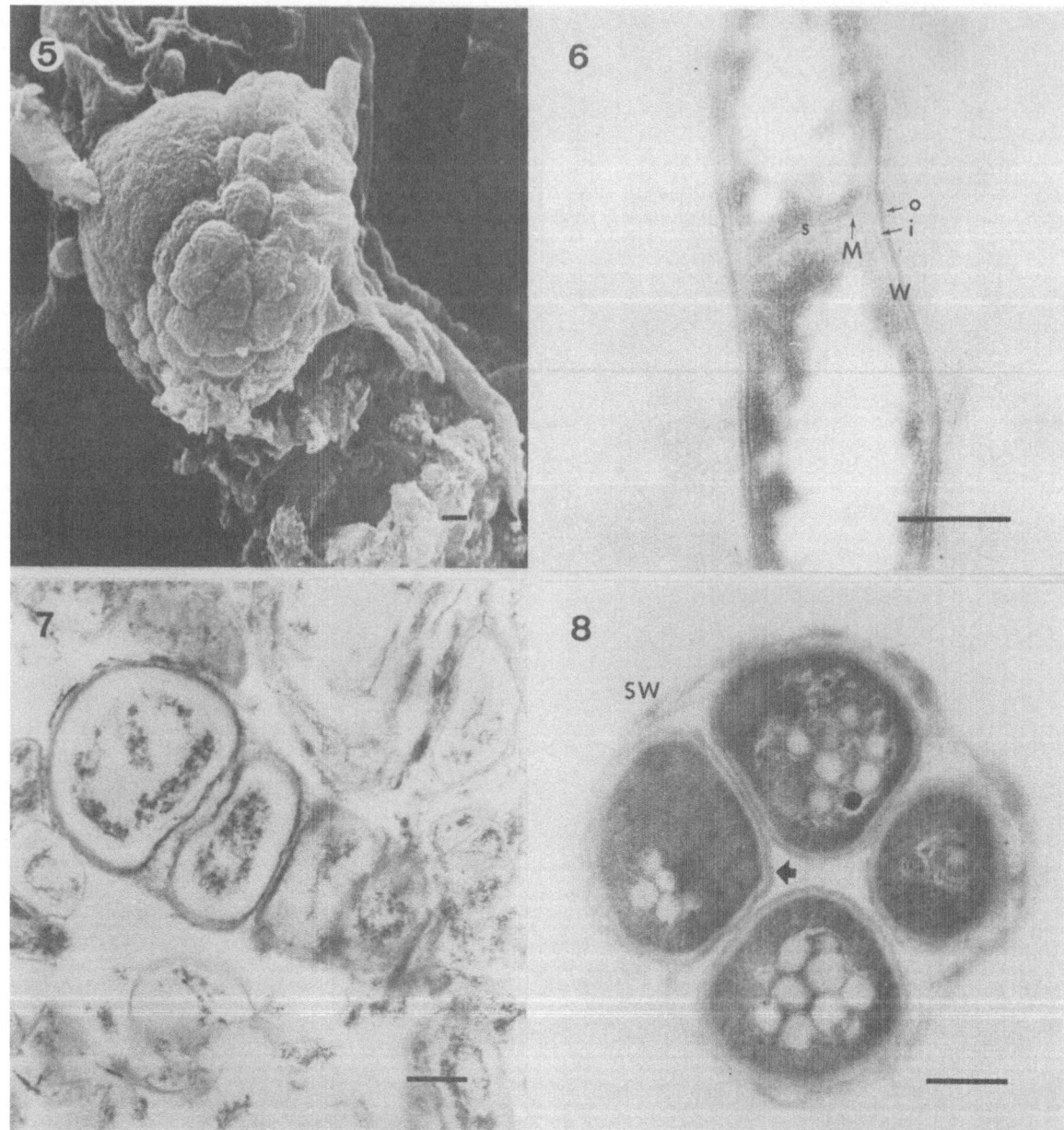

Fig. 5. SEM of a mature sporangium produced by *Frankia* strain DtI1 with the spores marking the sporangium wall. Bar = 1 μm.

Fig. 6. TEM of hypha of *Frankia* strain Dt1. Note the outer (o) and inner (i) layers of the cell wall (W), the plasma membrane (M) and septum (s). Bar = 0.5 μm.

Fig. 7. TEM of an early sporangium of *Frankia* strain DsI1 undergoing the initial transverse septation. Bar = 0.5 μm.

Fig. 8. TEM of a sporangium of *Frankia* strain ReI1. Note the two-layered spore wall (arrow) and the sporangium wall (SW). Bar = 0.5 μm.

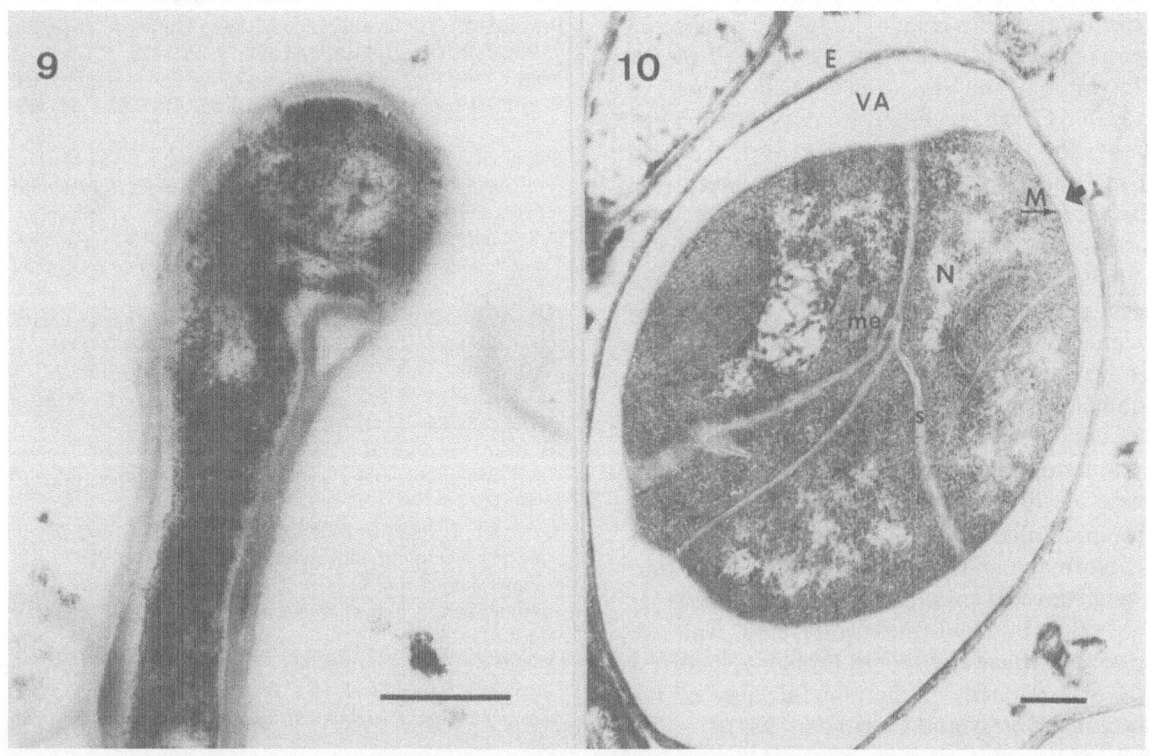

Fig. 9. TEM of an intrahyphal hypha produced in culture by *Frankia* strain DsI1. Bar = 0.5 μm.

Fig. 10. TEM of a vesicle of *D. trinervis* endophyte. Note the nuclear material (N), mesosomes (me), septa (s), the single-layered envelope (arrow) surrounding the plasma membrane (M), the void area (VA) and the electron-dense envelope (E). Bar = 0.5 μm.

development. Both the spore cell wall and sporangial wall show a two-layered structure, the inner layer being less electron-dense than the external layer (Fig. 8). Double tracked or multi-layered cell walls are characteristic of spores of *Frankia* grown endophytically and in culture (Baker *et al.*, 1980; Horriere *et al.*, 1983).

Growth of intrahyphal hypha was seen in cultures (Fig. 9). Intrahyphal growth has been reported in the wide hyphae of *Frankia* isolated from *Comptonia peregrina*, that are considered to be an early stage of sporangia formation (Newcomb *et al.*, 1979). The intrahyphal hyphae have a system of two cell walls, similar to those observed when the spores are in the sporangium, suggesting that they may be an early stage in the development of sporangia; the internal hypha would undergo septation to form the spores with two-layered walls and the wall of the external hypha would become the sporangium wall. However, intrahyphal hyphae have not been

related to sporangium formation in *Frankia* isolated from *Alnus rubra* (Horriere *et al.*, 1983).

In actinorhizal nodules, hyphae and vesicles are always enclosed in a common zone of low electron density that is separated from the host cytoplasm by the host cell membrane (Figs. 1 and 2). This zone is not present around the structures of the isolated strains (Figs. 4 and 6) and corresponds to the polysaccharide capsule synthesized by the host, and which has been described in all the *Frankia*-induced nodules (Lalonde and Quispel, 1977; Lalonde, 1978, 1979; Newcomb *et al.*, 1978; Baker *et al.*, 1980; Newcomb and Pankhurst, 1982; Newcomb and Heisey, 1984).

Most authors have designated as vesicle capsule the electron-dense layer enclosing the vesicle void area (Gardner, 1965; Lalonde and Quispel, 1977; Newcomb *et al.*, 1978; Lalonde, 1979; Newcomb and Pankhurst, 1982; Becking, 1984; Newcomb and Heisey, 1984), but since this structure is also present in our micrographs of vesicles produced by the free living strains (Fig. 4) it must be considered as part of the vesicle envelope synthesized by the microsymbiont, and the capsule itself surrounds it. Torrey and Callaham (1982) also detected, in TEM sections, this electron-dense layer over the void area in vesicles of a *Frankia* strain isolated from *Comptonia peregrina*. From freeze-etch preparations they concluded that it is the remnant of the lipidic multilayer structure that is part of the vesicle envelope itself, and that the void area is a consequence of the washing out of most of the inner lipid layers during preparation for transmission electron microscopy.

The vesicle cell wall has been located surrounding the cytoplasm (Lalonde and Quispel, 1977; Baker *et al.*, 1980; Newcomb and Pankhurst, 1982; Torrey and Callaham, 1982; Newcomb and Heisey, 1984; Normand and Lalonde, 1986), but the thin and single-layered structure of that envelope (Fig. 10) is different from the two-layered wall we observed in hyphae and spores. Probably, it corresponds to a remnant of the internal layer of the lipidic zone (Torrey and Callaham, 1982).

Our results suggest that in *Frankia* growing endophytically or *in vitro*, the external electron dense envelope that surround the vesicles, which has been commonly described as capsule, corresponds to the vesicle cell wall.

Acknowledgement

This paper is part of the research project No. 20.35.05 supported by the Dirección de Investigación de la Universidad de Concepción, Chile. Electron microscopy facilities were also provided by the Dirección de Investigación.

References

Baker D and O'Keefe D 1984 A modified sucrose fractionation procedure for the isolation of frankiae from actinorhizal root nodules and soil samples. Plant and Soil 78, 23–28.

Baker D, Newcomb W and Torrey J G 1980 Characterization of an ineffective actinorhizal microsymbiont, *Frankia* sp. EuI1 (Actinomycetales). Can. J. Bot. 26, 1072–1089.

Baker D, Torrey J G and Kidd G H 1979 Isolation by sucrose-density fractionation and cultivation *in vitro* of actinomycetes from nitrogen-fixing root nodules. Nature (London) 281, 76–78.

Becking J H 1984 Identification of the endophyte of *Dryas* and *Rubus* (Rosaceae). Plant and Soil 78, 105–128.

Benson D R 1982 Isolation of *Frankia* strains from alder actinorhizal root nodules. Appl. Environ. Microbiol. 44, 461–465.

Burggraaf A J P, Quispel A, Tak T and Valstar J 1981 Methods of isolation and cultivation of *Frankia* species from actinorhizas. Plant and Soil 61, 157–168.

Callaham D, Del Tredici P and Torrey J G 1978 Isolation and cultivation *in vitro* of the actinomycete causing root nodulation in *Comptonia*. Science 199, 899–902.

Diem H G and Dommergues Y 1983 The isolation of *Frankia* from nodules of *Casuarina*. Can. J. Bot. 61, 2822–2825.

Gardner I C 1965 Observation on the fine structure of the endophyte of the root nodule of *Alnus glutinosa* (L.) Gaertn. Arch. Mikrobiol. 51, 365–383.

Gauthier D L, Diem H G and Dommergues Y R 1984 Tropical and subtropical actinorhizal plants. Pesq. Agropec. Bras. 19, 119–136.

Horriere F, Lechevalier M P and Lechevalier H A 1983 *In vitro* morphogenesis and ultrastructure of *Frankia* sp. ArI3 (Actinomycetales) from *Alnus rubra* and a morphologically similar isolate (AirI2) from *Alnus incana* subsp. *rugosa*. Can. J. Bot. 61, 2843–2854.

Lalonde M 1978 Confirmation of the infectivity of a free-living actinomycete isolated from *Comptonia peregrina* root nodules by immunological and ultrastructural studies. Can. J. Bot. 56, 2621–2635.

Lalonde M 1979 Immunological and ultrastructural demonstration of nodulation of the European *Alnus glutinosa* (L.) Gaertn. host plant by an actinomycetal isolate from the North American *Comptonia peregrina* (L.) Coult. root nodule. Bot. Gaz. (Chicago) 140 (Suppl.), S35–S43.

Lalonde M and Quispel A 1977 Ultrastructural and immunological demonstration of the nodulation of the European *Alnus glutinosa* (L.) Gaertn. host plant by the North-American *Alnus crispa* var. *mollis* Fern. root nodule endophyte. Can. J. Microbiol. 23, 1529–1547.

Lechevalier M P and Ruan J S 1984 Physiology and chemical diversity of *Frankia* spp. isolated from nodules of *Comptonia peregrina* (L.) Coult. and *Ceanothus americanus* L. Plant and Soil 78, 15–22.

Newcomb W, Callaham D, Torrey J G and Peterson R L 1979 Morphogenesis and fine structure of the actinomycetous endophyte of nitrogen-fixing root nodules of *Comptonia peregrina*. Bot. Gaz. 140 (Suppl.), S22–S34.

Newcomb W and Heisey R M 1984 Ultrastructure of actinorhizal root nodules of *Chamaebatia foliolosa* (Rosaceae). Can. J. Bot. 62, 1697–1707.

Newcomb W and Pankhurst C E 1982 Ultrastructure of actinorhizal root nodules of *Discaria toumatou* Raoul (Rhamnaceae). N.Z.J. Bot. 20, 105–113.

Newcomb W, Peterson R L, Callaham D and Torrey J G 1978 Structure and host-actinomycete interactions in developing root nodules of *Comptonia peregrina*. Can. J. Bot. 56, 502–531.

Normand P and Lalonde M 1982 Evaluation of *Frankia* strains isolated from provenances of two *Alnus* species. Can. J. Microbiol. 28, 1133–1142.

Normand P and Lalonde M 1986 The genetics of actinorhizal *Frankia*: A review. Plant and Soil 90, 429–453.

Shipton W A and Burggraaf A J P 1982 A comparison of the requirements for various carbon and nitrogen sources and

vitamins in some *Frankia* isolates. Plant and Soil 69, 149–161.

Torrey J G and Callaham D 1982 Structural features of the vesicle of *Frankia* sp. CpI1 in culture. Can. J. Microbiol. 28, 749–757.

Session 3

Azolla and cyanobacteria

F. A. Skinner et al. (Eds.), Nitrogen fixation with non-legumes, 57–62.
© 1989 by Kluwer Academic Publishers.

Physiology and agronomy of *Azolla-Anabaena* symbiosis

I. WATANABE, C. LIN,[1] C. RAMIREZ, M. T. LAPIS, T. SANTIAGO-VENTURA and C. C. LIU[1]
*Department of Soil Microbiology, International Rice Research Institute, P.O. Box 933, Manila,
Philippines. [1]Fujian Academy of Agricultural Sciences, Fuzhou, Fujian, People's Republic of China.*

Key words: *Azolla*, *Anabaena*, N-fixation, P-nutrition, sporocarps

Abstract

To grow *Azolla* in sufficient quantity, input of some resources and good husbandry by farmers are required. In the nutrition of *Azolla*, P is most frequently a limiting factor for the growth of *Azolla*. The threshold value of plant P and flood water P concentrations were determined. Surveys of *Azolla* grown in the Philippines showed that about 80% of field-grown samples were deficient in P. Enrichment of *Azolla* with P in inoculum production plots was effective. *Azolla microphylla* from Paraguay was the most tolerant of high temperature and had the highest N content. By placing an indusium of megasporocarps onto *Anabaena*-free megasporocarps — either produced by cutting the apical portion of a megasporocarp or from *Anabaena*-free *Azolla* — the re-establishment of *Anabaena* to *Anabaena*-free *Azolla* was made. Thus, change of algal partner was achieved. *Azolla filiculoides* which had *Anabaena* from *A. microphylla* had a higher tolerance to 33°C (37° day/29° night) than the original *A. filiculoides*. The change of algal partner and sexual hybridization opens new ways of studying the physiology and agricultural application of *Azolla*.

Introduction

Azolla is a small aquatic fern, which grows in habitats such as ponds, canals, and flooded rice fields. The fern can grow without N compounds because N_2 is fixed by symbiotic *Anabaena azollae*. *Azolla* was not studied to any great extent until the 1970s, but since then, basic understanding of the physiology of the symbiosis has been greatly advanced by the work of Peters and his co-workers (Peters *et al.*, 1986). Other reviews on the environmental conditions needed for *Azolla* growth were made by Becking (1978, 1979). In this paper, the authors describe only important information relevant to the agronomic applications of the *Azolla–Anabaena* symbiosis.

For centuries, *Azolla* has been used as green manure for rice in China and northern Vietnam. Knowledge on its use in these countries has been extensively introduced into other areas since the late 1970s (Liu, 1979, 1984, 1987; Tuan and Thuyet, 1979; Li, 1984; Lumpkin and Plucknett,

1982; Lumpkin, 1985). Basic knowledge on the multiplication, environmental requirements, sexual reproduction, and fertilizer values of *Azolla* has been very much developed in China (Zhejian Academy of Agricultural Sciences, 1977; Liu, 1979, 1984, 1987; Lumpkin, 1985). Workshop proceedings on the use of *Azolla* are available (Silver and Schroeder, 1984; IRRI, 1987). At present, the volume of information is increasing. In China, *Azolla* is being used more for purposes other than fertilizer, because chemical N fertilizer is readily available (Wen, 1984).

The use of *Azolla* as green manure in wetland rice was summarized by Watanabe (1987) from internationally coordinated work by the International Network on Soil Fertility and Fertilizer Evaluation for Rice (INSFFER). Incorporation of one crop of *Azolla* gave an increased yield equivalent to that given by 30 kg urea N.ha^{-1}. From INSFFER trial data, the average yield of fresh biomass of *Azolla* before transplanting was only 15 t.ha^{-1} despite the application of P and insecticides in the experimental fields. This level of production of biomass was lower than expected. Biological constraints are thought to limit the growth of *Azolla*. These constraints are: low P fertility of soils, insect pest pressure and water excess or shortage. Various ways of eliminating these constraints have been studied (Zhejian Academy of Agricultural Sciences, 1977; Kannaiyan, 1987; Mochida, 1987).

IRRI started work on *Azolla* in 1975, and IRRI scientists have concentrated on finding solutions to these technological constraints. We have a collection of about 400 strains of all 6 species of *Azolla*.

The authors do not intend this review to cover most of the information to date, but to limit it to describing our research achievements at IRRI, particularly those done in collaboration with the National *Azolla* Research Center in Fuzhou, China.

Phosphorus nutrition of *Azolla*

Phosphorus supply and growth of Azolla

A handful of *Azolla* was brought to various areas of the Philippines in 1980. In Koronodal, South Cotabato, Mindanao, the growth of *Azolla* was sufficient for farmers to be convinced of its value as a green manure for rice. Farmers quickly adopted the technique of using *Azolla* and found it economically feasible (Watanabe, 1984; Kikuchi *et al.*, 1984). The success of *Azolla* technology in this area was due to high soil P fertility and a short dry season.

To evaluate the ability of soils to allow growth of *Azolla*, it was grown in small pots with various soils. The results revealed that the growth and P and N contents of *Azolla* were strongly related to the available P content of the soil. It was concluded that 25 mg.kg^{-1} available soil P (Olsen P) was required for better *Azolla* growth (Watanabe and Ramirez, 1984). Further studies also confirmed that the growth of *Azolla* in soils was primarily determined by its P nutrition. Dry matter production of *Azolla* was proportional to N content. Nitrogen content of *Azolla* was proportional to P content up to a content of 0.15% P. Above this level, the increase in P content was not accompanied by an increase in N content. The phosphorus % and N % of *Azolla* increased with the increase of

flood water P until it reached about 0.4 mg.l^{-1} P. At a flood water P content below 0.15 mg.l^{-1} *Azolla* suffered severely from P deficiency. In soils with Olsen P below 15 mg.kg^{-1}, flood water P content was below 0.15 mg.l^{-1} (Watanabe and Ali, 1986).

N and P contents of Azolla in the Philippines

Azolla samples (240) were taken from rice fields, canals and ponds from 12 Philippine provinces and analysed for their N and P contents. *Azolla* species, habitat, coverage, colour, healthiness, use by farmers, etc. were recorded. Soil samples were also taken from some fields. The surveys gave the following results.

1. Only 20% of the samples were sufficient in P ($> 0.4\%$) and N ($> 4\%$).
2. *A. microphylla* had higher N and P% (3.8% N and 0.35% P) than *A. pinnata* var. *imbricata* (3.22% N and 0.274% P). *A. pinnata* var. *pinnata* had lower N and P % (2.17% N and 0.14% P). N and P contents were highly correlated.
3. P contents of *Azolla* were higher in South Cotabato, Bicol, Southern Leyte, and Mountain Province.
4. Red *Azolla* had lower P and N content than green *Azolla*. When coverage of *Azolla* on the water surface was low ($< 10\%$) or high ($> 50\%$), N and P % of *Azolla* were higher than those with intermediate coverage.
5. No clear correlation was found between soil P status and P content of *Azolla*.

These surveys showed that *Azolla* could grow well in some areas in the Philippines. It may further be stated that the improvement of P nutrition of *Azolla* is a prerequisite for growing *Azolla* nationwide.

Improvement of P application and Azolla-P-enrichment in an inoculum pond

To grow *Azolla* well, P fertilizer application is needed. To make P application economically viable, the price of P fertilizer should be lower than the price of N fertilizer equivalent to the fixed N obtained by the enhanced growth of *Azolla*. At 1987 prices in the Philippines, 1 kg P applied as triple superphosphate was equivalent in its effect on

Table 1. Effect of application method to biomass production of *A. microphylla* (1986 dry and wet seasons)

Treatment	P applied		Biomass at 28 d					
	In the main field	Including inoculum plot	Dry season (Apr–May)			Wet season (Sep–Oct)		
			FW[a] (kg m^{-2})	N (kg ha^{-1})	N gain/P	FW (kg m^{-2})	N (kg ha^{-1})	N gain/P
Split application	3.93	3.93	2.2	45	8	2.3	54	11
Basal only	2.18	2.18	1.3	24	5.5	0.98	21	5
Basal + one split at 16 d	3.06	3.06	1.8	32	6	1.35	30	7.6
Basal + two splits at 16, 19 d	3.93	3.93	1.7	34	6	1.9	41	8
P-enriched Azolla	0	0.88	1.4	20	9	1.35	29	23
P-enriched + one split at 16 d	0.88	0.88	1.6	30	10	20	44	20
P-enriched + two splits at 16, 19 d	1.75	2.62	1.8	31	7	2.9	62	20
S.E.			0.10	1.7		0.25	1.7	

[a] FW, fresh weight. Control had N biomass of 12.3 in dry season and 9.4 kg N.ha^{-1} in wet season.

yield to 4.3–5 kg N applied as urea. Therefore, the application of 1 kg P to *Azolla* should encourage N accumulation by *ca.* 5 kg.

The application of P to the *Azolla* mat in an inoculum pond increased P content of *Azolla* considerably. In these conditions, the efficiency of P absorption by *Azolla* was more than 40% whereas the application of P to a small inoculum in the main field gave only 10% P absorption efficiency. Thus P-enriched *Azolla* (P content of 0.6–1.0%) could multiply several times after inoculation in the main field before it became P deficient (P content of 0.1–0.15%). Additional P applied once or twice could produce a biomass similar to that obtained by split application (Table 1); this procedure required less P and labour than was required by split application. The ratio of N gain to P applied was more than 5.

The problem is the unavailability of single P fertilizer. However, diammonium phosphate (DAP) (18% N — 46% P$_2$O$_5$) can be used without affecting greatly the growth of *Azolla*. Moreover, the smaller content of N in DAP is compensated for by the N gain in *Azolla*.

Strain difference in the ability to grow at a low P level

By continuous flow culture with 0.03 mg.l^{-1} P, *A. pinnata* var. *imbricata* strains could grow better

than *Euazolla* species (Subudhi and Watanabe, 1981). There is meagre information on the ability of different strains and species to grow in P-limited conditions.

Response to high temperature

Although *Azolla* is widely found in the tropics, growth is not always good and the plant often disappears during the hot and dry season (Gopal, 1967; Satapathy and Chand, 1984). At IRRI, *Azolla* growth was poorest during this period (April–May), when the average daily temperature exceeds 29°C and the maximum daily temperature exceeds 32°C (Watanabe *et al.*, 1981a, b). During this period, flood water temperature often exceeds 40°C near noon.

The ability to grow well above 30°C is an important screening criterion for finding *Azolla* strains suitable for the tropics. Screening of *Azolla* strains relatively tolerant of temperatures higher than 30°C was first made by growing them under controlled temperature in a Phytotron. Some of the promising strains were then grown in the field.

Differences among species

The growth of four species of *Azolla* (*A. pinnata* var. *imbricata*, *A. filiculoides*, *A. caroliniana* and *A. mexicana*) was compared at 22, 25, 29 and 33°C, with an 8°C difference between day and night temperature in a N-free medium (Watanabe and Berja, 1983). *Azolla filiculoides* was the most sensitive to high temperature. As the temperature increased, the biomass decreased. Some strains made fairly good growth at 33°C. When *A. microphylla* from Paraguay and Galapagos, and *A. pinnata* var. *pinnata* from Australia, were examined, the strains from Paraguay (IRRI Acc. no. 418 and 417) gave the best performance in terms of growth rate in both non-crowded and crowded conditions at 33(37/29)°C.

The effect of short time exposure to temperatures above 40°C on survival and growth was determined. A two-hour exposure at 40°C did not kill *Azolla*, but growth after such treatment was retarded. One- to 2-hour exposure at 44°C or a half-hour exposure at 48°C revealed differences in tolerance among strains. *Azolla filiculoides* was the most sensitive; *A. pinnata* var. *pinnata* and *A. microphylla* from Paraguay were the most tolerant to high temperature. Exposure at 52°C for half an hour killed all *Azolla*.

Response of Brazil collections and their growth in the field

Growth of *Azolla* strains (IRRI Acc. no. 307–315) taken from EMBRAPA, CNPAF (Brazil) was examined at controlled temperatures of 22(26-day/18-night), 29(33/25) and 33(37/29)°C. *Azolla* was grown up to 16 d. Except *Azolla* #315, all grew better than *A. filiculoides* #101 and *A. caroliniana* #301 at 29 and 33°C. Some strains like #308 and #309 produced biomass similar to that of *A. microphylla* #418 at 33°C. *Azolla* #309 and #310 are likely to be *A. mexicana*.

The growth of these strains in a field at IRRI was also studied. *Azolla* was grown continuously by removing 3/4 of the biomass whenever it exceeded 1 kg fresh weight/m^2. *Azolla* 309 and 310 performed best in terms of higher accumulated biomass and smaller variations from season to season.

Growth of Anabaena-*free and symbiotic* Azolla *at high temperature*

Temperatures higher than 30°C inhibit growth and nitrogen fixing ability. The reduction in nitro-

gen fixing activity was accompanied by a decreased heterocyst frequency in the symbiotic *Anabaena* (Tung and Watanabe, 1983). To determine whether growth retardation and nitrogen fixation are due only to the suppression of the nitrogen fixing activity, symbiotic *Azolla* and *Anabaena*-free *Azolla* were grown at 33(37/29)°C. *Anabaena*-free *Azolla* was grown with NH_4NO_3 and symbiotic *Azolla* was grown without combined nitrogen. *Anabaena*-free *Azolla* was obtained from decapitated megasporocarps (Lin and Watanabe, 1988) and in most cases, symbiotic *Azolla* strains and the corresponding *Anabaena*-free *Azolla* strains were compared. *Azolla* plants were also grown at 22(26/18) °C. Biomass production and N concentration of symbiotic *Azolla* at 22°C were higher than those of *Anabaena*-free *Azolla*. At 33°C, the growth of both symbiotic and non-symbiotic *A. filiculoides* was greatly reduced, but growth of *A. microphylla* and *A. pinnata* was not depressed as much as that of *A. filiculoides*. The growth of symbiotic *A. filiculoides* was more severely suppressed than that of the *Anabaena*-free plants. The nitrogen concentration of symbiotic *A. filiculoides* was similarly lower. This indicates that the low tolerance of *A. filiculoides* to high temperature is primarily determined by the host. Combined N improved the growth of *A. filiculoides* with *Anabaena*.

The alga of *A. filiculoides* per se is possibly sensitive to high temperature. To test this hypothesis, the growth of *A. filiculoides* which had *Anabaena* from *A. microphylla*, was examined. The method of obtaining *Azolla* in symbiosis with *Anabaena* from other species is described by Lin *et al.* (1988). At 33°C the growth of *A. filiculoides* with *Anabaena* from *A. microphylla* was better than that of *A. filiculoides* with its usual symbiont, but still less than that of the parent *A. microphylla*. This fact revealed that low tolerance of *A. filiculoides* to high temperature is determined both by the host and the symbiont.

Improvement of *Azolla*

Change of algal partner

Although many reports have claimed that *Anabaena azollae* Strasburger strains have been isolated in laboratory culture (Newton and Herman, 1979; Bai *et al.*, 1979; Venkataraman, 1962; Huneke,

1933). Koch's postulates were not satisfied. In fact, recent studies showed genetic and antigenic differences between laboratory-grown *Anabaena azollae* and the symbotic *Anabaena* (Franche and Cohen Bazire, 1985; Nierzwicki-Bauer and Haselkorn 1986; Ladha and Watanabe, 1982).

Anabaena cells are found in megasporocarps and establish symbiosis with a new sporophyte during embryo development (germination of megasporocarp) (Konar and Kapoor, 1974; Becking, 1978; Herd *et al.*, 1986). This fact suggests the possibility of making inoculation at the early stage of sporophyte germination. For this purpose, it was necessary to make *Anabaena*-free *Azolla*.

Anabaena cells reside beneath the indusium of the megasporocarps. The removal of the indusium resulted in the production of an *Anabaena*-free sporophyte. But some plants still contained *Anabaena*, because some cells were on the surface of the apical membrane in the pericollumella region at the time of indusium removal. The removal of indusium and apical membrane by cutting horizontally at the tip of the megaspore structure resulted in the complete removal of *Anabaena*. In the new sporophyte *Anabaena*-free *Azolla* produced sporocarps which were free from *Anabaena* and the germinated sporophytes remained free from it (Lin and Watanabe, 1988).

The removed indusium was used as an inoculum of the symbiont alga. The indusium removed from the megasporocarps of the same species or other species was placed in the decapitated (indusium and apical membrane removed) megaspore apparatus. The identification of *Anabaena* was made by a monoclonal antibody (Mc/Ab FC-16) specific to the *Anabaena* from *A. filiculoides*, prepared by the National *Azolla* Research Center, Fujian Academy of Agricultural Sciences. The *Anabaena* of *A. microphylla* did not react with the antibody, but the *Anabaena* of *A. microphylla* inoculated with the *Anabaena* of *A. filiculoides* (Amfa) reacted positively. Similarly, the *Anabaena* of *A. filiculoides* inoculated with the *Anabaena* from *A. microphylla* (Amfa) did not react with this antibody. These results confirmed the fact that the antigen for this monoclonal antibody was determined solely by the symbiont endophyte, and not by the host. The response to temperature as mentioned above also confirmed the change of algal partner.

A cell suspension of laboratory-grown *Anabaena* isolated by Newton and Herman (1979) and Bai *et*

al. (1979) was inoculated to decapitated *A. filiculoides* megaspore structures. In most new sporophytes, *Anabaena* disappeared as the *Azolla* grew. *Anabaena* was found in the leaf cavity in only a few plants. Interestingly, *Anabaena* cells were also present outside the leaf cavity. The *Anabaena* reacted positively with the antibody for *Anabaena azollae* but not with that for symbiotic algae. The plants exhibited nitrogen fixing activity but they could not grow in a N-free medium. These results indicated new ways of studying the *Anabaena-Azolla* symbiosis.

Sexual hybridization

Wei *et al.* (1986) reported the success of sexual hybridization between two species of *Azolla* (*A. filiculoides* and *A. microphylla*). The hybrid was confirmed by an esterase isozyme pattern. In some crosses, new sporophytes happened to be albinos and could not grow in N-free medium. One hybrid, now tested in Fujian, China, showed hybrid vigour and exhibited low temperature tolerance that was apparently inherited from one parent. Unluckily, the F_1 did not form megasporocarps (Wei, pers. comm.).

At IRRI and the University of the Philippines at Los Baños (P. Payawal pers. comm.) hybrids between *A. microphylla* and *A. filiculoides* were obtained, and the hybridization was confirmed biochemically.

A combination of hybridization and change of algal partner would be useful in changing *Azolla's* characteristics for wider applications.

References

Bai K Z, Yu, S L, Chen W L, Yang S Y and Cui C 1979 Isolation and pure culture of alga-free *Azolla* and *Anabaena azollae*. [in Chinese]. Kexue Tongbao 24, 664–666.

Becking J H 1978 Ecology and physiological adaptations of *Anabaena* in the *Azolla Anabaena* symbiosis. Ecol. Bull. (Stockholm) 26, 258–273.

Becking J H 1979 Environmental requirements of *Azolla* for use in tropical rice production. *In* Nitrogen and Rice, pp. 245–274. International Rice Research Institure, P.O. Box 933, Manila, Philippines.

Franche C and Cohen-Brazire 1985 The structural NIF genes of four symbiotic *Anabaena azollae* show a highly conserved physical arrangement. Plant Sci. 39, 125–131.

Gopal B 1967 Contribution of *Azolla pinnata* RBr to the

productivity of temporary ponds at Varanasi. Trop. Ecol. 8, 126–139.

Herd Y R, Cutter E G, and Watanabe I 1986 An ultrastructural study of postmeiotic development in the megasporocarps of A. microphylla. Can. J. Bot. 64, 822–833.

Huneke A 1933 Beiträge zur Kenntnis der Symbiose zwischen Azolla und Anabaena. Beitr. Biol. Pflanz. 20, 315–341.

International Rice Research Institute 1987 Azolla Utilization. P.O. Box 933, Manila, Philippines. 295 pages.

Kannaiyan S 1987 Use of Azolla in India. In Azolla Utilization. pp. 109–118 International Rice Research Institute, P.O. Box 933, Manila, Philippines.

Konar R N and Kapoor R K 1974 Embryology of Azolla pinnata. Phytomorphology 22, 211–223.

Kikuchi M, Watanabe I and Haws L D 1984 Economic evaluation of Azolla use in rice production. In Organic Matter and Rice. pp. 569–592. International Rice Research Institute, P.O. Box 933, Manila, Philippines.

Ladha J K and Watanabe I 1982 Antigenic similarity among Anabaena azollae separated from different species of Azolla. Biochem. Biophys. Res. Comm. 109, 675–677.

Li S H 1984 Azolla in paddy fields of Eastern China. In Organic Matter and Rice. pp. 179–192. International Rice Research Institute, P.O. Box 933, Manila, Philippines.

Lin C and Watanabe I 1988 A new method for obtaining Anabaena-free Azolla. New Phytologist 108, 341–344.

Lin C, Watanabe I, Tan L F and Liu C C 1988 Reestablishment of symbiosis to Anabaena-free Azolla. Zhunguo Kexue (Scientia Sinica) 30B.

Liu C C 1979 Use of Azolla in rice production in China. In Nitrogen and Rice. pp. 375–394. International Rice Research Institute, P.O. Box 933, Manila, Philippines.

Liu C C 1984 Recent advances on Azolla research. In Practical Application of Azolla for Rice Production. Eds. W S Silver and E C Schroeder. pp. 45–54. Martinus Nijhoff Publication, Dordrecht.

Liu C C 1987 Reevaluation of Azolla utilization in agricultural production. In Azolla Utilization. pp. 67–76. International Rice Research Institute, P.O. Box 933, Manila, Philippines.

Lumpkin T A 1985 Advances in Chinese research on Azolla. Proc. Roy. Soc. Edinburgh 86B, 161–167.

Lumpkin T A and Plucknett D A 1982 Azolla as a Green Manure: Use and management in Crop Production. Westview Press, Boulder, Colorado. 230 p.

Mochida O 1987 Insect pests of Azolla in the Philippines. In Azolla Utilization. pp. 207–221. International Rice Research Institute, P.O. Box 933, Manila, Phillipines.

Newton J W and Herman A L 1979 Isolation of cyanobacteria from the aquatic fern Azolla. Arch. Microbio. 120, 161–165.

Nierwicki-Bauer S A and Haselkorn R 1986 Differences in m RNA levels levels in Anabaena living freely or in symbiotic association with Azolla. EMBO J. 5, 29–35.

Peters G A, Toia R E Jr., Calvert H E, and Marsh B H 1986 Lichens to Gunnera — with emphasis on Azolla. In Nitrogen Fixation with Non-Legumes. Eds. F. A. Skinner and P. Uomala. pp. 17–34. Martinus Nijhoff Publishers, Dordrecht.

Satapathy, K B and Chand P K 1984 Studies on the ecology of Azolla pinnata RBr of Orissa. J. Indian Bot. Soc. 63, 44–52.

Silver W S and Schroeder E C (eds) 1984 Practical Application of Azolla for Rice Production. Martinus Nijhoff Publication, The Hague. 227 p.

Subudhi B P R and Watanabe I 1981 Differential phosphorus requirements of Azolla species in phosphorus limited continuous culture. Soil Sci. Plant Nutr. 27, 237–247.

Tuan D T and Thuyet T Q 1979 Use of Azolla in rice production in Vietnam. pp. 395–406. In Nitrogen and Rice. International Rice Research Institute, P.O. Box 933, Manila, Philippines.

Tung H G and Watanabe I 1983 Differential response of Azolla-Anabaena associations to high temperature and minus phosphorus treatments. New Phytologist 93, 423–431.

Venkataraman G S 1962 Studies on nitrogen fixation by blue green algae. III. Nitrogen fixation by Anabaena azollae. Indian J. Agric. Sci. 32, 22–24.

Watanabe I 1984 Use of symbiotic and free-living blue-green algae in rice culture. Outlook Agric. 13, 166–172.

Watanabe I 1987 Summary report of the Azolla Program of the International Network on Soil Fertility and Fertilizer Evaluation for Rice (INSFFER). In Azolla Utilization. pp. 197–205. International Rice Research Institute, P.O. Box 933, Manila, Philippines.

Watanabe I and Ali S 1986 Response of Azolla to P, K and Zn in different wetland soils in relation to chemistry of floodwater. Soil Sci. Plant Nutr 32, 239–252.

Watanabe I, Bai Keh-zhi, Berja N S, Espinas C R, Ito O and Subudhi B P R 1981a Azolla-Anabaena complex and its use in rice culture. IRRI Res Pap. Ser. 69. 11 p.

Watanabe I and Berja N S 1983 Growth of four species of Azolla as affected by temperature. Aquatic Bot. 15, 175–185.

Watanabe I, Berja N S and del Rosario D C 1981b Growth of Azolla in paddy field as affected by phosphorus fertilizer. Soil Sci. Plant Nutr. 26, 301–307.

Watanabe I and Ramirez C M 1984 Relationship between soil phosphorus availability and Azolla growth. Soil Sci. Plant Nutr. 30, 595–598.

Wei W X, Jin G Y and Zhang N 1986 Preliminary report of Azolla hybridization studies [in Chinese]. Bull. Fujian Acad. Agric. Sci. 1, 73–79.

Wen Q X 1984 Utilization of organic materials in rice production in China. In Organic Matter and Rice. pp. 45–46. The International Rice Research Institute, P. O. Box 933, Manila, Philippines.

Zhejian Academy of Agricultural Sciences, eds. 1977. Culture and utilization of Azolla [in Chinese]. Agricultural Press, Beijing. 127 p.

F. A. Skinner et al. (Eds.), Nitrogen fixation with non-legumes, 63–70.
© 1989 by Kluwer Academic Publishers.

Heterotrophic metabolism and diazotrophic growth of *Nostoc* sp. from *Cycas circinalis*

M. R. TREDICI, M. C. MARGHERI, L. GIOVANNETTI, R. DE PHILIPPIS and M. VINCENZINI
Centro di Studio dei Microrganismi Autotrofi, CNR and Istituto di Microbiologia Agraria e Tecnica, Università, Piazzale delle Cascine, 27, I-50144 Firenze, Italy

Key words: cyanobacteria, *Cycas*, heterotrophic metabolism, nitrogen-fixation, *Nostoc*, symbiosis

Abstract

The cyanobiont of *Cycas circinalis* (identified as *Nostoc* sp.) was isolated and its heterotrophic metabolism was studied in free culture under nitrogen-fixing conditions. Morphology, growth rate, nitrogenase activity, biochemical composition, efficiency of assimilation of organic carbon and molecular nitrogen were determined under different conditions of energy and carbon supply. The study has revealed the high potential of the heterotrophic metabolism in this symbiotic cyanobacterium. Although low rates of metabolic activities were attained under heterotrophic conditions, the efficiencies of organic carbon utilization (0.48 g cell-carbon per g glucose-carbon in chemoheterotrophy, from 0.65 to 0.74 under photoheterotrophy) and of N_2 assimilation (35.0 mg N_2 fixed per g glucose used in chemoheterotrophy, from 58.3 to 61.9 under photoheterotrophy) displayed by this organism were among the highest ever found in diazotrophically grown microorganisms. The isolate from *C. circinalis* was able to grow indefinitely in the dark under nitrogen-fixing conditions, maintaining a well balanced biosynthetic activity and the capacity to resume photosynthetic metabolism quickly. The significance of the heterotrophic potential of this symbiotic *Nostoc* is discussed.

Introduction

In recent years an ever-increasing number of cyanobacteria have been shown to be capable of growth at the expense of exogenous organic compounds (Khoja and Whitton, 1971; Rippka *et al.*, 1979), nevertheless the significance of heterotrophic metabolism under natural conditions in this group of organisms generally considered to be photoautotrophic, is still uncertain (Smith, 1982).

As stated by Smith (1983) it is likely that cyanobacterial heterotrophy is of importance only in specialized environments. Indeed in particular ecological niches, such as the coralloid roots of cycads, the cyanobacterium is exposed to conditions that allow exclusively a heterotrophic mode of carbon metabolism.

Cyanobacteria which develop in symbiosis are, with only a few unconfirmed exceptions, nitrogen-fixing heterocystous forms (Stewart *et al.*, 1983). It appears therefore of particular interest to study the relationships between heterotrophy and nitrogen fixation in heterocystous symbiotic cyanobacteria.

We have isolated the cyanobiont from the coralloid roots of *Cycas circinalis* and maintained it in axenic culture. The aim of this paper is to present the results of a study concerning the essential morphology and physiology of this organism grown heterotrophically under nitrogen-fixing conditions. It represents, at our knowledge, the first approach to the biology of a cycad symbiont grown on organic carbon and atmospheric nitrogen, the bulk of the studies having been carried out on the isolate from *Macrozamia lucida* (*Nostoc* sp. strain MAC) after it had lost its heterocystous habit (Bottomley and Van Baalen, 1978a, b; Hoare *et al.*, 1971; Ingram *et al.*, 1973).

Materials and methods

Organism

The endophyte of *Cycas circinalis* was isolated from the coralloid roots of a specimen collected at the Botanical Garden of S. Paulo (Brazil). It was obtained in axenic culture and identified as *Nostoc* sp. (*Nostoc* sp. strain Cc.)

Utilization of organic compounds

The ability of the isolate to use sugars and organic acids as sources of carbon and energy for growth was examined qualitatively under chemoheterotrophic and photoheterotrophic conditions. Cell suspensions were plated on $BG-11_0$ medium (Rippka *et al.*, 1979) supplemented with the carbon source and incubated both in the dark and under continuous illumination ($10\,\mu E\,m^{-2}\,s^{-1}$ of PAR) at 30°C. The following compounds were tested: glucose, fructose, sucrose, maltose, ribose (0.3% w/v), glycerol, acetate, pyruvate, succinate, malate, α-ketoglutarate, oxalacetate, citrate (0.1% w/v). Photoheterotrophic utilization of the carbon source was assayed in the presence of $2 \times 10^{-5}\,M$ 3-(3,4 dichlorophenyl)-1,1 dimethyl urea (DCMU). Growth was assessed by visual inspection after 40 days of incubation.

To determine the capacity of Krebs cycle intermediates to support nitrogenase activity under phototrophic conditions, 50 ml of cultures in $BG-11_0$ medium were supplemented with 0.1% (w/v) of the substance to be tested and incubated in a shaker at 30°C, under a photon flux of $50\,\mu E\,m^{-2}\,s^{-1}$ (PAR). After three hours of incubation, acetylene was added and ethylene production measured.

Growth parameters in phototrophic and chemotrophic cultures

To determine the main parameters characterizing phototrophic and chemotrophic cultures during the exponential growth phase the isolate was grown in 500-ml Erlenmeyer flasks, each containing 250 ml of culture. The medium ($BG-11_0$) was buffered with N-2 hydroxyethyl piperazine-N-2 ethansulphonic acid, sodium salt (HEPES) $0.5\,g\,l^{-1}$

(pH 7.5) and supplemented, when needed, with 0.25% (w/v) glucose. Photoheterotrophy was achieved by excluding CO_2 from the growth medium (a 40% NaOH solution trap was used in order to avoid any further contact with atmospheric CO_2) or by blocking its assimilation with DCMU ($2 \times 10^{-5}\,M$). Mixotrophic and autotrophic cultures were bubbled with 1% CO_2-enriched air. All cultures were incubated in an orbital shaker (Psycrotherm, New Brunswick Scientific Co-USA, model 627) at 30°C under continuous illumination provided by cool white fluorescent tubes which gave an incident photon flux rate of $95\,\mu E\,m^{-2}\,s^{-1}$ (PAR). Light was excluded from chemoheterotrophic cultures by wrapping the flasks with aluminium foil and black plastic.

Efficiency of organic carbon and molecular nitrogen assimilation

Cultures of *Nostoc* sp. strain Cc were grown in 500 ml flasks under the conditions previously described. The efficiencies of organic carbon utilization (expressed as gram cell-carbon produced per gram glucose-carbon used) and of nitrogen fixation (expressed as milligram N_2 fixed per gram glucose consumed) were determined from the growth yield, taking into account the elemental composition of the biomasses. When glucose in the medium was almost exhausted, the growth yields were calculated from the dry biomass produced and the amount of glucose consumed.

Analytical methods

Growth was followed by determining the cell dry weight on 10–20 ml volumes of the cultures. The cells were collected by filtration through a membrane filter of $5\,\mu m$ pore size, washed and dried at 70°C to constant weight. Cell protein content was determined according to the Lowry procedure as described by Herbert *et al.* (1971). Total cell carbohydrates were estimated by the phenol-sulphuric acid method (Dubois *et al.*, 1956), using glucose as a standard. Lipids were extracted in a Soxhlet apparatus with chloroform-methanol (2:1), dried at 75°C for 4 h and weighed. Pigments were deter-

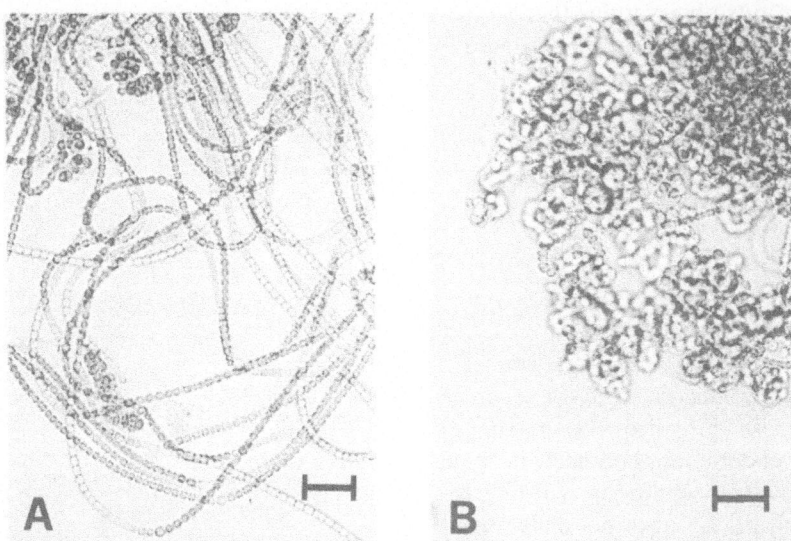

Fig. 1. Nostoc sp. strain Cc grown under photoautotrophic conditions. Free filaments and aseriate stage formation during the exponential growth phase (**A**); aseriate stage formation under C-deficient conditions (**B**). Bar markers = 20 μm.

mined as reported by Vincenzini *et al.* (1986). Nitrogenase activity was assayed by the acetylene reduction method (Turner and Gibson, 1980), as previously reported (Vincenzini *et al.*, 1986). Elemental composition (CHN) of the biomasses was determined by means of an automatic elemental analyzer (Carlo Erba Strumentazione-Italy, mod. 1106). Biomasses were lyophylized and dried at 110°C for 2 h before analysis. Acetanilide was used as a standard. Glucose concentration in the medium was determined by the phenol-sulphuric acid method (Dubois *et al.*, 1956).

Results and discussion

Morphological aspects of Nostoc sp. strain Cc

Under photoautotrophic conditions *Nostoc* sp. strain Cc cultures grew predominantly as long filaments of cylindrical vegetative cells with intercalary and terminal heterocysts (Fig. 1,A). The 'aseriate stage' consisted of irregularly coiled filaments within a firm envelope, and motile hormogonia were also produced. Under unfavourable conditions (light-stressed or C-deficient cultures) the aseriate stage predominated (Fig. 1,B). Akinetes were never observed and thus *Nostoc* sp. strain Cc maintains in culture a feature which is typical of the symbiotic life of *Nostoc* species (Grilli Caiola, 1980).

In medium BG-11_0 supplemented with glucose, either in the light or in darkness, *Nostoc* sp. strain Cc cultures consisted only of long heterocystous filaments. In contrast to the findings of Tel-Or *et al.* (1983) with *Anabaena azollae*, the heterocyst frequency did not increase upon addition of glucose. In fact it reached the maximum of 4% both in photoautotrophy and mixotrophy. In photoheterotrophic and chemoheterotrophic conditions fewer heterocysts were formed (Table 1).

Table 1. Growth parameters of *Nostoc* sp. strain Cc under different culture conditions for carbon and energy supply[a]

Growth conditions	Generation time (h)	Nitrogenase activity (nmol C_2H_4 mg cell d.w.$^{-1}$h^{-1})	Heterocyst frequency (%)
Photoautotrophy	21.5 ± 2	235 ± 30	4.0 ± 0.3
Photoheterotrophy	27.5 ± 3	185 ± 20	2.1 ± 0.2
Mixotrophy	16.8 ± 2.5	280 ± 30	4.0 ± 0.2
Chemoheterotrophy	70.0 ± 10	60 ± 10	1.3 ± 0.2

[a] Data were calculated during the exponential phase of growth.

It is evident from our observations that *Nostoc* sp. strain Cc, when grown on organic carbon, reproduces indefinitely by filament breakage without passing through the aseriate form. From this point of view this strain is different from *Nostoc muscorum* A which, in the dark, grows mostly as a mass of globose cells (Lazaroff, 1973).

Utilization of organic compounds

Heterotrophic growth of *Nostoc* sp. strain Cc was sustained only by glucose, fructose, sucrose and maltose. Glucose provided the maximum growth stimulation under both photoheterotrophic and chemoheterotrophic conditions. In this respect, *Nostoc* sp. strain Cc does not differ from *Nostoc* sp. strain MAC which grows in dim light and in the dark on the same sugars (Hoare *et al.*, 1971).

Table 2. Effect of Krebs cycle intermediates and glucose on nitrogenase activity of *Nostoc* sp. strain Cc under phototrophic conditions

Organic compound added	Nitrogenase activity (nmol C_2H_4 mg cell d.w.$^{-1}$ h^{-1})
None	108 ± 5
Glucose	248 ± 25
Pyruvate	183 ± 18
Succinate	175 ± 15
Acetate	162 ± 15
Malate	146 ± 10
Isocitrate	121 ± 5
α-Ketoglutarate	97 ± 5
Oxalate	92 ± 4

Although *Nostoc* sp. strain Cc was not able to utilize organic acids as carbon sources for growth, acetate, pyruvate, succinate, malate, and isocitrate enhanced the nitrogenase activity of carbon-limited (not air-CO_2 bubbled) cultures grown phototrophically (Table 2).

Generation time and nitrogenase activity

The generation times observed under phototrophic conditions (Table 1) were in the range of values displayed by most of the cyanobacteria so far tested (Smith, 1982). In chemoheterotrophy the generation time of *Nostoc* sp. strain Cc was among the highest found in cyanobacteria growing either

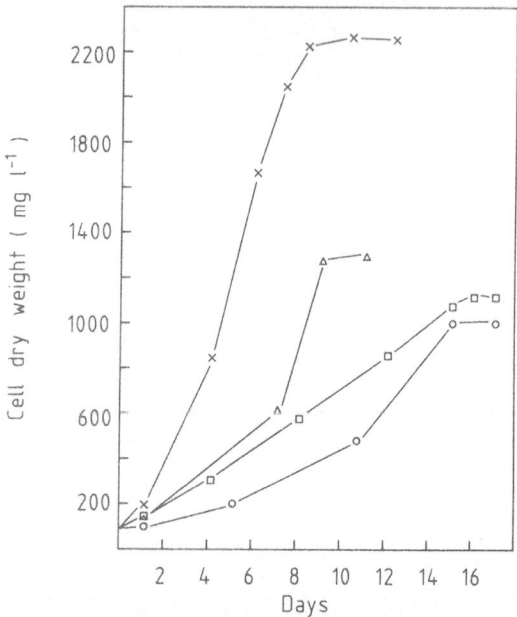

Fig. 2. Growth of *Nostoc* sp. strain Cc under different culture conditions for carbon and energy supply: x, mixotrophy; △, photoheterotrophy; □, autotrophy; O, chemoheterotrophy. Experimental conditions were as reported in Materials and Methods. Photoheterotrophy was achieved by blocking CO_2 assimilation with DCMU.

on nitrate and diazotrophically. It is worth mentioning, however, the capacity of the isolate to grow diazotrophically in the dark for long periods, retaining its full pigmentation even after about two years of dark cultivation.

It is generally reported that at light intensities near to saturation, organic compounds do not increase the growth rate of cyanobacteria. The growth response of *Nostoc* sp. strain Cc was different since exogenous glucose enhanced, to nearly 30%, the growth rate of photosaturated autotrophic cultures.

The generation time in photoheterotrophy was about one third of that observed in the dark, thus suggesting that energy (ATP) is the limiting factor during chemoheterotrophic growth.

Typical growth curves of *Nostoc* sp. strain Cc under both autotrophic and heterotrophic conditions are shown in Fig. 2. Growth in the dark on glucose was exponential almost to the point of glucose exhaustion, while in the light, both in the presence and absence of glucose, growth was soon limited (photolimited). Under mixotrophic and photoheterotrophic conditions growth ceased

Table 3. Chemical composition of *Nostoc* sp. strain Cc grown under different conditions (data expressed as % on organic fraction)

Growth conditions	Proteins	Carbohydrates	Lipids
Photoautotrophy	55.4	26.7	17.9
Mixotrophy	55.6	28.2	16.2
Photoheterotrophy	58.7	24.7	16.6
Chemoheterotrophy	43.6	44.2	12.2

abruptly when glucose in the medium was exhausted.

Nitrogenase activity appears related both to the specific growth rate and, to a lesser degree, to heterocyst frequency (Table 1). In fact the highest rates of acetylene reduction were found under mixotrophy and photoautotrophy, and the lowest under chemoheterotrophy. Acetylene reduction rates in chemoheterotrophically grown *Nostoc* sp. strain Cc were 20—25% of those attained photoautotrophically. Hence, our findings do not agree with those reported by Newton and Herman (1979) for *A. azollae*.

Nostoc sp. strain Cc showed relatively low nitrogenase activity in each of the conditions tested, compared both to other free-living and symbiotic cyanobacteria (Jensen, 1983; Newton and Herman, 1979; Stacey *et al.*, 1977).

Biochemical composition

No significant differences were noticed in the biochemical cell composition among the cultures grown under the different phototrophic conditions (Table 3). In the dark, proteins decreased from 55–59% of the organic fraction to 43.6%, and lipids from 16–18 to 12.2%, while the carbohydrate content nearly doubled, passing from 27–28 to 44.2%. Notwithstanding the increased carbohydrate content, the cyanobiont from *C. circinalis* presented, as *Nostoc* strain MAC grown on nitrate (Bottomley and Van Baalen, 1978a), a well balanced biosynthetic metabolism under chemoheterotrophic conditions. From this point of view the two *Nostoc* strains behave diversely from *Aphanocapsa* 6714 (Pelroy *et al.*, 1976). The dark heterotrophic metabolism of the latter organism is characterized by a near 90% reduction in the rate of protein, lipid and ribonucleic acid synthesis which, as observed by Smith (1982), would be in-

compatible with sustained growth under these conditions.

Pigment composition

The cell pigmentation was strongly influenced by growth conditions (Table 4). A general decrease in the chlorophyll *a* content was observed upon addition of glucose; this fact, also noted in other cyanobacteria and green algae (Haury and Spiller, 1981; Ogawa and Aiba, 1981; Vaara *et al.*, 1984), would simply suggest a diminished role of photosynthesis under photoheterotrophic and mixotrophic conditions as compared to photoautotrophy. Except for a reduced Chl *a* content, the pigment composition of chemoheterotrophically grown cells did not differ much from that of photoautotrophically grown cells. Thus, prolonged dark incubation does not remove the capacity for a quick resumption of photosynthetic metabolism. Under photoheterotrophic conditions total phycobiliproteins increased from 5.06% of photoautotrophically grown cells to 9.16%, on a dry weight basis. Two main explanations can be sought for this fact. Photoheterotrophic growth appears to be limited on the carbon side, *i.e.* it is nitrogen sufficient, as indicated by the high protein content of the cell, consequently phycobiliproteins, which have a well known role as nitrogen reserve polymers, accumulate. Otherwise the increase of phycobiliproteins could be due to the 'shade effect' of DCMU as reported in *Anacystis nidulans* (Koening, 1985).

Efficiency of conversion of sugar-carbon into cell-carbon

The efficiency of sugar-carbon assimilation in the dark was 0.48 g cell-carbon. g glucose-carbon^{-1}

Table 4. Pigment composition of *Nostoc* sp. strain Cc grown under different conditions (data expressed as % of dry weight)

Growth conditions	PC	APC	PE	Chl *a*
Photoautotrophy	3.10	0.69	1.27	2.07
Mixotrophy	1.79	0.62	0.48	1.14
Photoheterotrophy	5.29	1.55	2.32	0.87
Chemoheterotrophy	3.60	0.87	0.81	0.73

PC, phycocyanin; APC, allophycocyanin; PE, phycoerythrin; Chl *a*, chlorophyll *a*.

Table 5. Elemental composition of the biomass and efficiency of utilization of sugar-carbon and molecular nitrogen of *Nostoc* sp. strain Cc grown under heterotrophic conditions

Growth conditions	Elemental composition (% of the cell d.w.)			Efficiency of organic carbon assimilation (g cell C g glucose C^{-1})	Efficiency of nitrogen fixation (mg N fixed g glucose^{-1})
	C	N	H		
Photoheterotrophy (+ DCMU)	48.8	11.0	7.25	0.65	58.3
Photoheterotrophy (− CO$_2$)	50.0	10.5	7.32	0.74	61.9
Chemoheterotrophy	46.9	8.55	6.92	0.48	35.0

(Table 5). It is a value similar to or even higher than the levels obtained for heterotrophic bacteria (Payne, 1970), and higher than that displayed by *Nostoc* sp. strain MAC grown on nitrate (Bottomley and Van Baalen, 1978a). For *A. variabilis*, Wolk and Shaffer (1976) roughly calculated the efficiency of the conversion of fructose-carbon to cell-carbon at about 0.6, but Jensen (1983) found a more likely value of 0.38 and 0.52 in diazotrophically-grown and ammonia-grown *A. variabilis* cells, respectively. Therefore *Nostoc* sp. from *C. circinalis* has shown an efficiency of organic carbon conversion in the dark among the highest ever found, to our knowledge, for cyanobacteria grown chemoheterotrophycally on molecular nitrogen. The increase in efficiency from 0.48 for chemoheterotrophy to 0.65 for photoheterotrophy (plus DCMU) (Table 5) is due to the energy supply (ATP) by photosystem I activity that allows a greater share of glucose carbon to be channelled to anabolic reactions instead of being oxidized. Therefore light, as a source of ATP, does not suppress organic carbon catabolism but influences the efficiency of its assimilation.

When photoheterotrophic growth is obtained through the exclusion of CO_2 from the growth medium the efficiency of glucose carbon assimilation increases from 0.65 to 0.74. Two main explanations can be sought for such an effect. First, it is very likely that the supply of reducing equivalents by non-cyclic electron transport reduces the share of carbon used for producing energy; otherwise it is conceivable that a fraction of respiratory CO_2 is re-fixed. It is worth mentioning, however, that, despite the carbon deficiency, not all the CO_2 produced by glucose respiration is incorporated into the biomass. Hence some mechanism that limits CO_2 fixation in the presence of organic carbon assimilation must be hypothesized.

Nitrogen fixation efficiency

Under chemoheterotrophic conditions *Nostoc* sp. strain Cc fixed 35 mg of N_2 per g glucose consumed (Table 5). Thus it appears to be one of the most efficient nitrogen-fixing organisms, in terms of substrate utilization, when compared both to other heterocystous cyanobacteria and to heterotrophic bacteria (Alexander, 1977; Dalton and Postgate, 1969; Jensen, 1983).

Under photoheterotrophic conditions the efficiency of nitrogen incorporation rose to 58.3 (with DCMU) and 61.9 mg per g glucose used (without CO_2) (Table 5). This near doubling of efficiency was due not only to the higher growth yield but also to the higher nitrogen content of photoheterotrophically grown cells. This suggests that carbon, but not ATP or reductant, is a limiting factor under these conditions.

Conclusions

The data here presented attest the high potential of the heterotrophic metabolism in *Nostoc* sp. from *C. circinalis*. Although low rates of metabolic activity were attained under heterotrophic conditions as compared to those displayed by other free-living or symbiotic cyanobacteria, the efficiencies of organic carbon and N_2 assimilation were among the highest ever found in diazotrophically grown microorganisms. Besides, *Nostoc* sp. strain Cc was able to grow indefinitely in the dark under nitrogen-fixing conditions, maintaining a well balanced biosynthetic activity and the capacity to resume photosynthetic metabolism in a short time.

The slow but highly efficient heterotrophic metabolism displayed by this cyanobacterium seems consistent with its symbiotic role.

An assessment of the importance of cyanobacterial heterotrophy in natural environments must take into account the whole of the interactions between cyanobacteria and other organisms that occupy the same ecological niche (Smith, 1982). Under certain conditions (low light intensities or complete darkness) survival, if not growth, of free-living heterotrophic cyanobacteria will be largely dependent on their effective capacity to compete with bacteria for potential growth substrates.

The significance of heterotrophy in symbiotic cyanobacteria is, very likely, different. Once the association has been established, the cyanobacterium undergoes large morphological and physiological modifications whose first consequence is a reduced growth (Stewart *et al.*, 1983). In this condition, sheltered from soil microbe competition, the essential requisite for the endophyte is not a high growth rate, but the capacity to grow under low light or in complete darkness with a high efficiency of conversion of the supplied substrate in terms of fixed nitrogen.

Although we must be cautious in extrapolating microbial behaviour in nature from laboratory experiments, the study of the basic physiology of symbiotic cyanobacteria in culture will certainly help to clarify the significance of cyanobacterial heterotrophy and to establish the relationships among heterotrophy, nitrogen-fixation and the ability to live in symbiosis, in this group of prokaryotes.

References

Alexander M 1977 Introduction to Soil Microbiology. John Wiley and Sons, New York, 289 p.

Bottomley P J and Van Baalen C 1978a Characteristics of heterotrophic growth in the blue-green alga *Nostoc* sp. strain Mac. J. Gen. Microbiol. 107, 309–318.

Bottomley P J and Van Baalen C 1978b Dark hexose metabolism by photoautotrophically and heterotrophically grown cells of the blue-green alga (cyanobacterium) *Nostoc* sp. strain Mac. J. Bacteriol. 135, 888–894.

Dalton H and Postgate J R 1969 Growth and physiology of *Azotobacter chroococcum* in continuous culture. J. Gen. Microbiol. 56, 307–319.

Dubois M, Gilles K A, Hamilton J K, Rebers P A and Smith F 1956 Colorimetric method for determination of sugars and related substances. Anal. Chem. 28, 350–356.

Grilli Caiola M 1980 On the phycobionts of the cycad coralloid roots. New Phytol. 85, 537–544.

Haury J F and Spiller H 1981 Fructose uptake and influence on growth of and nitrogen fixation by *Anabaena variabilis*. J. Bacteriol. 147, 227–235.

Herbert D, Phipps P J and Strange R E 1971 Chemical analysis of microbial cells. *In* Methods in Microbiology. Eds. J R Norris and D W Ribbons. Vol. 5B, pp 209–344. Academic Press, London.

Hoare D S, Ingram L O, Thurston E L and Walkup R 1971 Dark heterotrophic growth of an endophytic blue-green alga. Arch. Mikrobiol. 78, 310–321.

Ingram L O, Calder J A, Van Baalen C, Plucker F E and Parker P L 1973 Role of reduced exogenous organic compounds in the physiology of the blue-green bacteria (algae): photoheterotrophic growth of a "heterotrophic" blue-green bacterium. J. Bacteriol. 114, 695–700.

Jensen B B 1983 Energy requirement for diazotrophic growth of the cyanobacterium *Anabaena variabilis* determined from growth yield in the dark. J. Gen. Microbiol. 129, 2633–2640.

Khoja T and Whitton B A 1971 Heterotrophic growth of blue-green algae. Arch. Mikrobiol. 79, 280–282.

Koening F 1985 Shade adaptation in the blue-green alga *Anacystis nidulans*. 5th Int. Symp. Photosynthetic Prokaryotes, Grindelwald 22–28 September.

Lazaroff N 1973 Photomorphogenesis and nostocacean development. *In* The Biology of Blue-green Algae. Eds. N G Carr and B A Whitton. pp. 279—319. Blackwell Scientific Publications, Oxford.

Newton J W and Herman A I 1979 Isolation of cyanobacteria from the aquatic fern *Azolla*. Arch. Microbiol. 120, 161–165.

Ogawa T and Aiba S 1981 Bioenergetic analysis of mixotrophic growth in *Chlorella vulgaris* and *Scenedesmus acutus*. Biotech. Bioeng. 23, 1121–1132.

Payne W J 1970 Energy yields and growth of heterotrophs. Annu. Rev. Microbiol. 24, 17–52.

Pelroy R A, Kirk M R and Bassham J A 1976 Photosystem II regulation of macromolecule synthesis in the blue-green alga *Aphanocapsa* 6714. J. Bacteriol. 128, 623–632.

Rippka R, Deruelles J, Waterbury J B, Herdman M and Stanier R J 1979 Generic assignments, strain histories and properties of pure cultures of cyanobacteria. J. Gen. Microbiol. 111, 1–61.

Smith A J 1982 Modes of cyanobacterial carbon metabolism. *In* The Biology of Cyanobacteria. Eds. N G Carr and B A Whitton. pp 47–85. Blackwell Scientific Publications, Oxford.

Smith A J 1983 Modes of cyanobacterial carbon metabolism. Ann. Microbiol. (Inst. Pasteur) 134 B, 93–113.

Stacey G, Van Baalen C and Tabita F R 1977 Isolation and characterization of a marine *Anabaena* sp. capable of rapid growth on molecular nitrogen. Arch. Microbiol. 114, 197–201.

Stewart W D P, Rowell P and Rai A N 1983 Cyanobacteria-eukaryotic plant symbioses. Ann. Microbiol. (Inst. Pasteur) 134 B, 205–228.

Tel-Or E, Sandovsky T, Kobiler D, Arad H and Weinberg R 1983 The unique properties of the symbiotic *Anabaena* in the water fern *Azolla*. *In* Photosynthetic Prokaryotes: Cell Differentiation and Function. Eds. G C Papageorgiou and L Packer. pp 303–314. Elsevier Scientific Publishing Company, New York.

Turner G L and Gibson A H 1980 Measurement of nitrogen fixation by indirect means. *In* Methods for Evaluating Biological Nitrogen Fixation. Ed. F J Bergersen. pp. 111–138. John Wiley and Sons, New York.

Vaara T, Sivonen K and Kurkela S 1984 Mixotrophic growth of *Nostoc* 268. *In* Advances in Photosynthesis Research. Ed. C

Sybesma. pp. 809–812. Nijhoff and Junk Publishers, Dordrecht.

Vincenzini M, De Philippis R, Ena A and Florenzano G 1986 Ammonia photoproduction by *Cyanospira rippkae* cells entrapped in dialysis tube. Experientia 42, 1040–1043.

Wolk C P and Shaffer P W 1976 Heterotrophic micro-and macrocultures of a nitrogen-fixing cyanobacterium. Arch. Microbiol. 110, 145–147.

F. A. Skinner et al. (Eds.), Nitrogen fixation with non-legumes, 71–76.

Light dependent nitrogen chlorosis in a heterocystous cyanobacterium

M. VINCENZINI, C. SILI, M. R. TREDICI and R. MATERASSI
Centro di Studio dei Microrganismi Autotrofi, CNR and Istituto di Microbiologia Agraria e Tecnica, Università, Piazzale delle Cascine, 27, I-50144 Firenze, Italy

Key words: cyanobacteria, heterocysts, nitrogen chlorosis, nitrogen fixation

Abstract

The effects of a sudden increase in the amount of available light energy (light shock) on the biosynthesis of the main cellular constituents in the heterocystous cyanobacterium *Cyanospira rippkae*, growing diazotrophically under light-dark cycles, are described. Light shocks of different magnitude were achieved by quickly preparing four experimental cultures at cell concentrations lower than that of a dense, highly pigmented, stock culture which served as inoculum. The new light regimens were maintained for 10 days by keeping the cultures at the proper dilution rate. The main modifications induced by the light shock were: (1) increase in heterocyst frequency; (2) accumulation of carbohydrates; (3) reduction of phycobiliproteins. The extent of these modifications correlated directly with the intensity of the light shock. Cell composition reached a new steady state 4-5 light-dark cycles after the initial increase in available radiant energy. Since the observed changes in cell composition and in trichome structure showed a noticeable analogy with the chlorosis induced in cyanobacteria by nitrogen starvation (nitrogen chlorosis), we referred to them as 'light dependent nitrogen chlorosis'. Possibly the photooxidative conditions resulting from the increase in the amount of radiant energy available for highly pigmented cells unbalance nitrogen metabolism by affecting the activity of some key enzymes.

Introduction

It is well known that cyanobacteria have a worldwide distribution, occurring even in extreme environments. Recently in our Research Centre two heterocystous cyanobacteria, for which the creation of the new genus *Cyanospira* was proposed (Florenzano *et al.*, 1985), were isolated from the alkaline soda Lake Magady in Kenya. The lake is characterized by wide fluctuations in its physical and chemical features, due to climatic variability, and by a very shallow euphotic zone with high incident solar radiation (Melack and Kilham, 1974). Nevertheless, the cyanobacterial population in this extreme aquatic habitat can reach high concentrations (Melack and Kilham, 1974; Tuite, 1981). It is evident that these organisms have developed some biological mechanism which enables them to overcome the environmental difficulties. Among these, light and salt stresses are undoubtedly of major importance.

In this paper we describe the metabolic and structural modifications that a sudden increase in the amount of available light energy (light shock) caused in *Cyanospira rippkae* Mag II 702 (ATCC 43194), the type strain of the new cyanobacterial genus, growing under light-dark cycles and N_2-fixing conditions.

Methods

Organism and culture conditions

Cyanospira rippkae Mag II 702 (ATCC 43194) was routinely maintained on a liquid mineral medium containing (g): $NaHCO_3$, 7.5; Na_2CO_3, 1.5; NaCl, 0.5; K_2SO_4, 0.95; K_2HPO_4, 0.25; $MgSO_4.7H_2O$, 0.12; $CaCl_2.2H_2O$, 0.025; ferric ammonium citrate, 0.006, and 1 ml trace metal solution A_5 + Co (Rippka *et al.*, 1979) per litre of distilled water. The stock culture was grown up to about 1 g cell dry weight l^{-1} in two small raceway ponds (0.13 sq. metres of illuminated surface and 6.8 litres of culture each) provided with paddle wheels circulating the culture suspension at a speed of about $10\,cm\,s^{-1}$. The cultures were run under 16:8 h light: dark cycles at $30 \pm 1°C$ during the light phase and at $18 \pm 1°C$ in the dark. Illumination was provided by 1000 W incandescent lamps giving a mean photosynthetic photon flux density of $250\,\mu E\,m^{-2}\,s^{-1}$ at the culture surface. Pure CO_2 was bubbled into the cultures in order to keep the pH in the range 9.0–9.3. Light shocks of different magnitude were achieved by quickly preparing, from the stock culture, four experimental cultures at lower cell concentration (240, 430, 640 and 870 mg cell dry weight l^{-1}). These cultures were immediately started under the same culture conditions as those described above for the stock culture. This procedure allowed us to increase the amount of radiant energy available for each cell in the cultures without increasing the light intensity at the culture surface up to values possibly dangerous for cell viability. Taking into account the Lambert-Beer law for light attenuation in cultures of photosynthetic microorganisms (Lee *et al.*, 1987) it may be calculated that the amount of average light energy available for the single cell in the most diluted experimental culture was about three times higher than in stock culture. The effects produced in *C. rippkae* cells by the different light shocks were investigated over ten light-dark cycles. Cell concentration in each experimental culture was kept constant over the whole experimental period by adopting a semicontinuous regimen: the daily production of biomass was removed from the cultures at the beginning of each light phase. The dilution rates were inversely related to the cellular concentration, being on the average about 0.44 day^{-1} for the culture at 240 mg (d.w.) l^{-1} and 0.07 day^{-1} for the culture at 870 $mg\,l^{-1}$.

It is worth mentioning that, notwithstanding the high value of light intensity at the culture surface, all the experimental cultures were under light-limited growth conditions.

Analytical procedures

Cell dry weight was determined by filtering 10–20 ml portions of the cultures through 0.45 μm pore HA membrane filters (Millipore Corp., Bedford,

Massachussets). Filters were dried at 110°C for 3 h and then weighed.

The protein content of the cells was determined with the Folin-Ciocalteau reagent as described by Herbert *et al.* (1971) using bovine serum albumin as standard. Carbohydrates were determined according to the phenol-sulphuric acid method of Dubois *et al.* (1956) with glucose as standard.

Carbohydrate and protein analyses were performed at the beginning and at the end of each light phase and the results were reported as activities (specific rate of synthesis or consumption) referred to the mean protein concentration in the cultures.

For the determinations of phycobiliprotein composition, cells were extracted with 0.01 M phosphate buffer (pH 7)–0.15 M NaCl after mortar grinding. The absorbance of the extracts, measured at 562, 615 and 652 nm, was used for calculating the amount of phycoerythrin, phycocyanin and allophycocyanin according to the equations of Bennett and Bogorad (1973) except that the extinction coefficient of Tandeau De Marsac (1977) was used for phycoerythrin. The results were expressed as mg of total phycobiliproteins per 100 mg protein.

Chlorophyll a content of the cells was determined spectrophotometrically after extraction with 90% acetone according to the method of Parson and Strickland (1963). Heterocyst and akinete frequencies were determined by counting at least 1000 cells by light microscopy. Cell counts were performed in a Thoma counting chamber and the frequencies of the differentiated cells were expressed as a percentage of the total cell population.

Photosynthetic photon flux density at the culture surface was determined with a quantum/radiometer/photometer model LI-185 B equipped with a quantum sensor model LI-190 SB (LI-COR Ltd., Lincoln, Nebraska).

Results

Carbohydrate and protein synthesis

Light shock caused remarkable changes in the main biosynthetic activities of *C. rippkae* growing diazotrophically at high irradiation with light-dark cycles. In the first light phase after the sudden increase in the amount of light energy available for the individual cell in the culture, the specific rates of

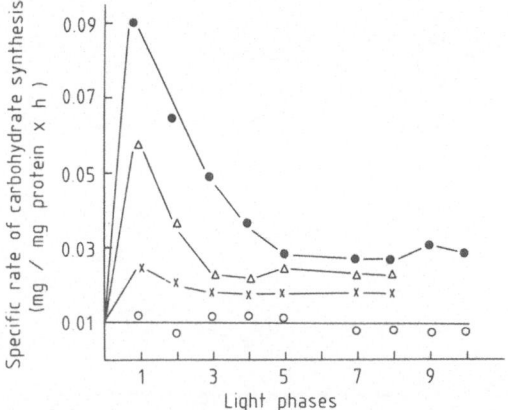

Fig. 1. Changes of the specific rate of carbohydrate synthesis in the light phases in *C. rippkae* cultures subjected to different light shocks (mg cell dry wt.l^{-1}): ●, 240; △, 430; ×, 640; ○, 870.

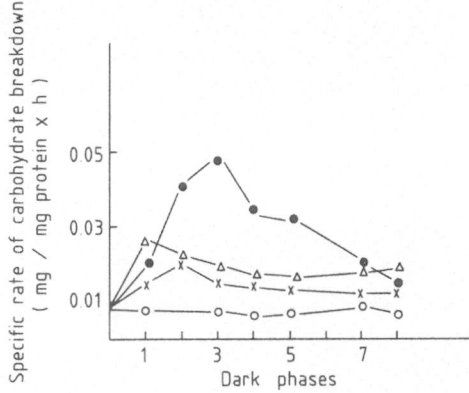

Fig. 3. Changes of the specific rate of carbohydrate consumption in the dark phases in *C. rippkae* cultures subjected to different light shocks (symbols as in Fig. 1).

carbohydrate synthesis increased according to the culture dilution (Fig. 1). The culture run at the highest cell concentration maintained a rate of carbohydrate synthesis unchanged compared to the corresponding rate of the stock culture while in the most diluted culture the rate of carbohydrate synthesis was nine times higher, reaching the value of 0.091 mg (mg protein)$^{-1}$h^{-1}. In successive light phases the specific rates of carbohydrate synthesis shown by the latter culture decreased regularly until a steady value was reached. The number of light-dark cycles required by the cultures to reach the steady state was directly related to the dilution rates. Indeed, five cycles were necessary for the culture growing at a cell concentration of 240 mg l^{-1} and only two cycles were required by the culture at 640 mg l^{-1}.

The increase in available radiant energy also caused an increase in the specific rate of protein synthesis but the kinetics of this increment were quite different compared to those of the synthesis of carbohydrates. In the cultures growing at the lowest cell concentrations the highest rate of protein synthesis was reached after three to four light-dark cycles and then remained substantially unchanged (Fig. 2). The difference in the time course of the two main biosynthetic activities was more pronounced where the intensity of the light shock was the highest: at the end of the first light phase the specific rate of carbohydrate synthesis in the most diluted culture was about four times the rate of protein synthesis while the values in the stock culture were about the same.

Carbohydrate consumption

Appreciable differences between the cultures were also found in the specific rate of carbohydrate consumption during the dark phases (Fig. 3). Dark carbohydrate catabolism was stimulated to a greater extent in the most diluted culture reaching the highest value during the third dark phase with a delay of two cycles for the highest activity of carbohydrate synthesis. In the successive dark phases the catabolic activity decreased and a steady state was reached as already seen for biosynthetic rates in the light.

As a consequence of the modifications of the specific rates of carbohydrate and protein synthesis

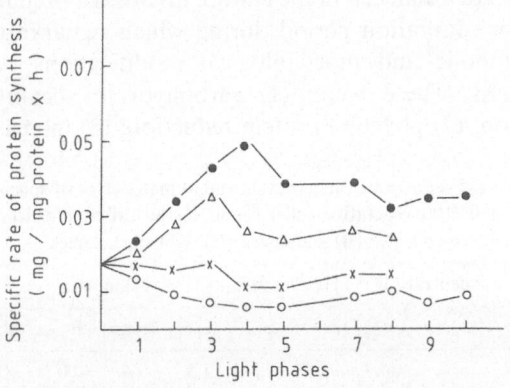

Fig. 2. Changes of the specific rate of protein synthesis in the light phases in *C. rippkae* cultures subjected to different light shocks (symbols as in Fig. 1).

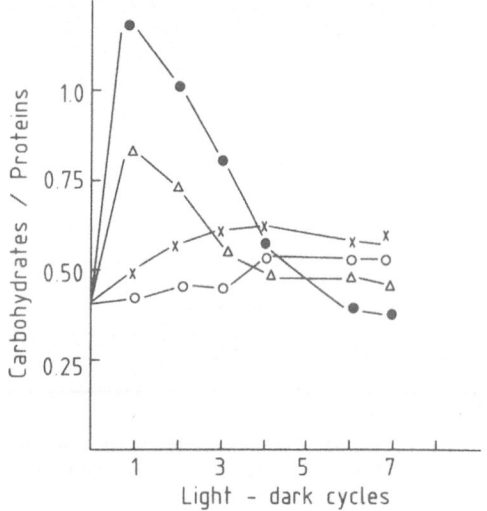

Fig. 4. Time course of the carbohydrate/protein ratio in *C. rippkae* cultures subjected to different light shocks (symbols as in Fig. 1).

in the light, and carbohydrate consumption in the dark, a remarkable change in the biochemical cell composition was observed in the cultures exposed to the most severe light shocks. In the most diluted culture the ratio between carbohydrate and protein concentrations (Fig. 4) rose by a factor of 3.4 after the first light-dark cycle while at the steady state it became similar to the starting value.

Phycobiliprotein level

Besides these biosynthetic alterations other significant modifications were observed following the increase in light availability for the cyanobacterial cells. After four light-dark cycles phycobiliprotein levels were markedly reduced in the most diluted cultures, changing, in the cells growing at the lowest concentration, from 16% to 4% of the protein content (Fig. 5). Hence in the cultures run at the highest dilution rates a substantial reduction in pigment content of the trichomes occurred although no variation in the level of chlorophyll *a* (on a protein basis) was observed.

Heterocyst frequency

The increase in available light energy also had a stimulatory effect on heterocyst differentiation: the

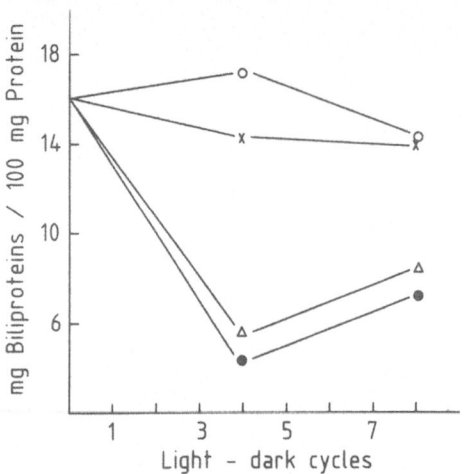

Fig. 5. Changes in the phycobiliprotein content of *C. rippkae* cultures subjected to different light shocks (symbols as in Fig. 1).

time courses of heterocyst frequency in the different cultures were the same as those observed for the other parameters, except for the specific rate of carbohydrate synthesis. Heterocyst frequency reached its maximal value during the third light-dark cycle in the culture growing at the highest dilution rate (Table 1). Quite different was the effect of the increased amount of light availability on the akinete differentiation which was substantially suppressed in the most diluted culture (Table 1).

Discussion

The response of *C. rippkae* cells to a sudden increase in the amount of light energy available for the individual cell in the culture involved a preliminary adaptation period during which remarkable metabolic and morphological modifications occurred. These were: (1) carbohydrate accumulation; (2) phycobiliprotein reduction; (3) increase

Table 1. Frequency of heterocysts and akinetes (No. of specialized cells/100 vegetative cells) found in the different cultures after three (A), five (B) and seven (C) light-dark cycles

Cell concentration (mg dry wt.l^{-1})	Heterocysts			Akinetes		
	A	B	C	A	B	C
240	9.1	7.6	7.5	0.4	0.0	0.2
430	8.5	7.0	6.9	4.2	8.4	8.8
640	7.3	6.7	6.5	8.5	9.5	8.5
870	6.4	6.4	6.1	8.5	8.0	11.4

in the heterocyst frequency. The extension of these modifications and the duration of the adaptation phase were directly related to the size of the increase in average light intensity in the culture.

Because of the strict analogy with the chlorosis induced in cyanobacterial cells by nitrogen starvation (De Vasconcelos and Fay, 1974; Wood and Haselkorn, 1980), the changes in cell composition and in trichome structure observed in *C. rippkae* during the adaptation period have been termed 'light dependent nitrogen chlorosis'. It is to be noted, however, that in contrast to true nitrogen chlorosis, the carbohydrate accumulation achieved by the light-shocked cells is accomplished through a different mechanism. Indeed, in the light dependent nitrogen chlorosis, both carbohydrates and proteins continue to be synthesized: carbohydrate accumulation is determined by a different time course of the two main biosynthetic activities with a net prevalence of carbohydrate over protein synthesis. On the other hand this divergence is not counterbalanced by an increase in the specific rate of carbohydrate consumption in the dark.

Since the phenomenon was not evident when urea was supplied as nitrogen source (data not shown), light-dependent nitrogen chlorosis might concern only cells growing on molecular nitrogen and originate from the heterocysts. It is well known that in heterocystous cyanobacteria growing under nitrogen fixing conditions protein synthesis in vegetative cells depends largely on carbohydrate catabolism in the heterocysts (Bothe, 1982). Carbohydrate breakdown in these specialized cells could be temporarily hindered by light shock and this would affect nitrogen metabolism. Indeed, the sudden increase in available radiant energy imposed on highly pigmented cells may cause a sudden rapid increase in photosynthetic oxygen evolution and establish conditions for generating intracellular oxidizing agents. These products may diffuse through microplasmodesmata into the heterocysts and interfere with the activation system, thioredoxin (Crawford *et al.*, 1984) and/or other reducing agents, of enzymes such as isocitrate dehydrogenase (Papen *et al.*, 1983) and glutamine synthetase (Bothe *et al.*, 1984), which regulate carbohydrate catabolism and ammonia incorporation. The consequence will be a rate of glutamine synthesis lower than the anabolic need and a delayed response of the specific rate of protein synthesis to the increase in average radiant energy in the culture.

The degradation of the phycobiliproteins observed in the light shocked cells could play the dual role of supplying additional nitrogen to cells growing under nitrogen limiting conditions and of preventing excess light harvesting by the photosynthetic apparatus. Furthermore, the induction of protective mechanisms against photooxidative damage and the possible feed-back inhibition of photosynthesis due to excess production of photosynthate (Post, 1986) could allow *C. rippkae* cells to overcome the imposed light shock and to grow as adapted cells under the new light conditions. From this point of view light-dependent nitrogen chlorosis may be regarded as a metabolic strategy evolved by *C. rippkae* to cope with the extreme conditions of its natural habitat. In this connection, cell photobleaching has been recently considered as a light adaptation process and not simply a photodamage phenomenon (Nultsch and Agel, 1986). On the other hand carbohydrate accumulation in cyanobacterial cells shifted from low to high photon flux densities is known to constitute a mechanism for regulating the light intensity experienced by the single cell (Kromkamp *et al.*, 1986).

References

Bennett A and Bogorad L 1973 Complementary chromatic adaptation in a filamentous blue-green alga. J. Cell. Biol. 58, 419–435.

Bothe H 1982 Nitrogen fixation. *In* The Biology of Cyanobacteria. Eds. N G Carr and B A Whitton. pp. 87–104. Blackwell Scientific Publications, Oxford.

Bothe H, Nelles H, Häger K P, Papen H and Neuer G 1984 Physiology and biochemistry of N_2-fixation by cyanobacteria. *In* Advances in Nitrogen Fixation Research. Eds. C Veeger and W E Newton. pp. 199–210. M Nijhoff and D R W Junk Pubs., The Hague/Boston/Lancaster.

Crawford N A, Sutton C W, Yee B C, Johnson T C, Carlson D C and Buchanan B B 1984 Contrasting modes of photosynthetic enzyme regulation in oxygenic and anoxygenic prokaryotes. Arch. Microbiol. 139, 124–129.

De Vasconcelos L and Fay P 1974 Nitrogen metabolism and ultrastructure in *Anabaena cylindrica*. I. The effect of nitrogen starvation. Arch. Microbiol. 96, 271–279.

Dubois M, Gilles R A, Hamilton J K, Robers P A and Smith F 1956 Colorimetric method for determination of sugar and related substances. Analyt. Chem. 28, 350–356.

Florenzano G, Sili C, Pelosi E and Vincenzini M 1985 *Cyanospira rippkae* and *Cyanospira capsulata* (gen. nov. and spp.nov.): New filamentous heterocystous cyanobacteria from Magadi lake (Kenya). Arch. Microbiol. 140, 301–306.

Herbert D, Phipps P J and Strange R E 1971 Chemical analysis of microbial cells. *In* Methods in Microbiology. Eds. J R Norris and D W Ribbons. Vol 5B, pp. 209–344 Academic Press, London/N.Y.

Kromkamp J, Konopka A and Mur L R 1986 Buoyancy regulation in a strain of *Aphanizomenon flos-aquae* (Cyanophyceae): The importance of carbohydrate accumulation and gas vesicle collapse. J. Gen. Microbiol. 132, 2113–2121.

Lee H Y, Erickson L E and Yang S S 1987 Kinetics and bioenergetics of light-limited photoautotrophic growth of *Spirulina platensis*. Biotechnol. Bioeng. 29, 832–843.

Melack J M and Kilham P 1974 Photosynthetic rates of phytoplankton in East African alkaline, saline lakes. Limnol. Oceanogr. 19, 743–755.

Nultsch W and Agel G 1986 Fluence rate and wavelength dependence of photobleaching in the cyanobacterium *Anabaena variabilis*. Arch. Microbiol. 144, 268–271.

Papen H, Neuer G, Refaian M and Bothe H 1983 The isocitrate dehydrogenase from cyanobacteria. Arch. Microbiol 134, 73–79.

Parson T R and Strickland J D H 1963 Discussion of spectrophotometric determination of marine plant pigments, with revised equations for ascertaining chlorophylls and carotenoids. J. Mar. Res. 21, 155–163.

Post A F 1986 Transient state characteristics of adaptation to changes in light conditions for the cyanobacterium *Oscillatoria agardhii*. I. Pigmentation and photosynthesis. Arch. Microbiol. 145, 353–357.

Rippka R, Deruelles J, Waterbury J B, Herdman M and Stanier R Y 1979 Generic assignments, strain histories and properties of pure cultures of cyanobacteria. J. Gen. Microbiol 111, 1–61.

Tandeau De Marsac N 1977 Occurrence and nature of chromatic adaptation in cyanobacteria. J. Bacteriol. 130, 82–91.

Tuite C H 1981 Standing crop densities and distribution of *Spirulina* and benthic diatoms in East African alkaline saline lakes. Fresh-water Biol. 11, 345–360.

Wood N B and Haselkorn R 1980 Control of phycobiliprotein proteolysis and heterocyst differentiation in *Anabaena*. J. Bacteriol. 141, 1375–1385.

F. A. Skinner et al. (Eds.), Nitrogen fixation with non-legumes, 77–82.

Isolation and characterization of the N_2-fixing symbiotic cyanobacterium *Anabaena azollae*

MINNI SRIVASTAVA, ANITA SHARMA and A. KUMAR
Centre of Advanced Study in Botany, Banaras Hindu University, Varanasi – 221005, India

Key words: *Azolla*, *Anabaena azollae*, heterocyst, nitrogenase, symbiosis

Abstract

Expression of nitrogenase activity (C_2H_2 reduction) has been studied in the symbiotic N_2-fixing cyanobacterium *Anabaena azollae* within the intact *Azolla* plant and after its isolation from the host. The cyanobacterium (blue-green alga) has been isolated successfully from the host plant by standard microbiological techniques. Optimum growth occurred in semi-solid (0.1% agar) Allen and Arnon's medium devoid of combined nitrogen. The alga has a generation time of 8–12 h when grown on atmospheric nitrogen. Addition of NO_3^- or NH_4^+ did not change the growth rate. Actively N_2-fixing filaments have a 15–18% heterocyst frequency. Under aerobic conditions, laboratory-grown *A. azollae* showed nitrogenase activity of 80–120 n mol C_2H_4 µg Chl $a^{-1}h^{-1}$. The activity was linear for 2 h in the light. The presence of high nitrogenase activity correlates with the observed high heterocyst frequency. Fresh intact *Azolla* plants showed a C_2H_2-reduction rate of 1.8–2 µmol C_2H_4 g (fresh wt)$^{-1}h^{-1}$ and the activity was almost equal in light and dark, but, after 2 h dark incubation, the activity began to decline. The repression of nitrogenase activity by combined nitrogen sources *viz.*, NO_3^- or NH_4^+ was more pronounced in isolated *A. azollae* than in the whole *Azolla* plants.

Introduction

Cyanobacteria fix atmospheric dinitrogen both under free-living and symbiotic conditions (Stewart, 1980). Symbiotic associations involving heterocyst-forming cyanobacteria are of interest because dinitrogen reduced by the cyanobacteria serves as nitrogen sources for growth of the eukaryotic host (Stewart and Rodgers, 1977; Stewart *et al.*, 1980; 1983). Although detailed studies have been carried out on growth, N_2 fixation and other physiological processes in various free-living cyanobacterial species (Stewart, 1980), little work has been done on symbiotic nitrogen fixation systems and their regulation (Meeks *et al.*, 1985; Stewart, 1980; Stewart *et al.*, 1980; 1983). The

water fern *Azolla* has attracted many workers due to its potential as biofertilizer (Stewart *et al.*, 1983; Zimmerman, 1987). *Azolla* is a genus of aquatic, heterosporous ferns and contains a symbiotic, heterocystous cyanobacterium, *Anabaena azollae*, within cavities formed in the aerial dorsal leaf lobes. The endophyte remains associated with the *Azolla* sporophyte in all stages of differentiation and development (Hill, 1975; 1977; Peters and Mayne, 1974; Peters, 1975; Peters *et al.*, 1977; 1978; 1980). Appreciable information is available on nitrogen fixation and related processes as well as for photosynthetic CO_2 fixation in the intact symbiotic association (Peters *et al.*, 1977; 1980; Ray *et al.*, 1978; 1979), but little or no information exists on growth and N_2 fixation of *Anabaena azollae* once isolated from *Azolla* (Stewart *et al.*, 1980), because attempts to isolate *A. azollae* have so far met with little success (Zimmerman, 1987). Here we report the successful isolation of *A. azollae* from *Azolla pinnata* and present evidence to show its persistent growth and active nitrogen fixation under asymbiotic conditions.

Materials and methods

Isolation of the endophyte

Anabaena azollae was isolated from its host *Azolla pinnata* which grows abundantly in various ponds near Banaras Hindu University campus. For isolation, *Azolla pinnata* was thoroughly washed with double-distilled water and thereafter incubated in 5-fold diluted AA$^-$-medium. After 72 h the *Azolla* fronds and medium were examined microscopically for free-living or epiphytic cyanobacteria. Healthy fronds not contaminated with epiphytic algae were selected and surface-sterilized by alcohol. After thorough washing in double-distilled water, these fronds were gently crushed in sterile AA$^-$-medium (Allen and Arnon, 1955) and the resulting suspension centrifuged at 1500 g for 5 min. The supernatant fraction was again centrifuged at 1500 g for 5 min to remove tissue and debris. Microscopic observation of the clear blue-greenish supernatant liquid showed the presence of short cyanobacterial filaments with heterocysts and many isolated heterocysts. Morphologically these filaments were indistinguishable from the *A. azollae* present in the host *Azolla*. This supernatant liquid was plated on 1% solid agar AA$^-$-medium devoid of combined nitrogen. After 3–4 days, blue-greenish filaments were spreading over the plates. These filaments were transferred both to solid and liquid AA$^-$ agar-medium. There was no growth in liquid plates but there was no visible growth in liquid medium. By repeated plating and transfers, the culture was made uni-algal and axenic on solid AA$^-$-medium. There was no growth in liquid medium but growth occurred in medium containing 0.1% agar; indeed, it was found that Allen and Arnon's medium containing 0.1% agar supported the best growth of *Anabaena azollae*. Cultures were grown routinely in this medium in a culture room at 27 ± 1°C illuminated with daylight fluorescent tubes (14.4 Wm^{-2} = 2400 lux) for 14 h d^{-1}.

Growth measurements

Growth was measured turbidimetrically at 660 nm in a Bausch and Lomb Spectronic-20 Spectrophotometer. Growth correlated with chlorophyll *a* content at chosen intervals of growth. Chlorophyll *a* was determined by extraction with acetone, employing the extinction coefficients given by Mackinney (1941). Heterocyst frequency was measured as a percentage of the total cells by counting at least 10 filaments (*ca.* 600–1000 cells).

Haemagglutination test

Haemagglutination tests were run with Dynatech microtiter V plates (Lis and Sharon, 1972) using rabbit trypsinized erythrocytes (1 mg trypsin, 1:250, per ml packed cells, 37°C for 1 h). Each well contained 25 μl algal extract diluted in phosphate-buffered saline (PBS, 0.01 M, pH 7.2), 25 μl PBS and 50 μl of the 2% erythrocyte suspension. The microtiter plate was covered and incubated for 2–3 h at room temperature. The positive (+) responses were scored when the bottoms of the wells were completely covered with a smooth mat of agglutinated cells and negative (−) responses when a definite compact button of non-agglutinated cells formed.

Nitrogenase activity

Nitrogenase activity was measured by the acetylene reduction technique (Stewart *et al.*, 1967). Two-ml *A. azollae* cultures, or the desired number of *Azolla* plants, were taken in 7-ml vacutainer tubes (Becton-Dickinson, Rutherford, N.J., U.S.A.) and acetylene was injected by a hypodermic syringe to maintain 10% final volume. All the assays were performed at 27°C and 14.4 Wm^{-2} light intensity. Dark treatment was given by wrapping the vials with aluminium foil. The ethylene formed was determined in a CIS (Baroda) gas chromatograph fitted with a Porapak R Column and flame ionization detector.

Results

The isolated cyanobacterium shows some morphological similarities with the endophyte associ-

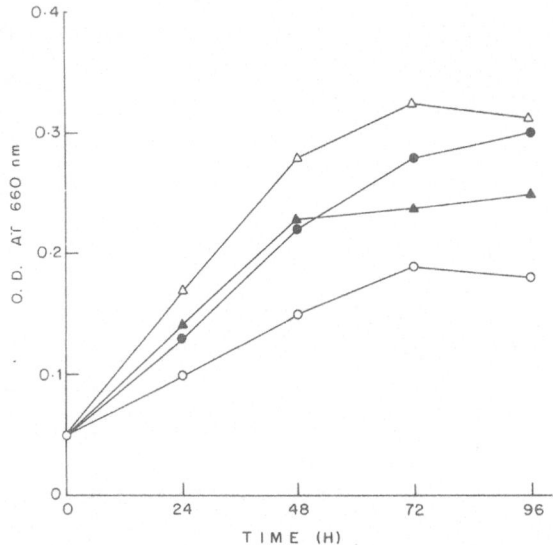

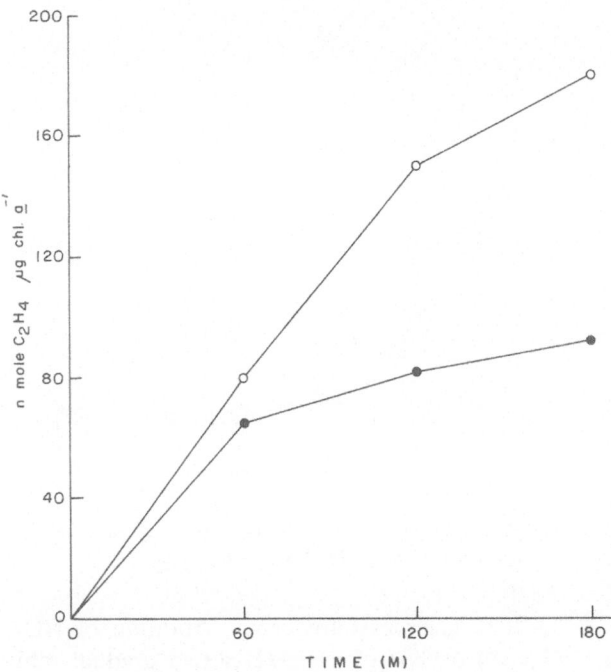

Fig. 1. Growth of *Anabaena azollae* in different nitrogen sources: △—△, AA + NH_4^+; ●—●, AA^-; ▲—▲, AA + NO_3^-; ○—○, AA^- but without agar.

Fig. 2. Nitrogenase activity by laboratory grown *A. azollae* under aerobic conditions: ○—○, light; ●—●, dark.

ated with the host *Azolla pinnata*. These include cell morphology, heterocyst frequency and sporulation. The cell size decreased markedly in the isolated *A. azollae* (length of vegetative cell; 20 μm; heterocyst; 27 μm) compared to the endophyte *in situ* (length of vegetative cell; 29 μm; heterocyst; 45 μm); however, morphology was similar in both cases. An individual filament of the cultured isolate was much longer than one living symbiotically. The spore was also smaller in laboratory-grown culture (39 μm length) than the symbiotic endophyte spore (57 μm). Again, both types of spore resembled each other. The heterocyst frequency was almost identical in the isolated (15–18%) and symbiotic alga (20–25%).

Apart from the similarity in heterocyst frequency, the above findings do not allow us to conclude that the isolated alga is indeed *A. azollae* and originating from *Azolla*. An extract of laboratory-grown cyanobacteria isolated from *Azolla* gave positive haemagglutination test results whereas free-living cyanobacteria *viz.*, *Anabaena doliolum*, *Nostoc muscorum* etc. failed to show any haemagglutination.

Extracts of freshly isolated endophyte (teased out from *Azolla* plants) or *Azolla* fronds also gave positive haemagglutination reactions. Furthermore, another symbiotic cyanobacterium, *A.*

cycadeae, isolated from *Cycas*, showed haemagglutination. These haemagglutination test results indicate that the isolated cyanobacterium is probably *A. azollae*.

Figure 1 shows the growth of *A. azollae* in simple Allen and Arnon's medium with or without combined nitrogen sources. The algae grew rapidly up to 72 h and thereafter more slowly. Best growth occurred in semi-solid 0.1% agar medium. After repeated transfer in liquid medium (without agar), the alga started growing but the yield was always half of that in the 0.1% agar-supplemented medium. Addition of NO_3^- (4 mM) or NH_4Cl (1 mM) did not affect growth rate. Under aerobic conditions, actively-growing cultures of *A. azollae* had an acetylene reduction rate of 80–120 n mol $C_2H_4 \mu g$ chl $a^{-1} h^{-1}$ (Fig. 2). A rate of 80 n mol $C_2H_4 \mu g$ chl $a^{-1} h^{-1}$ was routinely obtained. The nitrogenase activity was linear for 2 h in the light and similar in the dark initially but after 1 h there was a gradual decline of activity and only about 50% activity was left after 3 h. The rate of nitrogenase activity very much fits with the growth rate of this cyanobacterium. Furthermore, such high nitrogenase activity seems to be correlated with

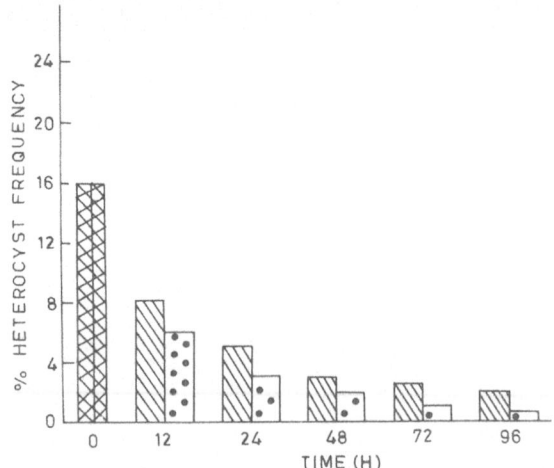

Fig. 3. Effect of NH$_4^+$ and NO$_3^-$ on heterocyst frequency; ⊠, AA$^-$; ◻, AA + NH$_4^+$; ◩, AA + NO$_3^-$.

15–18% heterocyst frequency. Addition of NO$_3^-$ (4 mM) or NH$_4$Cl (1 mM) caused gradual inhibition of heterocyst frequency (Fig. 3): the heterocyst frequency was below 4% within 24 h of NH$_4^+$ addition. However, neither NH$_4^+$ nor NO$_3^-$ at these concentrations could completely inhibit heterocyst formation even after 96 h (Fig. 3). Similarly, nitrogenase activity was typically repressed by the addition of NO$_3^-$ or NH$_4^+$ (Fig. 4). Inhibition of nitrogenase activity closely paralleled the decrease of heterocyst frequency (Fig. 4).

Intact fresh *Azolla* plants containing endophyte showed high nitrogenase activity both in the light and in the dark (Table 1). *Azolla* suspended in Allen and Arnon's medium elicited higher nitrogenase activity, i.e., 2.04 and 1.98 μmol C$_2$H$_4$ g (fresh wt)$^{-1}$ h^{-1} in light and dark, respectively. The specific activity of nitrogenase based on similar units in laboratory-grown *A. azollae* was 1.6–2.4 and 0.64–1.40 μmol C$_2$H$_4$ g (fresh wt)$^{-1}$ h^{-1}. Unlike isolated *A. azollae*, no immediate inhibition of nitrogenase activity was observed in intact *Azolla* plants by the addition of NO$_3^-$ or NH$_4^+$.

Discussion

Although the isolation of *A. azollae* and its growth under asymbiotic conditions has been reported (Gates *et al.*, 1980; Newton and Herman, 1979; Peters, 1975; Peters and Mayne, 1974; Ray *et*

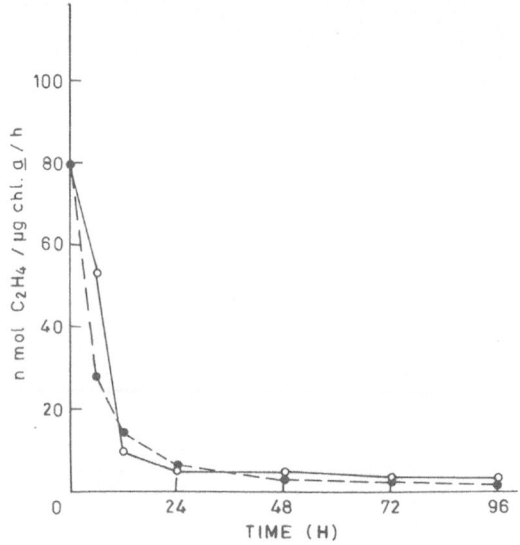

Fig. 4. Effect of combined nitrogen sources on nitrogenase activity: ●—●, AA + NH$_4^+$; ○—○, AA + NO$_3^-$.

al., 1979) there still exists confusion about the real identity of *A. azollae*. The alga isolated in the present study has been characterized as *A. azollae* on grounds of morphology, growth, agglutination reaction and nitrogenase activity. The shapes of vegetative cells, heterocysts and spores in the *in vitro* cultures were identical with those of the endophyte associated with the host but they were much smaller under non-symbiotic conditions, a fact noted by Newton and Herman (1979). The greatest similarity between laboratory-grown and symbiotic endophyte cells resides in heterocyst frequency. Except for *Anabaena* CA, no other free-living cyanobacterium has been reported to attain such a high (ca. 15–20%) frequency of heterocysts (Gotto *et al.*, 1979). These heterocysts are metabolically active and with high nitrogenase activity (Fig. 2 and Table 1). Further proof of the identity of the isolated cyanobacterium as *A. azollae* comes from the positive agglutination test. Our findings agree with earlier reports where the presence of lectin has been shown in *A. azollae* cultures freshly isolated from the host (Kobiler *et al.*, 1981). In our study, haemagglutination activity was specific for *A. azollae* or *A. cycadae* (another symbiotic cyanobacterium isolated from *Cycas* coralloid root) and all other free-living N$_2$-fixing species failed to show agglutination. The presence of agglutination activity in this alga leads us to conclude that the sugar-

Table 1. Comparison of nitrogenase activity between intact *Azolla* plants and isolated *A. azollae*

Conditions	Nitrogenase activity[a]	
	Light	Dark
Only *Azolla* plants	0.37	0.29
Azolla plants + double distilled water	1.01	0.96
Azolla plants + AA⁻ medium	2.04	1.98
Laboratory grown *A. azollae*	1.60 to 2.40[b]	0.64 to 1.40

[a] Nitrogenase activity expressed in μmol C_2H_4 g (fresh wt)$^{-1}$ h^{-1}.
[b] Values represent the range of activity obtained during the course of various estimations.

binding proteins (lectins) located on the cell surface of the alga may mediate the adherence of the *Anabaena* to the *Azolla*. We have not been able to perform a re-infection test with isolated *Anabaena azollae* but the results discussed above provide sufficient evidence that the isolated alga does in fact come from *Azolla*.

Nitrogenase activity and high heterocyst frequency separate this cyanobacterium from all other free-living cyanobacteria. The highest nitrogenase activity (108 n mol C_2H_4 μg chl a^{-1} h^{-1}) so far reported is from the marine cyanobacterium *Anabaena* CA (Gotto *et al.*, 1979) with optimum growth conditions. We have routinely detected 80–120 n mol C_2H_4 μg chl a^{-1} h^{-1} in this alga. Both heterocyst production and nitrogenase activity showed a typical but incomplete repression by the addition of NO_3^- or NH_4^+ (Figs 3 and 4). It seems that the repression of nitrogenase activity and heterocyst frequency by combined nitrogen sources are similar to those of free-living N_2-fixing cyanobacteria (Stewart, 1980).

Although the intact fresh *Azolla* plants containing the endophyte had high nitrogenase activity, the response of activity towards light/dark or combined nitrogen sources was different from the non-symbiotic isolated alga. Thus, similar nitrogenase activity was obtained both in light and in dark (Table 1); this is, presumably, due to the continued supply of reductant by the host to the endophyte. Probably, the host has a large pool of reductant that becomes available to the endophyte in the dark. There is no alternative source of reductant because the endophyte lacks the capacity to fix CO_2 even though it has normal photosynthetic pigments (Ray *et al.*, 1979; Tyagi *et al.*, 1980).

Similarly, unlike isolated *A. azollae*, no immediate inhibition of nitrogenase activity was observed in intact *Azolla* plants by the addition of nitrate or ammonium (NO_3^- or NH_4^+). Probably, the combined nitrogen sources are directly metabolized by the host and they do not reach the endophyte. Our results are consistent with earlier findings (Zimmerman, 1987). No doubt, the isolated *A. azollae* shows considerably high nitrogenase activity but if the rate is compared with intact *Azolla* plants (specific activity on g (fresh wt)$^{-1}$), it is evident that the alga loses appreciable activity (Table 1).

It is evident that *A. azollae* retains vital physiological features even in the free-living condition. The presence of high nitrogenase activity in intact *Azolla* plants, even in the presence of combined nitrogen sources, indicates that *Azolla* plants can fix nitrogen in nature despite the presence of high levels of combined nitrogen sources. This and other features make *Azolla* plants a suitable source of biofertilizer. The newly isolated *A. azollae* may help us understand the mechanism of association and nitrogen fixation process in greater detail.

Acknowledgement

This work has been supported by research grants from the Department of Non-Conventional Energy Sources (No. 103/3/85-NT and 102/16/86-NT), Govt. of India, New Delhi.

References

Allen M B and Arnon D I 1955 Studies on nitrogen-fixing bue-green algae. I. Growth and nitrogen fixation by *Anabaena cylindrica*. Plant Physiol. 30, 366–372.
Gates J E, Fisher R W, Goggin T W and Azrolan N I 1980 Antigenic differences between *Anabaena azollae* fresh from the *Azolla* fern leaf cavity and free-living cyanobacteria. Arch. Microbiol. 128, 126–129.
Gotto J W, Tabita F R and Van Baalen C 1979 Isolation and characterization of rapidly-growing marine, nitrogen-fixing strains of blue-green algae. Arch. Microbiol. 121, 155–159.
Hill D J 1975 The pattern of development of *Anabaena* in the *Azolla-Anabaena* symbiosis. Planta 122, 178–184.
Hill D J 1977 The role of *Anabaena* in the *Azolla — Anabaena* symbiosis. New Phytol. 78, 611–616.
Kobiler D, Cohen-Sharon A and Tel-or E 1981 Recognition between the N_2-fixing *Anabaena* and the water fern *Azolla*. FEBS Lett. 133, 157–160.

Lis H and Sharon N 1972 Soy bean (*Glycine max*) agglutinin. Methods Enzymol. 28, 360–368.

Mackinney G 1941 Absorption of light by chlorophyll solutions. J. Biol. Chem. 140, 315–322.

Meeks J C, Steinberg N, Joseph C M, Enderlin C S, Jorgensen P A and Peters G A 1985 Assimilation of exogenous and dinitrogen-derived $^{13}NH_4^+$ by *Anabaena azollae* separated from *Azolla caroliniana* Willd. Arch. Microbiol. 142, 229–233.

Newton J W and Herman A I 1979 Isolation of cyanobacteria from the aquatic fern, *Azolla*. Arch. Microbiol. 120, 161–165.

Peters G A 1975 The *Azolla–Anabaena azollae* relationship. III. Studies on metabolic capabilities and further characterization of the symbiont. Arch. Microbiol. 103, 113–122.

Peters G A and Mayne B C 1974 The *Azolla — Anabaena azollae* relationship. I. Initial characterization of the association. Plant Physiol. 53, 813–819.

Peters G A, Toia R E Jr and Lough S M 1977 The *Azolla–Anabaena azollae* relationship. V. $^{15}N_2$ fixation, acetylene reduction and H_2 production. Plant Physiol. 59, 1021–1025.

Peters G A, Toia R E Jr, Raveed D and Levine N J 1978 The *Azolla–Anabaena azollae* relationship. VI. Morphological aspects of the association. New Phytol. 80, 583–593.

Peters G A, Toia R E Jr, Evans W R, Crist D K, Mayne B C and Poole R E 1980 Characterization and comparisons of five N_2-fixing *Azolla–Anabaena* associations. I. Optimization of growth conditions for biomass increase and N content in a controlled environment. Plant Cell Environ. 3, 261–269.

Ray T B, Peters G A, Toia R E Jr and Mayne B C 1978 The *Azolla–Anabaena azollae* relationship. VII. Distribution of ammonia-assimilating enzymes, protein and chlorophyll between host and symbiont. Plant Physiol. 62, 463–467.

Ray T B, Mayne B C, Toia R E Jr and Peters G A 1979 The *Azolla–Anabaena azollae* relationship. VIII. Photosynthetic characterization of the association and individual partners. Plant Physiol. 64, 791–795.

Stewart W D P 1980 Some aspects of structure and function in N_2-fixing cyanobacteria. Annu. Rev. Microbiol. 34, 497–536.

Stewart W D P and Rodgers G A 1977 The cyanophyte-hepatic symbiosis. II. Nitrogen fixation and the interchange of nitrogen and carbon. New Phytol. 78, 459–471.

Stewart W D P, Fitzgerald G P and Burris R H 1967 *In situ* studies on N_2 fixation using acetylene reduction technique. Proc. Natl. Acad. Sci., U.S.A. 58, 2071–2078.

Stewart W D P, Rowell P and Rai A N 1980 Symbiotic nitrogen-fixing cyanobacteria. *In* Nitrogen Fixation. Eds. W D P Stewart and J R Gallon. pp 239–277. Academic Press, New York, London.

Stewart W D P, Rowell P and Rai A N 1983 Cyanobacteria-eukaryote plant symbiosis. Ann. Microbiol. (Inst. Pasteur) 134B, 205–228.

Tyagi V V S, Mayne B C and Peters G A 1980 Purification and initial characterization of phycobiliproteins from the endophytic cyanobacterium of *Azolla*. Arch. Microbiol. 128, 41–44.

Zimmerman W J 1987 Growth, nitrogen fixation and mass culture of isolated *Anabaena azollae*. Biotech. Lett. 9, 31–36.

F. A. Skinner et al. (Eds.), Nitrogen fixation with non-legumes, 83–88.

Bacteria in the *Azolla–Anabaena* symbiosis

C. FORNI[1], M. GRILLI CAIOLA[1] and S. GENTILI[2]
[1]*Dipartimento di Biologia and* [2]*Dipartimento di Medicina Sperimentale, II Università degli Studi di Roma, Via O. Raimondo, I-00173 Rome, Italy*

Key words: *Arthrobacter* spp., *Azolla* mucilage production, symbiosis

Abstract

The leaf cavities of the fern *Azolla* contain the N_2-fixing cyanobacterium *Anabaena azollae* Strasb. and bacteria. These bacteria have been isolated from five *Azolla* species, grown in the Botanic Garden of Naples, Italy; they were aerobic, Gram-positive and with a rod-coccus life cycle. Moreover they neither fixed nitrogen nor produced gas, H_2S or indole, and they were catalase positive. Nutritionally non-exacting strains grew on mineral salts medium with an ammonium salt and a carbon source. Most of the isolates produced mucilage. The bacteria were identified as *Arthrobacter globiformis* Conn and Dimmick, *A. nicotianae* Giovannozzi-Sermanni, *A. aurescens* Phillips and *A. cristallopoietes* Ensign and Rittenberg or *A. pascens* Lochhead and Burton. Among the species isolated *A. globiformis* is the most common. It is likely that the mucilage produced by *A. globiformis* and *A. aurescens* fulfils the function of retaining bacteria and *Anabaena* in the leaf cavities; moreover the bacteria, alone or together with the endophyte, may contribute to the formation of the envelope, which contains bacteria and algal cells.

Introduction

The leaf cavities of the aquatic fern *Azolla* contain the symbiotic N_2-fixing cyanobacterium *Anabaena azollae* Strasb. together with bacteria. In the literature there are several reports concerning the presence of these bacteria. The first description of them by means of transmission electron microscopy (TEM) observations was made by Grilli (1964) and later confirmed by Peters *et al.* (1978). The presence of the bacteria has been also detected by other methods (Newton and Cavins, 1976; Peters and Mayne, 1974; Gates *et al.*, 1980; Forni and Grilli Caiola, 1986; Wallace and Gates, 1986). These bacteria were first identified as *Pseudomonas* (Bottomley, 1920); later, Newton and Herman (1979) assigned to *Caulobacter fusiformis* and *Alcaligenes faecalis* the bacteria found associated with the phycobiont culture. Gates *et al.* (1980) reported the presence of coryneform bacteria in the fern leaf cavities. Recently, Wallace and Gates (1986) have identified them as eubacteria belonging to the genus *Arthrobacter* Conn and Dimmick.

The aim of this study was the taxonomic identification and metabolic characterization of the bacteria isolated from five *Azolla* species, grown in the Botanic Garden of Naples, Italy, in order to determine if more than one bacterial genus was present in the fern leaf cavities and to try to understand whether such bacteria are merely occasional occupants of the *Azolla* leaf cavities or if they play a role in the *Azolla–Anabaena* symbiosis as a constant component.

This study provides information on the enzymatic activities and the nutritional characteristics of the bacteria isolated.

Materials and methods

Fern species

Azolla plants were kindly provided by Dr. Moretti, Botanic Garden, University of Naples, Italy.

The bacteria were isolated from the leaf cavities of the following *Azolla* species: *Az. caroliniana*

Table 1. Characteristics of the bacteria isolated from the leaf cavities of five *Azolla* species

Characteristic	Isolate											
	A1	A2	A3	A4	A5	A6	B1	B2	C1	C2	D1	E1
Gram staining	+	+	+	+	+	+	+	+	+	+	+	+
Rod-coccus cycle	+	+	+	+	+	+	+	+	+	+	+	+
Colony colour												
white	+	+	+	+	+	+	+	+	−	−	−	+
pale buff	−	−	−	−	−	−	−	−	−	−	+	−
bright yellow	−	−	−	−	−	−	−	−	+	+	−	−
Growth at 30°C	+	+	+	+	+	+	+	+	+	+	+	+
Growth at 45°C	−	−	−	+	+	+	W^a	W^a	−	−	−	+
Mucilage production	+	+	+	+	+	+	+	+	−	−	+	−

^{a} W, weak growth

Willd., *Az. filiculoides* Lam., *Az. mexicana* Presl., *Az. microphylla* Kaulf. and *Az. pinnata* R. Brown.

Isolation of bacteria

Algal packets, isolated by manual dissection (from 5th to 10th leaves), were surface sterilized following the method of Gates *et al.* (1980), except that sterile distilled water was used instead of sterile buffered water (APHA, 1971). Various dilutions of the sterilized algal packets were plated on TRN medium containing (g.l^{-1}): tryptone (Difco), 10; NaCl, 5; agar (Oxoid), 15; pH 7. The Petri dishes were incubated at 30°C. Bacterial colonies were detected after 4–5 days.

Single colony isolates were obtained using routine bacteriological procedures.

Identification of the bacteria

Gram classification was made with the KOH test (Gregersen, 1978; Wallace and Gates, 1986), *Pseudomonas stutzeri* was used as negative control.

Bacterial identification was based on Bergey's Manual of Systematic Bacteriology vol. 2 (Sneath *et al.*, 1986).

Culture media and growth studies

Bacteria were isolated and then maintained on TRN medium.

The presence of anaerobic bacteria was tested for with thioglycollate broth (Difco), solidified with 1.5% agar, pH 7. The thioglycollate agar plates were incubated in a Brewer anaerobic chamber with a CO_2 and H_2 generating GasPak (Baltimore Biological Laboratories).

TSI (Triple Sugar Iron) medium was purchased from Baltimore Biological Laboratories.

Growth studies were conducted to determine the nutritional requirements of the bacteria isolated. Basal medium was the nitrogen-free M 9 containing (g.l^{-1}): KH_2PO_4, 3; $MgSO_4$, 0.2; Na_2HPO_4, 6; pH 7. Agar concentrations were 1.5% for solid and 0.5% for semi-solid medium. Ammonium chloride (20 m*M* was added where desired. Solutions (2% or 10% w/v) of different carbohydrates or organic

Table 2. Enzymatic activities present in the bacterial isolates

Enzymatic test	Isolate											
	A1	A2	A3	A4	A5	A6	B1	B2	C1	C2	D1	E1
Catalase	+	+	+	+	+	+	+	+	+	+	+	+
Oxidase	+	+	+	+	+	+	+	+	−	−	+	−
Urease	−	−	−	−	W^a	−	W^a	W^a	−	−	−	−
Arginine dehydrolase	−	−	−	−	−	−	−	−	−	−	−	+
Nitrate reduction	−	−	−	−	−	−	−	−	−	−	−	+
Nitrogenase	−	−	−	−	−	−	−	−	−	−	−	−

^{a} W, weakly positive

acids (the acids were neutralized with NaOH to pH 7) were sterilized separately and added to the medium to give final concentrations of 1% (w/v) or 0.5% (w/v). Growth factor solutions (biotin or riboflavin 10 mg. 100 ml^{-1}) were filter-sterilized (Millipore filter, 0.45 μm pore size) and added when needed (0.1 μg.ml^{-1}).

N_2-fixation

N_2-fixation was measured by the acetylene reduction assay (ARA). Acetylene (10%) was added under aerobic, microaerobic (2 or 5% O_2 in argon) and anaerobic conditions in cultures grown in semi-solid nitrogen-free M 9 medium with 1% glucose as carbon source. The cultures were incubated at 30°C for one week.

Ethylene formation was followed with a Perkin-Elmer 8310 gas-chromatograph provided with a silica gel column.

Enzyme assays

Catalase activity was determined following the Smibert and Krieg method (1981); *Staphylococcus epidermidis* was used as positive control.

The oxidase test was made by means of oxidase sticks (Oxoid BR 64).

API enzyme tests (Ayerst Italiana) were also used to determine some biochemical activities.

Results

Bacterial isolates

The bacteria isolated were aerobic; no anaerobic bacteria were detected.

Twelve single colonies were selected at random from the highest dilutions of the algal packets of five *Azolla* species; they formed white (isolates A, B and E) or pale-buff (isolate D 1) or bright yellow (isolates C1 and C2) colonies on TRN medium. All the bacteria grew well at 30°C, but some of them could also grow at 45°C (Table 1).

The Gram-positive bacteria were rods of varying sizes (1–5 μm in length and 0.5–0.6 μm in diameter), depending on the growth phase. 'V' formations were observed during the log phase. A rod-coccus cycle was detected for all the isolates (Table 1).

Cocci, produced during stationary or senescent phases, were *ca.* 0.6 μm in diameter. Sometimes rods were still visible after 15–20 days of growth in TRN medium, when the first Gram-positive cocci appeared. The length of the life cycles of the various isolates showed small differences (data not shown).

Twentyfour-hour-old cultures of the bacteria produced a viscous suspension when treated with 3% KOH; *Pseudomonas stutzeri* was used as a negative control.

Nine of twelve isolates produced mucilage (Table 1).

Enzyme assays (Table 2)

All isolates were catalase positive and all except three (C1, C2 and E1) were oxidase positive.

A weak urease activity was detected for three isolates (A5, B1 and B2). Only the isolate E1 showed arginine dehydrolase activity and reduced nitrate to nitrite.

Nitrogenase activity as determined by acetylene reduction, was not detected under the conditions tested.

Metabolic characteristics (Table 3)

The six A isolates hydrolyzed gelatin. None of the isolates produced gas, H_2S or indole.

Some isolates could produce acids from glucose (A3, A4, A5, A6 and C2) or lactose (E1), but only after two days or more of growth on TSI medium.

All the isolates could grow in the presence of 2 or 5% NaCl, and with the exception of E1, they also grew in 10% NaCl.

Only D1 required biotin as organic growth factor (Table 3), while for the other isolates growth occurred in mineral salts medium M9 with NH_4Cl as nitrogen source and a sugar or an organic acid as carbon source (Table 4).

Table 3. Metabolic characteristics of the bacterial isolates

Characteristic	Isolate											
	A1	A2	A3	A4	A5	A6	B1	B2	C1	C2	D1	E1
Gelatin hydrolysis	+	+	+	+	+	+	−	−	−	−	−	−
Production of:												
Gas	−	−	−	−	−	−	−	−	−	−	−	−
H$_2$S	−	−	−	−	−	−	−	−	−	−	−	−
Indole	−	−	−	−	−	−	−	−	−	−	−	−
Acid from:												
Glucose	−	−	+	+	+	+	−	−	−	+	−	−
Lactose	−	−	−	−	−	−	−	−	−	−	−	+
Growth in:												
2% NaCl	+	+	+	+	+	+	+	+	+	+	+	+
5% NaCl	+	+	+	+	+	+	+	+	+	+	+	+
10% NaCl	W[a]	+	−	+	+	W[a]	+	+	W[a]	W[a]	+	−
Vitamin requirement:												
Biotin	−	−	−	−	−	−	−	−	−	−	+	−

[a] W, weak growth

Identification of the bacteria

Based on morphological and biochemical data (Tables 1, 2, 3 and 4) and on the descriptive information in Bergey's Manual of Bacteriology (Sneath *et al.*, 1986), we identified the bacteria isolated as belonging to the genus *Arthrobacter* Conn and Dimmick. Furthermore the presence of different *Arthrobacter* species was detected.

The species *A. globiformis* Conn and Dimmick (isolates A1, A2, A3, A4, A5 and A6) was identified by the following characteristics: (1) the colonies on TRN showed no distinctive pigmentation (Table 1) (Keddie *et al.*, 1986); (2) the isolates were nu-

Table 4. Biochemical characteristics of the nutritionally non-exacting isolates

Characteristic	Isolate										
	A1	A2	A3	A4	A5	A6	B1	B2	C1	C2	E1
Utilization as carbon source of:											
Arabinose	+	+	+	+	+	+	−	−	−	−	−
Fructose	+	+	+	+	+	+	+	+	+	+	+
Galactose	+	+	+	+	+	+	+	+	+	+	+
Glucose	+	+	+	+	+	+	+	+	+	+	+
Glycerol	+	+	+	+	+	+	+	+	+	+	W[a]
Maltose	+	+	+	−	−	−	+	−	−	−	+
Mannitol	+	+	+	+	+	+	+	+	+	+	+
Ribose	+	+	+	+	+	+	+	+	+	+	+
Sorbitol	+	+	+	−	+	+	+	+	+	+	+
Sucrose	+	+	+	+	+	+	+	+	+	+	+
N-acetyl-glucosamine	−	+	−	+	+	+	+	−	−	−	+
Gluconate	+	+	+	+	+	+	+	+	−	−	+
Adipate	−	−	−	−	−	−	−	+	+	−	−
Malate	+	+	+	+	+	+	−	+	−	−	−
Citrate	+	+	+	+	+	+	+	+	+	+	+
Acetate	+	+	+	−	W[a]	+	+	+	+	−	+
Succinate	+	+	+	+	+	+	+	+	+	+	+
Caprate	−	−	−	−	−	−	−	−	−	−	−
Ascorbic acid	+	+	+	+	+	+	W[a]	W[a]	−	−	+

[a] W, weak

tritionally non-exacting, moreover they utilized several compounds as carbon sources (Table 4) (Keddie *et al.*, 1986); (3) all the isolates hydrolysed gelatin and did not produce indole (Table 3) and they were urease negative or weakly positive (Table 2) (Wallace and Gates, 1986). It is noteworthy that all the *A. globiformis* strains produced mucilage (Table 1).

The isolates B1 and B2 were probably *A. globiformis*, although they differed from the isolates A in gelatin hydrolysis (Table 3).

The species *A. nicotianae* Giovannozzi-Sermanni (isolates C1 and C2) was identified by the production of bright lemon-yellow colonies on TRN medium (Table 1); moreover the isolates did not require a vitamin as growth factor and utilized several carbon sources, particularly glycerol, ribose, arabinose and galactose (Table 4) (Keddie *et al.*, 1986).

Arthrobacter aurescens Phillips (isolate D1) produced pale buff colonies on complex medium, when incubated in the dark (Sneath *et al.*, 1986; Grainger *et al.*, unpublished data) and required biotin as a growth factor (Table 3). When supplied with biotin, growth occurred on M9 medium with an ammonium salt as nitrogen source and different sugars as carbon source (data not shown). Moreover, it produced mucilage (Table 1).

Lastly the colonies of the isolate E1 on TRN medium had no distinctive pigmentation (Table 1); furthermore the isolate could grow on M9 medium with an ammonium salt and a carbon source (Table 4), but it differed from *A. globiformis* strains because of the oxidase test and gelatin hydrolysis (Tables 2 and 3). Moreover it possessed arginine dehydrolase activity and reduced nitrate to nitrite (Table 2). Since the biochemical characteristics of E1, so far examined, were very similar to two different species, this isolate might be identified as *A.*

cristallopoietes Ensign and Rittenberg or *A. pascens* Lochhead and Burton (Sneath *et al.*, 1986).

Among the *Arthrobacter* species, *A. globiformis* was the most common. In fact it has been isolated eight times from four of five *Azolla* species (Table 5); while *A. nicotianae* (two isolates), *A. aurescens* (one isolate) and *A. cristallopoietes* or *A. pascens* (one isolate) have been isolated, respectively, only from *Azolla caroliniana*, *Az. filiculoides* and *Az. pinnata* (Table 5).

Discussion

A bacterial population was present in all leaf cavities of the *Azolla* species so far examined. These observations agree with those of Petro and Gates (1987). Moreover, the number of bacteria inside the leaf cavities increased as the leaf aged (Petro and Gates, 1987; Albertano and Grilli Caiola, 1988).

The life cycle of the bacteria, *i.e.* the formation of rods and 'V' forms during the log phase and the production of cocci, predominating in the senescent phase, indicated that the bacteria belonged to the genus *Arthrobacter*. These data were confirmed also by the biochemical characteristics so far determined. Considering the controversy which exists in the classification of this genus we have identified the following species: *A. globiformis*, *A. nicotianae*, *A. aurescens* and *A. cristallopoietes* or *A. pascens* based on the description in Bergey's Manual (Keddie *et al.*, 1986). *Arthrobacter globiformis* is the most common species isolated, having been found in four of five *Azolla* species. Moreover, Wallace and Gates (1986) reported the presence of only *A. globiformis* in their isolates. Since the other species have been found in more than one *Azolla* species, we cannot yet establish any correlation between species of *Arthrobacter* and species of *Azolla*. The

Table 5. Distribution of the *Arthrobacter* spp. isolated from the leaf cavities of five species of *Azolla*

Azolla spp.	A. globiformis	A. nicotianae	A. aurescens	A. cristallopoietes or A. pascens
A. caroliniana	4	2	–	–
A. filiculoides	1	–	1	–
A. mexicana	2	–	–	–
A. microphylla	1	–	–	–
A. pinnata	–	–	–	1

fact that our isolates, and those of Wallace and Gates (1986), were all of one genus, *Arthrobacter*, suggests that these bacteria may be a third component of the symbiosis (Petro and Gates, 1987) and not occasional invaders coming from the external environment. Moreover, bacteria have also been found in microsporocarps of *Azolla filiculoides*, suggesting that like the endophyte *Anabaena* (Peters and Calvert, 1982) these bacteria are retained throughout the sexual cycle of the fern.

No isolate fixed nitrogen. Therefore the bacteria do not contribute to the nitrogen metabolism of the symbiosis via ammonia production. However, one strain, i.e. the supposed *A. cristallopoietes* or *A. pascens*, reduced nitrate to nitrite. Wallace and Gates (1986) reported N_2 gas production for some of their bacterial isolates. It has not been demonstrated that these activities contribute to the nitrogen metabolism of the association. Apart from *A. aurescens*, the strains do not require growth factors.

So far the nutritional versatility of almost all the strains, which can use a wide and varied range of substrates as sole or major carbon + energy sources, suggests that inside the leaf cavities these bacteria may easily adapt their metabolism to the environmental circumstances. As yet, however, it is not known whether the bacteria benefit the *Anabaena* or not.

Nevertheless one may assume that the mucilage production detected in all the *A. globiformis* strains and in the *A. aurescens* strain might fulfil the function of retaining bacteria and *Anabaena* in the leaf cavities. Although some authors (Duckett *et al.*, 1975; Robins *et al.*, 1986) suggest that the envelope, which contains bacteria and algal cells, is produced by *Anabaena*, it is likely that the bacteria, alone or together with the endophyte, may contribute to the formation of the envelope.

Acknowledgement

Research work supported by CNR, Italy, Special grant I.P.R.A., Subproject 1. Paper N. 1671.

References

Albertano P and Grilli Caiola M 1988 Ultrastructural study of *Azolla microphylla* Kaulf. Giorn. Bot. Ital. 122, *In press*

American Public Health Association (APHA) 1971 Standard Methods for the Examination of Water and Freshwater, 13th edn. American Public Health Association, Inc., Washington D.C.

Bottomley W B 1920 The effect of organic matter on the growth of various water plants in culture solution. Ann. Bot. 39, 353–365

Duckett J G, Toth R and Soni S L 1975 An ultrastructural study of the *Azolla, Anabaena azollae* relationship. New Phytol. 75, 111–118

Forni C and Grilli Caiola M 1986 Studio in coltura dei batteri isolati dalle cavita fogliari di *Azolla*. Giorn. Bot. Ital. 120, n. 1–2, Suppl. 2, 100

Gates J E, Fisher E W and Candler R A 1980 The occurrence of the coryneform bacteria in the leaf cavity of *Azolla*. Arch, Microbiol. 127, 163–165

Gregersen T 1978 Rapid method for the distinction of the Gram-negative from the Gram-positive bacteria. Eur. J. Appl. Microbiol, Biotechnol. 5, 123–177

Grilli M 1964 Infrastrutture di *Anabaena azollae* vivente nelle foglioline di *Azolla caroliniana*. Ann. Microbiol. 14, 69–90

Keddie R M, Collins M D and Jones D 1986 Genus *Arthrobacter* Conn, and Dimmick. *In* Bergey's Manual of Systematic Bacteriology. Eds. P H A Sneath, N S Mair, M E Sharpe and J G Holt. pp 1288–1301. Williams & Wilkins, Baltimore and London

Newton J W and Cavins J F 1976 Altered nitrogenous pools induced by the *Azolla-Anabaena* symbiosis. Plant Physiol. 58, 798–799

Newton J W and Herman H I 1979 Isolation of cyanobacteria from the aquatic fern *Azolla*. Arch. Microbiol. 120, 161–165

Peters G A and Mayne B C 1974 The *Azolla, Anabaena azollae* relationship. II. Localization of nitrogenase activity as assayed by acetylene reduction. Plant Physiol. 53, 820–824

Peters G A, Toia R E Jr, Raveed D and Levine N J 1978 The *Azolla-Anabaena azollae* relationship. VI. Morphological aspects of the association. New Phytol. 80, 583–593

Peters G A and Calvert H E 1982 The *Azolla-Anabaena* symbioses. *In* Advances in Agricultural Microbiology. Ed. N S Subba Rao. pp. 191–218. Oxford & IBH Publishing Company, New Delhi, Bombay, Calcutta

Petro M J and Gates J E 1987 Distribution of *Arthrobacter* sp. in the leaf cavities of four species of the N-fixing *Azolla* fern. Symbiosis 3, 41–48

Robins R J, Hall D O, Shi D-J, Turner R J and Rhodes M J C 1986 Mucilage acts to adhere cyanobacteria and cultured plant cells to biological and inert surfaces. FEMS Microbiol. Lett. 34, 155–160

Smibert R M and Krieg N R 1981 General characterization. *In* Manual of Methods for General Bacteriology. Eds. P Gerhards, R G E Murray, R N Costilow, E W Nester, W A Wood, N R Krieg and G B Phillips. pp. 409–443. American Society for Microbiology, Washington D.C.

Sneath P H A, Mair N S, Sharpe M E and Holt J G (Eds) 1986 Bergey's Manual of Systematic Bacteriology, Vol. 2. Williams and Wilkins, Baltimore and London, 1599 pp.

Wallace W H and Gates J E 1986 Identification of eubacteria isolated from leaf cavities of four species of the N-fixing *Azolla* fern as *Arthrobacter* Conn and Dimmick. Appl. Environ. Microbiol. 52, 425–429

F. A. Skinner et al. (Eds.), Nitrogen fixation with non-legumes, 89–94.

New data on the *Azolla-Anabaena* symbiosis
I. Morphological and histochemical aspects

F. CARRAPIÇO and R. TAVARES
Departamento de Biologia Vegetal, Faculdade de Ciências de Lisboa, P-1294 Lisboa Codex, Portugal

Key words: *Azolla-Anabaena* bacteria symbiosis, electron light microscopy, histochemistry, transfer hairs

Abstract

Light and electron microscopy studies were made on dorsal leaves of *Azolla filiculoides* at different stages of development that corresponded to three different levels of nitrogen fixation in the plant. The morphological and histochemical studies referred to in this work showed a complex relationship between the three partners (*Azolla*, *Anabaena*, bacteria), with special emphasis on the biological recognition of these organisms in the symbiosis. The bacteria, present in all stages of leaf development, are in close association with the *Azolla* cavity cells and follow a location pattern identical to that of the cyanobacteria.

Introduction

The symbiotic association *Azolla-Anabaena* was described for the first time by Strasburger in 1873. Since then and particularly over the last few years, several studies have been made to elucidate the morphological and functional relationship between these two partners in the symbiosis (Grilli, 1964; Hill, 1975; Lumpkin and Plucnett, 1980; Neumuller and Bergman, 1981; Peters and Mayne, 1974; Peters and Calvert, 1983 and Sevillano *et al.*, 1984). Until recently it was generally accepted that this symbiosis was formed by two partners. After the isolation and identification of one type of bacteria in the leaf cavity of this fern (Wallace and Gates, 1986) the idea of a third symbiotic partner was implicitly suggested, but was not shown to be a permanent member of the prokaryotic colony at different stages of leaf development.

In this work, light and electron microscopy studies were made on dorsal lobe leaves of *Azolla filiculoides* at three different stages of development, with special attention to the relationship between the fern and the prokaryotic colony. Thus, the purpose of this work was to contribute to a better understanding of the symbiotic relationship

between the three partners (*Azolla*, *Anabaena*, bacteria) and to increase knowledge of some metabolic processes that occur in the symbiosis.

Materials and methods

The *Azolla filiculoides* plants were collected in Alcochete (70 km from Lisbon) and maintained in natural conditions in the Botanical Garden of Lisbon.

Light and electron microscopy studies were made on dorsal lobe leaves of this fern at different stages of development (apical meristem, full mature leaf and senescent leaf) that correspond to three different levels of nitrogen fixation in the plant.

For electron microscopy studies, leaves were submitted to a double step fixation process by 4% glutaraldehyde in 0.05 M cacodylate buffer (pH 7.2) and in 1% OsO_4 in 0.05 M cacodylate buffer (pH 7.4) with or without ruthenium red (Colombo and Rascio, 1977). The samples were dehydrated in an acetone series and embedded in Epon–Araldite (Mollenhauer, 1964). Semi-thin sections (800 nm) of this material without ruthenium red, were used in lipid histochemistry by sudan black B staining

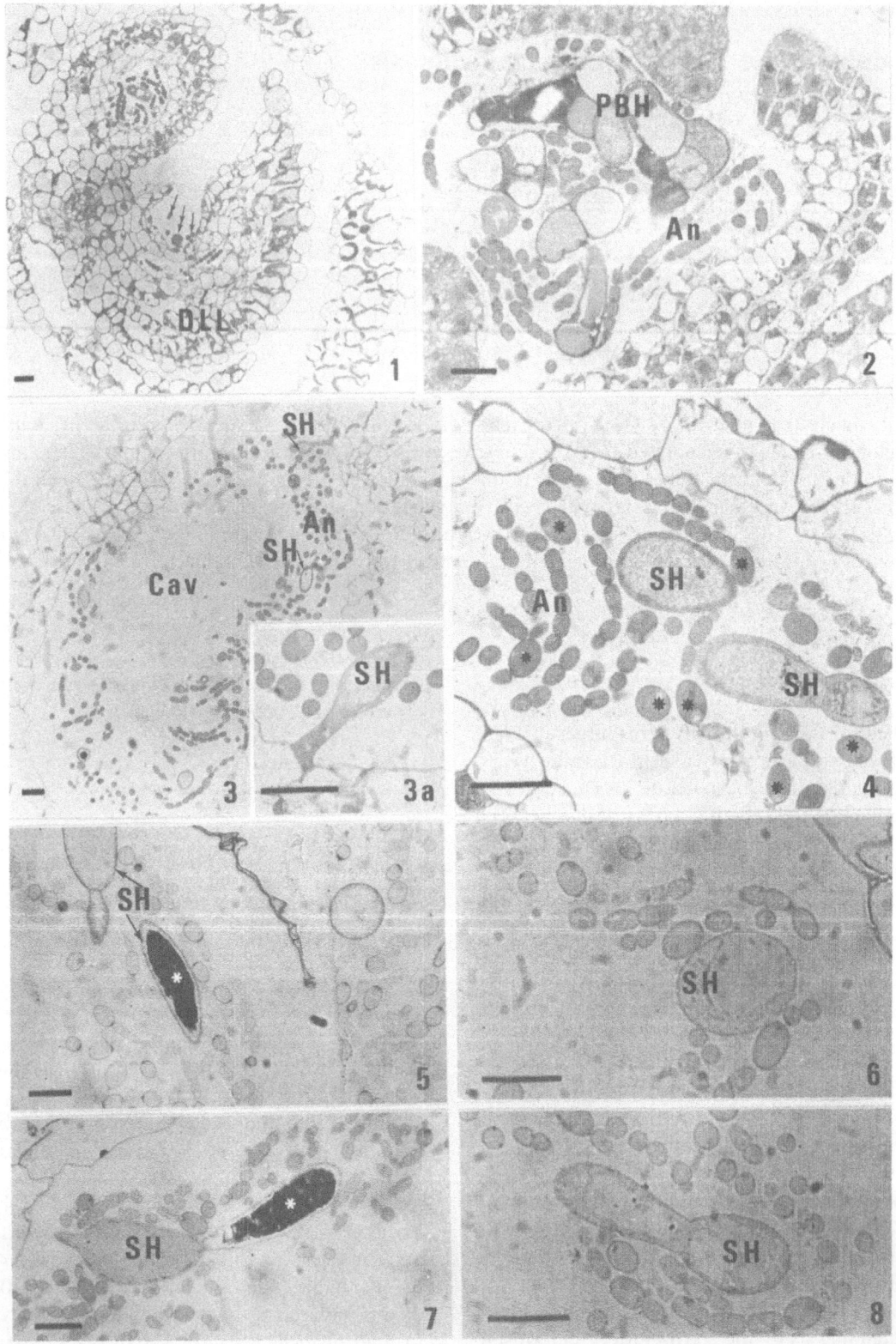

(Bronner, 1975) and in the detection of insoluble polysaccharides by the periodic acid Schiff (PAS) technique (Feder and O'Brien, 1968). For further electron microscopic morphological data, impregnation with Kl-OsO$_4$ mixture was made (Carrapiço et al., 1984).

Thin sections (60–80 nm) of the specimens were cut with a glass knife and Porter-Blum MT-2 ultramicrotome and mounted on copper grids. The sections were post-stained (except KI sections) and the observations were made in a Hitachi-12 electron microscope operating at an accelerating voltage of 75 Kv.

Results and discussion

In early stages of cavity formation, at the apical meristem, several filaments of undifferentiated cells of *Anabaena* and bacteria were seen in the proximity of epidermal cells or near the primary branched hair (PBH) cells (Figs. 1, 2, 9, 10, 12).

The blue-green alga (cyanobacterium) filaments were mainly characterized by the absence of heterocysts (Figs. 2, 9) and the PBH cells showed a typical transfer hair morphology (Fig. 10).

The next stages of leaf development (mature and senescent leaves) were characterized by the presence of other types of transfer hairs, namely the simple hairs (SH), which also showed a transfer cell morphology (Figs. 3, 3a, 11). These trichomes appeared at different stages of development in the

same leaf cavity and they seemed not to be related to the age of the leaf. The prokaryotic colonies (cyanobacteria and bacteria) appeared to be more numerous and the blue-green alga was characterized by the presence of a great number of heterocysts in the filaments (Fig. 4). A middle-age trichome revealed the presence of labyrinthine cell wall ingrowths, whose contours were closely followed by the plasmalemma (Fig. 11); a large central vacuole with deposits (Figs. 4, 5, 7); other different cytoplasmic organelles, in particular, a poorly developed endomembrane system revealed by Kl-O$_s$O$_4$ impregnation (Figs. 13, 14); some plastids with rudimentary tylakoid system; numerous mitochondria with developed cristae, and a few peroxisomes (Fig. 11). It seems that no significant morphological difference exists between these two levels of leaf development, with the main exception that the number of heterocysts in the blue-green alga present in senescent leaves decreases.

The presence of some compounds in the cell vacuoles of simple hairs are clearly evident in electron and light microscopy observations. The use of PAS and sudan black B tests in semi-thin sections revealed the polysaccharidic and lipophilic nature of these substances accumulated in the terminal cells of these trichomes (Figs. 5, 6, 7, 8). This accumulation may be the result of an autolytic or a secretion process. It is not clear whether the same hair can accumulate these two kinds of substance

Fig. 1. Early stage of leaf development. A depression, which corresponds to the cavity before closing, can be seen in the ventral side of the dorsal leaf lobe (DLL). This depression contains several filaments of *Anabaena* vegetative cells (arrows) and the primary branched hair (double arrow).

Fig. 2. The apical colony of *Anabaena* cells (An) is maintained in the proximity of the primary branched hairs (PBH). In this stage of leaf development no heterocyst are observed in the cyanobacteria.

Figs. 3–4. Several aspects of a cavity (Cav) in a mature leaf. Sections of simple hairs (SH) can be seen. The presence of a great number of heterocysts in *Anabaena* filaments is evident (asterisks). Fig. 3a shows a magnification of a simple hair in the cavity.

Fig. 5. Histochemical localization of insoluble polysaccharides (PAS test) in simple hair cells (SH). The reaction product is observed in vacuole sap of the terminal cells of these trichomes (asterisk).

Fig. 6. Control of the PAS test. No reaction product is observed in simple hair cells.

Fig. 7. Histochemical localization of lipids (sudan black B test). The reaction product is only seen in the vacuole of the terminal cells of simple hairs (asterisk).

Fig. 8. Control of the sudan black B test. No reaction is observed in the terminal cells of the simple hair (SH).
In Figures 1–8 bar represents 20 μm.

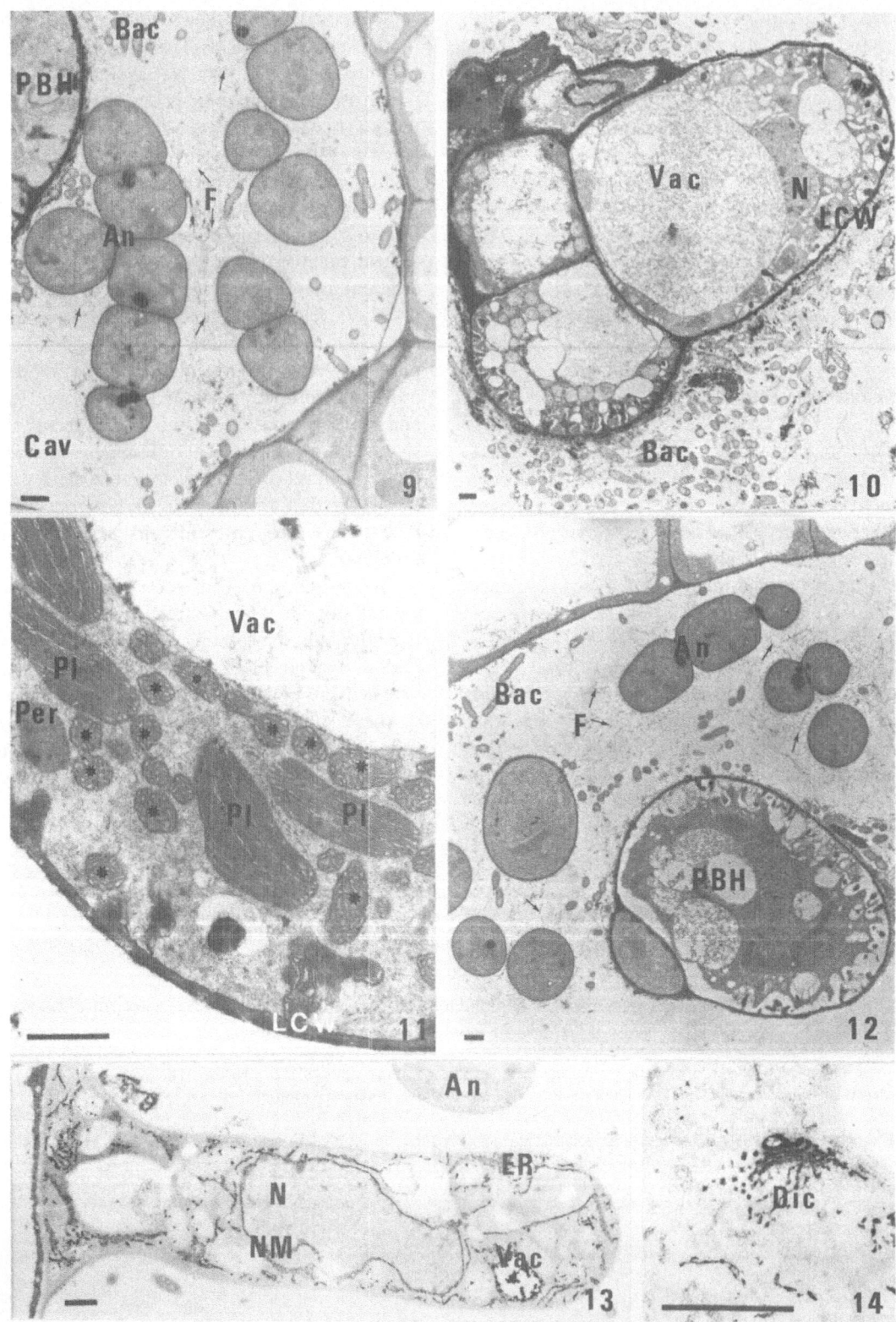

or if, on the contrary, different types of transfer hairs exist with specialized metabolic functions. These compounds may be involved in the protection against foreign organisms outside the symbiotic partners. Further research is needed to clarify this matter.

One of the more interesting features of this association is the existence of a mucilaginous fibrillar network, revealed by ruthenium red and present in all stages of leaf development, which fills the cavity and in which the blue-green algae and bacteria are immersed (Figs. 9, 12). This technique shows the polysaccharadic nature of the *Anabaena* mucilaginous cell wall and agrees with the results of Bergman *et al.*, (1985) in the free-living species *Anabaena cylindrica*. The mucilaginous fibrillar network observed in the *Azolla* cavity may play a part in the recognition process between the fern cells, cyanobacteria and bacteria. The presence of this mucilaginous network in the first steps of cavity formation, strongly suggests that its formation is not dependent on the fern, but related to the metabolic activity of the blue-green algae or the bacteria. These data support the hypothesis formulated by Duckett *et al.*, (1975) in *Anabaena*-free *Azolla*, that any mucilaginous compounds were formed in the cavity of the fern, which indicates that these substances are produced by the *Anabaena* cells.

Numerous bacteria with a typical coryneform or coccal morphology, probably of the genus *Arthrobacter* (Forni, 1987, personal communication; Wallace and Gates, 1986), are present in all stages of leaf development, as a member of the prokaryotic colony, in close association with the transfer hairs and in the proximity of *Azolla* epidermal cells. The metabolic function of these bacteria in the symbiosis is unknown.

In conclusion, we suggest that a bacterium, probably *Arthrobacter*, is the third permanent partner of the symbiosis. The bacteria are present at all stages of leaf development in close association with the primary branched hairs, simple hairs or *Azolla* epidermal cells, following a location pattern identical to that of *Anabaena*.

A mucilaginous fibrillar network revealed by ruthenium red and present at all stages of leaf development, fills the cavity in which blue-green algae and bacteria are contained. This network may play a part in the recognition process between the three partners of the symbiosis.

The apical cells of simple hairs accumulate substances of lipophilic and polysaccharidic nature. The function of these compounds is unknown, but we suggest that they may be involved in the defence against foreign organisms other than the symbiotic partners.

References

Bergman B, Lindblad P, Pettersson A, Renstrom E and Tiberg A 1985 Immuno-gold localization of glutamine synthethase in a nitrogen-fixing cyanobacterium (*Anabaena cylindrica*). Planta 166, 329–334

Bronner R 1975 Simultaneous demonstration of lipids and starch in plant tissues. Stain Technol. 50, 1–4

Carrapiço F, Pais MS and Madalena-Costa F 1984 Impregnation of biological material by the $ZnI_2 - OsO_4$, $KI-OsO_4$ and $NaI-OsO_4$ mixtures for electron microscopic observations. Chemical interpretation of the reaction. J. Microsc. 134, 193–202

Colombo PM and Rascio N 1977 Ruthenium red staining for

Figs. 9 and 12. Ultrastructural aspects of the early stages of cavity formation. The *Anabaena* filaments (An) and bacteria (Bac) are in close proximity with the primary branched hair (PBH). A fibrillar network (F), revealed by ruthenium red (arrows) can be observed, surrounding the *Anabaena* cells and bacteria.

Fig. 10. Section of a primary branched hair. The cells of this hair present a labyrinthine cell wall (LCW) and are highly vacuolated (Vac). A great number of bacteria (Bac) are in close association with this hair.

Fig. 11. Detail of a simple hair cell. These cells also show a transfer cell morphology. A great number of mitochondria (asterisks) and a few plastids (Pl) are observed. The number of peroxisome (Per) sections are, on the contrary, very low. These cells show, frequently, a large vacuole (Vac) containing fibrillar material and electron-dense deposits.

Figs. 13–14. Aspects of a simple hair cell treated with the $KI-OsO_4$ mixture. The reactivity is observed in the endoplasmic reticulum (ER), dictyosome (Dic), nuclear membrane (NM) and in the sap of some vacuoles (Vac).
In figures 9 to 14 bar represents 1 μm.

electron microscopy of plant material. J. Ultrastruct. Res. 60, 135–139

Duckett JG, Toth R and Soni SL 1975 An ultrastructural study of the *Azolla-Anabaena azollae* relationship. New Phytol. 75, 111–118

Feder N and O'Brien TP 1968 Plant microtechnique: some principles and new methods. Am. J. Bot. 55, 123–144

Grilli M 1964 Infrastrutture di *Anabaena azollae* vivente nelle foglioline di *Azolla caroliniana* Ann. Microb. Enzim. XIV, 69–90

Hill DJ 1975 The pattern of development of *Anabaena* in the *Azolla-Anabaena* symbiosis. Planta 122, 179–184

Lumpkin TA and Plucnett DL 1980 *Azolla*: Botany, physiology and use as green manure. Econ. Bot. 34, 111–153

Mollenhauer HH 1964 Plastic embedding for use in electron microscopy. Stain Technol. 39, 111–114

Neumuller M and Bergman B 1981 The ultrastructure of *Anabaena azollae* in *Azolla pinnata*. Physiol. Plant. 51, 69–76

Peters GA and Mayne BC 1974 The *Azolla-Anabaena azollae* relationship. I. Initial characterization of the association. Plant Physiol. 53, 813–819

Peters GA and Calvert HE 1983 The *Azolla-Anabaena azollae* symbiosis. *In* Algal Symbiosis. Ed. LJ Goff. pp 109–145 Cambridge University Press, New York

Sevillano F, Subramanian P and Rodriguez-Barrueco C 1984 La association simbiotica fijadora de nitrogenio atmosferico *Azolla-Anabaena* An CEBA Vol II

Wallace WH and Gates JE 1986 Identification of eubacteria isolated from leaf cavities of four species of the N-fixing *Azolla* fern as *Arthrobacter* Conn and Dimmick. Appl. Environ. Microbiol. 52, 425–429

F. A. Skinner et al. (Eds.), Nitrogen fixation with non-legumes, 95–100.
© 1989 by Kluwer Academic Publishers.

New data on the *Azolla-Anabaena* symbiosis
II. Cytochemical and immunocytochemical aspects

F. CARRAPIÇO and R. TAVARES
Departamento de Biologia Vegetal, Faculdade de Ciências de Lisboa, P-1294 Lisboa Codex, Portugal

Key words: *Azolla-Anabaena*, ATPase, catalase, glycollate oxidase, peroxisomes, cytochemistry, RuBisCO, immunocytochemistry, transfer hairs

Abstract

Cytochemical and immunocytochemical studies were made on the *Azolla-Anabaena* symbiosis to try to understand how metabolites are interchanged between the prokaryotic colony and the fern. The detection of ribulose-1,5-bisphosphate carboxylase/oxygenase (RuBisCO) by immuno-gold staining was considered to be important in order to understand the behaviour of *Anabaena* cells in the association and the role of the plastids present in *Azolla* transfer hairs and mesophyll cells. We have detected the presence of catalase and glycollate oxidase activity in peroxisomes of *Azolla*. The relationship between these two sets of data is discussed.

Introduction

The physiology and biochemistry of nitrogen fixation in different plants has been studied intensively during the last few years, particularly in the *Rhizobium*-legume symbiosis (Reynolds *et al.*, 1982; Sprent, 1984). These studies have been extended to other plant associations, *e.g.* the *Azolla-Anabaena* symbiosis (Lumpkin and Plucnett, 1980; Peters, 1975; Ray *et al.*, 1978). The most important characteristic of this symbiotic association is that a cyanobacterium (blue-green alga) *Anabaena azollae*, lives in a cavity of the dorsal lobe of the fern leaf, where the ecological conditions stimulate high heterocyst frequency in the cyanobacterium.

The exchange of metabolites between the two organisms is assumed to be an active process of photosynthate translocation from the host to the algae and translocation of fixed nitrogen compounds from the algae to the host.

In this work, cytochemical and immunocytochemical research has been developed, with special attention to the relation between the catalase and glycollate oxidase activities and the presence of RuBisCO in the *Azolla* cells. The detection of this

enzyme in the *Anabaena* cells will be reported elsewhere.

Materials and methods

Cytochemistry

Catalase. (Carrapiço *et al.*, 1985; Silverberg and Sawa, 1973). Leaves of *Azolla filiculoides* were fixed with 4% (v/v) glutaraldehyde in $100\,mM$ cacodylate buffer (pH 7.4) overnight at 4°C. After several washes in the same buffer, the material was incubated for 1 h, at 37°C, in a mixture containing: $50\,mM$ glycine/NaOH buffer pH10, 5 ml; DAB (3,3′diaminobenzidine), 10 mg; 3% H_2O_2, 0.1 ml. The final pH of the incubated solution was adjusted to 9.5. The control mixture contained $20\,mM$ 3-amino-1,2,4-triazole (inhibitor of the catalase activity) added to the initial solution.

Glycollate oxidase. (Carrapiço *et al.*, 1985; Thomas and Trelease, 1981). The material was pre-fixed in an ice-cold fixative consisting of formaldehyde-glutaraldehyde (2%–2.5%) in $50\,mM$ PIPES buffer

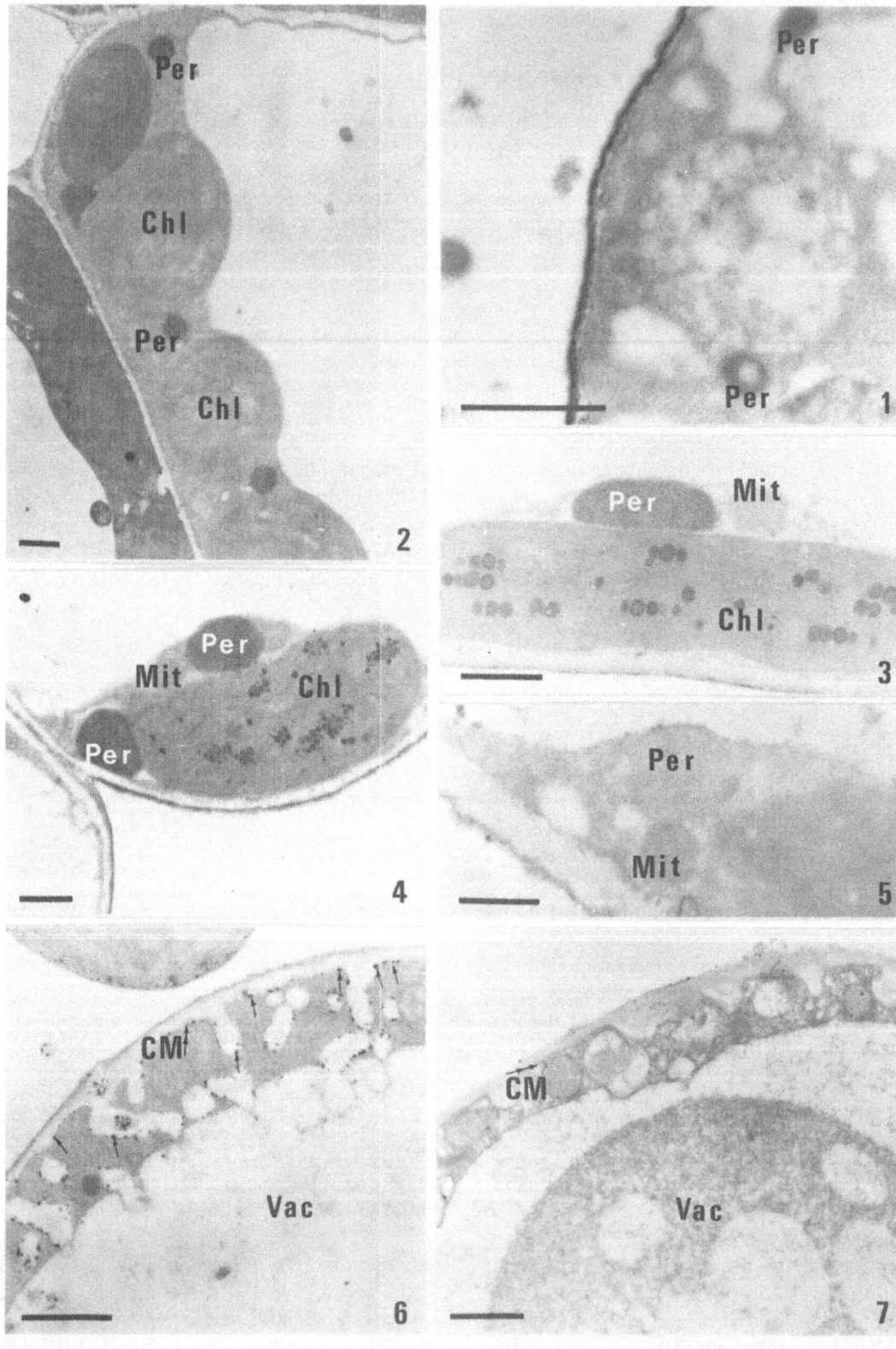

(pH 7.4) for 15 min at 4°C. The formaldehyde was freshly prepared from paraformaldehyde immediately before use.

After three washes (10 min each) in the same fixative buffer, the material was pre-incubated overnight, at 4°C, in 100 mM tris-maleate buffer (pH 7.8), containing 3 mM $CeCl_3$ and 50 mM 3-amino-1,2,4-triazole. After this step, the samples were incubated at 37°C, for 3 h, in an aerated mixture, containing 100 mM Tris-maleate buffer (pH 7.8), 50 mM 3-amino-1,2,4-triazole, 20 mM cerium chloride ($CeCl_3$) and 60 mM sodium glycollate. The final pH of the mixture was adjusted to 7.8. The control of the reaction was made with the same medium, but without sodium glycollate.

ATPase. (Hulstaert *et al.*, 1983). *Azolla* leaves fixed in the same way as for glycollate oxidase detection, were pre-incubated overnight at 4°C in a solution consisting 100 mM Tris-maleate buffer (pH 7.2) and 1 mM $CeCl_3$. After this step the material was incubated for 45 min at 37°C in a solution consisting 100 mM Tris-maleate buffer (pH 7.2), 1 mM $CeCl_3$, 2.5 mM ATP and 4 mM $MgSO_4$. The final pH of the mixture was adjusted to 7.2. The control was done without substrate (ATP). All the solutions containing $CeCl_3$ were made with boiled double-distilled water to prevent carbonate precipitation.

After incubation, the specimens were washed several times in 50 mM cacodylate buffer (pH 6.0) to remove non-specific cerium precipitates, and fixed overnight at 4°C in the same pre-fixative mixture. The samples were washed in 50 mM PIPES buffer (pH 7.4) (10 min each) and post-fixed in 1%

O_sO_4 in 50 mM cacodylate buffer (pH 6.8) for 2 h at room temperature. All the samples were dehydrated in an acetone series and embedded in Epon-Araldite (Mollenhauer, 1964).

Immunocytochemistry

RuBisCO (Ribulose-1,5-biphosphate carboxylase/ oxygenase). (Vaughn, 1987). *Azolla* leaves were fixed in an ice-cold fixative consisting of formaldehyde-glutaraldehyde (2%–2.5%) in 50 mM PIPES buffer (PH 7.4) for 3 h at 4°C and washed in three changes of the same buffer (10 min each). The samples were dehydrated through an ethanol series and embedded in LR White's resin. Thin sections (60–80 nm) of the specimens were cut with a glass knife and Porter-Blum Mt-2 ultramicrotome and mounted on nickel grids.

After this step the grids were floated on 1% (w/s) bovine serum albumin (BSA) in 50 mM sodium phosphate (pH 7.4) with 0.85% sodium chloride (=BSA–PBS) for 30 min to block non-specific antibody sticking. The grids were then transferred to a drop of anti-RuBisCO (antitobacco) diluted 1:20 (v/v) in BSA–PBS for 3 h. After several washes in the same buffer (BSA–PBS) the grids were floated on a 1:30 (v/v) dilution of protein A-colloidal gold (15 nm gold, EY Laboratories, San Mateo, CA) in BSA–PBS for 60 min. After several washes in phosphate-buffered saline (PBS) and water the grids were post-stained in 2% (w/v) uranyl acetate and Reynold's lead citrate. The controls were made with pre-immune sera. The observations were made in a Hitachi-12 electron microscope at 75 Kv.

Fig. 1. Localization of catalase activity in a simple hair cell. The stained organelles inside the cell are peroxisomes (Per).

Figs. 2–4. Cytochemical localization of catalase activity in the mesophyll cells of the dorsal leaf lobe (fig. 2 and 3) and in epidermal cells which surround the cavity (fig. 4). The reaction product is only observed in the peroxisomes (Per). No stain is seen in chloroplasts (Chl) and mitochondria (Mit).

Fig. 5. Control of the DAB catalase cytochemical test made by the addition of aminotriazole to the incubation medium. No reaction product is observed in peroxisomes.

Fig. 6. Cytochemical localization of adenosine triphosphatase (ATPase) activity in simple hair cells. The reactivity (arrows) is mainly located in the inner face of the cell membrane (M) (double arrow).

Fig. 7. Control of ATPase by omission of substrate (ATP) from the incubation medium. No reaction product is observed in the cell membrane (CM) (double arrow).

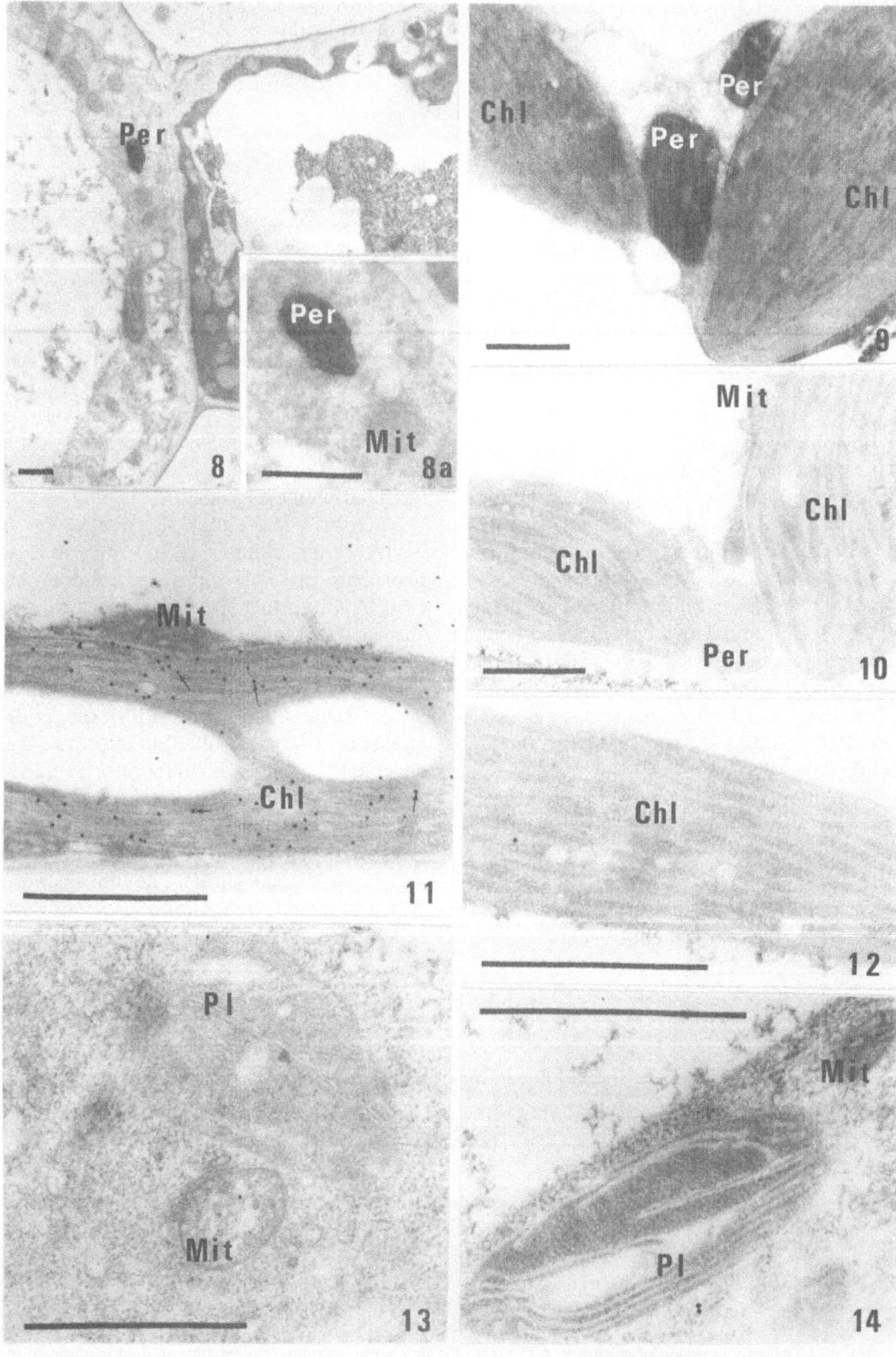

Results and discussion

After incubation of the *Azolla* leaves in a DAB medium to detect catalase activity and in a $CeCl_3$ glycollate solution to detect glycollate oxidase activity, stained peroxisomes were observed both in mesophyll and cavity epidermal cells of the fern in close association with chloroplasts and mitochondria (Figs. 2, 3, 4, 9). These two organelles did not show any reaction to these tests (Figs. 3, 8, 8a). Stained peroxisomes were also present in the cells of simple hairs (Figs. 1, 8, 8a). The controls of these two tests gave no reaction in peroxisomes (Figs. 5, 10).

The material submitted to the ATPase test revealed the presence of electron-dense deposits of cerium phosphate in the inner face of the cell membrane in simple hair cells (Fig. 6). The control gave a negative reaction in the plasmalemma (Fig. 7).

The immuno-gold staining of RuBisCO was apparent only in the stroma of chloroplasts present in the mesophyll cells (Fig. 11) and no colloidal gold particles were seen in plastids of simple hairs (Fig. 13). The control gave no labelling in these organelles (Figs. 12, 14).

The presence of reactive peroxisomes to catalase and glycollate oxidase tests showed that these organelles contained these two enzymes and are associated with an active metabolic pathway, especially for degradation of glycollate. This product (glycollate), formed in the chloroplasts by the action of RuBisCO's oxygenase activity, is transferred to the peroxisomes, where by the action of glycollate oxidase, it is converted into glyoxylate and initiates the photorespiration process (Tolbert, 1982). The glycollate oxidase activity generates H_2O_2 which is broken down by the action of catalase, also present in the peroxisomes. In these conditions, the results obtained by the glycollate oxidase and catalase cytochemistry tests in *Azolla* mesophyll cells suggest that these organelles are involved in glycollate degradation and are probably involved in the photorespiration process (Peters *et al.*, 1982).

RuBisCO is an important enzyme in the photosynthetic carbon reduction pathway (Grey and Kekwick, 1974; Vaughn, 1987) and its immunodetection in the chloroplasts of the *Azolla* mesophyll cells confirms the capacity of these organelles to fix CO_2. This result is normal, but the absence of this enzyme of the plastids in the transfer hair cells was not expected. These results suggest that in simple hair cells, plastids are not involved in CO_2 fixation and in glycollate synthesis. The presence of a few starch granules in these orgnelles could be explained by a flow of sugars through these hairs, where an excess of monosaccharides produced in the mesophyll cells could participate in starch formation. The detection of ATPase activity in the cell membrane of the transfer hairs confirms this idea and probably suggests the existence of a proton co-transport of sugars from the mesophyll cells into the mature leaf cavity. This enzyme may be involved in ion movements and could account for the stimulation of sugar loading (Malek and Baker, 1977; Peters and Calvert, 1983).

If the presence of the glycollate oxidase activity in the mesophyll peroxisomes is normal, the existence of this enzyume in the peroxisomes of hair cells is unexpected. Since there is no RuBisCO in the plastids of these cells, no glycollate can be formed there. In these conditions, the presence of glycollate oxidase in hair cell peroxisomes could

Figs. 8–9. Cytochemical localization of glycollate oxidase activity. Electron dense deposits of cerium perhydroxide are only observed in peroxisomes (Per) from simple hair cells (figs. 8, 8a) or mesophyll cells (fig. 9). No reaction product is observed in mitochondria (Mit).

Fig. 10. Control of glycollate oxidase test by omission of glycollate from the incubation medium. No reaction product is seen in peroxisomes (Per).

Fig. 11. Immuno-gold staining of RuBisCO in mesophyll cells. Colloidal gold particles are mainly restricted to the chloroplasts (Chl). No particles are observed in mitochondria (Mit).

Fig. 13. Immuno-gold staining of RuBisCO in simple hair cells. No colloidal gold particles are seen in the plastids (Pl) of these cells.

Figs. 12 and 14. Control of immunocytochemical localization of RuBisCO, by treatment with the pre-immuno serum. No colloidal gold particles are observed in the chloroplasts of mesophyll cells (fig. 12) or in the plastids of simple hair cells (fig. 14). Bar in figures = 1 μm indicate that a flow of glycollate occurs from the mesophyll cells or from another source to the simple hairs. Another possible explanation of these results may be found in the affinity of this enzyme to other substrates in addition to the glycollate, such as L-lactate and other hydroxy acids (Thomas and Trelease, 1981).

In conclusion, we suggest that the simple hair cells are involved in a process of carbohydrate translocation to the cavity and that sugars are formed in the chloroplasts of the mesophyll cells. The plastids of the transfer hairs are probably involved in the formation of lipophilic compounds found in the vacuoles of these cells.

Acknowledgements

The authors are grateful to Drs Richard N Trelease (Arizona State University) and Kevin C Vaughn (USDA-Stoneville) for their support and constructive comments. Dr Vaughn also generously supplied the RuBisCO antisera used in this work.

References

Carrapiço F, Diamond J and Tavares R 1985 The metabolic function of peroxisomes in the leaves of *Azolla filiculoides*. Eur. J. Cell Biol. Suppl. 14 (41), 9

Gray JC and Kekwick RGO 1974 An immunological investigation of the structure and function of ribulose 1,5-bisphosphate carboxylase. Eur. J. Biochem. 44, 481–489

Hulstaert CE, Kalicharan DE and Hardonk MJ 1983 Cytochemical demonstraion of phosphatases in the rat liver by the cerium-based method in combination with osmium tetroxide and potassium ferrocyanide postfixation. Histochemistry 78, 71–79

Lumpkin TA and Plucnett DL 1980 *Azolla* botany, physiology and use as green manure. Econ. Bot. 34, 111–153

Malek AS and Baker DA 1977 Proton co-transport of sugars in phloem loading. Planta 135, 297–299

Mollenhauer HH 1964 Plastic embedding for use in electron microscopy. Stain Technol. 39, 111–114

Peters GA 1975 The *Azolla-Anabaena azollae* relationship. III Studies on metabolic capabilities and a further characterization of the symbiont. Arch. Microbiol. 103, 113–122

Peters GA and Calvert HE 1983 The *Azolla-Anabaena azollae* symbiosis. *In* Algal Symbiosis. Ed. LJ Goff. pp. 109–145 Cambridge University Press, New York

Peters GA, Calvert HE, Kaplan D, Ito O and Toia RE, Jr. 1982 The *Azolla-Anabaena* symbiosis: Morphology, physiology and use. Israel J. Bot. 31, 305–323

Ray TB, Peters GA, Toia, RE, Jr and Mayne BC 1978 *Azolla-Anabaena* relationship. VII. Distribution of ammonia-assimilating enzymes, protein and chlorophyll between host and symbiont. Plant Physiol. 62, 463–467

Reynolds PHS, Blevins DG, Boland MJ, Schubert KR and Randall DD 1982 Enzymes of ammonia assimilation in legume nodules: A comparison between ureide- and amide-transporting plants. Physiol. Plant. 55, 256–260

Silverberg BA and Sawa T 1973 A structural and cytochemical study of microbodies in the genus *Nitella* (Characeae). Can. J. Bot. 51, 2025–2032

Sprent JI 1984 Nitrogen fixation. *In* Advanced Plant Physiol. Ed MB Wilkins. pp. 249–256. Pitman Pub. Inc., London

Thomas J and Trelease RN 1981. Cytochemical localization of glycolate oxidase in microbodies (glyoxysomes and peroxisomes) of higher plant tissues with the CeCl₃ technique. Protoplasma 108, 39–53

Tolbert NE 1982 Leaf peroxisomes Ann N.Y. Acad. Sci. 386, 254–268

Vaughn KC 1987 Two immunological approaches to the detection of ribulose-1,5-bisphosphate carboxylase in guard cell chloroplasts. Plant Physiol. 84, 188–196

Session 4

Isolation and identification of
plant-associated N_2-fixing bacteria

F. A. Skinner et al. (Eds.), Nitrogen fixation with non-legumes, 103–108.

Isolation and identification of root associated diazotrophs

JOHANNA DÖBEREINER
EMBRAPA-UAPNPBS, Km 47, Seropédica, 23851, Rio de Janeiro, Brazil

Key words: bacteria, diazotrophs, rhizosphere

Abstract

Diazotrophs have been isolated from the rhizosphere or roots of plants by many workers. To recognize a certain diazotroph as the most abundant bacterium at a certain site or as the principal agent responsible for N_2-fixation is much more difficult. It is probable that many diazotrophs, including possibly the most efficient ones, have not been identified yet. The use of proper selective media which simulate the environment of the various diazotrophs *in situ* has led to the discovery of 10 new root-associated diazotrophs, three of them during 1986/1987 (*Azospirillum halopraeferans*, *Herbaspirillum seropedicae* and the recently proposed *Acetobacter diazotrophicus*). The importance of using a variety of carbon substrates in the growth media with pH indicators, and the use of N-free semi-solid media, is discussed. Recognition of plant-bacteria interactions requires, in addition to the identification of the bacteria, the demonstration of effects of the plant on the bacteria and of the bacteria on the plant. Confirmation of the identity of diazotrophs responsible for response of plants to inoculation must be made in experiments with strains labelled with antibiotic resistance or other markers. If establishment of the inoculated strain is demonstrated in plants grown in ^{15}N-labelled soil, the ^{15}N enrichment of the plants will reveal if any observed responses in N yield are due to N_2-fixation or increased soil/fertilizer-N uptake.

Introduction

Associations of plants with microorganisms range from those with pathogens and harmless saprophytes to beneficial associations of great complexity as in the symbioses of legumes with diazotrophs, or mycorrhizal symbioses. The plant is the prime source of nutrients for microorganisms in soil, providing them indirectly from exudates or dead tissues, or directly when microorganisms colonize the interior of plant roots or other organs. It is understandable that microorganisms should have developed various ways of interacting with plants in order to gain access to the nutrients provided by them (Beringer and Johnston, 1984). The elucidation of the mechanisms and especially the identification of the responsible microorganisms in such associations is still in its infancy. It is considerably more advanced with pathogenic associations than with those of cereals or grasses with diazotrophs. Various bacteria have been isolated from roots or from the rhizosphere and were described as diazotrophic but later have not been confirmed as such (*e.g.* Hill and Postgate, 1969). More recently, several new actively N_2-fixing microaerobic bacteria have been identified and their association with cereals and grasses demonstrated (Döbereiner and Pedrosa, 1987).

In the present paper I shall try to answer the frequently-posed question of how our group proceeded in order to identify seven new root-associated diazotrophs which have all been confirmed by modern methods and subsequently been re-isolated by many authors from various parts of the world.

Aerobic ecto-rhizosphere diazotrophs

For almost 20 years Fred and Waksman's (1928) book on soil microbiological methods and Winogradsky's (1949) textbook recommended the use of N-free agar or silica-gel plates as the only media for the isolation of aerobic diazotrophs. On such plates few microorganisms grow and colonies are very characteristic. Silica-gel plates with sucrose inoculated with soil particles from the rhizosphere or root surface of various tropical plants almost always yielded typically gummy *Beijerinckia indica* colonies. Completely different, small dry colonies, which often occurred on the same plates, attracted our attention. They could be easily isolated and at least 20 strains were obtained from various sources. Kjeldahl analyses of liquid cultures confirmed their capacity to fix N_2 in pure culture. Based on a few physiological tests and microscopical observations we proposed our first root-associated new species, *Beijerinckia fluminensis* (Döbereiner and Ruschel, 1958). There were so few aerobic diazotrophs at the time that just by comparing them phenotypically left little doubt that this was a new organism.

Very similar procedures, except that N-free agar media recommended for *Azotobacter* were used, led to the discovery of *Azotobacter paspali* (Döbereiner, 1966). With both species selective enrichment in root surface soil was demonstrated (Döbereiner, 1961; 1970) which in the case of *A. paspali*, was specific for one ecotype of *Paspalum notatum*: to date this association remains the most specific among all diazotrophic symbioses, being even more specific than the legume-*Rhizobium* symbioses. Both species have since been found in many other parts of the world and have been confirmed as distinct species (Krieg and Holt, 1984). There has been no questioning of the inclusion of *B. fluminensis* within the genus *Beijerinckia* and the distinct phenotypic characteristics have been confirmed by numerical taxonomy (Thompson and Skermann, 1979). The genus *Beijerinckia*, including *B. fluminensis*, is clearly distinguishable from the *Azotobacter/Azomonas* group which belongs to a different rRNA superfamily (De Smedt *et al.*, 1980).

Azotobacter paspali could be separated from all other Azotobacter spp. by DNA homology experiments (De Ley, 1968). In fact Thompson and Skermann (1979) proposed a new genus *Azorhizosphilus* for it, based on numerical taxonomy but this was opposed by De Smedt *et al.* (1980) who found that the rRNA cistron of *A. paspali* is almost identical to that of the other *Azotobacter* spp. Immunoelectrophoresis confirmed the similarity of that species to *Azotobacter* spp. (Tchan *et al.*, 1983).

Microaerobic endorhizosphere diazotropohs

No further true aerobic N_2 fixing bacteria (which fix N_2 under air) have been found in association with plant roots indicating that the observed nitrogenase activity must occur on sites where O_2 access is restricted either by heavy colonization of various microorganisms in the mucigel layer on the root surface or by mechanical barriers offered by the root tissue itself. The microaerobic *Spirillum lipoferum* described by Beijerinck in 1925 was found to be unable to fix N_2 in pure culture (Schröder, 1932), presumably because free O_2 in the nutrient solution inactivated its nitrogenase. It was rediscovered as a true diazotroph due to the introduction of N free semi-solid media which came to solve, very simply and practically, the rather complicated problem of microaerobic diazotrophs. All the organisms found so far possess more or less pronounced aerotactic attraction mechanisms which permit them to move rapidly within the semisolid medium towards the site where their respiration requirements equilibrate with the O_2 diffusion rate and no O_2 accumulates, and where they find optimal conditions for N_2-fixation and N_2-dependent growth. In addition, pH indicators such as bromothymol blue were included in the now universally-used NFb medium to give information on the use of various carbon substrates and consequent acid or alkali production which help to characterize the various diazotrophs. From enrichment cultures in completely N-free semi-solid media isolation is achieved after one or two transfers into new N-free semisolid vials followed by streaking out of 24 h-old cultures on agar plates. The same medium is used, supplemented with low concentrations (0.002–0.005%) of yeast extract, to permit colony formation under air but eliminate growth of many other bacteria. Due to the simplicity of these media they became the key for the discovery of almost all of the more recently-described root-associated diazotrophs: four *Azospirillum* spp. (Tarrand *et al.*, 1978; Magalhães *et al.*, 1983; Reinhold *et al.*, 1987), *Campylobacter nitrofigilis* (McClung and Patriquin, 1980), *Herbaspirillum seropedicae* (Baldani *et al.*, 1986a), various unidentified *Pseudomonas* spp. (Haahtela *et al.*, 1981; Bally *et al.*, 1983; Barraquio *et al.*, 1983), the recently proposed *Saccharobacter* (now *Acetobacter*) (Cavalcante and Döbereiner, 1987) and several others. Even the highly efficient facultative N_2-fixing *Bacillus azotofixans* (Seldin *et al.*, 1984; Seldin and Dubnau, 1985) was isolated originally from a semi-solid medium.

The first attempts to identify microaerobic diazotrophs in roots which show nitrogenase activity used N-free, semi-solid media with various carbon substrates, and the introduction of semi-solid malate medium (NFb) was based on experiments where the ARA of individual 1 cm long root pieces of grasses and later maize were evaluated within syringes which were then placed in various media. The only medium which consistently showed significant correlations of root piece ARA with enrichment culture ARA was the N-free semi-solid malate medium (von Bülow and Döbereiner, 1975; Döbereiner and Day, 1976) used until today and *Azospirillum lipoferum* and *A. brasilense* were obtained with this medium. The modification of this medium according to the ecological requirements, *e.g.* saline soils which frequently show high temperatures, led to the discovery of *A. halopraeferans* (Reinhold *et al.*, 1987). Other bacteria, such as *A. amazonense* or *Herbaspirillum seropedicae* were initially observed in NFb media and were isolated following observation of the frequent occurrence of certain colony types or growth habits. The identities of all these organisms were confirmed by DNA/DNA and RNA/RNA homologies (Tarrand *et al.*, 1978; Falk *et al.*, 1985a, 1986; Reinhold *et al.*, 1987).

Isolation of a new N_2-fixing *Acetobacter* species

After it was shown that certain sugar cane cultivars could obtain up to 60% of their N requirements through associations with diazotrophs (Lima *et al.*, 1987) the search for responsible bacteria was intensified. Vials of semi-solid pure sugar cane juice inoculated with serial dilutions of sugar cane roots or stems showed high nitrogenase activity (above 50 nmoles $C_2H_4 h^{-1}$ per culture) in dilutions of up to 10^{-7} or occasionally 10^{-8}. By diluting the sugar cane juice to 1% but maintaining the sugar concentration at 10% by adding sucrose, nitrogenase activities of such enrichment cultures were increased. Replication into N-free mineral sucrose (10%) medium with bromothymol blue showed very strong acid production and no growth

at pH above 6.0. Growth and nitrogenase activity continued for several days at pH below 3.0 and thick, orange, surface pellicles formed. Streaking out these enrichment cultures on cane juice agar and on N-free mineral sucrose (10%) agar with bromothymol blue supplemented with 0.005% yeast extract, yielded easily distinguishable colony types on the latter. The only colony type which consistently yielded nitrogenase-active (> 200 nmoles $C_2H_4 h^{-1}$ per culture) cultures when transferred into N-free semi-solid sucrose medium acidified with acetic acid to pH 4.5, were dark orange colonies (due to assimilation of the bromothymol blue). When purified on potato agar with 10% sucrose, dark brown colonies formed.

With all these unusual characteristics it seemed clear that another new diazotroph had been found. Sugar cane root and stem samples from various varieties were obtained from four distinct regions of Brazil and organisms with the same characteristics were isolated from all of them. The extreme acid-tolerance and high sugar requirements (growth on 10% sucrose or glucose was faster than on 5% and there was still growth with 30% sucrose) indicated that the new organism is related to the group of acetic bacteria. Contacts with N. R. Krieg, J. De Ley and M. Gillis confirmed our suspicion and type strains of *Acetobacter aceti*, *Gluconobacter oxidans* and *Frateuria aurantia* were obtained by courtesy of Dr. J. De Ley, and the phenotypic characteristics were compared. Apart from the capacity to fix N_2 and grow on it as sole N source a number of other characteristics distinguished the new organism from the generic characteristics given by Krieg and Holt (1984). Flagellation (lateral), over-oxidation of glucose, orange colonies on sugar-agar plates with bromothymol blue and brown colonies on potato plates (10% sugar) distinguished it from all three genera while over-oxidation of ethanol was as observed with *Acetobacter*. Production of a brown soluble pigment from glucose, formation of H_2S and growth at 30% glucose showed its similarity to *Frateuria*. Once more, phenotypic characteristics seemed sufficiently distinct to justify the proposal of a new bacterial taxon (Cavalcante and Döbereiner, 1988). DNA homology experiments confirmed that the organism is not a *Frateuria aurantia* and RNA/DNA hybridization experiments carried out by M. Gillis (pers. comm.) have shown it to be an *Ac-*

etobacter. Comparison with the other three species of this genus have shown the new organism to be distinct, and DNA/DNA homology experiments have confirmed it as a new species, now renamed *Acetobacter diazotrophicus*.

Identification of plant-bacteria interactions

In order to speak of plant-diazotroph associations it is not sufficient to isolate a certain bacterium from the root surface or even from the interior of roots: plant-bacteria interactions have to be demonstrated. Most information in this direction is available for *Azospirillum* spp. Koch's postulates, which require a bacterium to be isolated, purified and re-inoculated into the host, have been satisfied. Albrecht *et al.* (1981) were the first to demonstrate nitrogenase activity in sterile maize plants inoculated with *A. brasilense*. Since then many experiments all over the world showed inoculation effects under sterile conditions, pot experiments and in the field (Okon, 1985; Boddey *et al.*, 1986). Likewise, enrichment in the ecto- and endorhizosphere of many plants has been shown not only for *Azospirillum* spp. but also for *Herbaspirillum* and various less common or less well defined diazotrophs as in the case of *Campylobacter nitrofigilis* (McClung and Patriquin, 1980) or *Pseudomonas* spp. (Baraquio *et al.*, 1983) or a *Bacillus* sp. (Neal and Larson, 1976). The effect of certain *A. brasilense* strains on N incorporation by wheat was shown by the correlation of *A. brasilense* numbers in the endorhizosphere with total N in the plants, and which was not significantly correlated with numbers in the ectorhizosphere (root surface) (Baldani *et al.*, 1983). In field experiments the establishment of certain inoculated *Azospirillum* spp. strains in the endorhizosphere and of others in the ectorhizosphere (Baldani *et al.*, 1986b, 1987) shows the existence of recognition mechanisms. Chemotatic attraction by root exudates which also can be strain specific (Reinhold *et al.*, 1985) is another plant-bacteria interaction. Strain-specific root hair deformations (Patriquin *et al.*, 1983) and inoculation effects on root elongation and nutrient assimilation (Okon, 1985) are other examples.

Conclusion

From this very condensed history of isolation and identification of root-associated diazotrophs in our laboratory, some apparently logical and very simple recommendations can be made:

1. In order to identify new organisms one should have experience with as many diazotrophs as possible.
2. Only after at least 10 isolates from different origins have been obtained which, phenotypically and microscopically, are similar among themselves but distinct from other species, should one think of it as possibly a new bacterium.
3. In order to stress phenotypic differences several media should always be used containing indicators or stains which emphasize differences between growth habits.
4. Cultures should be observed daily and growth habits recorded.
5. Observations of wet mounts by phase contrast microscopy reveal much about the shape and motility of the organism at different growth stages.
6. Enrichment media should be almost (or preferably, completely) N-free, and the presence of important diazotrophs should be suspected only if nitrogenase activities of at least 50 nmoles $C_2H_4h^{-1}$ per culture are observed during log phase.
7. Enrichment and isolation media should be adapted to the environmental conditions from where the diazotrophs are to be isolated.
8. Once a large number of isolates is available which continue fixing N_2 after repeated transfers into various media, they have to be compared with type strains of similar organisms.
9. Test kits, especially those prepared for certain groups of bacteria such as the Enterobacteriaceae, can be very misleading when used with other groups.
10. Once a new species or genus is suspected it has to be confirmed by one or more of the modern taxonomic analyses based on DNA and rRNA sequences.

References

Albrecht S L, Okon Y, Lonquist J and Burris R H 1981 Nitrogen fixation by corn-*Azospirillum* associations in a temperate climate. Crop Sci. 21, 302–306.

Baldani V L D, Baldani J I and Döbereiner J 1983 Effects of *Azospirillum* inoculation on root infection and nitrogen incorporation in wheat. Can. J. Microbiol. 29, 924–929.

Baldani J I, Baldani V L D, Seldin L and Döbereiner J 1986a Characterization of *Herbaspirillum seropedicae* gen. Nov., sp. Nov.: A root-associated nitrogen-fixing bacterium. Int. J. Syst. Bacteriol. 36, 86–93.

Baldani V L D, Alvarez M A B, Baldani J I and Döbereiner J 1986 b Establishment of inocculated *Azospirillum* spp in the rhizosphere and in roots of field grown wheat and sorghum. Plant and Soil 90, 35–46.

Baldani V L D, Baldani J I and Döbereiner J 1987 Inoculation of field-grown wheat (*Triticum aestivum*) with *Azospirillum* spp. in Brazil. Biol. Fertil. Soil 4, 37–40.

Bally R, Thomas-Bauzon D, Heulin Th, Balandreau J, Richard C and De Ley J 1983 Determination of the most frequent N_2-fixing bacteria in a rice rhizosphere. Can. J. Microbiol. 29, 881–887.

Barraquio W L, Ladha J K and Watanabe J 1983 Isolation and identification of N_2-fixing *Pseudomonas* associated with wetland rice. Can. J. Microbiol. 29, 867–873.

Beijerinck M W 1925 Über ein Spirillum, welches freien Stickstoff binden kann? Centralbl. Bakt. II Abt 63, 353–357.

Beringer J E and Johnston A W B 1984 Concepts and terminology in plant-microbe interactions. *In* Plant-Microbe Interactions: Molecular and Genetic Perspectives. VI. Eds. T Kosuge and E W Nester. pp 3–18. Macmillan Publishing, New York.

Boddey R M, Baldani V L D, Baldani J I and Döbereiner 1986 Effects of inoculation of *Azospirillum* spp. on the nitrogen assimilation of field grown wheat. Plant and Soil 95, 109–121.

Bülow J F N von and Döbereiner J 1975 Potential for nitrogen fixation in maize genotypes in Brasil. Proc. Natl. Acad. Sci. 72, 2389–2393.

Cavalcante V and Döbereiner J 1988 A new acid-tolerant nitrogen-fixing bacterium associated with sugar-cane. Plant and Soil 108, 23–31.

De Ley J 1986 DNA base composition and classification of some more free-living nitrogen-fixing bacteria. Antonie van Leeuwenhoek 34, 66–70.

De Smedt J, Bauwens M, Tijtgat R and De Ley J 1980 Intra- and intergeneric similarities of ribosomal ribonucleic acid cistrons of free-living, nitrogen fixing bacteria. Int. Syst. Bacteriol 30, 106–122.

Döbereiner J 1961 Nitrogen-fixing bacteria of the genus *Beijerinckia* Derx in the rhizosphere of sugar cane. Plant and Soil 15, 211–216.

Döbereiner J 1966 *Azotobacter paspali* sp. n., uma bactéria fixadora de nitrogênio na rizosfera de *Paspalum*. Pesq. agropec. bras. 1, 357–365.

Döbereiner J 1970 Further research on *Azotobacter paspali* and its variety: Specific occurrence in the rhizosphere of *Paspalum notatum* Flügge. Zentralbl. Bakteriol. Parasitenkd. Infektionskr. Hyg. Abt II 124, 224–230.

Döbereiner J and Day J M 1976 Associative symbioses in tropical grasses: Characterization of microorganisms and dinitrogen fixing sites. *In* Proceedings of the 1st International Symposium on N_2 Fixation. Eds. W E Newton and C J Nyman. pp 518–538. Washington State University Press, Pullman.

Döbereiner J and Pedrosa F O 1987 Nitrogen-Fixing Bacteria in Nonleguminous Crop Plants. 155pp. Brock/Springer Series in Contemporary Biosciences. Science Tech Publishers, Madison.

Döbereiner J and Ruschel A P 1958 Uma nova espécie de *Beijerinckia*. Rev. Biol. 1, 261–272.

Falk E C, Döbereiner J, Johnson J L and Krieg N R 1985 Deoxyribonucleic acid homology of *Azospirillum amazonense* Magalhães *et al.* 1984 and emendation of the description of the Genus *Azospirillum*. Int. J. Syst. Bacteriol. 35, 117–118.

Falk E C, Johnson J L, Baldani V L D, Döbereiner J and Krieg N R 1986 Deoxyribonucleic and ribonucleic acid homology studies of the genera *Azospirillum* and *Conglomeromonas*. Int. J. Syst. Bacteriol. 36, 80–85.

Fred E B and Waksman S A 1928 Laboratory Manual of General Microbiology. McGraw-Hill Book Co. Inc., New York, London.

Haahtela K, Wartiovaara T, Sundman V and Skujins J 1981 Root-associated N_2 fixation (acetylene reduction) by Enterobacteriaceae and *Azospirillum* strains in cold-climate spodosols. Appl. Environ. Microbiol. 41, 203–206.

Hill S and Postgate J R 1969 Failure of putative nitrogen-fixing bacteria to fix nitrogen. J. Gen. Microbiol. 58, 277–285.

Krieg N R and Holt J G (Eds.) 1984 Bergey's Manual of Systematic Bacteriology. VI. Williams and Wilkins, Baltimore. 964p.

Lima E, Boddey R M and Döbereiner J 1987 Quantification of biological nitrogen fixation associated with sugar cane using a ^{15}N aided nitrogen balance. Soil Biol. Biochem. 19, 165–170.

Magalhães F M M, Baldani J I, Souto S M, Kuykendall J R and Döbereiner J 1983 A new acid tolerant *Azospirillum* species. An. Acad. Bras. Ciên. 55, 417–430.

McClung C R and Patriquin D G 1980 Isolation of a nitrogen-fixing *Campylobacter* species from the roots of *Spartina alterniflora* Loisel. Can. J. Microbiol. 26, 881–886.

Neal J L Jr and Larson R I 1976 Acetylene reduction by bacteria isolated from the rhizosphere of wheat. Soil Biol. Biochem. 8, 151–155.

Okon Y 1985 *Azospirillum* as a potential inoculant for agriculture. Trends Biotechnol. 3, 223–228.

Patriquin D G, Döbereiner J and Jain D K 1983 Sites and processes of association between diazotrophs and grasses. Can. J. Microbiol. 29, 900–915.

Reihold B, Hurek T, Fendrik I, Pot B, Gillis M, Kersters K, Thielemans S and De Ley J 1987 *Azospirillum halopraeferens* sp. nov., a nitrogen-fixing organism associated with roots of kallar grass (*Leptochloa fusca* (L.) Kunth.). Int. J. Syst. Bacteriol. 37, 43–51.

Schröder M 1932 Die Assimilation des Luftstickstoffs durch einige Bakterien. Zentralbl. Bakt. Parasitenkd. 85, 178–212.

Seldin L, Van Elsas J D and Penido E G C 1984 *Bacillus azotofixans* sp nov. a nitrogen fixing species from Brazilian soils and grass roots. Int. J. Syst. Bacteriol. 34, 451–456.

Seldin L and Dubnau D 1985 Deoxyribonucleic acid homology among *Bacillus polymyxa, Bacillus macerans, Bacillus az-*

otofixans, and other nitrogen-fixing *Bacillus* strains. Int. J. Syst. Bacteriol. 35, 151–154.

Tarrand J J, Krieg N R and Döbereiner J 1978 A taxonomic study of the *Spirillum lipoferum* group, with descriptions of a new genus, *Azospirillum* gen. nov. and two species. *Azospirillum lipoferum* (Beijerinck) comb. nov. and *Azospirillum brasilense* sp. nov. Can. J. Microbiol. 24, 967–980.

Tchan Y T, Wyszomirska-Dreher Z, New P B and Zhoy J C 1983 Taxonomy of the *Azotobacteraceae* determined by using immunoelectrophoresis. Int. J. Syst. Bacteriol. 33, 147–156.

Thompson J P and Skerman V B D 1979 *Azotobacteraceae*: The Taxonomy and Ecology of the Aerobic Nitrogen-Fixing Bacteria. Academic Press, London.

Winogradsky S 1949 Microbiologie du sol. Masson et Cie, Paris. 861p.

F. A. Skinner et al. (Eds.), Nitrogen fixation with non-legumes, 109–114.
© 1989 by Kluwer Academic Publishers.

Occurrence and survival of *Azospirillum* spp. in temperate regions

KATLEEN DE CONINCK, S. HOREMANS[1], SUNIETTA RANDOMBAGE and K. VLASSAK
Laboratory of Soil Biology and Soil Fertility, Faculty of Agriculture, K.U. Leuven, Kard. Mercierlaan 92, B-3030 Leuven, Belgium. [1]S.E.S., Industriepark 15, B-3300 Tienen, Belgium

Key words: *Azospirillum*, identification, isolation, population dynamics, seed inoculation

Abstract

In order to evaluate the suitability of *Azospirillum* spp. as a crop inoculant in temperate regions, the natural occurrence, distribution and survival of *Azospirillum* after seed inoculation in Belgian agricultural soils was studied. *Azospirillum* was present in most of the fields examined, but concentrations never exceeded 1000 cfu per g soil or per g roots. Under field conditions none of the known species was found to be localized inside the roots of barley, wheat, rye, maize or grasses. Also, the distribution of *A. brasilense* SpBr14 within the root system of hydroponic-grown wheat was studied by immunofluorescence. From the rhizosphere samples of the field crops investigated, a number of microaërophilic, diazotrophic bacteria were isolated and identified as *A. lipoferum*, found only on maize and grass roots, and *A. brasilense*, present under all crops. In contrast to *A. brasilense*, *A. lipoferum* was able to use different amino-acids and some derivatives as sole carbon and nitrogen sources. Use of a peat-based seed inoculant resulted in the establishment of the *Azospirillum* spp. in the rhizosphere of field-grown winter barley and winter wheat. The established population survived during winter without appreciable change in numbers, but there was no indication of active growth during spring or summer.

Introduction

To evaluate the use of *Azospirillum* as a potential inoculant in temperate regions we studied its natural occurrence in Belgian agricultural soils cultivated with winter barley, winter wheat, maize, rye and grasses.

Using MPN counts we tried to estimate the absolute population densities of *Azospirillum* and to locate them within different sites at the soil-root interface. The colonization of the internal root tissues of sterile wheat seedlings was investigated with the fluorescent antibody technique.

From the field samples in which microaërophilic growth and nitrogenase activity was detected, a total of 28 endogenous diazotrophs was isolated, purified and characterized.

In addition to the distribution study, the survival of the applied inoculum and its distribution within the rhizosphere as a function of time and depth, was studied.

Materials and methods

Media

The basic medium for growth of the *Azospirillum* isolates was a nitrogen-free malate stock solution (NfM) of the following composition (g.l^{-1} of distilled water): DL-malic acid, 50; KH_2PO_4, 6; K_2HPO_4, 2; $MgSO_4 \cdot 7H_2O$, 2; NaCl, 1; $Na_2MoO_4 \cdot 2H_2O$, 0.02; KOH, 49; $FeCl_3 \cdot 6H_2O$, 0.15; $CaCl_2 \cdot 2H_2O$, 0.25; the final pH was adjusted to 6.8.

All nitrogenase activity assays were performed in NfM semi-solid medium which is prepared by diluting the stock solution in distilled water (1/9 v/v) adding 4 ml l^{-1} of a bromothymol blue solution (0.5% in ethanol), 100 mg l^{-1} of yeast extract and 2 g l^{-1} of agar to the basic medium.

Culture medium was prepared by adding 1 g l^{-1} of NH_4Cl to the tenfold dilution of NfM stock solution. TAE buffer was a 50-fold dilution of a stock solution containing 242 g Trizma base (Sigma) and 37.2 g disodium EDTA per litre. The pH was adjusted to 7 with acetic acid and 7.5 g NaCl was added per litre.

Phosphate buffer saline (PBS: pH 7.2) was prepared by dissolving 1.15 g K_2HPO_4, 0.2 g KH_2PO_4 and 7.5 g NaCl per litre of distilled water. All media were sterilized by autoclaving.

Organic substances as sole sources of carbon and nitrogen to identify the diazotrophs were used according to Kersters (Kersters *et al.*, 1984).

These tests were performed with the API CH 50, API AA 50 and API AO 50 galleries (API System, La Balme Les Grottes, Montalieu, France).

The glycine solution needed for the staining procedure was prepared by dissolving 7.5 mg of glycine in 10 ml of 0.1 *M* sodium phosphate buffer (pH 7.2).

Strains used for inoculation

Survival study. A mixture of four *A. brasilense* spp., all originating from Belgian soils, was used for seed inoculation: 85W5, 85W16 and 85M82 were isolated and identified during this project and S631 was obtained from Dr. L. Reynders. *Azospirillum brasilense* 85W5, 85W16 and S631 originate from soils under winter wheat cultivation whereas 85M82 was isolated from a calcareous field under maize cultivation. The peat inoculant was prepared by mixing 100 g of finely sieved, sterilized peat, adjusted to pH 6.8, with 125 ml of a two-day-old *A. brasilense* culture.

After one week at 30°C the peat inoculant was stored at 4°C since preliminary experiments showed no loss of viability up to six weeks of storage.

Just before planting, the seeds were coated, using 10 g of peat inoculant and 4 g of gum arabic for 100 g of seeds.

Sterile seedling experiment. Azospirillum brasilense SpBr14 was used to study the colonization and the effect of inoculation on some root characteristics of sterile-grown wheat seedlings (var. Bastion). One ml of the young *Azospirillum* culture was added to 100 ml of modified Hoagland solution just after planting.

Sampling and enumeration

Distribution study. To study the distribution of *Azospirillum* spp. in Belgian soils, about 15 plants originating from different sites in a particular field were sampled. All samples were taken between 0 and 15 cm deep and stored at 4°C up to the time of analysis, maximally 24 h later. A distinction was made between the 'non-rhizosphere soil', the 'rooted soil' *i.e.* the soil hanging loosely between the roots and the 'rhizosphere soil' *i.e.* the soil firmly attached to the roots. The 'soil' and 'rooted soil' samples, as well as the roots with the firmly adhering soil, were suspended in 10 ml of TAE buffer per g of material. All suspensions were shaken vigorously for 20 min and the homogenate was serially diluted in 10-fold steps with PBS. The washed roots were sterilized 4 min in ethanol, crushed, shaken again for 20 min in TAE, allowed to settle and the supernatant fluid was serially diluted. This sample was designated 'roots'. Five replicate 0.1 ml aliquots of each dilution were added to 14 ml tubes each containing 5 ml of NfM semi-solid agar. Tubes exhibiting both micro-aërophilic growth and acetylene reduction activity (ARA) were scored as positive for the presence of *Azospirillum* spp. and counted by the most probable number (MPN) method.

Survival study. Fifteen plants from each of the six treated plots were taken at various growth stages by excavating the root system (15 cm deep) and were used for estimates of inoculant establishment. To study the dispersion of the inoculum, the root system was excavated down to 80 cm. The survival study was performed at two field sites, one loamy soil and one sandy loamy soil, and with two different crops, i.e. winter barley and winter wheat. Soil not influenced by the roots was taken at twenty different sites between the plots. The sampling and counting procedures were identical to those described in the distribution study except that the washed roots were crushed immediately without sterilization.

Sterile seedling experiment. The hydroponically-grown spring wheat plants were harvested after two weeks of growth. Root pieces of c. 1.0 cm were collected from different parts of the inoculated and non-inoculated plants. The colonization of the internal root tissues was investigated using the indirect fluorescent antibody technique. First the root sections, prepared with a freezing microtome, were treated with a glycine solution for 10 min in order to block all the free aldehyde groups which can interfere with fluorescent staining. The sections were then washed twice with the 0.1 *M* sodium phosphate buffer, each time for 5 min. In order to block all the unspecific binding sites, the sections were treated with normal sheep serum for 20 min after which they were divided into two sets. One set was treated with the specific antiserum of *Azospirillum brasilense* SpBr14, which was diluted 20 times in normal sheep serum in order to reduce the non-specific binding of the antiserum. The other set of root sections was treated with normal rabbit serum which was diluted in the same manner as the specific antiserum. The treatment was carried out for 30 min. The sections treated with the normal rabbit serum were used as controls to compare the non-specific and the specific fluorescent staining. Both sets were again treated with the normal sheep serum for 5 min. Then the samples were rinsed three times (5 min) in the 0.1 *M* sodium phosphate buffer to remove the excess of antiser. After washing, the root sections were treated with anti-rabbit Ig biotinylated species-specific whole antibody (diluted 50 times in the buffer) for 30 min to increase the number of binding sites for the fluorescent material. After washing, (three times, for 5 min each, in the buffer) the sections were treated with the fluorescent Texas Red compound (diluted 50 times in the 0.1 *M* phosphate buffer to reduce non-specific staining) for 30 min, after which they were rinsed again in the buffer to remove the excess of fluorescent material. Then the sections were carefully transferred to cleaned glass

slides, mounted with 90% buffered glycerol and the observations were made under a light microscope equipped with a filter of 520 nm excitation wavelength together with a cut-off filter.

Results and discussion

Distribution study

Total numbers of sampling sites and the numbers of sampling sites positive for the presence of *Azospirillum* spp. according to the criteria given in the Materials and methods section, are shown in Table 1. Results are presented for the different crops and soil textures investigated.

Azospirillum spp. were present in 81% of the samples investigated and no correlation with the type of the crop or the soil texture was evident. As can be seen from Table 1, *Azospirillum* spp. were found in 16 of 18 maize fields, 11 of 13 winter wheat fields, 12 of 16 winter barley fields, 1 of 2 rye fields and 13 of 16 grasslands. For winter barley, winter wheat, rye and grasses the absolute population densities of *Azospirillum* spp. in the rooted soil, rhizosphere soil and within the roots were estimated as well (data not shown).

The numbers of *Azospirillum* spp. per gram of rooted soil or rhizosphere soil ranged from 3 to 1163, similar densities to those found in Germany (Markus and Kramer, 1987). Other bacteria, also capable of growing in the NfM medium, were present in much higher numbers (10^4 to 10^6 per gram of 'rooted' or 'rhizosphere' soil).

MPN counts revealed that in 9 of 37 positive samples the numbers of *Azospirillum* spp. were at least ten times higher in the rhizosphere samples compared to the rooted soil samples. However, due to the heterogeneous nature of the rhizosphere

Table 1. Total number of sampling sites (a) and number of samples positive for microaërophilic growth and nitrogenase activity (b) for each type of crop and soil texture analysed

Soil texture	Corn		Winter barley		Winter wheat		Rye		Grassland	
	a	b	a	b	a	b	a	b	a	b
Sandy	5	5	2	0	–	–	2	1	5	4
Sandy loam	1	1	2	1	2	1	–	–	2	1
Clay	6	6	6	5	6	5	–	–	5	4
Loamy	3	1	6	6	5	5	–	–	4	4
Calcareous	3	3	–	–	–	–	–	–	–	–

Table 2. Numbers of *Azospirillum* sp. in different soil/root zones after seed inoculation of winter barley with a mixture of 85M82, 85W5, 85W16 and S631

Treatment	Sample	Numbers of *Azospirillum* spp.* at different sampling dates			
		21/12/85	11/03/86	26/05/86	26/06/86
Non-inoculated	soil	2	2	4	0
	rooted soil	2	20	70	0
	rhizosphere soil + root surface	81	62	183	3
	roots	68	0	15	0
Inoculated	rooted soil	220	330	93	39
	rhizosphere soil + root surface	1.1×10^5	6.5×10^4	1.5×10^4	5.0×10^2
	roots	2.9×10^3	2.05×10^3	6.0×10^2	1.8×10^3

* Numbers of *Azospirillum* spp. are expressed per gram of soil for the 'soil' samples; for the other counts results are expressed per gram of fresh root.
Data evaluated by MPN counts are from the field site in Tienen (loamy soil).

sample (*i.e.* the closely adhering soil and the root surface), we can not conclude that rhizosphere enrichment indeed occurred here. Finally, *Azospirillum* was never detected within surface-sterilized roots of the crops investigated.

This field survey clearly indicates that although *Azospirillum* is widely distributed in Belgian agricultural soils, the absolute densities are rather low compared to other bacteria also capable of growing in NfM semi-solid medium.

Identification of the diazotrophs

From the samples showing microaërophilic growth and nitrogenase activity we isolated a total of 28 diazotrophs: 12 from maize, 7 from winter wheat, 3 from barley, 5 from grasses and one from rye. Based on 201 physiological, biochemical and nutritional characteristics, eight were identified as *Azospirillum lipoferum*, isolated only from maize fields and grasslands, 18 as *A. brasilense*, recovered from all five crops investigated, and one strain with intermediate characteristics. One strain, responsible for the *Azospirillum*-like activity found within surface-sterilized grass roots, represented, for us, an unknown diazotrophic species.

The results obtained with the different API-galleries showed that in contrast to *A. brasilense*, *A. lipoferum* was able to use different amino-acids and some derivatives as a sole carbon and nitrogen source. The strains classified as *A. lipoferum* all grew well on suberate, azelate, adipate, pimelate,

sarcosine, ethanolamine, l-serine, phenylalanine, tyrosine, arginine, threonine, 2-ketoglutarate, D-alanine, berrilinate, aspartate, citrate and histidine.

All the *A. brasilense* strains could use itaconate, mesaconate and xylose. The former two compounds were not used by *A. lipoferum*, and xylose was utilized only by half of the *A. lipoferum* strains investigated.

Survival study

Since the results obtained were very similar for the different field locations and for the two crops investigated (winter barley and winter wheat) only one of the experiments will be discussed here.

Between 2×10^4 and 10^6 living bacteria per seed were recovered at the moment of planting. MPN counts of the inoculant establishment in the different soil/root zones are shown in Table 2. During the whole growing season the control plots con-

Table 3. Incidence of *A. brasilense* in the rhizosphere soil of winter barley as a function of depth after seed inoculation with 85M82, 85W5, 85W16 and S631

Depth	Roots fresh weight as a % of total root weight in 0–80 cm	Number of *Azospirillum brasilense* per gram soil
0–10 cm	72%	1.7×10^3
10–30 cm	21%	0
30–60 cm	5.4%	0
60–80 cm	1.8%	0

Data, evaluated by MPN counts, are from a field experiment in Tienen.

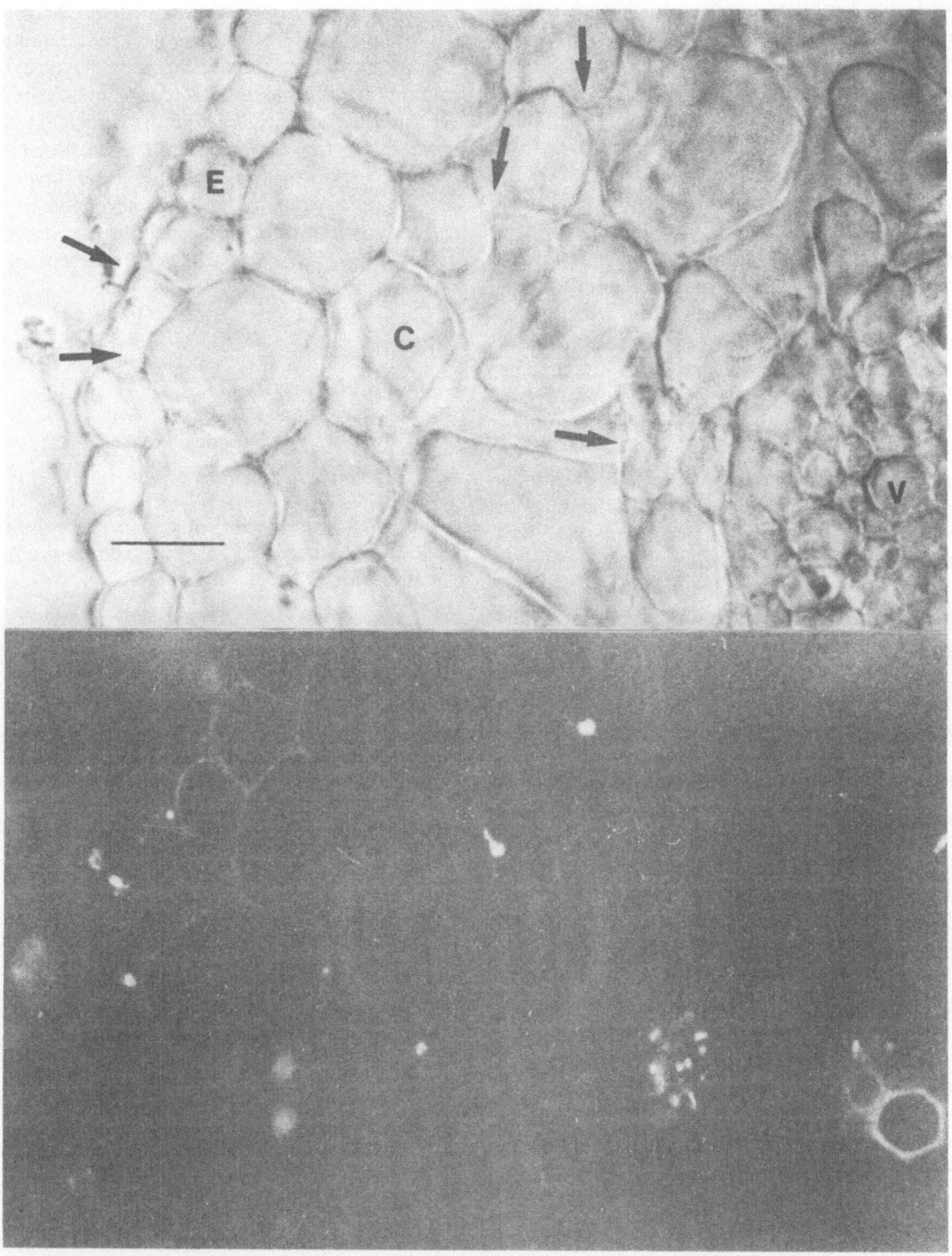

Fig. 1. Immunofluorescence indicating the presence of *A. brasilense* SpBr14 in cross-sections of inoculated wheat roots. Fluorescent spots in the photograph at the bottom correspond to the sites indicated by arrows in the photograph at the top. E, epidermis; C, cortex; V, vascular system. Magnification bar = 50 μm.

tained only very low numbers of microaërophilic diazotrophs, confirming the earlier distribution study. In all experiments the numbers of *A. brasilense* recovered from the rhizosphere samples were much higher in the inoculated plots as compared to the control plots. Actual numbers at the moment of tillering ranged from 10^3 to 5×10^4 living bacteria per gram of rhizosphere soil indicating a quantitative survival of the established inoculum during the winter period. During the following spring, however, no increase in actual numbers was observed. On the contrary, in all four experiments the numbers of *Azospirillum* spp. decreased slowly to reach a level between 30 and 10^3 living bacteria per gram of rhizosphere soil or per gram of roots.

These densities are probably insufficient to have any significant effect on the plant performance through the production of auxins and cytokinins (Horemans *et al.*, 1986). Probably the applied inoculum does not migrate away from the roots since actual numbers in the rooted soil were always very low. In addition the *Azospirillum* population was only recovered from the upper 10 cm of the root system, as shown in Table 3.

Sterile seedling experiment

Photographs of the immunofluorescence of the root cross sections are given in Fig. 1. Although some autofluorescence of the xylem vessels and weak background non-specific fluorescence was noticed, the bright red fluorescence due to the specific staining could clearly be distinguished from the non-specific staining. No such bright fluorescence was noticed in control sections.

Transverse sections of the inoculated roots showed few fluorescent spots in the cortical tissue, indicating the possible occurrence of *Azospirillum* SpBr14 in the internal root tissues. Any evidence of the presence of bacteria in the vascular tissues as suggested by Patriquin and Döbereiner (1978) was not obtained in this experiment.

Similar results were obtained by Bashan and Levanony (1987) studying wheat root colonization by *A. brasilense* Cd during primary stages using immunogold labelling. Bacteria were neither detected in the endodermis nor in the vascular system of wheat, but they were present in intercellular spaces of the cortex as far as the endodermis layer.

As already reported we were never able to detect *Azospirillum* within surface-sterilized roots of field grown maize, winter barley, winter wheat, rye and grasses.

Acknowledgements

This study was supported by the Instituut voor Wetenschappelijk Onderzoek in Nijverheid en Landbouw. We wish to thank the Nationaal Fonds voor Wetenschappelijk Onderzoek, S Demarsin for skillful assistance and M J Struyven for typing this manuscript.

References

Bashan Y and Levanony H 1987 Interaction between *Azospirillum brasilense* Cd and wheat root cells during early stages of root colonization. *In Azospirillum* 4, Genetics, Physiology, Ecology. Ed. W Klingmüller, Springer Verlag.

Horemans S, De Coninck K, Neuray J, Hermans R and Vlassak K 1986 Production of plant growth substances by *Azospirillum* spp. and other rhizosphere bacteria. Symbiosis 2, 341–346.

Kersters K, Hinz K H, Hertle A, Segers P, Lievens A, Siegman O and Deley J 1984 *Bordetella avium* sp. nov., isolated from the respiratory tracts of turkeys and other birds. Int. J. Syst. Bacteriol. 34, 57–70.

Markus P and Kramer J 1987 Importance of non-symbiotic nitrogen-fixing bacteria in organic farming systems. *In Azospirillum* 4, Genetics, Physiology, Ecology. Ed. W Klingmüller, Springer Verlag.

Patriquin D G and Döbereiner J 1978 Light microscopy observations of tetrazolium-reducing bacteria in the endorhizosphere of maize and other grasses in Brazil. Can. J. Microbiol. 24, 734–742.

F. A. Skinner et al. (Eds.), Nitrogen fixation with non-legumes, 115–120.

Klebsiella sp. *NIAB-I:* A new diazotroph, associated with roots of kallar grass from saline sodic soils

JAVED A. QURESHI, YUSUF ZAFAR and KAUSER A. MALIK
Nuclear Institute for Agriculture and Biology, P.O. Box 128, Jhang Road, Faisalabad, Pakistan

Key words: diazotroph, kallar grass, *Klebsiella*

Abstract

A nitrogen fixing organism containing a plasmid has been isolated from the rhizosphere fraction of *Leptochloa fusca* (L) Kunth (kallar grass) growing on saline soils in the Punjab area. This bacterium can grow aerobically in a medium containing 1 M NaCl and can fix nitrogen efficiently under microaerobic conditions on semi-solid medium with glucose or sucrose as a carbon source. Maximum N_2-fixation in batch cultures occurred with 100 mM NaCl at pH 8.0 and 35°C. DNA hybridization and analysis of the protein pattern were carried out to establish its taxonomic position. On the basis of protein electrophoretic pattern, physiological characteristics, DNA relatedness, and better growth in the presence of high NaCl concentration, we regard this strain as a new species of *Klebsiella*.

Introduction

Diazotrophs are ubiquitous. Several genera of dinitrogen fixing bacteria are reported to be associated with the roots, stems and leaves of various plants (Ladha *et al.*, 1983; Dart, 1986; Bilal and Malik, 1987). Kallar grass (*Leptochloa fusca* L. Kunth) a salt-tolerant grass, is a primary colonizer of salt-affected lands in Pakistan (Sandu and Malik, 1975). Nitrogenase activity (acetylene reduction) associated with the roots of this grass has been reported (Malik *et al.*, 1980, 1982; Zafar 1985) and the grass roots are infected by nitrogen fixing bacteria (Zafar *et al.*, 1986, Reinhold *et al.*, 1987). One of these isolates, NIAB-I, resembled *Klebsiella* in morphology, biochemical characteristics and DNA base composition (Zafar *et al.*, 1987). This paper describes further characteristics of this isolate (NIAB-I) and its comparison with the known N_2-fixing wild type *Klebsiella pneumoniae* strain M5A1. Furthermore, on the basis of physiological behaviour, protein electrophoretic pattern and DNA—DNA hybridization we propose this strain as a new species of *Klebsiella*.

Materials and methods

Bacterial strains

Two wild type *Klebsiella* strains M5A1 and NIAB-I, were used as experimental organisms. M5A1, known as *Klebsiella pneumoniae* NCIB 12204 (identified in European Collections as *Klebsiella oxytoca*) was supplied by Dr. David Lowe, AFRC Unit of N_2–Fixation, University of Sussex, England. NIAB-I strain was isolated form the roots of kallar grass (Zafar, 1985). Both organisms were kept on nutrient agar slants and subcultured monthly.

Media and growth conditions

The minimal medium (MM) described by Yoch and Pengra (1966) was used with slight modification, containing (g.1^{-1}) Na$_2$ HPO$_4$, 6.25; KH$_2$PO$_4$, 0.75; MgSO$_4$.7H$_2$O, 0.2; Na$_2$MoO$_4$, 0.01; FeSO$_4$. 7H$_2$O, 0.01, glucose, 20.0. pH and sodium chloride concentration were adjusted as required. This medium was routinely prepared by autoclaving phosphates and the remaining ingredients separately in distilled water. $FeSO_4.7H_2O$ was added after filter-sterilization.

Organisms were grown at 35°C under an atmosphere of either N_2 or N_2 + air in a 6 l fermenter (Eyla laboratory fermenter. Model No. M-160 Tokyo, Japan). The pH of the growth medium was stabilized with $2N$ NaOH added by an automatic bench pH controller (Model No. FC-1 Eyla, Tokyo, Japan). The 6 l fermenter was inoculated with 300 ml of a culture grown aerobically for 24 h on minimal medium with 300 mg ammonium sulphate per litre.

Preparation of cell extracts for protein analysis

Whole cells protein samples, each from a 5 ml overnight culture of a bacterial strain, were prepared by SDS, protease inhibitor (phenylmethylsulphonyl fluoride) and lysozyme treatment as described earlier (Shavikumar *et al.*, 1986). SDS polyacrylamide gel electrophoresis (PAGE) was performed by the method of Laemmli (1970).

Molecular studies on bacterial strains M5A1 and NIAB-I

Labelled DNA was extracted from cells which had been grown in a medium containing 1mCi of [^{3}H] thymidine per 300 ml of Luria broth. The extraction and purification of labelled DNA from M5A1 strain and unlabelled DNA from NIAB-I was performed according to the method of Marmour (1961). The concentration and purity of DNA was determined by measuring the absorbance at 260 nm and 280 nm.

Hybridization procedure

The denaturation of unlabelled DNA was carried out by boiling the DNA (10 μg ml^{-1} of 0.1 × Sodium Saline Citrate Buffer; SSC) three times at 95°C in closed pyrex tubes in a water bath for 2 min. SSC contains $0.15M$ NaCl and $0.15M$ sodium citrate (pH 7.0). The buffers 0.1 × SSC and 2 × SSC (see below) contain one tenth of, and twice,

these concentrations, respectively. After denaturation, the tubes were placed in an ethanol ice bath for 10–15 min. The procedure for DNA–DNA hybridization followed that of McConaughy et al., (1969) with slight modifications. The denatured suspensions were poured on pre-soaked nitro-cellulose filters and allowed to flow under gravity. After 24 to 32 h the filters were removed, placed in sterile Petri-dishes (glass) and baked for 2 h at 80°C in a vacuum oven (this denaturation enables the DNA to stick to the filters). The immobilized DNA filters were then placed in vials with different concentrations of labelled DNA (in 2 × SSC) and incubated for 18–24 h at 67°C. The hybridization reactions were terminated by washing the filters with 2 × SSC solution (thrice) and then incubated in another vial containing DNAase for 1 h at 37°C.

DNA filters were again washed three times with cold (2 × SSC), dried in a vacuum oven at 70°C and placed in 10 ml of scintillation fluid. The amount of input DNA bound to the immobilized DNA was estimated by measuring the radioactivity of the nitrocellulose filter in a liquid scintillation counter (Packed Tricarb 4388). The percent hybrids was determined for each sample relative to the control.

Nitrogenase assay and protein estimation

The nitrogenase activity (acetylene reduction) of 5 ml samples from batch cultures was measured in

Table 1. Effect of NaCl and pH on the N_2ase activity of *Klebsiella* strains M5A1 and NIAB-I in batch cultures at 35°C

Klebsiella strain	pH	Conc. of NaCl (mM)	Max. N_2ase activity at (h)[b]	Max. N_2ase activity (nmoles $C_2H_4^{-1}$ min^{-1} mg protein^{-1} ml^{-1})
M5A1	6	100	32	180
	7[a]	100	24	290
	7	34.2	26	278
	7	100	24	316
	7	250	38	140
	8	100	44	270
NIAB-I	7	100	30	233
	8	34.2	26	320
	8	100	28	418
	8	250	38	270
	8[a]	100	24	350
9	10 0	34	250	

[a] Strains were grown under anaerobic N_2-fixing conditions.
[b] Maximum N_2-ase activity observed after inoculation.

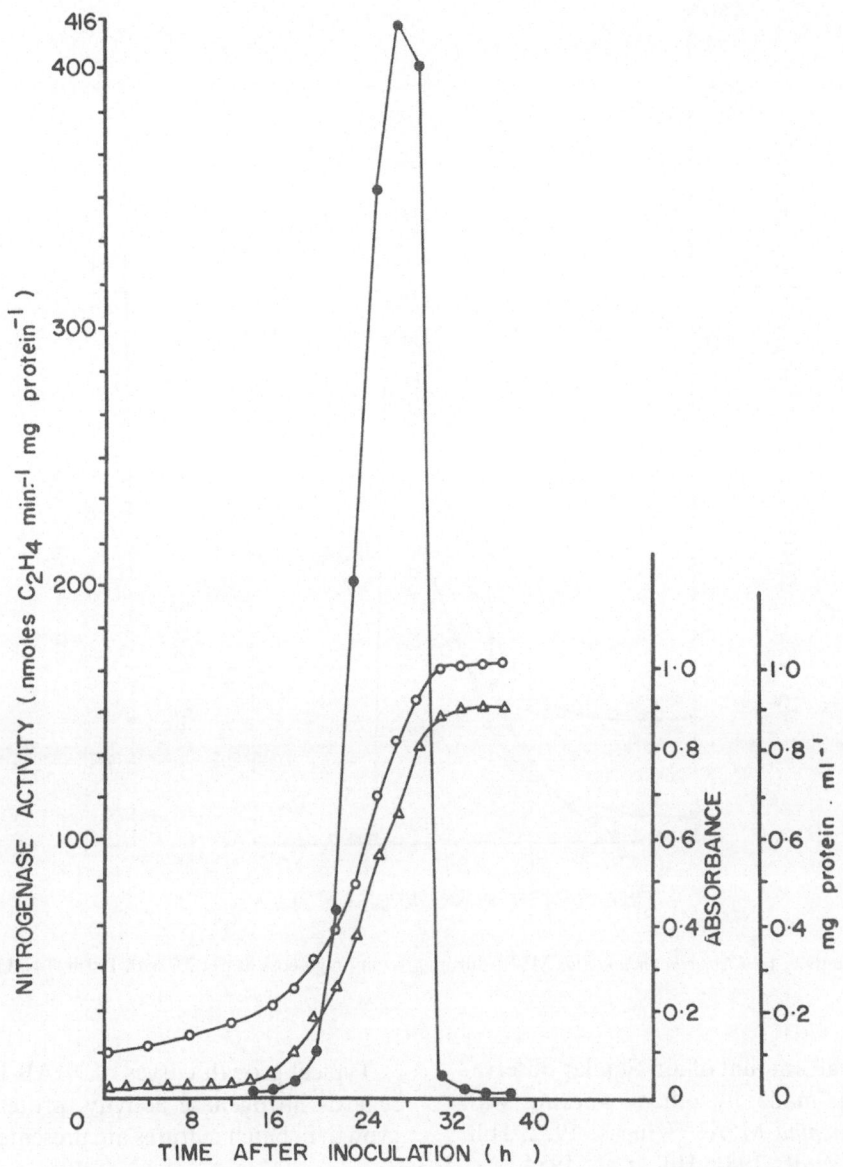

Fig. 1. Nitrogenase activity by *Klebsiella* sp. NIAB-I during growth on glucose at pH 8.0, with 100 mM NaCl and temperature 35°C.

13 ml serum vials as described by Hill (1976). Total protein of whole cells was determined by the method of Gowa (1953).

Results and discussion

The comparison of the nitrogenase activities of two wild type *Klebsiella* strains M5A1 and NIAB-I in batch cultures at various pH values and NaCl concentrations is summarized in Table 1. *Klebsiella* strains are facultative anaerobes and N_2-fixation in these organisms is mainly associated with anaerobic metabolic processes but our results provide evidence that a gas mixture (3% air and 97% N_2) and 100 mM NaCl in the growth medium improves the physiological state of the cells and nitrogen fixing ability of the two experimental organisms. This is an indication that the two organisms (M5A1 and NIAB-I) grow and fix N_2 actively in the

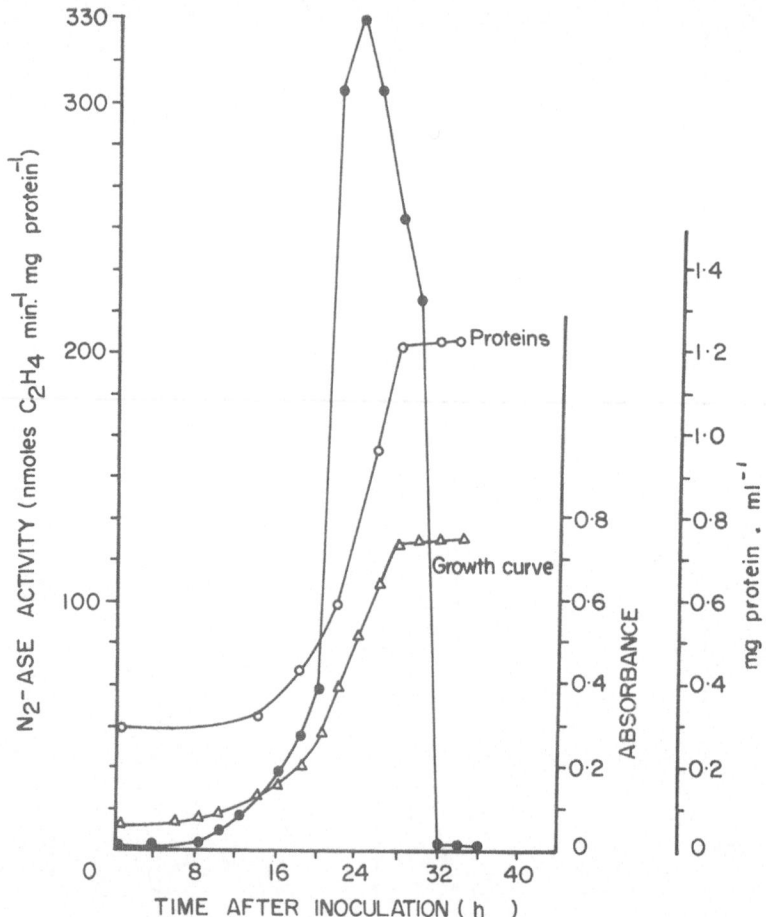

Fig. 2. Nitrogenase activity by *Klebsiella pneumoniae* M5A1 during growth on glucose at pH 7.0 with 100 m*M* NaCl and temperature 35°C.

presence of a small amount of air. Similar observations have been made by others working with *Klebsiella pneumoniae* M5A1 (Klucas, 1972; Hill, 1976; Bergersen *et al.*, 1981; Hill *et al.*, 1984.)

It is of interest that NaCl concentration and pH of the growth medium also has an important effect on the growth rate and on the nitrogenase activity of the intact organisms; 100 m*M* NaCl in the growth medium increases the N_2ase activity by 12% in M5A1 and about 24% in NIAB-I at the optimal pH value. The pH optimum for NIAB-I was 8.0 and for M5A1, 7.0. A higher concentration of salt (NaCl), however, decreases the nitrogen fixing ability of the cells. Thus, with 250 m*M* NaCl in the growth medium, M5A1 and NIAB-I showed a 55 and 35% decrease, respectivley, in N_2ase activity.

Typical growth curves of NIAB-I and M5A1 as regards nitrogenase activity, protein content and growth in batch cultures are presented in Figs 1 and 2.

The aerobic growth patterns of M5A1 and NIAB-I in Luria broth medium supplemented with different NaCl concentrations are presented in Fig. 3a,b. It can be seen that *Klebsiella* sp NIAB-I grows well even at 1*M* NaCl; M5A1 does not grow at this NaCl concentration.

Comparison of electrophoretic pattern of proteins

SDS–PAGE patterns of some diazotrophs are shown in Fig. 4. The three genera differ from each other in this respect. The two *Azospirillum*

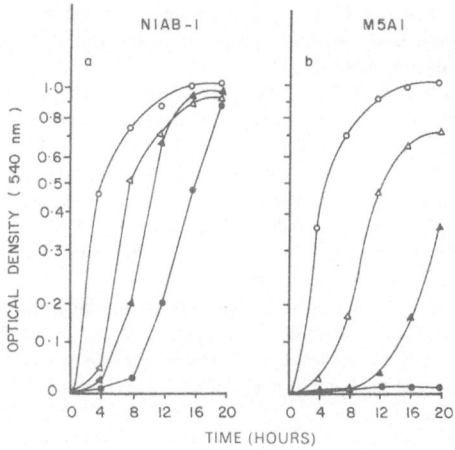

Fig. 3. Growth behaviour of (a) *Klebsiella* sp. NIAB-I and (b) *Klebsiella pneumoniae* M5A1 in Luria broth medium supplemented with different NaCl concentrations. The optical density is plotted as a function of incubation time. Open circle represents control; closed circle, 1*M* NaCl; closed triangle, 0.8*M* NaCl; open triangle, 0.6*M* NaCl.

brasilense strains 1690 and 2915 have identical protein patterns whereas the three *Klebsiella* strains M5A1, NIAB-I and 681 show differences in their electrophotograms. Bacteria with identical protein patterns have a high genomic similarity (Kersters and De Ley 1975, 1980). Izard *et al.*, (1981) identified four groups of *Klebsiella* spp. by the electrophoretic behaviour of proteins.

DNA–DNA hybridization

Because of the protein electrophoretic differences between the two *Klebsiella* spp. studied, it became important to investigate their DNA homology. Binding of 60–70% between the DNAs from M5A1 and NIAB-I was found (Fig. 5). At lower values of DNA (in μg) the binding is 100% but as the amount of DNA is increased the binding is decreased indicating that, genetically, the two species are different.

There are reports (Steigerwalt *et al.*, 1976; Izard *et al.*, 1981; Reinhold *et al.*, 1987) that DNA–DNA homology values within the genus range between 20 and 100%.

Klebsiella pneumoniae M5A1 is indole positive whereas our isolate is indole negative and has a G + C content 56.9 (Zafar *et al.*, 1987). Bagley *et al.*, (1981) proposed a new species, *Klebsiella planticola*, on biochemical and physiological

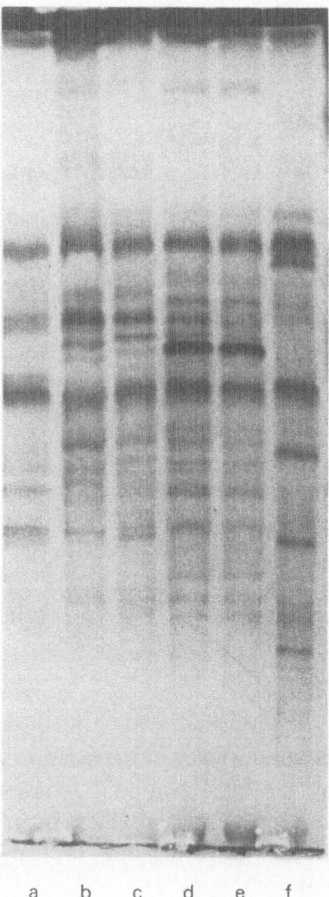

Fig. 4. SDS–PAGE Pattern of (a) *Bacillus polymyxa* 372, (b) *Azospirillum brasilense* 1690, (c) *Azospirillum brasilense* 2915, (d) *Klebsiella pneumoniae* 681, (e) *Klebsiella* sp. NIAB-I and (f) *Klebsiella pneumoniae* M5A1. Strain 681 is a non-N_2-fixer from the German Culture Collection whereas M5A1 strain is a N_2-fixer. M5A1 strain is now held with the National Collection of Industrial Bacteria as *Klebsiella pneumoniae (oxytoca)* NCIB 12204.

characteristics. Similarly Ladha *et al.* (1983) have identified new species on biochemical and immunological characteristics. *Klebsiella terrigena* was also identified as a new species using electrophoretic and DNA hybridization technique (Izard *et al.*, 1981). Recently, Reinhold *et al.* (1987) and Kersters and De Ley (1980) have used the SDS–PAGE method to classify bacteria. The present studies on the physiological behaviour, protein pattern and molecular studies on *Klebsiella* spp. NIAB-I provide evidence that our isolate is a new species of *Klebsiella*. Recently we have found that NIAB-I contains a plasmid which confers salt tolerance whereas M5A1 has no plasmid.

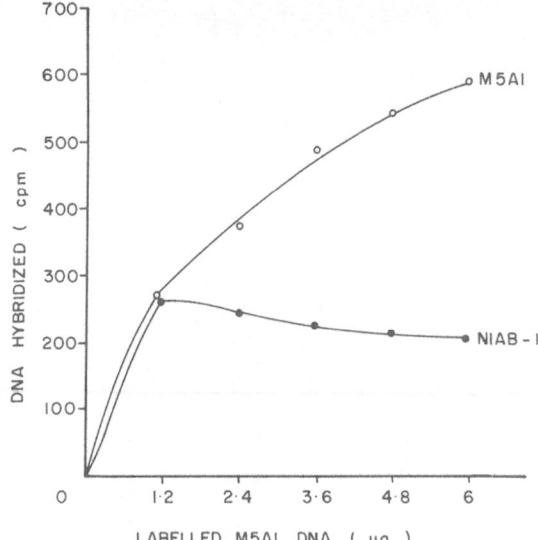

Fig. 5. DNA–DNA hybridization of *Klebsiella* sp. NIAB-I and *Klebsiella pneumoniae* M5A1.

Acknowledgement

Financial support for this research was partly provided by the U.S. National Academy of Sciences/National Research Council by means of a grant from the U.S. Agency for International Development.

References

Bagley ST, Seidler RJ and Brenner DJ 1981 *Klebsiella planticola* sp. nov.: A new species of Enterobacteriacae found primarily in non-clinical environments. Curren. Microbiol. 6, 105–109.

Bergersen FJ, Kennedy C and Hill S 1981 Influence of low oxygen concentration on depression of nitrogenase in *Klebsiella pneumoniae*. J. Gen. Microbiol. 128, 909–915.

Bilal R and Malik KA 1987 Isolation and identification of a nitrogen fixing zoogloea forming bacterium from Kallar grass histoplane. J. Appl. Bacteriol. 62, 289–294.

Dart PJ 1986 Nitrogen fixation associated with non-legumes in agriculture. Plant and Soil 90, 303–334.

Gowa J 1953 A micro biuret method for protein determination. Scan. J. Clin. Lab. Invest. 5, 218–222.

Hill S 1976 Influence of oxygen concentration on acetylene reduction and efficiency of nitrogen fixation in intact *Klebsiella pneumoniae*. J. Gen. Microbiol. 93, 335–345.

Hill S, Turner GL and Bergersen FJ 1984 Synthesis and activity of nitrogenase in *Klebsiella pneumoniae* exposed to low concentrations of oxygen. J. Gen. Microbiol. 130, 1061–1067.

Izard D, Ferragut F, Kersters K, De Ley J and Leclerc H 1981 *Klebsiella terrigina*: A new species from soil and water. Int. J. Syst. Bacteriol. 31, 116–127.

Kersters K and De Ley J 1975 Identification and grouping of bacteria by numerical analysis of their electrophoretic patterns. J. Gen. Microbiol. 87, 333–342.

Kersters K and De Ley J 1980 Classification and identification of bacteria by electrophoresis of their proteins. *In* Microbiological Classification and Identification. Eds. M. Goodfellow and RG Board. pp. 273–297. Soc. Appl. Bacteriol. Symp. Ser. 8. Academic Press, London.

Klucas R 1972 Nitrogen fixation grown in the presence of oxygen. Can. J. Microbiol. 18, 1845–1850.

Laemmli UK 1970 Cleavage of structural proteins during the assembly of the bacteriophage T4. Nature (London) 27, 680–685.

Ladha JK, Barraquio WL and Watanabe I 1983 Isolation and identification of nitrogen-fixing *Enterobacter cloacae* and *Klebsiella planticola* with rice plants. Can. J. Microbiol. 29, 1301–1308.

Malik KA, Zafar Y and Hussain A 1980 Nitrogenase activity in the rhizosphere of Kallar grass. Biologia 26, 107–112.

Malik KA, Zafar Y and Hussain A 1982 Associative dinitrogen fixation in *Diplachne fusca* (Kallar grass). *In* Biological Nitrogen Fixation Technology for Tropical Agriculture. Eds. PH Graham and SC Harris. pp. 503–507. Cali, Colombia. CIAT.

Marmour J 1961 A procedure for the isolation of deoxyribonucleic acid from micro-organisms. J. Mol. Biol. 3, 208–218.

McConaughy BC, Laird CD and McCarthy BJ 1969 Nucleic acid reassociation in formamide. Biochemistry 18, 3289–3295.

Reinhold B, Hurek T, Fendrik I, Pot B, Gills M, Kersters K, Thielemans S and De Ley J 1987 *Azospirillum halopraeferens* sp. nov.: A nitrogen fixing organism associated with roots of Kallar grass (*Leptochloa fusca* (L) Kunth.) Int. J. Sys. Bacteriol. 37, 43–51.

Sandhu GR and Malik KA 1975 Plant succession, key to utilization of salt affected soils. Nucleus 12, 35–38.

Shavikumar AG, Grundling GJ, Benson TA, Casuto D, Miller MF and Spear BB 1986 Vegetative expression of endotoxin genes of *Bacillus thuringiensis* sub sp. Kurstaki in *Bacillus subtilus* J. Bacteriol. 116, 194–204.

Steigerwalt AG, Fanning GR, Fife-Ausbury MA and Brenner DJ 1976 DNA relatedness among species of *Enterobacter* and *Serratia*, Can. J. Microbiol. 22, 121–127.

Yoch DC and Pengra RM 1966 Effect of amino acids on the nitrogenase system of *Klebsiella pneumoniae* J. Bacteriol. 92, 618–622.

Zafar Y 1985 Studies on the diazotrophic biocoeonosis in Kallar grass (*Leptochloa fusca*) (L) Kunth). Ph.D thesis. Quaid-i-Azam University, Islamabad, Pakistan.

Zafar Y, Ashraf M and Malik KA 1986 Nitrogen fixation associated with roots of Kallar grass (*Leptochloa fusca* (L) Kunth). Plant and Soil 90, 102–105.

Zafar Y, Malik KA and Niemann EG 1987 Studies on N_2-fixing bacteria associated with the salt tolerant grass, *Leptochloa fusca* (L) Kunth. MIRCEN J. Appl. Bacteriol. Biotech. 3, 45–56.

Session 5

Physiology of *Azospirillum* and
other diazotrophs

F. A. Skinner et al. (Eds.), Nitrogen fixation with non-legumes, 123–136.

Ecophysiological aspects of growth and nitrogen fixation in *Azospirillum* spp.

A. HARTMANN

Lehrstuhl für Mikrobiologie, Universität Bayreuth, Postfach 101251, D-8580 Bayreuth, FRG

Key words: amino acids, *Azospirillum brasilense, A. lipoferum, A. amazonense, A. halopraeferens,*
              growth regulation, nitrogen fixation, osmotolerance, siderophores

Abstract

The nitrogenase activity of *Azospirillum* spp. is efficiently regulated by environmental factors. In *A. brasilense* and *A. lipoferum* a rapid 'switch off' of nitrogenase activity occurs after the addition of ammonium chloride. As in photosynthetic bacteria, a covalent modification of nitrogenase reductase (Fe-protein) is involved. In *A. amazonense*, a non-covalent mechanism causes only a partial inhibition of nitrogenase activity after ammonium chloride is added. In anaerobic conditions, nitrogenase reductase is also 'switched off' by a covalent modification in *A. brasilense* and *A. lipoferum*. Short-time exposure of *Azospirillum* to increased oxygen levels causes a partially reversible inhibition of nitrogenase activity, but no covalent modification is involved. *Azospirillum* spp. show variations in their oxygen tolerance. High levels of carotenoids confer a slightly improved oxygen tolerance. Certain amino acids (*e.g.* glutamate, aspartate, histidine and serine) affect growth and nitrogen fixation differently in *Azospirillum* spp. Amino acids may influence growth and nitrogen fixation of *Azospirillum* in the association with plants. *Azospirillum brasilense* and *A. halopraeferens* are the more osmotolerant species. They utilize most amino acids poorly and accumulate glycine betaine, which also occurs in osmotically stressed grasses as a compatible solute to counteract osmotic stress. Nitrogen fixation is stimulated by glycine betaine and choline. Efficient iron acquisition is a prerequisite for competitive and aerotolerant growth and for high nitrogenase activity. *Azospirillum halopraeferens* and *A. amazonense* assimilate iron reasonably well, whereas growth of some *A. brasilense* and *A. lipoferum* strains is severely inhibited by iron limitation and by competition with foreign microbial iron chelators. However, growth of certain iron-limited *A. brasilense* strains is stimulated by the phytosiderophore mugineic acid. Thus, various plant-derived substances may stimulate growth and nitrogen fixation of *Azospirillum*.

Introduction

The rhizosphere is a challenging habitat for microorganisms and a challenge for ecological studies. At the interface of root and soil, the growing plant provides macronutrients as well as micronutrients and the soil contributes its structure, mineral composition, physicochemical properties and water regime. Thus, the stage is set for a competitive and balanced growth of soil microorganisms, which can themselves change the habitat (Brock, 1985; Curl and Truelove, 1986; Whipps and Lynch, 1986). Growth and multiplica-

tion, despite competition for nutrients and the effects of antagonistic agents, give the best chance to escape predators, such as grazing amoebae or bdellovibrios (Brock, 1985; Curl and Truelove, 1986). Microorganisms, which succeed in colonizing the root surface and are able to reach the endorhizosphere, may be regarded as fortunate winners of the game. At these sites, more substrates are available and competition may be less severe. This is particularly important for the nitrogen-fixing ability of plant-associated bacteria, because nitrogen fixation is an energy-demanding process (Whipps and Lynch, 1983).

Table 1. Influence of amino acids on growth and nitrogen fixation of *Azospirillum* spp.

(A) Amino acids as sole sources of carbon and nitrogen:

Amino acid (10 mM)	Stimulation of growth[a]: A. amazonense	A. lipoferum	A. brasilense	A. halopraeferens
glutamate	19.2	22.3	9.3	3.1
proline	25.7	28.7	9.9	2.2
serine	2.7	10.8	1.6	1.3
alanine	9.5	15.0	8.9	17.6
histidine	3.7	14.2	1.1	1.2

(B) Effects of amino acids on nitrogen fixation in the presence of 0.5% malate (Y1: sucrose):

Amino acid (10 mM)	Relative nitrogen fixation activity[a]: A. amazonense	A. lipoferum	A. brasilense	A. halopraeferens
glutamate	0.14	0	0.45	0.73
proline	0.01	0.14	0.25	0.86
serine	0.05	0.16	0.73	0.73
alanine	0.15	0.01	0.08	0
histidine	0	0	1.47	0.90

[a] The figures are ratios of growth (A_{560}) (A) and nitrogen fixation (B) in the presence and absence of amino acids.
The tests were performed in semisolid minimal medium (Albrecht and Okon, 1980) without nitrogen and/or carbon sources added. Growth (A_{560}) was measured after 3 days incubation at 30°C (*A. halopraeferens*: 41°C). Nitrogenase activity (acetylene reduction) was determined 1 or 2 days after inoculation.

Numerous factors limit or enhance nitrogen-fixing activity. A comparison of the different responses of *Azospirillum* spp. to environmental conditions may help us to recognize the characteristics of different species and strains. The identification of specific factors involved in the plant-*Azospirillum* interaction may lead to a better understanding of how the plant host and associated bacteria may be physiologically interdependent, and so help to improve the competitiveness of inoculant strains and their ability to fix nitrogen.

This article summarizes recently published and unpublished results on the influence of combined nitrogen sources (ammonium and amino acids), oxygen, osmotic stress and iron limitation on growth and nitrogen fixation of different *Azospirillum* species and closely related bacteria. The reader is also referred to previous reviews on the physiology (Okon, 1985a), ecology (Balandreau, 1986; Okon and Kapulnik, 1986; Patriquin *et al.*, 1983) and applied aspects (Elmerich, 1984; Okon, 1985b) of *Azospirillum*.

Growth substrates and their influence on nitrogen fixation

Organic acids and sugars. The utilization of substrates available as plant cell constituents, muci-

lages and in root exudates is a prerequisite for growth in the rhizosphere, colonization of the root and finally, nitrogen fixation in association with the root. The chemotaxis towards components of root exudates and mucilage (Heinrich and Hess, 1985; Mandimba *et al.*, 1986; Reinhold *et al.*, 1985) and the utilization of organic acids and sugars by *Azospirillum* is well documented (Magalhaes *et al.*, 1983; Reinhold *et al.*, 1987; Tarrand *et al.*, 1978). There are species-specific differences in the utilization of sugars; thus *A. lipoferum* readily utilizes glucose and *A. amazonense*, sucrose, whereas *A. brasilense* and *A. halopraeferens* cannot grow on these substrates efficiently. The physiological basis for these differentiating characteristics are the abilities to take up mono- and disaccharides and utilize them along different pathways (Martinez-Drets *et al.*, 1984; Martinez-Drets *et al.*, 1985).

Amino acids. Azospirillum spp. differ in the utilization of amino acids. *Azospirillum lipoferum* and *A. amazonense* readily used many amino acids as the sole sources of carbon, nitrogen and energy (Table 1A). Nitrogen fixation measured in the presence of malate or sucrose was drastically inhibited at high concentrations of amino acids. At low concentration of glutamate, nitrogen fixation in *A. amazonense* was slightly stimulated (Fig. 1). In contrast, *A. brasilense* and *A. halopraeferens* grew poor-

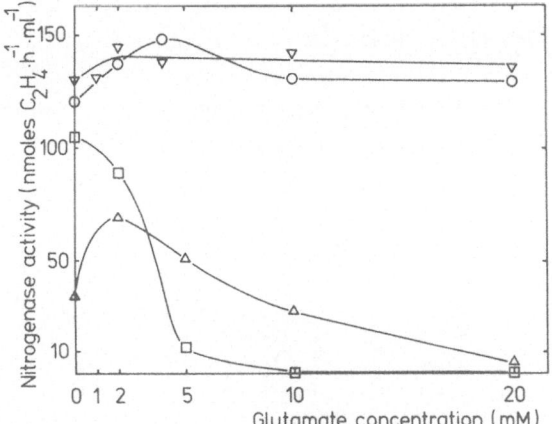

Fig. 1. Effect of different concentrations of glutamate on the total nitrogen fixation activity of *A. amazonense* Y1 (△), *A. lipoferum* SpRG20a (□), *A. brasilense* Sp7 (○) and *A. halopraeferens* (▽). Semisolid minimal medium, prepared according to Albrecht and Okon (1980), was inoculated with 2% (v/v) overnight cultures grown with ammonium as combined nitrogen source. After 24 h at 30°C (*A. halopraeferens*: 41°C) the acetylene reduction activity (Burris, 1972) was measured.

ly or not at all with amino acids as sole source of carbon, nitrogen and energy (exception: alanine), and nitrogen fixation was not inhibited at even high concentrations of amino acids, *e.g.* glutamate (Table 1, Fig. 1). Accordingly, glutamate is a good nitrogen source for *A. lipoferum* and *A. amazonense*, but a poor nitrogen source for *A. brasilense* and *A. halopraeferens* (Hartmann *et al.*, 1988). The physiological basis for the different efficiency of the utilization of glutamate is probably the different activities for glutamate uptake, glutamate dehydrogenase and glutamate transaminases (Hartmann *et al.*, 1988). The ability of *Azospirillum* spp. to grow on amino acids as sole sources of carbon and nitrogen may be helpful as additional taxonomic characteristics and to isolate certain species specifically (*e.g. A. lipoferum* with histidine). This ability may also be relevant for their growth behaviour in the rhizosphere of particular plants and for the establishment of associative nitrogen fixation.

Polymer-substrates and C_1-compounds. *Azospirillum* can grow and fix nitrogen using polymer substrates, such as straw and xylan (Halsall *et al.*, 1985) and hemicellulose (Ladha *et al.*, 1986). In co-culture with cellulose degrading bacteria, *Azospirillum* can fix nitrogen (Halsall and Gibson, 1985). The ability for autotrophic growth is widely distributed among *Azospirillum* (Malik and

Schlegel, 1981; Tilak *et al.*, 1986). *Azospirillum* spp., including the recently described *A. halopraeferens*, are able to grow on methanol or other C_1-compounds and to fix nitrogen (Sampaio *et al.*, 1982; Hartmann, unpublished results). *Azospirillum* may also be able to use the methyl groups of pectin, a major constituent of plant cell walls. Cell-wall degrading enzymes are present in *Azospirillum* at rather low activities (Tien *et al.*, 1981).

Regulation of nitrogenase activity by ammonium and oxygen

The expression of nitrogen-fixing activity is under genetic control. Possibly a *nifA*-type regulation, which activates the transcription of the *nif*-genes at nitrogen limiting conditions (Postgate, 1982), is operating in *A. brasilense* (Pedrosa and Yates, 1984). In addition, the nitrogenase activity itself is regulated by environmental factors. An effective regulatory control of nitrogen fixation may have evolved because of the high energy costs of nitrogen fixation (Postgate, 1982).

Ammonium. After the addition of very low concentrations of ammonium, nitrogen fixation is rapidly and reversibly inhibited in *Azospirillum* spp. (Gallori and Bazzicalupo, 1985; Hartmann *et al.*, 1986; Haian Fu, personal communication). A covalent modification of the nitrogenase reductase of *A. brasilense* and *A. lipoferum* was demonstrated using the quick filtration and extraction procedure (Kanemoto and Ludden, 1984) in combination with the immuno-blotting technique (Hartmann *et al.*, 1986). The modification resembles the situation in the photosynthetic bacterium *Rhodospirillum rubrum*, where a 'switch off' of nitrogenase activity is caused by an ADP-ribosylation of nitrogenase reductase (Pope *et al.*, 1985). In *A. amazonense* (Hartmann *et al.*, 1986; Song *et al.*, 1985) and *Herbaspirillum seropedicae* (Haian Fu, personal communication) no evidence for an involvement of covalent modification in the 'ammonium switch off' was apparent. Probably, the less complete inhibitory effect is mediated by a noncovalent inhibitory mechanism in these bacteria. The rapid inhibition of nitrogenase activity by ammonium and other nitrogen compounds is widely distributed in N_2-fixing bacteria (Heda and Madigan, 1986; Kush *et al.*, 1985; Reich *et al.*, 1986; Turpin *et al.*, 1984; Yoch and Whiting, 1986; Zumft, 1985/

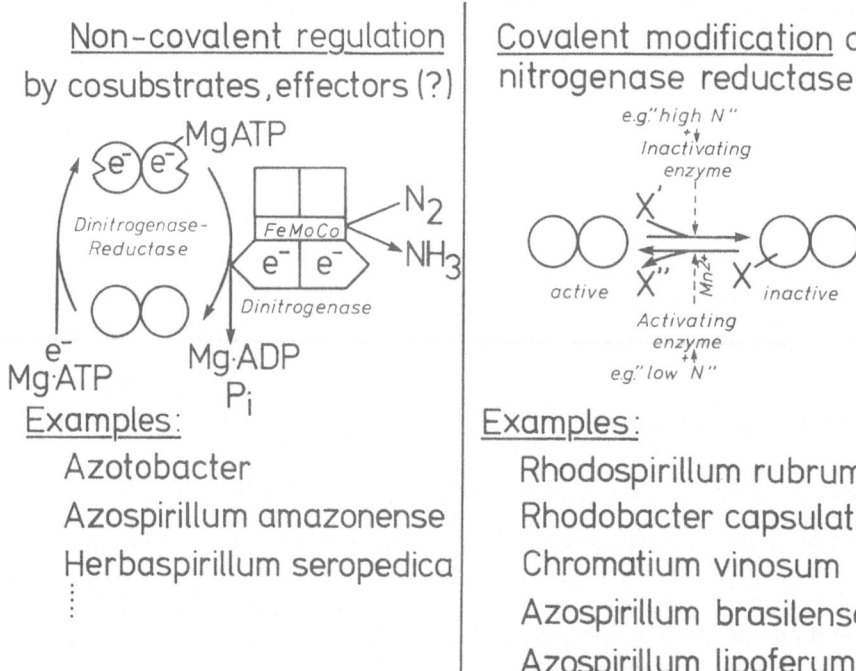

Fig. 2. Covalent and non-covalent regulation of the nitrogenase activity (schematic overview).

review). However, the appearance of a covalently modified nitrogenase reductase during 'switch off' has clearly been demonstrated only in a few species (Gotto and Yoch, 1985; Hartmann *et al.*, 1986; Jouanneau *et al.*, 1983; Kanemoto and Ludden, 1984) (Fig. 2). The involvement of an activating (Saari *et al.*, 1984) and inactivating enzyme (Lowery *et al.*, 1986) in the regulation of nitrogenase reductase probably improves the fine tuning and efficiency of nitrogenase regulation. *Azospirillum brasilense* and *A. lipoferum* have an activating enzyme, which activates modified nitrogenase reductase of *Rhodospirillum rubrum* (Ludden and Burris, 1976; Ludden *et al.*, 1978; Hartmann and Fu, unpublished). The regulatory system of nitrogenase in *R. rubrum*, a close relative to *Azospirillum*, appears to be very similar to *A. brasilense* and *A. lipoferum*.

Oxygen. After a shift to anaerobic conditions, nitrogen fixation of *A. brasilense* and *A. lipoferum* was inhibited and simultaneously the nitrogenase reductase was modified. However, during the rapid inhibition of nitrogenase activity by high oxygen

levels, no covalent modification occurred (Hartmann and Burris, 1987). The decrease of nitrogenase activity during anaerobiosis was not accompanied by an increase of the glutamine pool, as in the 'ammonium switch off', and no alteration in the activity of glutamine synthetase occurred (Hartmann and Burris, 1987). Therefore, a signal independent of the nitrogen metabolism can also trigger the inactivating enzyme which catalyzes the covalent modification of nitrogenase (Fig. 2). A decrease in the oxoglutarate level could also provide the signal, because this would raise the glutamine/oxoglutarate ratio, which is of general importance in metabolic control of nitrogen metabolism. Thus details in the regulation of nitrogenase activcity by the modification system are not yet understood.

Release of nitrogen compounds by Azospirillum

In an effectively nitrogen fixing association of mutual benefit, a carbon-nitrogen interchange should occur. The association of *Peltigera* with the cyanobacterium *Nostoc* provides a good example. Sarcosine was found to act as a regulator which inhibits glutamine synthetase and stimulates nitrogenase activity and ammonium release in *Nostoc* (Hällborn, 1984). For rhizocoenosis with heterotrophic diazotrophs, a similar physiological interaction has not yet been described.

At low levels of glutamate or aspartate, small amounts of ammonium are released from N_2-fixing cultures of *A. brasilense* and *A. amazonense* (Hartmann *et al.*, 1988). Glutamate and aspartate, or metabolities derived from it, might reduce the activities of ammonium assimilatory enzymes and ammonium uptake and foster ammonium release. It would be very interesting to find plant-derived effectors and the N_2-fixing bacterial partner, which respond by releasing substantial amounts of ammonium, amino acids or other nitrogen compounds. A release of nitrogen compounds in laboratory batch cultures of *A. lipoferum* (Volpon *et al.*, 1981), *A. amazonense* and *A. halopraeferens* (Hurek *et al.*, 1987b) has been described. The nature of the compounds and the mechanism involved are not known. Under certain environmental conditions, certain strains of *Azotobacter chroococcum* release ammonium (Narula and Gupta, 1986). Further studies on the regulatory properties of glutamine synthetase (Gauthier and Elmerich, 1977), glutamate synthase (Ratti *et al.*, 1985), glutamate dehydrogenase (Maulik and Ghosh, 1986) and ammonium permease (Hartmann and Kleiner, 1982; Kleiner and Castorph, 1982) are needed in strains isolated from effectively nitrogen fixing associations.

Energy starved cultures of *Azospirillum* release small amounts of ammonium due to an inactivation of the ammonium permease and a decrease in the activities of ammonium assimilatory enzymes (Hartmann *et al.*, 1984). After the addition of malate, ammonium uptake is rapidly reactivated and ammonium is taken up again (Hartmann, unpublished results). This mechanism of nitrogen release might operate in rhizocoenoses, when the energy supply is transiently reduced due to environmental effects on the plant (Whiting *et al.*, 1986) or at later growth stages, when the root becomes a minor sink for photosynthates (Curl and Truelove, 1986).

Ammonium release and constitutive N_2-fixation in

mutants. Glutamine auxotrophic mutants of *A. brasilense*, defective in glutamine synthetase (Gauthier and Elmerich, 1977) or glutamate synthase (Bani *et al.*, 1980), release ammonium under N_2-fixing conditions; these mutants have no ammonium uptake activity (Hartmann *et al.*, 1984). Recently, prototrophic mutants of *A. brasilense* which fix nitrogen constitutively were isolated (Fischer *et al.*, 1986). In prototrophic mutants of *A. brasilense* resistant to methionine sulphoximine (MSX), constitutive N_2-fixation occurred but no ammonium release was observed (Hartmann, 1982; Hartmann *et al.*, 1983). MSX-resistant mutants of *Anabaena variabilis* (Spiller *et al.*, 1986) and *Nostoc muscorum* (Singh *et al.*, 1983) release ammonium while fixing nitrogen. Certainly more research on this topic is needed. *Azotobacter* mutants with enhanced nitrogen fixation activity also release some ammonium (Gordon and Jacobson, 1983). It may be speculated that mechanisms for the release of nitrogen compounds are present in bacteria from N_2-fixing associations, but this property is under the physiological control of plant factors.

Inhibition of nitrogen fixation by oxygen and oxygen protective mechanisms

Species and strain specific differences in oxygen tolerance. All *Azospirillum* species are characterized as microaerobically nitrogen fixing bacteria (Krieg and Döbereiner, 1984; Okon *et al.*, 1977). The narrow range of oxygen tolerance obviously limits nitrogen fixation in the association with roots (Zuberer and Alexander, 1986). A comparison of the oxygen tolerance of nitrogen fixation in a well mixed chamber (Hochman and Burris, 1981), which allowed simultaneous measurements of dissolved oxygen, respiration activity and acetylene reduction, revealed an increased oxygen tolerance in the order *A. brasilense* Sp7, *A. lipoferum* SpRG20a and *A. amazonense* Y1 (Hartmann *et al.*, 1985). Using the same experimental device, this order of oxygen tolerance was also found, when the depression of nitrogen fixation in cultures pregrown on ammonium was performed at various dissolved oxygen concentrations (Fig. 3). Strain specific differences of the efficiency of nitrogen fixation at different oxygen concentrations were also found in chemostat cultures (Hurek *et al.*, 1987a;

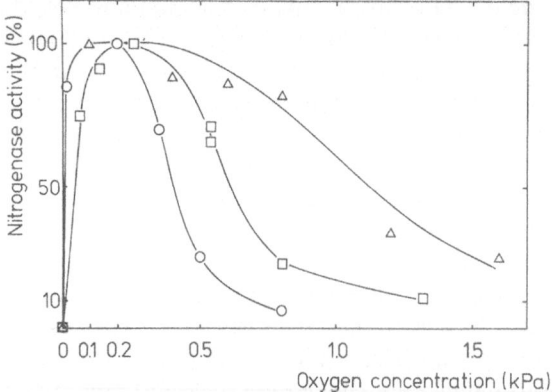

Fig. 3. Different oxygen tolerance of nitrogen fixation in *A. brasilense* Sp7 (O), *A. lipoferum* SpRG20a (□) and *A. amazonense* Y1 (△).

Cells growing with ammonium chloride as nitrogen source were centrifuged and resuspended in nitrogen free minimal medium. Derepression of nitrogen fixation was examined at different constant dissolved oxygen concentrations in a rapidly stirred chamber equipped with an oxygen electrode (Hartmann and Burris, 1986). After the acetylene reduction activity reached a constant rate the speific activity was determined. For comparison reasons the highest acetylene reduction rate in each strain was set to 100%. The lag period until the acetylene reduction rate reached a fairly constant rate was considerably prolonged at increasing dissolved oxygen concentrations. The dissolved oxygen concentration in equilibrium with 1 kPa oxygen equals about $12 \mu M$.

Nelson and Knowles, 1978). The physiological basis for this variation in oxygen tolerance is not known. Recently, an oxygen resistant hydrogenase was found in *A. amazonense* Y1, which may contribute to the oxygen tolerance; *A. brasilense* and *A. lipoferum* lacked this activity (Fu and Knowles, 1986).

Oxygen protection by carotenoids. Carotenoids contribute to the oxygen tolerance due to their ability to quench radical reactions which are initiated by toxic oxygen metabolites, thereby preventing autooxidation reactions (Burton and Ingold, 1984). Inhibition of carotenoid biosynthesis by diphenylamine in the red-coloured strain. *A. brasilense* Cd reduced the oxygen tolerance of its nitrogen fixation process (Nur *et al.*, 1981). Carotenoid overproducing mutants, obtained from the slightly pink wild type *A. brasilense* Sp7, grew and fixed nitrogen at oxygen stress conditions, whereas the wild type failed to grow (Hartmann *et al.*, 1983). In oxygen controlled, nitrogen-free continuous cultures an improved nitrogen fixation activity occurred due to carotenoids at oxygen stress conditions ($12 \mu M$ dissolved oxygen), although growth and nitrogen fixation was greatly inhibited (Hartmann and Hurek, unpublished results). It is concluded that a high level of carotenoids do not extend the optimum for nitrogen fixation to higher oxygen levels, but provide some limited protection against oxygen damage.

Reversible 'oxygen switch off'. A rapid, and partially reversible, inhibition of nitrogenase activity by oxygen was observed in *Azospirillum* spp. (Hartmann and Burris, 1987; Hartmann *et al.*, 1985). A similar type of reversible inhibition of nitrogenase activity by oxygen was demonstrated in *Azotobacter* spp. (Dingler and Oelze, 1985; Yates and Jones, 1974), photosynthetic bacteria (Hochman and Burris, 1981), cyanobacteria (Stal and Krumbein, 1985), *Klebsiella pneumoniae* (Goldberg *et al.*, 1987) and anaerobic N_2-fixers (Postgate *et al.*, 1985). The 'oxygen switch off' is probably a general phenomenon of nitrogen fixation and does not depend on the presence of the so-called oxygen protection protein of *Azotobacter* (Robson and Postgate, 1980; Yates and Jones, 1974). Using antiserum against purified Fe_2S_2-protein (Shetnaprotein) of *A. vinelandii* (kindly provided by Dr. F. Simpson, Madison WI), no immunologically crossreactive protein could be detected by the immunoblotting technique in *Azospirillum* spp., *R. rubrum* and *K. pneumoniae* (Hochman *et al.*, 1987). Goldberg *et al.*, (1987) suggested that at high oxygen levels the limited pool of electrons is diverted towards respiration, thus leaving the nitrogenase without a sufficient supply of reduction equivalents.

Respiratory protection. Nitrogen fixation in mutants of the aerobically nitrogen-fixing *A. chroococcum*, which have a reduced respiratory capacity due to a lesion in citrate synthase, was restricted to microaerobic conditions (Ramos and Robson, 1985). It is reasonable that respiratory protection is only efficient with a highly active tricarboxylic acid cycle. A comparison of the activities of citrate synthase and other TCA-cycle enzymes of *Azospirillum* (Martinez-Drets *et al.*, 1984) with *Azotobacter* (Ramos and Robson, 1985) reveals that respiration protection of *Azospirillum* may indeed be limited by the supply of reduction equivalents. Therefore, attempts to improve oxygen tolerance of nitrogen fixation should focus on an improvement of respiratory capacity, so that enough electrons are available to support high respiration rates and high nitrogenase activity simultaneously. This may in part be accomplished by providing *Azospirillum* with an optimum iron supply, because iron is an important constituent in many of the enzymes involved. In laboratory minimal medium cultures, aerotolerant growth of *A. brasilense* was improved by the addition of dihydroxyphenyl iron-binding compounds (Das and Mishra, 1984). Similar effects were found in other microaerophilic bacteria (Bowdre *et al.*, 1976). Oxygen tolerance of nitrogen fixation can be regarded as a co-operative property of the cell metabolism, where many different mechanisms contribute (Krieg and Hoffman, 1986). Therefore, protection against oxygen damage may be further improved by increased levels of superoxide dismutase or peroxidase, which were demonstrated in *Azospirillum* (Clara and Knowles, 1985; Nur *et al.*, 1982). The recent isolation of an aerotolerant mutant of *A. brasilense* SpCd (Del Gallo *et al.*, 1987) which was selected in a chemostat under oxygen stress conditions, demonstrates that improvements in oxygen tolerance of *Azospirillum* are feasible. However, as expected, the efficiency of nitrogen fixation was very low due to an increased consumption of malate.

Osmotolerance and osmoregulatory properties

Metabolic activities of microorganisms are greatly influenced by the availability of water (Brown, 1976). Soil bacteria have to cope with water stress in salt affected soils or when soil becomes dry. Low water potentials frequently occur in the rhizosphere of plants growing in a dry climate, because of the high water need of the transpiring plant. In microorganisms which are able to adapt to changing water potentials in the environment, compatible solutes like amino acids, betaines and sugars play an important role (Imhoff, 1986; Ken Dror *et al.*, 1986; Yancey *et al.*, 1982). The intracellular accumulation of these solutes prevents water loss from the cell and is compatible with or even preserves enzymatic functions. In drought or osmotic stress conditions various plants, including grasses, accumulate betaines and/or proline (Wyn Jones and Storey, 1981). Probably these substances are available for microorganisms in the rhizosphere.

In *Azospirillum*, osmotolerance is a species-specific character and declines in the order *A. halopraeferens*, *A. brasilense*, *A. lipoferum* and *A. amazonense* (Hartmann, 1987; Reinhold *et al.*, 1987; Tarrand *et al.*, 1978). Nitrogenase activity is more sensitive towards salt stress than cellular growth on combined nitrogen (Rao and Venkateswarlu, 1985). Half-maximal *in vitro* nitrogenase activity of *A. amazonense* was obtained at 110 mM NaCl, as was the case with whole cell nitrogen fixation (Hartmann, 1987; Hartmann, unpublished results). In osmotic stress conditions, growth of *A. brasilense* and *A. halopraeferens* was stimulated by glutamate or proline, whereas no effect was found with *A. lipoferum* or *A. amazonense* (Hartmann, 1987). The latter two species efficiently use glutamate and proline as carbon and nitrogen sources (see Table 1). Therefore, *A. lipoferum* and *A. amazonense* may not be able to use amino acids as compatible solutes during osmotic stress. Betaine glycine (N,N,N-trimethyl glycine) stimulated nitrogen fixation in *A. brasilense* and *A. halopraeferens* at otherwise inhibitory osmotic stress (Table 2). Even low concentrations of glycine betaine (0.1 mM) were sufficient to improve osmotolerance. *Azospirillum lipoferum* grew efficiently with choline and glycine betaine as sole source of carbon and nitrogen; accordingly, *A. lipoferum* could not use them as osmoprotectants (Table 2). In contrast, *A. halopraeferens* had a high affinity uptake for choline (k$_M$: 16 μM), and possibly oxidized it to glycine betaine, the most potent osmolyte (Imhoff, 1986; Yancey *et al.*, 1982). Obviously, osmotolerant strains scavenge potential osmoprotectants such as proline, betaine or choline, originating from the plant. To take advantage of this relationship, the ability to synthesize betaines and proline by the plant partner should be better explored and improved cultivars used if possible. The halophyte kallar grass (*Leptochloa fusca*), where *A. halopraeferens* resides in the rhizoplane, accumulates high levels of betaine and proline at high salinity (Sandhu *et al.*, 1981). *Azospirillum halopraeferens* is characterized by its sodium requirement (Hartmann, 1987; Reinhold *et al.*, 1987). As an adaptation to the sodium-rich alkaline soil, certain cellular processes (such as transport of amino acids) may require sodium ions.

Table 2. Effect of glycine betaine and choline on growth, nitrogen fixation and osmotolerance of *Azospirillum* spp.

	A. amazonense Y1	A. lipoferum Sp59b	A. brasilense Sp7	A. halopraeferens Au4
a) Growth stimulation by [a]				
betaine glycine	0.95	27	0.96	1.1
choline	0.95	37	0.96	1.0
b) Effect on nitrogen fixation activity of [b]				
betaine glycine	1.02	0	0.97	0.84
choline	0.91	0.18	0.97	0.98
c) Effect on nitrogen fixation activity at osmotic stress conditions [c]				
without additions	2.5	2.0	1.5	2.0
with 1 mM betaine glycine	4.0	4.0	73	40
with 1 mM N,N'-dimethylglycine	n.t.	n.t.	1.5	20
with 1 mM choline	5.0	4.0	26	20

[a] Growth with betaine glycine or choline (10 mM) as sole carbon and nitrogen sources was measured in semisolid minimal medium after 3 days incubation at 30°C (Au4: 41°C). Growth stimulation was calculated as the ratio of A_{560} in the presence and absence of betaine or choline.
[b] Nitrogen fixation activity (NA) was determined as acetylene reduction in semisolid minimal media with malate or sucrose (Y1) as carbon sources. The figures were calculated as the ratios of NA in the presence and absence of betaine or choline (10 mM).
[c] The following NaCl concentrations were applied:
A. amazonense Y1, *A. lipoferum* Sp59b: 0.3 M NaCl; *A. brasilense* Sp7: 0.4 M NaCl; *A. halopraeferens* Au4: 0.6 M NaCl. The specific nitrogenase activity (nmoles ethylene h^{-1} ml^{-1} (A = 1)) was measured after 24 hours incubation.

Iron uptake properties

Iron is essential for aerobic life and also for biological nitrogen fixation. It is an essential component of nitrogenase and many enzymes involved in energy metabolism and reactions with oxygen. However, ferric iron (Fe^{3+}) is extremely insoluble in an aerobic environment; the solubility constant of ferric oxyhydroxy polymers is 10^{-38}. Therefore, most microorganisms produce and excrete low molecular weight molecules (e.g. catechols, hydroxamates), which can efficiently complex and solubilize iron (Neilands, 1984). These chelators (siderophores) are taken up by the microorganisms in specific high affinity uptake systems, thereby assimilating the iron (Braun, 1985; Hider, 1984). It became apparent from research with human pathogens that the ability to scavenge iron is crucial for competitiveness and infectivity. Also in the rhizosphere the availability of iron is a major pathogenicity factor and exerts a selective pressure on the composition of the microbial community (Kloepper *et al.*, 1980; Leong, 1986; Neilands and Leong, 1986). Siderophores can be extracted from soil (Bossier and Verstraete, 1986; Powell *et al.*, 1980) and it is therefore probable that competition for available iron is also a factor regulating growth of microorganisms in soil. In addition, roots of grasses excrete phytosiderophores, such as mugineic acid and avenic acid at iron-limiting conditions (Römheld and Marschner, 1986; Sugiura and Nomoto, 1984).

Different iron acquisition abilities in Azospirillum. The degree of establishment of a potentially beneficial microorganism, like *Azospirillum*, in the rhizosphere is expected to be influenced by its vigour for efficient iron supply in competition and/or co-operation with other microorganisms and the plant. The potential for high affinity iron uptake of *Azospirillum* spp. was tested under severe iron limiting conditions, using high phosphate mineral medium (Albrecht and Okon, 1980) without any iron added and supplemented with the artificial iron chelator αα'-dipyridyl. Under these conditions, *E. coli* K12 grew as well as with high iron supply, because it excretes the high affinity siderophore enterochelin (red colour) to scavenge and assimilate iron (Hider, 1984), but *Azospirillum* spp. were inhibited to various degrees (Fig. 4). In *A. brasilense*, two groups with different iron assimilation efficiency existed. *Azospirillum brasilense* Sp245 (Fig. 4) and Nig5a (not shown) grew reasonably well at iron stress conditions, whereas *A. brasilense* Sp7 (Fig. 4) and SpCd (not shown) were relatively poor iron scavengers. The siderophores

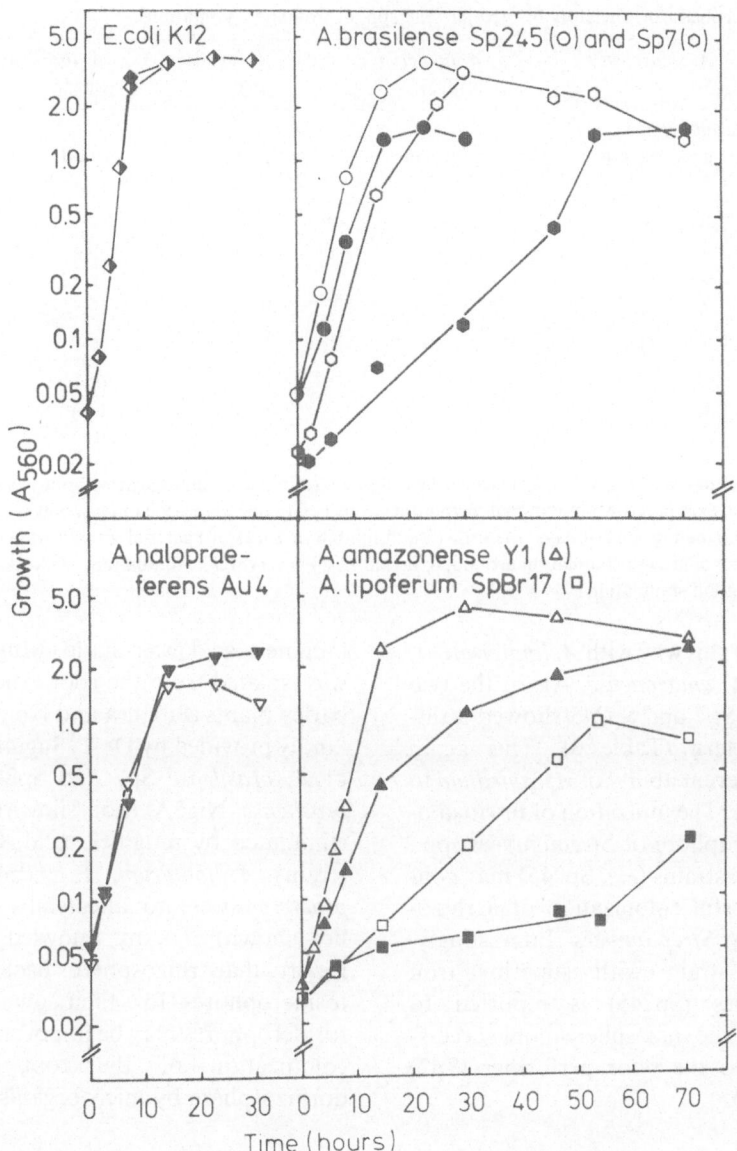

Fig. 4. Growth at iron limiting conditions of *E. coli* K12 (◇), *A. brasilense* Sp245 (o) and Sp7 (O), *A. halopraeferens* Au4 (▽), *A. lipoferum* SpBr17 (□) and *A. amazonense* Y1 (△).

At iron stress conditions (closed symbols), FeCl$_3$ was omitted from the minimal medium (Albrecht and Okon, 1980) and the artificial iron chelator αα'-dipyridyl (100 μ*M*) was added. Growth was performed in 20 ml cultures with shaking at 30°C (Au4: 41°C) and was measured turbidimetrically at 560 nm.

of *A. lipoferum* and *A. brasilense* were investigated recently and amino acid conjugates of the phenolate siderophores 2,3-dihydroxybenzoic acid and 3,5-dihydroxybenzoic acid were characterized (Bachhawat and Gosh, 1987; Saxena *et al.*, 1986).

Utilization of microbial siderophores. Iron chelators, excreted by other microorganisms, can be used efficiently as iron carriers by, for example, *E. coli* (Braun, 1985), *Rhizobium* (Smith and Neilands, 1984) and *Agrobacterium* (Page and Dale, 1986) to overcome iron deficiency. The effect of foreign fungal and bacterial siderophores on growth of *Azospirillum* spp. was tested on iron-depleted minimal-medium plates containing αα'-dipyridyl (Table 3). Microbial siderophores could

Table 3. Utilization of microbial iron chelators by *Azospirillum* spp. at iron stress conditions.

Conditions	A. amazonense Y1	A. lipoferum SpBr17	A. brasilense Sp7	A. brasilense Sp245	A. halopraeferens Au4
a) *Without additions of siderophores*:					
	good growth	very poor growth	poor growth	good growth	good growth
b) *With the additions of siderophores**:					
ferrichrome	−	0	−/+	+++	0
ferrichrysin	−	0	−/+	+++	0
ferrioxamine B	−	0	0	++	0
arthrobactin	−	0	(+)	(+)	0
coprogen	−	0	+	++	0
fusigen	−	0	0	0	0
schizokinen	−	(+)	(+)	(+)	(+)
aerobactin	−	0	0	0	0

* Microbial iron chelators (5 μg) were added in the centre of the test plates (minimal medium agar without $FeCl_3$ but containing 100 μm αα′-dipyridyl). The bacteria were pregrown in minimal medium without $FeCl_3$ and were added in a top agar overlay (0.1 ml culture in 2.5 ml 0.5% agar). After incubating for 2–4 days, the growth and inhibition zones of bacterial growth were recorded. −: inhibition, −/+: inhibition with a zone of growth stimulation around, 0: no effect, (+): very weak stimulation, +: weak stimulation, ++: good stimulation, +++: very good stimulation.

usually not be used in this way with *A. lipoferum*, *A. halopraeferens* and *A. amazonense*. Again the two *A. brasilense* strains Sp7 and Sp245 showed a different response pattern (Table 3). This again demonstrates the different ability of *Azospirillum* to face the iron problem. The utilization of ferrioxamine B, the main siderophore of *Streptomyces* spp., by some *A. brasilense* strains (*e.g.* Sp245) may contribute to the successful colonization of a rhizosphere dominated by *Streptomyces*. Interestingly, the *A. brasilense* strain with superior iron acquisition properties (Sp245) is reported to become estabished in the rhizosphere more successfully as compared to the poor performer (Sp7) (Baldani *et al.*, 1986).

Utilization of phytosiderophores. An investigation to determine whether *Azospirillum* spp. could utilize iron complexed in the phytosiderophore

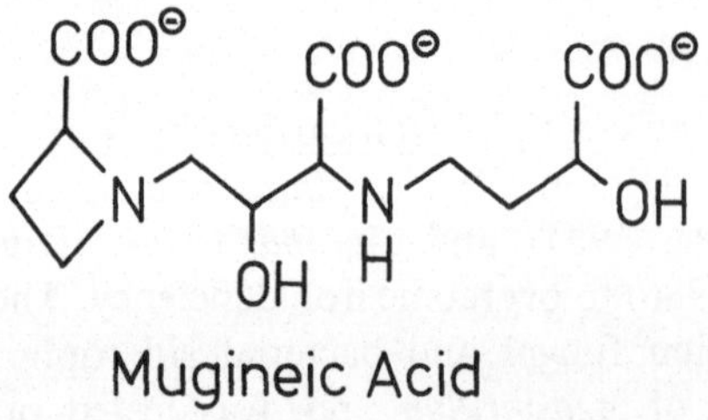

Mugineic Acid

Fig. 5. Structure of mugineic acid, a phytosiderophore isolated from exudates of iron deficient barley plant roots.

mugineic acid was made. Mugineic acid (Fig. 5) was isolated from the root exudate of iron-limited barley plants (Sugiura and Nomoto, 1984) and was kindly provided by Dr Y. Sugiura (Kyoto). Growth of *A. brasilense* Sp7 and Sp245 (Fig. 6) and *A. brasilense* Nig5A (not shown) was considerably stimulated by mugineic acid. Other *A. brasilense* strains, *A. lipoferum*, *A. amazonense* and *A. halopraeferens* were not affected by mugineic acid (data not shown). To my knowledge, this is the first report that rhizosphere bacteria can use phytosiderophores for their own iron needs. Phytosiderophores may be important for the successful colonization of the root surface and endorhizosphere by microorganisms.

Conclusions

In addition to the availability of substrates for growth, various substances of plant or microbial origin may influence the growth of *Azospirillum* in the rhizosphere by specific stimulatory or inhibitory effects. Plant-derived amino acids can enhance and stimulate, or repress, nitrogenase activity of certain *Azospirillum* spp.. Proline and betaines can, almost like catalysts, support growth and nitrogen fixation of osmotolerant strains of *A. brasilense* and *A. halopraeferens* by improving their osmoprotection. Siderophores of plant and microbial origin

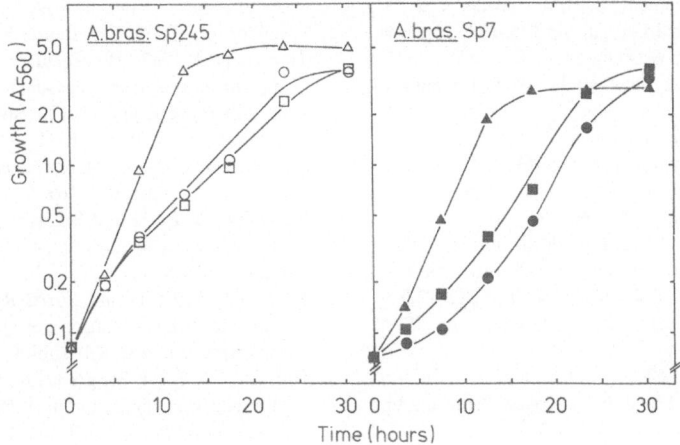

Fig. 6. Stimulation of iron-limited growth by mugineic acid in *A. brasilense* Sp245 (open symbols) and *A. brasilense* Sp7 (closed symbols). Minimal medium, extracted for iron by the 8-hydroxyquinoline method (Waring and Werkman, 1943), was inoculated (2% v/v) with iron-limited overnight cultures. Growth was performed in culture volumes of 10 ml in Erlenmeyer flasks with 100 rev. min^{-1} rotary shaking. Conditions: No $FeCl_3$ added (O); 1.3 μM $FeCl_3$ in 0.02 M HCl solution added (□); 1.3 μM Fe^{3+}-mugineic acid added (△).

can enhance or abolish growth of *Azospirillum* spp. in the rhizosphere. The ability of *Azospirillum* to respond to these interactions possibly determines whether it will proliferate further or will die out. Depending on the environmental conditions (climate, soil type *e.g.*) these interactive processes may determine the microbial community structure to various extents and may also regulate the *Azospirillum*/plant association, its actual nitrogen fixation activity and its contribution to plant nutrition. The plant/microbe and microbe/microbe interactions should be exploited by, for example, the proper choice of plant/bacterial partners in a given environmental/soil situation.

Acknowledgements

I thank Professor RH Burris (Madison, WI) for generous support and enlightening comments. The major part of my own work presented was performed during a two year post-doctoral appointment in his laboratory. I am also very grateful to my colleague Haian Fu (Madison, WI) for sharing unpublished results and to Professor PW Ludden and his coworkers for stimulating discussions and provision of antisera. I also greatly appreciate the generous support by Professor Dr. H Stolp (Bayreuth, FRG) for the investigations on osmotolerance and iron acquisition and Mrs H Ram-

ming for skilful technical assistance. I thank Drs Y Sugiura, K Hantke and P Fiedler for gifts of siderophores and Drs J Döbereiner, F Favilli and B Reinhold for providing *Azospirillum* strains. This work was in part supported by the Deutsche Forschungsgemeinschaft (Ha 1241/2 and SFB 137).

References

Albrecht S L and Okon Y 1980 Cultures of *Azospirillum*. Methods Enzymol. 69, 740–749.
Bachhawat A K and Gosh S 1987 Iron transport in *Azospirillum brasilense*: Role of the siderophore spirilobactin. J. Gen. Microbiol. 133, 1759–1765.
Balandreau J 1986 Ecological factors and adaptive processes in N_2-fixing bacterial populations of the plant environment. Plant and Soil 90, 73–92.
Baldani V L D, Alvarez MA de B, Baldani JI and Döbereiner J 1986 Establishment of inoculated *Azospirillum* spp. in the rhizosphere and in roots of field grown wheat and sorghum. Plant and Soil 90, 35–46.
Bani D C, Barberio M, Favili F, Gallori E and Polsinelli M 1980 Isolation and characterization of glutamate synthase mutants of *Azospirillum brasilense*. J. Gen. Microbiol. 119, 239–244.
Bossier P and Verstraete W 1986 Detection of siderophores in soil by a direct bioassay. Soil Biol. Biochem. 18, 481–486.
Bowdre J H, Krieg N R, Hoffman P S and Smibert R M 1976 Stimulatory effect of dihydroxyphenyl compounds on the aerotolerance of *Spirillum volutans* and *Campylobacter fetus* subspecies jejuni. Appl. Environ. Microbiol. 31, 127–133.
Braun V 1985 The unusual features of the iron transport systems of *Escherichia coli*. Trends Biochem. Sci. 10, 75–78.
Brock T D 1985 Procaryotic population ecology. *In* Engineered

Organisms in the Environment: Scientific Issues. Eds. HO Halvorson, D Pramer and M Rogul. pp 176–179. American Society for Microbiology, Washington, D.C.

Brown A D 1976 Microbial water stress. Bact. Reviews 40, 803–846.

Burris R H 1972 Nitrogen fixation — Assay methods and techniques. Methods Enzymol. 24B, 415–431.

Burton G W and Ingold K U 1984 β-Carotene: An unusual type of lipid antioxidant. Science 224, 569–573.

Clara R W and Knowles R 1985 Superoxide dismutase, catalase and peroxidase in ammonium-grown and nitrogen-fixing *Azospirillum brasilense*. Can. J. Microbiol. 30, 1222–1228.

Curl E A and Truelove B 1986 The Rhizosphere, pp 55–92. Springer-Verlag, Berlin.

Das A and Mishra A K 1984 Aerotolerant growth in *Azospirillum brasilense* induced by dihydroxyphenyl iron-binding compound. Current Microbiol. 11, 313–316.

Del Gallo M, Gratani L and Morpurgo G 1987 Selection at the chemostat of *Azospirillum brasilense* Cd N_2-fixing at high oxygen pressure. In *Azospirillum* IV, Genetics, Physiology, Ecology. Ed. W Klingmüller. pp. 75–82. Springer-Verlag, Berlin.

Dingler Ch and Oelze J 1985 Reversible and irreversible inactivation of cellular nitrogenase upon oxygen stress in *Azotobacter vinelandii* growing in oxygen controlled continuous culture. Arch. Microbiol. 141, 80–84.

Elmerich C 1984 Molecular biology and ecology of diazotrophs associated with non-leguminous plants. Bio/Technology 2, 967–978.

Fischer M, Levy E and Geller T 1986 Regulatory mutation that control *nif* expression and histidine transport in *Azospirillum brasilense*. J. Bacteriol. 167, 423–426.

Fu Ch and Knowles R 1986 Oxygen tolerance of uptake hydrogenase in *Azospirillum* spp. Can. J. Microbiol. 32, 897–900.

Gallori E and Bazzicalupo M 1985 Effect of nitrogen compounds on nitrogenase activity in *Azospirillum brasilense*. FEMS Microbiol. Lett. 28, 35–38.

Gauthier D and Elmerich C 1977 Relationship between glutamine synthetase and nitrogenase in *Spirillum lipoferum*. FEMS Microbiol. Lett. 2, 101–104.

Goldberg I, Nadler V and Hochman A 1987 Mechanism of nitrogenase switch-off by oxygen. J. Bacteriol. 169, 874–879.

Gordon J K and Jacobson M R 1983 Isolation and characterization of *Azotobacter vinelandii* mutant strains with potential as bacterial fertilizer. Can. J. Microbiol. 29, 973–978.

Gotto J W and Yoch D C 1985 Regulation of nitrogenase activity by covalent modification in *Chromatium vinosum*. Arch. Microbiol. 141, 40–43.

Hällborn L 1984 Sarcosine: A possible regulatory compound in the *Peltigera praetextata-Nostoc* symbiosis. FEMS Microbiol. Lett. 22, 119–121.

Halsall D M and Gibson A H 1985 Cellulose decomposition and associated nitrogen fixation by mixed cultures of *Cellulomonas gelida* and *Azospirillum* species or *Bacillus macerans*. Appl. Environ. Microbiol. 50, 1021–1026.

Halsall D M, Turner G L and Gibson A H 1985 Straw and xylan utilization by pure cultures of nitrogen-fixing *Azospirillum* spp. Appl. Environ. Microbiol. 49, 423–428.

Hartmann A 1982 Antimetabolite effects on nitrogen metabolism of *Azospirillum* and properties of resistant mutants. In

Azospirillum, Genetics, Physiology, Ecology. Ed. W Klingmüller. pp 59–68. Experientia Suppl. 42, Birkhäuser, Basel.

Hartmann A 1987 Osmoregulatory properties of *Azospirillum* spp. In *Azospirillum* IV, Genetics, Physiology, Ecology. Ed. W Klingmüller. pp 122–130. Springer-Verlag, Berlin.

Hartmann A and Burris R H 1987 Regulation of nitrogenase activity by oxygen in *Azospirillum brasilense* and *Azospirillum lipoferum*. J. Bacteriol. 169, 944–948.

Hartmann A, Fu H and Burris R H 1986 Regulation of nitrogenase activity by ammonium chloride in *Azospirillum* spp. J. Bacteriol. 165, 864–870.

Hartmann A, Fu H and Burris R H 1988 Influence of amino acids on nitrogen fixation ability and growth of *Azospirillum* spp. Appl. Environ. Microbiol. 54, 87–93.

Hartmann A, Fu H, Song S-D and Burris R H 1985 Comparison of nitrogenase regulation in *A. brasilense*, *A. lipoferum* and *A. amazonense*. In *Azospirillum* III, Genetics, Physiology, Ecology. Ed. W Klingmüller. pp. 116–126. Springer-Verlag, Berlin.

Hartmann A, Fußeder A and Klingmüller W 1983 Mutants of *Azospirillum* affected in nitrogen fixation and auxin production. In *Azospirillum* II, Genetics, Physiology, Ecology. Ed. W Klingmüller. pp 78–88. Experientia Suppl. 48, Birkhäuser-Verlag, Basel.

Hartmann A and Kleiner D 1982 Ammonium (methylammonium) transport by *Azospirillum* spp. FEMS Microbiol. Lett. 15, 65–67.

Hartmann A, Kleiner D and Klingmüller W 1984 Ammonium uptake and release by *Azospirillum*. In Advances in Nitrogen Fixation Research. Eds. C Veeger and W E Newton. p. 227. Nijhoff/Junk/Pudoc Publishers, The Hague, Wageningen, The Netherlands.

Heda G D and Madigan MT 1986 Aspects of nitrogen fixation in *Chlorobium*. Arch. Microbiol.143, 330–336.

Heinrich D and Hess D 1985 Chemotactic attraction of *Azospirillum lipoferum* by wheat roots and characterization of some attractants. Can. J. Microbiol. 31, 26–31.

Hider R C 1984 Siderophore mediated absorption of iron. Structure Bonding 58, 26–87.

Hochman A and Burris R H 1981 Effect of oxygen on acetylene reduction by photosynthetic bacteria. J. Bacteriol. 147, 492–499.

Hochman A, Goldberg I, Nadler V and Hartmann A 1987 Reversible inhibition of nitrogen fixation by oxygen. In Aspects of Nitrogen Metabolism. Eds. W R Ullrich, P J Aparicio, P J Syrett and F Castillo. pp 173–174. Springer-Verlag, Berlin.

Hurek T, Reinhold B, Fendrik I and Niemann E-G 1987a Root-zone-specific oxygen tolerance of *Azospirillum* spp. and diazotrophic rods closely associated with Kallar grass. Appl. Environ. Microbiol. 53, 163–169.

Hurek T, Reinhold B, Niemann E-G and Fendrik I 1987b N_2-dependent growth of *Azospirillum* spp. in batch cultures at low concentrations of oxygen. In *Azospirillum* IV, Genetics, Physiology, Ecology. Ed. W Klingmüller. pp 115–121. Springer-Verlag, Berlin.

Imhoff J F 1986 Osmoregulation and compatible solutes in eubacteria. FEMS Microbiol. Rev. 39, 57–66.

Jouanneau Y, Meyer C M and Vignais P M 1983 Regulation of nitrogenase activity through iron protein interconversion into an active and an inactive form in *Rhodospeudomonas capsulata*. Biochim. Biophys. Acta 749, 318–328.

Kanemoto R H and Ludden P W 1984 Effect of ammonia darkness and phenazine methosulfate on whole-cell nitrogenase activity and Fe protein modification in *Rhodospirillum rubrum*. J. Bacteriol. 158, 713–720.

Ken Dror S, Preger R and Avi-Dor Y 1986 Role of betaine in the control of respiration and osmoregulation of a halotolerant bacterium. FEMS Microbiol. Rev. 39, 115–120.

Kleiner D and Castorph H 1982 Inhibition of ammonium (methylammonium) transport in *Klebsiella pneumoniae* by glutamine and glutamine analogues. FEBS Lett. 146, 201–203.

Kloepper J W, Leong J, Teintze M and Schroth MN 1980 Enhanced plant growth by siderophores produced by plant growth-promoting rhizobacteria. Nature 286, 885–886.

Krieg N R and Döbereiner J 1984 Genus *Azospirillum. In* Bergey's Manual of Systematic Bacteriology, Vol. 1. Eds. NR Krieg and J G Holt. pp. 94–104. Williams & Wilkins, Baltimore, London.

Krieg N R and Hoffman P S 1986 Microaerophily and oxygen toxicity. Annu. Rev. Microbiol. 40, 107–130.

Kush A, Elmerich C and Aubert J P 1985 Nitrogenase of *Sesbania Rhizobium* strain ORS 571: Purification, properties and 'switch off' by ammonia. J. Gen. Microbiol. 131, 1765–1777.

Ladha J K, Triol A C, Daroy M L, Nayak D N, Caldo G and Watanabe I 1986 Degrading straw as source of carbon and energy for nitrogen-fixing organisms in wetland rice soil. Proceedings of the Fourth International Symposium on Microbial Ecology 1986, Ljubljana (*In press*).

Leong J 1986 Siderophores: Their biochemistry and possible role in the biocontrol of plant pathogens. Annu. Rev. Phytopathol. 24, 187–209.

Lowery R G, Saari L L and Ludden P W 1986 Reversible regulation of nitrogenase iron protein from *Rhodospirillum rubrum* by ADP-ribosylation *in vitro*. J. Bacteriol. 166, 513–518.

Ludden P W and Burris R H 1976 Activating factor for the iron protein of nitrogenase from *Rhodospirillum rubrum*. Science 194, 424–426.

Ludden P W, Okon Y and Burris R H 1978 The nitrogenase system of *Spirillum lipoferum*. Biochem. J. 173, 1001–1003.

Magalhaes R M, Baldani J I, Souto SM, Kuykendall J R and Döbereiner J 1983 A new acid-tolerant *Azospirillum* species. An. Acad. Brasil. Cienc. 55, 417–430.

Malik K A and Schlegel H G 1981 Chemolithoautotrophic growth of bacteria able to grow under N$_2$-fixing conditions. FEMS Microbiol. Lett. 11, 63–67.

Mandimba G, Heulin T, Bally R, Guckert A and Balandreau J 1986 Chemotaxis of free-living nitrogen-fixing bacteria towards maize mucilage. Plant and Soil 90, 129–139.

Martinez-Drets G, Fabiano E and Cardona A 1985 Carbohydrate catabolism in *Azospirillum amazonense*. Appl. Environ. Microbiol. 50, 183–185.

Martinez-Drets G, del Gallo M, Burpee C L and Burris R H 1984 Catabolism of carbohydrate and organic acids by the azospirilla. J. Bacteriol. 159, 80–85.

Maulik P and Ghosh S 1986 NADPH/NADH-dependent cold labile glutamate dehydrogenase of *Azospirillum brasilense*. Eur. J. Biochem. 155, 595–602.

Narula N and Gupta K G 1986 Ammonia excretion by *Azoto-

bacter chroococcum* in liquid culture and soil in the presence of manganese and clay minerals. Plant and Soil 93, 205–209.

Nelson L M and Knowles R 1978 Effect of oxygen and nitrate on nitrogen fixation and denitrification by *Azospirillum brasilense* grown in continuous culture. Can. J. Microbiol. 24, 1395–1403.

Neilands J B 1984 Methodology of siderophores. Structure Bonding 58, 1–24.

Neilands J B and Leong S A 1986 Siderophores in relation to plant growth and disease. Annu. Rev. Plant Physiol. 37: 187–208.

Nur I, Okon Y and Henis Y 1982 Effect of dissolved oxygen tension on production of carotenoids, poly-β-hydroxybutyrate, succinate oxidase and superoxide dismutase by *Azospirillum brasilense* Cd grown in continuous culture. J. Gen. Microbiol. 128, 2937–2943.

Nur I, Steinitz Y L, Okon Y and Henis Y 1981 Carotenoid composition and function in nitrogen-fixing bacteria of the genus *Azospirillum*. J. Gen. Microbiol. 122, 27–32.

Okon Y 1985a The physiology of *Azospirillum* in relation to its utilization as inoculum for promoting growth of plants. *In* Nitrogen Fixation and CO$_2$ Metabolism. Eds. P W Ludden and J E Burris. pp. 165–174.Elsevier, New York.

Okon Y 1985b *Azospirillum* as a potential inoculant for agriculture. Trends Biotechnol. 3, 223–228

Okon Y, Houchins J P, Albrecht S L and Burris R H 1977 Growth of *Spirillum lipoferum* at constant partial pressures of oxygen, and the properties of its nitrogenase in cell-free extracts. J. Gen. Microbiol. 98, 87–93.

Okon Y and Kapulnik Y 1986 Development and function of *Azospirillum*-inoculated roots. Plant and Soil 90, 3–16.

Page W J and Dale P L 1986 Stimulation of *Agrobacterium tumefaciens* growth by *Azotobacter vinelandii* ferri-siderophores. Appl. Environ. Microbiol. 51, 451–454.

Patriquin D G, Döbereiner J and Jain D K 1983 Sites and process of association between diazotrophs and grasses. Can. J. Microbiol. 29, 900–915.

Pedrosa F O and Yates M G 1984 Regulation of nitrogen fixation (*nif*) genes of *Azospirillum brasilense* by *nif*A and *ntr* (*gln*) type gene products. FEMS Microbiol. Lett. 23, 95–101.

Pope M R, Murrel S A and Ludden P W 1985 Covalent modification of the iron protein of nitrogenase from *Rhodospirillum rubrum* by adenosine diphosphoribosylation of a specific arginyl residue. Proc. Natl. Acad. Sci. USA 82, 3173–3177.

Postgate J R 1982 The Fundamentals of Nitrogen Fixation, pp 60–102. Cambridge University Press, Cambridge.

Postgate J·R, Kent H M, Hill S and Blackburn H 1985 Nitrogen fixation by *Desulfovibrio gigas* and other species of *Desulfovibrio. In* Nitrogen Fixation and CO$_2$ Metabolism. Eds. P W Ludden and J E Burris. pp 225–234. Elsevier Science Publishers, Amsterdam.

Powell P E, Cline G R, Reid C P P and Szaniszlo P J 1980 Occurrence of hydroxamate siderophore iron chelators in soils. Nature 287, 833–834.

Ramos J L and Robson RL 1985 Lesions in citrate synthase that affect aerobic nitrogen fixation by *Azotobacter chroococcum*. J. Bacteriol. 162, 746–751.

Rao A V and Venkateswarlu B 1985 Salt tolerance of *Azospirillum brasilense*. Acta Microbiol. Hung. 32, 221–224.

Ratti S, Curti B, Zanetti G and Galli E 1985 Purification and characterization of glutamate synthase from *Azospirillum brasilense*. J. Bacteriol. 163, 724–729.

Reich S, Almon H and Böger P 1986 Short-term effect of ammonia on nitrogenase activity of *Anabaena variabilis* (ATCC 29413). FEMS Microbiol. Lett. 34, 53–56.

Reinhold B, Hurek T and Fendrik I 1985 Strain-specific chemotaxis of *Azospirillum* spp. J. Bacteriol. 162, 190–195.

Reinhold B, Hurek T, Fendrik I, Pot B, Gillis M, Kersters K, Thielemans S and De Ley J 1987 *Azospirillum halopraeferens* sp. nov., a nitrogen-fixing organism associated with roots of Kallar grass (*Leptochloa fusca* (Linn.) Kunth). Int. J. Syst. Bacteriol. 37, 43–51.

Robson R L and Postgate J R 1980 Oxygen and hydrogen in biological nitrogen fixation. Annu. Rev. Microbiol. 34, 183–207.

Römheld V and Marschner H 1986 Evidence for a specific uptake system for iron phytosiderophores in roots of grasses. Plant Physiol. 80, 175–180.

Saari L L, Triplett E W and Ludden P W 1984 Purification and properties of the activating enzyme for iron protein of nitrogenase from the photosynthetic bacterium *Rhodospirillum rubrum*. J. Biol. Chem. 259, 15502–15508.

Sampaio M J A M, Pedrosa F O and Döbereiner J 1982 Growth of *Derxia gummosa* and *Azospirillum* spp. on C_1-compounds. An. Acad. brasil. Cienc. 54, 457–458.

Sandhu G R, Aslam Z, Salim M, Sattar A, Qureshi R H, Ahmad N and Wyn Jones RG 1981 The effect of salinity on the yield and composition of *Diplachne fusca* (Kallar grass). Plant Cell Environ. 4, 177–181.

Saxena B, Modi M and Modi V V 1986 Isolation and characterization of siderophores from *Azospirillum lipoferum* D-2. J. Gen. Microbiol. 132, 2219–2224.

Singh H N, Singh R K and Sharma R 1983 An L-methionine-D,L-sulfoximine-resistant mutant of the cyanobacterium *Nostoc muscorum* showing inhibitor-resistant δ-glutamyl-transferase, defective glutamine synthetase and producing extracellular ammonia during N_2-fixation. FEBS Lett. 154, 10–14.

Smith M J and Neilands J B 1984 Rhizobactin, a siderophore from *Rhizobium meliloti*. J. Plant Nutr. 7, 449–458.

Song S D, Hartmann A and Burris R H 1985 Purification and properties of the nitrogenase of *Azospirillum amazonense*. J. Bacteriol. 164, 1271–1277.

Spiller H, Latorre C, Hassan M E and Shanmugam KT 1986 Isolation and characterization of nitrogenase-derepressed mutant strains of Cyanobacterium *Anabaena variabilis*. J. Bactriol. 165, 412–419.

Stal L J and Krumbein W E 1985 Oxygen protection of nitrogenase in the aerobically nitrogen fixing, nonheterocystous cyanobacterium *Oscillatoria* sp. Arch. Microbiol. 143, 72–76.

Sugiura Y and Nomoto K 1984 Phytosiderophores, structure and properties of mugineic acids and their metal complexes. Structure Bonding 58, 107–135.

Tarrand J J, Krieg N R and Döbereiner J 1978 A taxonomic study of the *Spirillum lipoferum* group, with descriptions of a new genus, *Azospirillum* gen. nov. and two species, *Azospirillum lipoferum* (Beijerinck) comb. nov. and *Azospirillum brasilense* sp. nov. Can. J. Microbiol. 24, 967–980.

Tien T M, Diem H G, Gaskins M H and Hubbell D H 1981 Polygalacturonic acid transeliminase production by *Azospirillum* species. Can. J. Microbiol. 27, 426–431.

Tilak K V B R, Schneider K and Schlegel H G 1986 Autotrophic growth of nitrogen-fixing *Azospirillum* species and partial characterization of hydrogenase from strain CC. Current Microbiology 13, 291–297.

Turpin D H, Edie S A and Canvin D T 1984 *In vivo* nitrogenase regulation by ammonium and methylamine and the effect of MSX on ammonium transport in *Anabaena flos-aguae*. Plant Physiol. 74, 701–704.

Volpon A G T, De-Polli H and Döbereiner J 1981 Physiology of nitrogen fixation in *Azospirillum lipoferum* Br 17 (ATCC 29709). Arch. Microbiol. 128, 371–375.

Whiting G J, Gandy E L and Yoch D C 1986 Tight coupling of root-associated nitrogen fixation and plant photosynthesis in the salt marsh grass *Spartina alterniflora* and carbon dioxide enhancement of nitrogenase activity. Appl. Environ. Microbiol. 52, 108–113.

Waring W S and Werkman C H 1943 Growth of bacteria in an iron-free medium. Arch. Biochem. 1, 303–310.

Whipps J M and Lynch J M 1986 The influence of the rhizosphere on crop productivity. *In* Advances in Microbial Ecology 9. Ed. KC Marshall. pp 187–244. Plenum Press, New York.

Whipps J M and Lynch J M 1983 Substrate flow and utilization in the rhizosphere of cereals. New Phytol. 95, 605–623.

Wyn Jones R G and Storey R 1981 Betaines. *In* Physiology and Biochemistry of Drought Resistance in Plants. Eds. L G Paleg and D Aspinall. pp 171–204. Sydney, Academic Press Australia.

Yancey P H, Clark M E, Hand SC, Bowlus R D and Somero G N 1982 Living with water stress: Evolution of osmolyte systems. Science 217, 1214–1222.

Yates M G and Jones C W 1974 Respiration and nitrogen fixation in *Azotobacter*. *In* Advances in Microbial Physiology 11. Eds. AH Rose and D W Tempest. pp 97–135. Academic Press, New York.

Yoch D C and Whiting G J 1986 Evidence for NH_4^+ switch-off regulation of nitrogenase activity by bacteria in salt marsh sediments and roots of the grass *Spartina alterniflora*. Appl. Environ. Microbiol. 51, 143–149.

Zuberer D A and Alexander D B 1986 Effects of oxygen partial pressure and combined nitrogen on N_2-fixation (C_2H_2) associated with *Zea mays* and other gramineous species. Plant and Soil 90, 47–58.

Zumft W G 1985 Regulation of nitrogenase activity in the anoxygenic phototrophic bacteria. *In* Nitrogen Fixation Research Progress. Eds. H J Evans, P J Bottomley and W E Newton. pp 551–557. Martinus Nijhoff Publishers, Dordrecht, The Netherlands.

F. A. Skinner et al. (Eds.), Nitrogen fixation with non-legumes, 137–145.
© 1989 by Kluwer Academic Publishers.

The phytohormonal interactions between *Azospirillum* and wheat

W. ZIMMER and H. BOTHE*
Botanisches Institut, Universität Köln, Gyrhofstr. 15, D-5000 Köln 41, FRG

Key words: *Azospirillum*, bacteria-plant association, cytokinin, gibberellin, indoleacetic acid, phytohormones

Abstract

A simple model system was designed to detect positive effects of *Azospirillum* on the root growth of cereals. Cultures of *A. brasilense* Sp7 and *A. lipoferum* Sp59 did not excrete gibberellins and cytokinins in the logarithmic and in the early stationary growth phase. Indoleacetic acid (IAA) was formed, however, only in the stationary phase of the cultures. The addition of D,L-tryptophan to the medium enhanced the formation of IAA. A further, still unidentified substance was produced by *Azospirillum* under denitrifying conditions in the logarithmic growth phase. The substance was almost twice as active as IAA in increasing the wet weight of wheat root segments. It is suggested that this unidentified substance is the major stimulus affecting the growth of cereals.

Introduction

There is currently much interest in soil bacteria of the genus *Azospirillum*. *Azospirillum* is not a major constituent of soils of the temperate zone where the genera *Pseudomonas* and *Enterobacter* are more dominant (Lindberg and Granhall, 1984; Kleeberger *et al.*, 1983). For tropical soils, however, numbers as high as 10^4–10^7 *Azospirillum* cells g^{-1} dry weight of roots have been determined. *Azospirillum* lives in association with grasses (wheat, corn, rice, sugar cane, *Sorghum*). It seems to colonize the inner cortex of the roots, the root cap and/or a zone a few mm away from the root surface (Döbereiner, 1983b; Elmerich, 1984; Okon, 1985; Whallon *et al.*, 1985).

Azospirillum is thought to improve the nitrogen supply to cereals by means of N_2-fixation and thereby to enhance crop productivity. However, *Azospirillum* can perform N_2-fixation only under microaerobic conditions. A direct transfer of products of N_2-fixation (NH_4^+, glutamate, glutamine or other compounds) from *Azospirillum* to plants has never been demonstrated. *Azospirillum*

* Dedicated to Professor E.-G. Niemann, Hannover, on the occasion of his 60th birthday.

could have a shorter life span than the cereals, and could perhaps decompose before the grain-filling stage and so improve crop productivity by its mineralization. However, most investigators agree that the number of *Azospirillum* cells in roots, though considerable, is too small to enhance crop yield significantly by means of N_2-fixation and mineralization (Zimmer *et al.*, 1987).

Azospirillum also deserves special attention because of its denitrification properties (Bothe *et al.*, 1981; Nelson and Knowles, 1978). When oxygen is limiting in cultures, *Azospirillum* can utilize either nitrate, nitrite or nitrous oxide as terminal respiratory electron acceptors. Experimental conditions were established for both batch and continuous cultures where the bacterium anaerobically reduced nitrate to nitrite, nitrous oxide or molecular nitrogen, nitrite to nitrous oxide or molecular nitrogen, and nitrous oxide to molecular nitrogen. Growth yield measurement indicated that the energy yields are comparable between O_2 and N_2O as respiratory electron acceptors. They are 1/3 lower with nitrite and 2/3 lower with nitrate than with O_2 (Danneberg *et al.*, 1985; Zimmer *et al.*, 1987). Model experiments with wheat indicated that *Azospirillum* performs either nitrogen fixation

(C_2H_2-reduction) or denitrification (N_2O-formation) depending on the assay conditions (Neuer *et al.*, 1985). *Azospirillum* may contribute significantly to NO_x production in soils particularly when the concentration of nitrate is high and that of O_2 is limiting.

Azospirillum also produces phytohormones and this production may enable *Azospirillum* to enhance plant growth even when occuring in small quantities in roots. Tien *et al.*, (1979) showed that *Azospirillum* produced indoleacetic acid (IAA) and indolelactic acid when the medium contained tryptophan. The authors also stated that the bacterium produced small amounts of gibberellins and three cytokinin-like compounds. These compounds remained to be identified. Tien *et al.*, (1979) showed that the phytohormones produced by *Azospirillum* stimulated the formation of additional root hairs and of lateral roots with pearl millet (*Pennisetum americanum*). Hartmann *et al.*, (1983) isolated mutants of *A. brasilense* Cd which formed thirty times more IAA than the wild strain. Some mutants also produced IAA independently of tryptophan in the medium. The effect of *Azospirillum* on the formation of root hairs and lateral roots is due not only to IAA but, probably, to still unidentified phytohormones. Such a conclusion follows from the experiments of Kapulnik *et al.*, (1985) who showed that the overproducing mutants of *A. brasilense* Cd had the same effects on root hair and lateral root formation as the wild strain.

In the current study, cultures of the two type strains, *A. brasilense* Sp7 and *A. lipoferum* Sp59, were assayed for phytohormone production. The medium in which *Azospirillum* had lived was tested for gibberellin, cytokinin and indoleacetic acid content using classical plant assays for these phytohormones. In addition, a simple assay was designed to detect stimulatory effects of bacteria on the root growth of cereals.

Materials and methods

Organisms and growth

The two *Azospirillum* strains used, *Azospirillum brasilense* Sp7 (ATCC 29145) and *A. lipoferum* Sp59 (ATCC 29707), were grown in 100 ml and 1 l flasks in a shaking water bath (60 strokes min^{-1}) at

30°C for 24 h under batch culture conditions and in a medium containing (gl^{-1} water): K_2HPO_4, 0.78; KH_2PO_4, 0.61; $MgSO_4 \cdot 7H_2O$, 0.2; $FeSO_4 \cdot 7H_2O$, 0.00625; EDTA (= Titriplex III), 0.0093; $Na_2MoO_4 \cdot 2H_2O$, 0.02; $MnSO_4 \cdot H_2O$, 0.01; malate (adjusted to pH 7.0 with KOH), 5.0; KNO_3, 2.0. In the cultures grown with NH_4^+ as the nitrogen source, KNO_3 was replaced by 0.265 gl^{-1} NH_4Cl. The *A. lipoferum* cultures were supplemented with a sterile solution of biotin (10^{-5} gl^{-1}). The cultures had an optical density of 1.0–1.5 at 560 nm after 24 h of growth under sterile conditions.

Assay to test the stimulatory effects of Azospirillum *on root growth*

A surface sterilized (see Neuer *et al.*, 1985) grain of spring wheat (variety Ralle) was placed in an Eppendorf plastic tube of which the tip had been excised. The tube was placed in a 25 ml test tube on top of 10 ml of the wheat medium described by Neuer *et al.*, (1985) containing 3.5 mM $NaNO_3$, but no agar in this case. In the experiments with *Azospirillum* (see Fig. 1), the test tubes were supplemented with 10 μl of a 1-day-old culture of *A. brasilense* Sp7 with an optical density of 1.43 at 560 nm. All manipulations were performed under sterile conditions. After incubating at 23°C for 10 d (14 h in the light, 10 h in the dark per day), the dry weight of the excised roots was determined.

Phytohormone tests

Assay for cytokinins. The *Amaranthus* test used is based on the cytokinin-induced formation of beta-cyanins in the cotyledons and hypocotyls of this plant incubated in the dark in the presence of tyrosine, as described by Biddington and Thomas (1973). Seeds of *Amaranthus caudatus* (C. Walz Seed Co., Stuttgart-Feuerbach) were sown in 10 cm × 2 cm Petri dishes on a double layer of filter paper moistened with 5 ml distilled H_2O and incubated for 4 d in the dark at 25°C. The seed coats and the roots were then removed, and the upper part of each germinated plant, consisting of the hypocotyl plus the cotyledons (10 pieces), were placed on a double layer of filter paper in a 25 ml Erlenmeyer flask. The filter paper was moistened with 2 ml of a solution containing 13.3 mM potassium phosphate buffer pH 8.0, 2 mg tyrosine and the cytokinin solution to be assayed. The Erlenmeyer flasks were incubated at 25°C for 18 h in the dark and the *Amaranthus* samples were then removed and transferred to 3 ml of distilled water. The samples were frozen and re-thawed twice, and the quantity of betacyanin released was determined by calculating the differences between the absorbances at 542 nm and 620 nm which correlate with the cytokinin concentration. The detection limit was 0.5 ng cytokinin ml^{-1} *Azospirillum* culture.

Assay for auxins. The cell elongation of wheat root segments was shown to be dependent on the presence of indoleacetic acid and can be taken as a test for determining the amount of auxin in a solution (Libbert, 1957). Spring wheat grains (var. Ralle) were watered with 10 ml H$_2$O for 24 h at 4°C in a Petri dish. For germination, they were placed on two layers of filter paper moistened with 6 ml H$_2$O in a Petri dish and incubated for 72 h at 26°C in the dark. Twelve wheat root segments of 5 mm length, taken from the zone 2–7 mm away from the tip, were transferred to a Petri dish containing one layer of filter paper moistened with 2.5 ml of the solution to be assayed for auxin content. The segments were incubated for 24 h at 26°C in the dark, and the wet weight of the root segments was then determined. This test is a direct assay of the effect of auxins on the growth of roots with a detection limit of 1 pg indoleacetic acid ml^{-1} of culture and gives statistically sound data.

Assay for gibberellins. The method described by Jones and Varner (1967) was used as the bioassay for gibberellins. Barley seeds were cut in halves transversely. The embryo was removed and the halves were sterilized with an 1% NaOCl solution for 5 min under vacuum, followed by three washings in sterile distilled water. The sterilized halves were incubated for 4 d in sterile Petri dishes with 4 ml H$_2$O at 4°C in the dark. Ten half seeds for each vessel were transferred to autoclaved 7.2 ml Fernbach flasks each containing 2 ml 1 mM sodium acetate buffer pH 4.8, 10 mM CaCl$_2$ and the gibberellin solution to be tested. After incubation of this mixture for 48 h at 30°C in a shaking water bath, the medium was decanted into centrifuge tubes, and the Fernbach flasks were washed with

2 ml distilled water. The combined solution (medium and washing) was centrifuged (10 min, 10,000 g). The assay for α-amylase activity in the supernatant could be performed in the same way as by Jones and Varner (1967), but it was easier to adopt the procedure described in the Sigma diagnostic leaflet no. 576. In this, 0.6 ml of the Sigma reaction mixture containing mainly *p*-nitrophenyl-D-maltoheptaoside as substrate and 0.2 ml of the supernatant were mixed. After 1 min, the formation of *p*-nitrophenol was monitored by reading the absorbance changes at 420 nm. This gibberellin-induced release of α-amylase allows a determination of 25 pg gibberellin ml^{-1} of culture as the detection limit.

Separation of the auxins by HPLC

The separation was performed by the use of a Novapak C$_{18}$ (0.8 × 10 cm) reversed phase column with 5 μm particle size using a Waters model 510 HPLC. The solvent was methanol/1% H$_3$PO$_4$ in water = 40/60(v/v), the flow rate was 1 ml min^{-1} and the injection volume was 25 μl. The peak areas (u.v.-detector at 278 nm) were calculated by using an arbitrary unit for the unidentified substance (see Results). One peak area unit corresponded to 1/4096 optical density unit at 278 nm × 1/20 sec. The unidentified substance had a retention time of 162 sec and IAA had 306 sec under these conditions. IAA could be quantified by standards.

Results

A new test to assay the effects of Azospirillum on root growth

Tien *et al.*, (1979) reported that the root system of pearl millet responded to inoculation of *A. brasilense* Sp13t SR2 by forming additional root hairs and lateral roots. The same effect can also be seen in a simple and convenient test system described in detail under Materials and methods. Figure 1 indicates that 10-d-old wheat plants had formed much longer roots, and more root hairs and lateral roots after the incubation with *Azospirillum* in a medium containing nitrate. These assays were performed with either *A. brasilense* Sp7 or *A.*

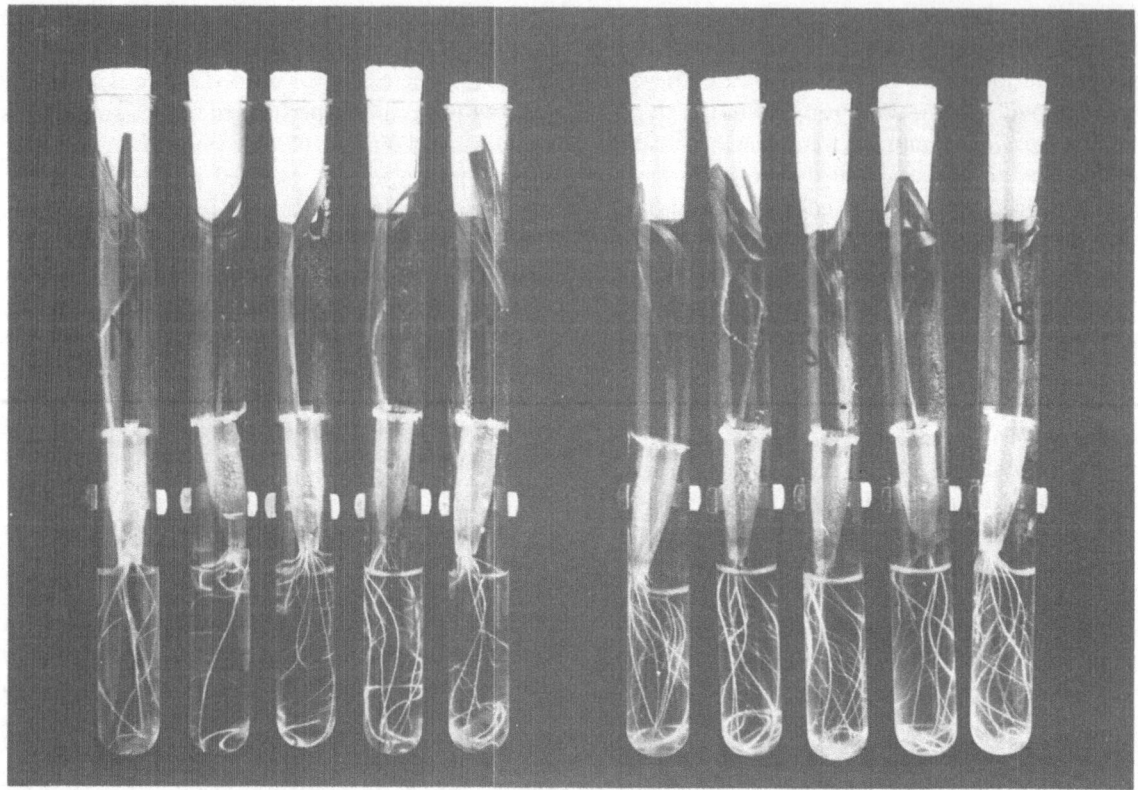

Fig. 1. The stimulatory effect of *Azospirillum* on the growth of wheat roots.
A wheat plant was grown for 10 d in the presence of *A. brasilense* Sp7 (5 test tubes at the right side of the photo). The controls without *A. brasilense* are the 5 tubes at the left side. For experimental details see Materials and methods.

Table 1. Test for cytokinins in the supernatant of *Azospirillum*

Assay conditions	Absorbance difference $E_{542}-E_{620}$
1. control without cytokinin	0.024
2. control with	
a) 0.2 ng kinetin/assay	0.036
b) 200 ng kinetin/assay	0.060
c) 2000 ng kinetin/assay	0.123
3. supernatant of *A. brasilense* Sp7	
a) of an 1-day-old culture	0.023
b) of a 3-days-old culture	0.024
4. supernatant of *A. lipoferum* Sp59	
a) of an 1-day-old culture	0.024
b) of a 3-days-old culture	0.022

The test for cytokinin was performed by measuring the production of betacyanin in seedlings of *Amaranthus caudatus*. The standard was kinetin (6-furfurylaminopurine, Sigma no. K 2751). Both *Azospirillum* cultures had O.D.$_{560\,nm}$ = 1.3 after 1 d and 1.5 after 3 d. The cultures were centrifuged, and 0.2 ml of the supernatant was assayed for cytokinin content. The method would have allowed detection of 100 pg cytokinin 0.2 ml^{-1}. Thus the cultures contained less than 0.5 ng cytokinin ml^{-1}.

lipoferum Sp59 with the same results. The dry weight of roots excised afterwards gave 8.55 ± 1.74 mg for 5 plants incubated in the presence of *Azospirillum brasilense* for 10 d, and 6.30 ± 0.74 mg for 5 plants grown without the bacteria for the same time. Using the Student *t*-test, the confidence in the difference of these data was 98.9%.

Assays for cytokinins and gibberellins

The *Amaranthus* assay was chosen for determining the amount of cytokinins by *Azospirillum*. Cytokinin induces the formation of betacyanin in the cotyledons and hypocotyls of this plant. Betacyanin can easily be extracted and measured quantitatively with high sensitivity. Table 1 indicates that betacyanin formation by the plants was, indeed, dependent on the addition of low amounts of kinetin (6-furfurylaminopurine) as a control.

Table 2. Test for gibberellins in the supernatant of *Azospirillum*

Assay conditions	$\Delta E_{420\,nm}/min$
1. control without gibberellins	0.000
2. control with 10 pg GA_3/assay	0.015
3. control with 1000 pg GA_3/assay	0.273
4. supernatant of *A. brasilense* Sp7	
of an 1-day-old culture	0.000
of a 3-days-old culture	0.000
5. supernatant of *A. lipoferum* Sp59	
of an 1-day-old culture	0.000
of a 3-days-old culture	0.000

The test for gibberellins was performed with 10 sterilized endosperm halves of barley seeds. Both *Azospirillum* cultures were grown aerobically with nitrate as the nitrogen source and had an $O.D._{560\,nm}$ of 1.3 after 1 d and of 1.5 after 3 d. The cultures were centrifuged, and 0.2 ml of the supernatant was assayed for gibberellin content. The method would have allowed to detect 5 pg gibberellin 0.2 ml^{-1}. Thus the cultures contained less than 25 pg gibberellin ml^{-1}.

There were, however, no indications for the release of cytokinins into the medium by *Azospirillum brasilense* Sp7 and *A. lipoferum* Sp59 when either 1- or 3-d-old cultures were assayed (Table 1).

The production of α-amylase in barley endosperm is a sensitive bioassay for gibberellins. This test has the advantage of correlating the concentration of gibberellin with the activity of an enzyme which is released specifically upon the addition of this phytohormone. The test is also insensitive to solvents (Jones and Varner, 1967). Table 2 indicates that the addition of GA_3 as a control, indeed resulted in the formation of high α-amylase activity. In contrast, the medium of either 1- or 3-d-old *Azospirillum* cultures did not contain detectable levels of gibberellins.

Auxin formation by Azospirillum

Azospirillum brasilense Sp7 had a generation time of 5.5 h under the growth conditions of the present study and the stationary phase was reached within 1 d (Fig. 2). An excretion of IAA was not detectable until 3 d after inoculation. The maximal production was 15 μM IAA after 4–5 d in this culture which then had an optical density of 1.5 at

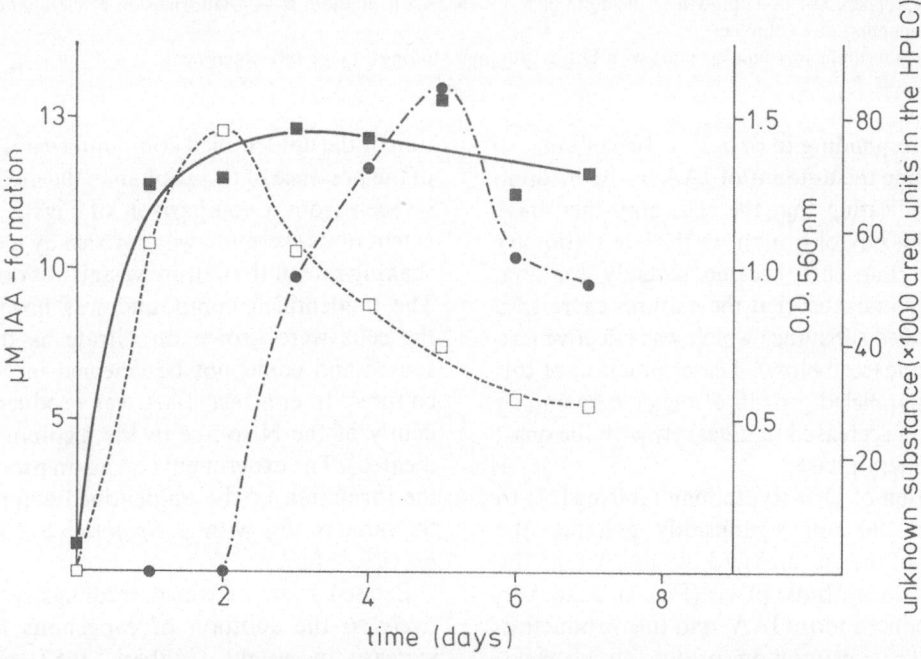

Fig. 2. The formation of IAA and of the unidentified substance by *A. brasilense* Sp7 under batch culture conditions.
■——■ growth of the culture ($O.D._{560\,nm}$); ●–·–·–● IAA-formation; □– – – –□ formation of the unknown substance. The experimental details are described under Materials and methods.

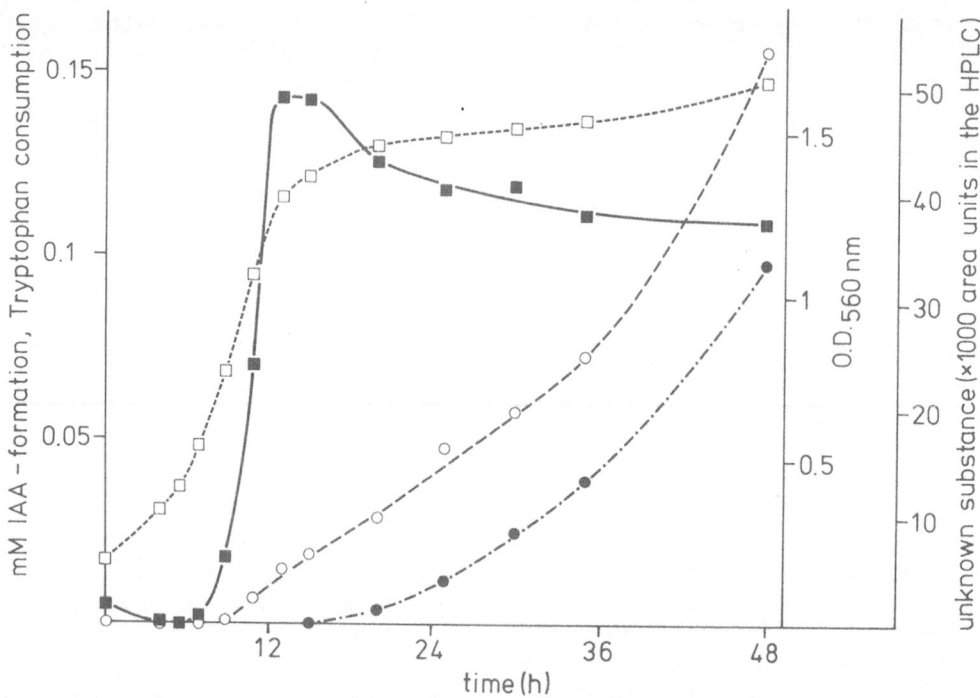

Fig. 3. The formation of IAA and of the unidentified substance by *A. brasilense* Sp7 under batch culture conditions grown in the presence of tryptophan. □----□ growth of the culture; ●-·-·-● IAA-formation; ■——■ formation of the unknown substance. ○———○ consumption of tryptophan
The *Azospirillum* medium was supplemented with D,L tryptophan (100 mg l⁻¹) for this experiment.

560 nm (corresponding to *ca.* 8.3×10^8 cells ml⁻¹). After that time the amount of IAA in the medium decreased indicating that the cells may then have re-utilized this phytohormone in the late stationary phase when nutrients became severely limiting. Figure 2 also indicates that the cultures excreted a still unidentified substance which was effective as a phytohormone (see below). The production of this compound paralleled growth of the culture and its concentration decreased significantly with the onset of the stationary phase.

The addition of D,L-tryptophan (100 mg l⁻¹) to the medium did not significantly enhance the growth rate of the cultures and did not extend the length of the logarithmic phase (Fig. 3). However, the cells produced more IAA, and this production paralleled the consumption of the amino acid. After 2 d, the medium contained 0.1 m*M* IAA and thus 6–8 times more than was produced maximally by a culture grown without tryptophan. The forma-

tion of the unidentified compound was independent of the presence of tryptophan in the medium, as can be seen from a comparison of Figs 2 and 3. The extent of its excretion was reduced by increasing the shaking rate of the culture vessel (not documented). The unidentified compound was produced when the cells were grown on nitrate as the nitrogen source and could not be detected in NH₄⁺ grown cultures. In contrast, IAA was produced independently of the N-source in the medium (not documented). The experiments on auxin production and the formation of the unidentified compound gave the same results with *A. lipoferum* Sp59 as with *A. brasilense* Sp7.

Excised roots of wheat seedlings respond positively to the addition of exogenous IAA by an increase in weight (Libbert, 1957) and by the formation of additional lateral roots (Blakely *et al.*, 1986). The addition of IAA largely enhanced the wet weight of wheat root segments under the con-

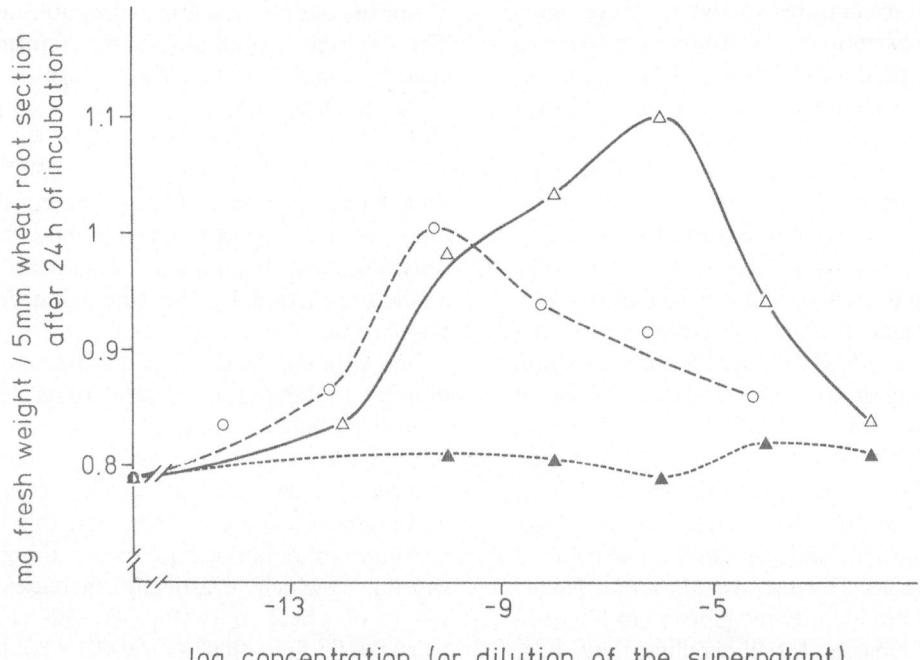

Fig. 4. Effects of IAA and of the unknown substance on wheat root segments.
o— — —o increase of the wet weight by IAA (IAA concentration in *M*); △———△ increase by the supernatant (obtained by growing *A. brasilense* Sp7 for 1 d in the *Azospirillum* medium and in the presence of NO_3^- and by removing the cells by centrifugation); ▲– – – –▲ supernatant of a culture grown in the presence of NH_4^+.

ditions employed in the present study. A curve with a pronounced optimum and with higher concentrations being inhibitory is typical for a phytohormonal response, and this was also obtained in Fig. 4 by using either IAA or the unidentified substance. Remarkably, the unidentified compound was almost twice as active as IAA in this assay. It was formed only when the cells were grown with nitrate as the nitrogen source in the medium. Supernatant from NH_4^+-grown cells completely failed to stimulate the growth of wheat root segments (Fig. 4).

Discussion

The simple test designed in the present study to detect stimulatory effects of *Azospirillum* on the growth of wheat roots gives impressive results. Profound differences between plants inoculated with and without bacteria can be seen even by eye after a week. Thus this test confirms the reports about positive effects of *Azospirillum* on the growth of plants published by others (Döbereiner, 1983; Elmerich, 1984; Kapulnik *et al.*, 1985; Okon, 1985). This rapid test may also replace more complicated assays described earlier (Tien *et al.*, 1979; Thomas-Bauzon *et al.*, 1982; Neuer *et al.*, 1985). It might also be suitable for the assay of many kinds of associations between plants and bacteria.

The present study shows that *A. brasilense* Sp7 and *A. lipoferum* Sp59 do not produce gibberellins and cytokinins in the logarithmic and early stationary growth phase. The bioassays used here allowed the detection of 0.5 ng cytokinin ml^{-1} and 25 pg gibberellin ml^{-1} culture. It could well be that *Azospirillum* had produced the phytohormones in concentrations under these detection limits. The radioimmunoassay for phytohormones as a more sensitive assay (Weiler, 1984) was not tried in the present study, because very low production might be insufficient to have a stimulatory effect on the root growth of cereals even when the population of bacteria in the roots is dense. In the late stationary phase or when the cultures are dying, *Azospirillum* and any other bacterium may produce cytokinin as a degradation product of DNA or RNA and may possibly also form gibberellins. This was not investigated in this study, because such conditions can hardly be controlled in any association between bacteria and cereals.

The results of this communication are contradictory to those of Tien *et al.*, (1979) who reported the excretion of gibberellins and cytokinins by *Azospirillum*. Several facts may explain this discrepancy. Tien *et al.*, (1979) used a different strain which grew very slowly. The stationary phase of their culture was not reached before 10 d (see Fig. 7 in Tien *et al.*, 1979), and the cell density was only 1/3 of that obtained with the cultures used here. The formation of phytohormones by *Azospirillum* may be strain specific and growth-phase dependent as noted for auxins (Hartmann *et al.*, 1983). Tien *et al.*, (1979) did not identify the gibberellin-like substance. As the bioassay for gibberellins, they chose the increase of the growth of lettuce hypocotyls which was described to be gibberellin specific (Frankland and Warening, 1960). However, elongations of hypocotyls are often also stimulated by auxins. In the present study, therefore, the gibberellin dependent release of α-amylase activity in barley endosperms was chosen as specific assay where gibberellin cannot be substituted by other phytohormones. Tien *et al.*, (1979) also did not identify the cytokinins, and almost all *n*-butanol fractions were stimulatory in the chlorophyll retention assay which they chose for assaying these plant growth substances. The tests used in the present study allowed us to detect 1/2000 of the gibberellin and 1/2 of the cytokinin found in the medium of *A. brasilense* Sp13t SR2 by Tien *et al.*, (1979). Since we could not find any cytokinin and gibberellin in our cultures, we did not attempt to purify these compounds. The tests in the present study were performed directly with the medium in which *Azospirillum* had lived. Control experiments indicated that the salts of the medium did not interfere with the bioassays.

The present communication confirms that *A. brasilense* Sp7 produces IAA. Significant amounts are, however, only formed in the late stationary growth phase. Within a detection limit of 1 pg ml^{-1} culture, there was no evidence for the phytohormone production in the logarithmic phase (Fig.

2) nor in continuous cultures (unpublished results). The present communication confirms earlier reports that *Azospirillum* forms IAA in larger quantities when the medium is supplemented with tryptophan (Tien *et al.*, 1979; Reynders and Vlassak, 1979; Hartmann *et al.*, 1983). We suspect that the production of IAA in the cultures of the late stationary growth phase comes from dying cells releasing tryptophan. This may stimulate auxin production by the still living cells in the population.

The stimulus involved in the formation of additional root hairs and lateral roots needs to be identified. Auxin produced by *Azospirillum* may particularly cause the formation of lateral roots as observed for several plants (see Blakely *et al.*, 1986 and references therein). However, this laboratory attributes more importance to the still unidentified substance which drastically increases the wet weight of wheat roots (Fig. 4). This is concluded from the findings that *Azospirillum* could exert its effects on the wet weight increase of roots when the medium contained nitrate but not ammonia as the only N-source in the assay system used in Fig. 1 (not documented). *Azospirillum* excretes this compound only when grown on nitrate, and oxygen suppresses its formation. Denitrifying conditions apparently favour its formation. A publication describing its characteristics is in preparation.

Acknowledgement

This work was kindly supported by a grant from the Bundesministerium für Forschung und Technologie (no. 0318961 A6 of the Projektträger Biologie, Ökologie und Energie der KFA Jülich).

References

Biddington NL and Thomas TH 1973 A modified *Amaranthus* betacyanin bioassay for the rapid determination of cytokinins in plant extracts. Planta (Berl.) 111, 183–186.

Blakely LM, Blakely RM and Galloway CM 1986 Effects of dimethyl sulfoxide and pH on indoleacetic acid induced lateral root formation in the radish seedling root. Plant Physiol. 80, 790–791.

Bothe H, Klein B, Stephan MP and Döbereiner J 1981 Transformations of inorganic nitrogen by *Azospirillum* spp. Arch. Microbiol. 130, 96–100.

Danneberg G, Zimmer W and Bothe H 1985 Some physiological and biochemical properties of denitrification by *Azospirillum brasilense. In* Azospirillum III, Genetics, Physiology, Ecology, Ed. W Klingmüller. pp 127–138, Springer, Berlin, Heidelberg.

Döbereiner J 1983a Dinitrogen fixation in the rhizosphere and phyllosphere associations. *In* Inorganic Plant Nutrition, A. Läuchli and RL Bieleski (eds.), Encycl. Plant Physiol., New Series, Vol. 15A, pp 33–35, Springer, Berlin, Heidelberg, New York.

Döbereiner J 1983b Ten years of *Azospirillum. In* Azospirillum II, Genetics, Physiology, Ecology. Ed. W Klingmüller. pp. 9–23, Birkhäuser, Basel, Boston, Stuttgart.

Elmerich C 1984 Molecular biology and ecology of diazotrophs associated with non-leguminous plants. Bio/Technology 2, 967–978.

Frankland B and Warening PF 1960 Effect of gibberellic acid on hypocotyl growth of lettuce seedlings. Nature (London) 185, 225–226.

Hartmann A, Singh M and Klingmüller W 1983 Isolation and characterization of *Azospirillum* mutants excreting high amounts of indoleacetic acid. Can. J. Microbiol. 29, 916–923.

Jones RL and Varner JE 1967 The bioassay of gibberellins. Planta (Berl.) 72, 155–161.

Kapulnik Y, Okon Y and Henis Y 1985 Changes in root morphology of wheat caused by *Azospirillum* inoculation. Can. J. Microbiol. 31, 881–887.

Kleeberger A, Castorf H and Klingmüller W 1983 The rhizosphere microflora of wheat and barley with special reference to Gram negative bacteria. Arch. Microbiol. 136, 306–311.

Libbert E 1957 Die Regulation des Wurzelwachstums durch synthetische und endogene Inhibitoren. Planta (Berl.) 5, 25–40.

Lindberg T and Granhall U 1984 Isolation and characterization of dinitrogen fixing bacteria from the rhizosphere of temperate cereals and forage grasses. Appl. Environ. Microbiol. 48, 683–689.

Nelson LM and Knowles R 1978 Effect of oxygen and nitrate on nitrogen fixation and denitrification by *Azospirillum brasilense* grown in continuous culture. Can. J. Microbiol. 24, 1395–1403.

Neuer G, Kronenberg A and Bothe H 1985 Denitrification and nitrogen fixation by *Azospirillum* III. Properties of a wheat *Azospirillum* association Arch. Microbiol. 141, 364–370.

Okon Y 1985 *Azospirillum* as a potential inoculant for agriculture. Trends Biotechnol. 3, 223–228.

Reynders L and Vlassak K 1979 Conversion of tryptophan to indoleacetic acid by *Azospirillum* sp. Soil Biol. Biochem. 11, 547–548.

Thomas-Bauzon D, Weinhard P, Villecourt P and Balandreau J 1982 The spermosphere model: Its use in growing, counting and isolating N_2-fixing bacteria from the rhizosphere of rice. Can. J. Microbiol. 28, 922–928.

Tien TM, Gaskin MH and Hubbell DH 1979 Plant growth substances produced by *Azospirillum brasilense* and their effect on the growth of pearl millet (*Pennisetum americanum* L.). Appl. Environ. Microbiol. 37, 219–226.

Weiler EW 1984 Immuno assay of plant growth regulators. Annu. Rev. Plant Physiol. 35, 85–95.

Whallon JH, Acker G and El-Khawas H 1985 Electron microscopy of young wheat roots inoculated with *Azospirillum. In* Azospirillum III. Genetics, Physiology, Ecology. Ed. W Klingmüller. pp 222–229, Springer, Berlin, Heidelberg.

Zimmer W, Roeben K, Danneberg G and Bothe H 1987 The bacterial genus *Azospirillum* and its potential applications. *In* Inorganic Nitrogen Metabolism. Ed. W Ullrich. pp 177–182. Springer, Berlin, Heidelberg.

F. A. Skinner et al. (Eds.), Nitrogen fixation with non-legumes, 147–152.
© 1989 by Kluwer Academic Publishers.

Investigations on growth, nitrogen fixation and ammonium assimilation by a bacterium isolated from rice soil

C. FRITZSCHE and E.-G. NIEMANN
Institut für Biophysik, Universität Hannover, Herrenhäuser Str. 2, D-3000 Hannover 21, FRG

Key words: ammonia metabolism, nitrogen fixation, oxygen, rice

Abstract

Isolate R7, a gram negative nitrogen fixing rod isolated from a paddy field at Suweon, Korea, was grown in continuous culture at several dilution rates and several oxygen concentrations under nitrogen fixing conditions. Growth parameters ($\mu_{max} = 0.17\,h^{-1}$, $Ks < 2\,\mu mol$ L-malate l^{-1} and $Y_{max} = 195\,mg$ dry weight g^{-1} L-malate) and the oxygen tolerance were determined. Enzyme activities of glutamine synthetase, glutamate synthase (GOGAT), and glutamate dehydrogenase were measured for all steady states in continuous culture. The base composition is 66% G + C, but no definite identification could be obtained.

Introduction

Nitrogen fixing bacteria have been demonstrated by many researchers to occur in the soil of paddy fields. In addition to azospirilla and enterobacters many isolates have been referred to *Pseudomonas* sp. but little is known of their systematics and physiology. (Barraquio *et al.*, 1983).

Isolate R7 is a gram negative rod obtained from a soil sample from a rice (*Oryza sativa*) field in Suweon, Republic of Korea. Enrichment was performed in continuous culture at a low dilution rate ($D = 0.0125\,h^{-1}$) where Isolate R7 reached 80% of the total population. At the same time 75–80% of total fixed N was found in the supernatant fluid (Kloss *et al.*, 1986; in this paper Isolate R7 was named Isolate R). These characteristics made Isolate R7 an interesting subject for further investigations of general growth parameters as well as of ammonia metabolism.

The determination of growth parameters and oxygen tolerance is important for the evaluation of bacteria which live near or inside the rhizosphere of plants, because the roots are exuding carbon sources and vitamins, and in the case of rice growing in anaerobic soil, oxygen and nitrogen are supplied to rhizosphere organisms through the roots (Yoshida and Broadbent, 1975).

Glutamine synthetase (GS), glutamate synthase (GOGAT) and glutamate dehydrogenase (GDH) are the major enzymes for converting ammonia into amino acids and proteins.

The GS/GOGAT pathway has a greater affinity for ammonia and is active in most bacteria under nitrogen fixing conditions whereas GDH is active when they are growing on ammonia as a nitrogen source (Kleiner *et al.*, 1981).

Materials and methods

Bacterial strain

R7 was purified by Kloss from colonies on double layer agarose plates. It is able to fix nitrogen under microaerophilic conditions, and requires biotin and niacinamide as vitamins (Kloss *et al.*, 1986).

For further characterization DNA was prepared as described by Reinhold *et al.*, (1987), modified from Marmur (1961). The DNA base composition was measured by thermal denaturation and calculated by the equation of De Ley (1970). Quinones were extracted in acetone from sonicated cells and separated by chromatography (Kroppenstedt,

1982). Several other tests for characterization were done by standard procedures.

Medium

All experiments were carried out in a nitrogen-free medium (below $1 \, mg \, N \, l^{-1}$) containing $(g \, l^{-1})$: L-malate, 2.5 (neutralized with KOH); K_2HPO_4, 0.2; KH_2PO_4, 0.3; $MgSO_4 \cdot 7 \, H_2O$, 0.2; NaCl, 0.1; $CaCl_2$, 0.02; $MnSO_4 \cdot H_2O$, 0.01; $Na_2MoO_4 \cdot 2H_2O$, 0.002; $FeSO_4 \cdot 7H_2O$, 0.01; and $(mg \, l^{-1})$: biotin, 0.1; niacinamide, 0.5; pH 6.8.

Continuous culture

Cultures were grown in a Biostat M and a Biostat V fermenter (B. Braun, Melsungen, FRG) with culture volumes of 1,350 ml and 1,000 m and stirring rates of 800 rev. min^{-1} and 400 rev. min^{-1}, respectively. Growth temperature was kept constant at 37°C, and the pH was measured electrometrically and maintained at pH 6.8 by the automatic addition of 4.25% (w/v) orthophosphoric acid. Dissolved oxygen levels were measured by autoclavable oxygen electrodes (Clark Typ, Ingold). To maintain a constant oxygen concentration in the vessel an appropriate amount of synthetic air (20% O_2, 80% N_2) regulated by an oxygen control unit (B. Braun, Melsungen, FRG) was added to a constant flow of N_2 (230 ml min^{-1} and 130 ml min^{-1} for the larger and smaller culture volumes, respectively). Media were supplied by piston pumps with autoclavable pumping heads.

For each set of culture conditions four to five culture volumes were passed through the vessel until steady states were reached. Samples were taken by collecting the effluent in autoclaved containers kept in ice bath. Samples (150 ml each) were used for the determination of dry weight by freeze drying. Optical density was measured at 578 nm. Ten ml of the suspension were stored at $-20°C$ for the determination of total protein by the micro Goa method (Bergersen, 1980) and total nitrogen by Kjeldahl digestion (Bergersen, 1980); the ammonia formed was estimated by a colorimetric method (Mueller-Beisenhirtz and Keller, 1965). Supernatant fractions of 20 ml were sterile-filtered and

stored at $-20°C$ for determination of protein by Coomassie Brilliant Blue G-250 micro assay (Bio Rad, Munich FRG), of ammonia by the Urea/Ammonia Test (Boehringer, Mannheim, FRG), of nitrogen by Kjeldahl digestion, and of L-malate by the L-malate Test (Boehringer). Bovine serum albumin was used as reference standard for all protein assays. Two hundred and fifty ml of the bacterial suspension were centrifuged at 18,500 g at 4°C for 10 min, washed twice in Tris buffer (50 mM, pH 7.0), and homogenized in a cell homogenizer (B. Braun, Melsungen, FRG) with glass beads of 0.10–0.11 mm diameter and cooled with liquid CO_2 (Merkenschlager *et al.*, 1957). After centrifugation at 22.800 g at 4°C for 10 min the supernatant fraction was used for glutamine synthetase (transferase hydroxamate, pH 7.0), NADPH-dependent glutamate synthase (pH 7.5), and NADH-dependent glutamate dehydrogenase (pH 7.5) assays (Farnden and Robertson, 1980. Enzyme activities are given as $1 \, U = 1 \, nmol \, NAD(P)H$ oxidized $min^{-1} \, mg^{-1}$ protein (GDH, GOGAT) and $1 \, U = 1 \, nmol \, \gamma$-glutamyl hydroxamate formed $min^{-1} \, mg^{-1}$ protein (GS). The protein content was determined by the Coomassie Brilliant Blue G-250 standard assay.

After taking the last subsample, 0.5 g of neutralized L-malate was added to the vessel to test for carbon limitation of the culture.

Results

Characterization

R7 is a gram negative rod with a base composition of 66% G + C and Q_{10} as the main quinone component. Catalase, oxidase, and urease reactions are positive, nitrate reduction and denitrification negative; O/F test: oxidation with light alkalization, fermentation negative; no PHB content; these tests and a characterization with API 20 NE (Bio Merieux, Nürtingen) did not lead to a definite identification. The results indicate that R7 belongs to the RNA superfamily IV, as *Rhizobium*, *Azospirillum*, *Beijerinckia*, and *Agrobacterium* do, because to our knowledge, no organism with ubiquinone Q_{10} has been found which belongs to a group other than RNA superfamily IV.

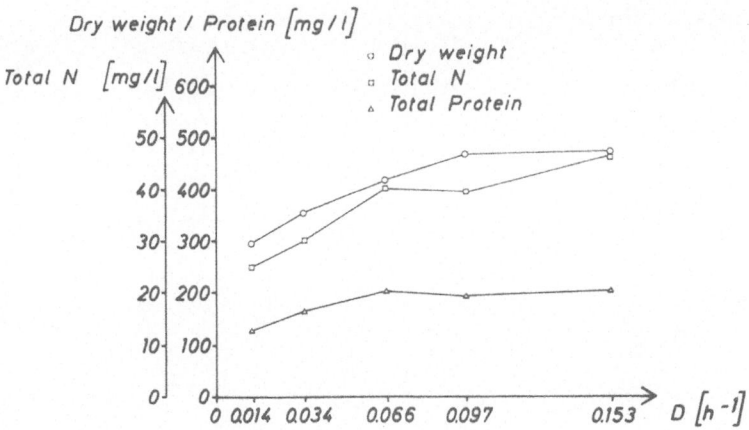

Fig. 1. Variations of growth parameters from different steady
states at a constant oxygen concentration of 8 μmol O_2 l^{-1}

General growth parameters

The dilution rates were varied between 0.014 h^{-1}
and 0.153 h^{-1} (Fig. 1). The oxygen concentration
was kept constant at 8 μmol O_2 l^{-1}. For each dilution rate carbon limitation occurred and almost no
ammonium, nitrogen, or L-malate could be detected in the supernatant liquid (< 0.5 mg l^{-1},
1 mg l^{-1}, 2 mg l^{-1}, respectively). Protein concentrations in the supernatant fractions of all experiments
fell between 1.4 and 2.1 mg l^{-1} which equals about
1% of total protein. For determination of μ_{max} the

wash-out method (Jannasch, 1969) was used (data
not shown). Under the given conditions $\mu_{max} =$
0.17 h^{-1} and $Ks < 2$ μmol L-malate l^{-1}. Maximum
yield and maintenance requirements were calculated (Pirt, 1975) as: $Y_{max} = 195$ mg dry weight g^{-1}
L-malate ($= 16$ mg N g^{-1} L-malate) and $m =$
0.049 g L-malate g^{-1} dry weight h^{-1}.

At D $= 0.064$ an additional steady state was set
with non-nitrogen fixing conditions by adding 0.5 g
NH_4Cl to the medium. The following results were
obtained: 585 mg l^{-1} dry weight, 260 mg l^{-1} protein, and 124 mg l^{-1} nitrogen; in the supernatant

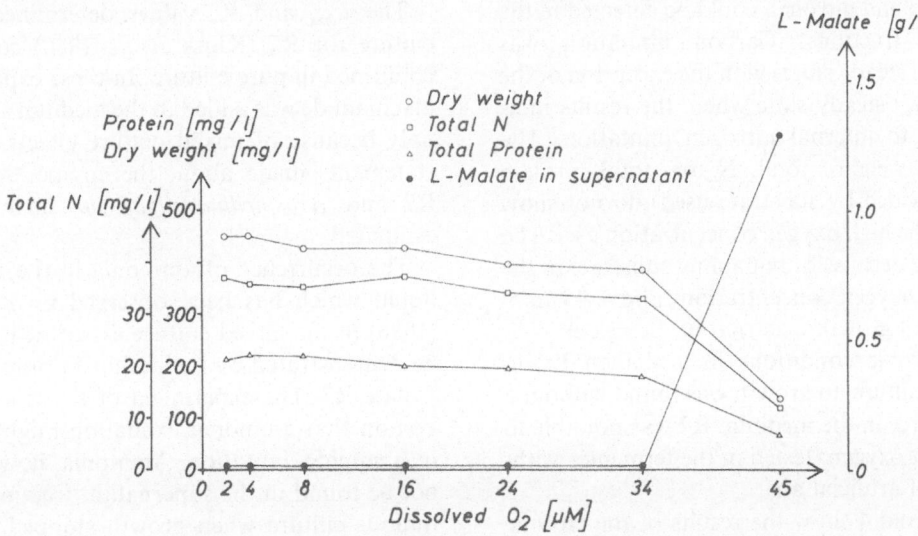

Fig. 2. Variations of growth parameters at different oxygen
concentrations from steady states at D $= 0.075$ h^{-1}

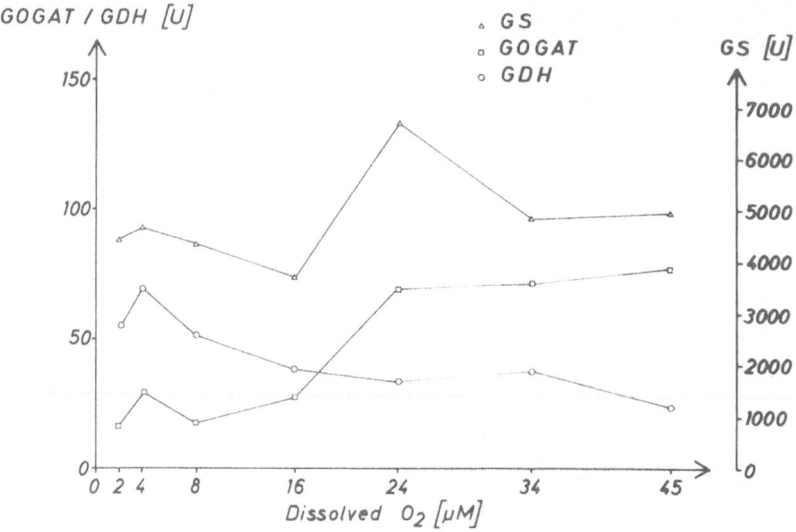

Fig. 3. Enzyme activities at different oxygen concentrations from
steady states at $D = 0.075\,h^{-1}$

fraction 70 mg l^{-1} ammonia and 58 mg l^{-1} nitrogen
were found; enzyme assays were 300 U (GS), 35 U
(GOGAT), and 116 U (GDH).

Oxygen tolerance

At a constant dilution rate of $D = 0.075\,h^{-1}$
steady states were established at different oxygen
concentrations between 2 and 45 μmol l^{-1} (Fig. 2).
No ammonia and nitrogen could be detected in the
supernatant fractions. Carbon limitation was
shown for all steady states with the exception of the
45 μmol O$_2$ l^{-1} steady state where the results indicate a shift to internal nitrogen limitation. The
yields (dry weight, total N or total protein
produced divided by substrate used) do not show
the drop at the high oxygen concentration as can be
seen in Fig. 2 because of remaining substrate in the
medium. At oxygen concentrations above 45 μmol
l^{-1} heavy wall growth occurred in the vessel.

Under aerobic conditions in a shaken Erlenmeyer flask culture no growth was found without a
nitrogen source in the medium. R7 was not able to
grow without oxygen (tested in the fermenter without supply of artificial air).

Figures 3 and 4 show the results of the enzyme
assays. Data for GS activity showed a high standard deviation within parallel experiments. No
NADPH-depending GDH activity and no NADH-

depending GOGAT activity was found. When the
cell homogenate was used for enzyme tests without
final centrifugation (so that it included larger particles from cell breakdown) activities did not
change significantly compared to those of the centrifuged extract.

Discussion

The μ_{max} and K_s values determined in mixed
culture for R7 (Kloss *et al.*, 1986) could not be
confirmed in pure culture. In these experiments no
niacinamide was added to the medium and R7 grew
only because of mixed culture effects. Therefore,
statements made about the competition between
R7 and *Azospirillum lipoferum* have to be re-
evaluated.

The occurrence of ammonia in the supernatant
liquid which has been observed by Kloss *et al.*,
(1986) in the mixed culture experiments could not
be demonstrated by the author in pure cultures of
Isolate R7. The experiment of Kloss led to a suggestion that ammonia exudation might be due to
niacinamide limitation. Ammonia, however, could
not be found in the supernatant fraction of a continuous culture when growth stopped because of
niacinamide limitation (an established steady state
culture was supplied with fresh medium without the
vitamin; data not shown).

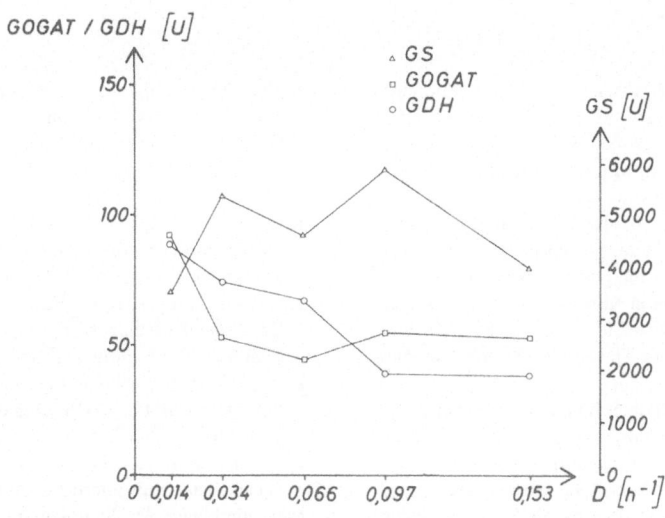

Fig. 4. Enzyme activities from steady states at a constant oxygen concentration of 8 μmol O_2 l^{-1}

Only little information on physiological growth parameters under controlled oxygen concentration culture is available. The following values have been found: for *Azospirillum brasilense* μ_{max} = 0.146 h^{-1}, K_s = 72 μmol l^{-1}, Y_{max} = 7.9 mg N g^{-1} L-malate (Kloss *et al.*, 1983); for *Arthrobacter giacomelli* μ_{max} = 0.076 h^{-1} and K_s = 83 μmol l^{-1} (Cacciari *et al.*, 1986); and for *Xanthobacter autotrophicus*(formerly *Corynebacterium autotrophicum*) μ_{max} = 0.14 h^{-1} (Berndt *et al.*, 1976). Compared to these results R7 has a high μ_{max} (= 0.17 h^{-1}) and a very high substrate affinity (K_s < 2 μmol L-malate l^{-1}). Simulation of growth by computer models show that R7 outgrows *A. brasilense* very quickly, even at low dilution rates because of its high substrate affinity, but for many reasons predictions of competition in the natural habitat are very difficult.

Under nitrogen fixing conditions R7 shows a ratio between dry weight, protein, and nitrogen content of 12:6:1 changing to 10:4.5:1 at the highest dilution rates (D = 0.153 h^{-1}).

From different yields under nitrogen fixing and nonfixing conditions the energy requirement of the nitrogen fixing system (including oxygen protection etc.) can be estimated as 29% of the total energy need at the 8 μmol $O^2 l^{-1}$ and D = 0.064 h^{-1} steady state.

The yield decreased linearly with increasing oxygen concentration. This leads to the conclusion that a respiratory mechanism is involved in the oxygen protection of the nitrogenase. For *Azotobacter* the possible existence of such a mechanism, which lowers the actual O_2 concentration at the site of the nitrogenase if diffusion of O_2 towards the nitrogenase is limited, is discussed (Bergerson, 1984). The protein yield of *Azotobacter vinelandii* decreases between 2 and 45 μmol O_2 l^{-1} by more than 55% (Post *et al.*, 1983) whereas the yield of R7 decreases only by 35%. A shift from carbon limitation to internal nitrogen limitation occurs between 34 and 45 μmol O_2 l^{-1}. Hurek found that oxygen tolerance is strain- and root zone-specific for two azospirilla and two diazotrophic rods isolated from kallar grass (*Leptochloa fusca*) (Hurek *et al.*, 1987). Compared to their results Isolate R7 has a high oxygen tolerance.

GDH and GOGAT activities vary with changing growth conditions significantly. There is no indication whether these changes are due to some kind of regulation or to inhibition by intracellular compounds (Kleiner *et al.*, 1981), or other reasons, but effects of different binding of the enzymes to membranes can be excluded.

In paddy soils there are, as well as azospirilla, many other microaerophilic nitrogen-fixing bacteria with interesting features for mutualistic interactions. High substrate affinity and high oxygen tolerance seem to favour isolate R7 in the presence of other competing bacteria.

References

Barraquio WL, Ladha JK and Watanabe I 1983 Isolation and identification of N₂-fixing *Pseudomonas* associated with wetland rice. Can. J. Microbiol 29, 867–873.

Bergersen FJ 1980 Measurement of nitrogen fixation by direct means. *In* Methods for Evaluating Biological Nitrogen Fixation. Ed. FJ Bergersen. pp. 65–110. John Wiley & Sons, Chichester, New York, Brisbane, Toronto.

Bergersen FJ 1984 Oxygen and the physiology of diazotrophic microorganisms. *In* Advances in Nitrogen-Fixation Research. Eds. C Veeger and WE Newton. pp. 171–181. Martinus Nijhoff/Dr W Junk Publishers, The Hague, Boston, Lanchester.

Berndt H, Ostwal K-P, Lalucat J, Schumann C, Mayer F and Schlegel HG 1976 Identification and physiological characterisation of the nitrogen fixing bacterium *Corynebacterium autotrophicum* GZ 29. Arch. Microbiol. 108, 17–26.

Cacciari I, Del Gallo M, Ippoliti S, Pietrosanti T and Pietrosanti W 1986 Growth and Survival of *Azospirillum brasilense* and *Arthrobacter giacomelli* in binary continuous culture. Plant and Soil 90, 107–116.

De Ley J 1970 Reexamination of the association between the melting point, buoyant density, and chemical base composition of deoxyribonucleic acid. J. Bacteriol. 101, 738–754.

Farnden KFJ and Robertson JG 1980 Methods for studying enzymes involved in metabolism related to nitrogenase. *In* Methods for Evaluating Biological Nitrogen Fixation. Ed. FJ Bergersen. pp. 265–314. John Wiley & Sons, Chichester, New York, Brisbane, Toronto.

Jannasch HW 1969 Estimation of bacterial growth in natural water. J. Bacteriol. 99, 156–160.

Hurek T, Reinhold B, Fendrik I and Niemann E-G 1987 Rootzone-specific oxygen tolerance of *Azospirillum* spp. and diazotrophic rods closely associated with Kallar grass. Appl. Environ. Microbiol. 53, 163–169.

Kleiner D, Phillips S and Fitzke E 1981 Pathways and regulatory aspects of N₂ and NH₄⁺ assimilation in N₂-fixing bacteria. *In* Biology of Inorganic Nitrogen and Sulfur. Eds. H Bothe and A Trebst. pp. 131–140. Springer-Verlag, Berlin, Heidelberg, New York.

Kloss M, Iwannek K-H and Fendrik I 1983 Physiological properties of *Azospirillum brasilense* SP7 in a malate limited chemostat. J. Gen. Appl. Microbiol. 29, 447–457.

Kloss M, Iwannek K-H, Fendrik I and Niemann E-G 1986 Enrichment of diazotrophic bacteria from rice soil in continuous culture. Plant and Soil 90, 151–164.

Kroppenstedt RM 1982 Anwendung chromatographischer HP-Verfahren (HPTLC, HPLC) in der Bakterientaxonomie. GIT Lab. Med. 5, 266–275.

Marmur J 1961 A procedure for the isolation of deoxyribonucleic acid from microorganisms. J. Mol. Biol 3, 208–218.

Merkenschlager M, Schlossmann K and Kurz W 1957 Ein mechanischer Zellhomogenisator und seine Anwendbarkeit auf biologische Probleme. Biochem. Zeitschr. 329, 332–340.

Mueller-Beisenhirtz W and Keller H 1965 Zur Bestimmung des Blutammoniaks. Klin Wochenschr. 43, 43–49.

Pirt SJ 1975 Principles of Microbe and Cell Cultivation. John Wiley & Sons, New York.

Post E, Kleiner D and Oelze J 1983 Whole cell respiration and nitrogenase activities in *Azotobacter vinelandii* growing in oxygen controlled continuous culture. Arch. Microbiol. 134, 68–72.

Reinhold B, Hurek T, Fendrik I, Pot B, Gillis M, Kersters K, Thielemans S and De Ley J 1987 *Azospirillum halopraeferens* spec. nov. a nitrogen-fixing organism associated with roots of Kallar grass (*Leptochloa fusca* (L.) Kunth). Int. J. Sys. Bacteriol. 37, 43–51.

Yoshida T and Broadbent FE 1975 Movement of atmospheric nitrogen in rice plants. Soil Science 120, 288–291.

Session 6

Genetics of *Azospirillum* and other diazotrophs

F. A. Skinner et al. (Eds.), Nitrogen fixation with non-legumes, 155–163.

Regulation of *nif* genes expression in *Azospirillum brasilense* and *Herbaspirillum seropedicae*

F.O. PEDROSA, E.M. DE SOUZA, H.B. MACHADO, L.U. RIGO and S. FUNAYAMA
Department of Biochemistry, Universidade Federal do Paraná, C. Postal 19046, 81504 – Curitiba, PR, Brazil

Key words: ammonium transport, *Azospirillum brasilense*, *Herbaspirillum seropedicae*, *nif*A, Nif constitutive mutant, *nif* regulation

Abstract

Several ethylenediamine (EDA)-resistant mutants of *A. brasilense* were isolated and found to depress nitrogenase in the presence of high NH_4^+ concentrations. These mutants were defective in NH_4^+ uptake and capable of excreting NH_4^+ in N-free semi-solid medium. A gene library of *H. seropedicae* strain HZ78 was constructed in cosmid pVK102. A recombinant plasmid pEMS1, capable of complementing a *nif*A^- mutant of *A. brasilense*, was isolated and is being characterized.

Introduction

Azospirillum brasilense (Tarrand *et al.*, 1978) and *Herbaspirillum seropedicae* (Baldani *et al.*, 1986) are free-living aerobic diazotrophs found in soils and associated with the roots of several grasses. N_2-dependent growth of both organisms occurs only under limiting NH_4^+ and O_2 concentrations, as in all other diazotrophs (see Postgate, 1982). In *Azospirillum brasilense*, nitrogenase activity and synthesis is regulated by NH_4^+ ions (Okon *et al.*, 1976, Gauthier and Elmerich, 1977, Tarrand *et al.*, 1978). Nitrogenase undergoes reversible inactivation (switch off/on) in the presence of NH_4^+, glutamine or anaerobic conditions (Pedrosa and Yates, 1984, Hartmann *et al.*, 1985, 1986). This involves covalent modification of Fe-protein as deduced from ^{32}P incorporation experiments (Hartmann *et al.*, 1985, 1986). The mechanism of NH_4^+ switch-off of nitrogenase in *A. brasilense* thus resembles that described for *Rhodopseudomonas capsulata* (Jouanneau *et al.*, 1983) and *Rhodospirillum rubrum* (Kanemoto and Ludden, 1984). Reactivation (switch-on) of *A. brasilense* nitrogenase occurs *in vivo* under NH_4^+ limitation and is carried out by a membrane-bound activating enzyme resembling

that of *Rhodospirillum rubrum* (Okon *et al.*, 1977, Ludden *et al.*, 1978, Pedrosa and Yates 1983, 1984). The modifying group is probably removed in the process of activation (see Zumft, 1985).

Nitrogenase activity is not observed in the presence of high NH_4^+ concentrations suggesting effective repression of nitrogen fixation (*nif*) genes in *Azospirillum brasilense* as reported for other diazotrophs (Eady *et al.*, 1978, Postgate, 1982). A number of regulatory mutants of *A. brasilense* have been isolated (Gauthier and Elmerich, 1977, Bani *et al.*, 1980. Hartmann *et al.*, 1983, Pedrosa and Yates, 1984, Funayama *et al.*, 1985, Fischer *et al.*, 1986). These mutants are important for an understanding of the regulation of *nif* gene expression in this diazotroph. Genetic evidence indicates that nitrogenase synthesis in *Azospirillum brasilense* strains Sp7 (Pedrosa and Yates, 1984) and Sp245 (Funayama *et al.*, 1985) involves *nif* A- and *ntrC*-like genes, as positive regulators, possibly by a mechanism analogous to that described for *Klebsiella pneumoniae* (Merrick, 1983, Ow and Ausubel, 1983). Recently, *nif*A-: Tn5 insertion mutants of *Azospirillum* have been reported (Singh and Klingmüller, 1987). Assimilation of NH_4^+ by diazotrophically grown *Azospirillum brasilense* most probably occurs via glutamine synthetase (GS) and glutamate synthase (GOGAT) activities as indicated by physiological and genetic studies (Okon *et al.*, 1976, Gauthier and Elmerich, 1977, Bani *et al.*, 1980). Glutamine auxotrophic (Gln⁻, Gauthier and Elmerich, 1977) and glutamate synthase negative (GOGAT⁻, Bani *et al.*, 1980) mutants failed to grow diazotrophically. Glutamine auxotrophs were GS⁻ and displayed a nitrogen fixation negative (Nif⁻) or nitrogen fixation constitutive (Nifᶜ) phenotype. These opposite phenotypes led the authors to suggest a regulatory role for GS in nitrogenase expression in *Azospirillum brasilense* (Gauthier and Elmerich, 1977). Restoration of normal control of nitrogen fixation to both classes of mutants by plasmids containing only the *glnA* gene of *A. brasilense* supports a regulatory role for glutamine synthetase (Bozouklian *et al.*, 1986). In *Klebsiella pneumoniae*, active GS has been shown to be essential to elicit nitrogenase repression by NH₄⁺ through the generation of a metabolic signal. There is, however, no evidence for a direct role of the GS protein in positive control of nitrogen-regulated operons (see Dixon, 1984). Nif-GS-mutants of *K. pneumoniae* are polar mutants in *glnA* affecting expression of *ntrB,C* downstream in the *glnAntrBC* operon (Dixon, 1984). In *Azospirillum brasilense*, however, recent evidence suggests that *ntrBC* and *glnA* genes are unlinked, thus excluding the possibility of a polar *glnA* mutation giving rise to a Nif⁻ phenotype (Bozouklian *et al.*, 1986). Inhibition of GS by methionine sulphoximine (MSX) caused derepression of nitrogenase in *A. brasilense* cultures growing in the presence of NH₄⁺ (Okon *et al.*, 1976), stressing the role of active GS in nitrogenase repression.

Glutamine synthetase activity has also been implicated in *in vivo* NH₄⁺ switch-off of nitrogenase (Pedrosa and Yates, 1983, Hartmann *et al.*, 1986) and NH₄⁺ transport (Kleiner *et al.*, 1983) in *A. brasilense* and *A. lipoferum*. Glutamine has been suggested as the metabolic signal for repression of nitrogen-regulated operons in *A. brasilense* and *A. lipoferum* (Hartmann *et al.*, 1986).

Mutants deficient in glutamate synthase were Nif⁻, and unable to grow on poor N-sources (Bani *et al.*, 1980). High intracellular glutamine concentration could explain the Nif⁻ phenotype. The inability to grown on poor N-sources is probably related to NH₄⁺ assimilation.

Three other regulatory mutants (BTG 182, 200 and 1130) of *Azospirillum brasilense* were Nifᶜ, histidine uptake constitutive, displayed normal control and levels of GS, and were suggested to carry *ntr*-type (*ntrC*) mutations (Fischer *et al.*, 1986).

In *A. brasilense*, NH₄⁺ transport (uptake) is carried out by an inducible, energy-dependent system (Hartmann and Kleiner, 1982). This transport system is repressed by ammonia, but synthesized in the presence of glutamate.

In this paper we describe novel Nif constitutive mutants of *Azospirillum brasilense* and present evidence for a *nifA*-like gene function in *Herbaspirillum seropedicae*.

Materials and methods

Organisms and growth conditions

Azospirillum brasilense strain FP2 (Sp7 Nalʳ Smʳ, wild type, Pedrosa and Yates, 1984), and its Nif constitutive derivatives (HM14, HM26, HM53 and HM210, this work), and *A. brasilense* strain FP10 (*nif*A⁻-like) were grown in NfbHP medium (Pedrosa and Yates, 1984). *Herbaspirillum seropedicae* strain HZ78 (Baldani *et al.*, 1986), a gift from Dr. J. Döbereiner, was also grown in NFbHP medium. *Escherichia coli* strains HB101 (*pro, leu; thi, lacY, hsd*R, *hsd*M, *rec*A; Ditta *et al.*, 1980), BHB 2688 and 2690 were grown in Luria broth (LB) or agar (LA) (Maniatis *et al.*, 1982). Antibiotics were added to growth media as required. All bacterial strains were grown at 30 °C.

Plasmids

Plasmid pGE10 (R68.45, Tra⁺, Kmʳ, Tcʳ, Apʳ and IncP, *glnAntrBC* from *Klebsiella pneumoniae*) was a gift from M. Merrick (Espin *et al.*, 1981). The cosmid vector pVK102 (IncP, Kmʳ, Tcʳ, Tra⁻, lambda *cos*) was from Knauf and Nester (1982). The helper plasmid pRK2013 (Kmʳ, Tra⁺) was from Ditta *et al.* (1980). Plasmid R68.45 (Tra⁺, Kmʳ, Tcʳ, Apʳ, IncP) was used in surinfection experiments to eliminate plasmids of the same incompatibility group. Recombinant plasmid pEMS1 contains a *Herbaspirillum seropedicae nifA*-like gene cloned into the *SalI* site of pVK102 and is Kmʳ, Tcˢ (this work).

Conjugation

Biparental and triparental matings were carried out as described (Pedrosa and Yates, 1984).

Mutant isolation

Spontaneous mutants of *A. brasilense* strains FP2 resistant to EDA (0.05–0.2%) were isolated in NFbHP-plates containing glutamate (5 mM) incubated at 30 °C. Resistant colonies were purified and tested for nitrogenase activity in semi-solid NFbHP medium containing 0, 5 and 20 mM NH$_4$Cl at 30 °C. Fifty percent of the isolates were Nif^c. Four Nif^c isolates, HM14, HM26, HM53 and HM210, were further characterized.

Enzymatic assays

Nitrogenase activity (acetylene reduction) was determined in cultures grown in N-free semi-solid medium (Pedrosa and Yates, 1984), at 30 °C for 1 h. Ethylene was determined by GLC. Glutamine synthetase (GS) biosynthetic and transferase activities were determined as described by Goldberg and Hanau (1979). NH$_4^+$ uptake activity was measured in washed, glutamate-grown cells, resuspended in N-free liquid NFbHP medium (O.D.$_{540nm}$ = 1.0). Cell suspensions were incubated for 30 min at 30 °C and 130 rev. min^{-1} to deplete residual N-sources, and the reaction was initiated by the addition of NH$_4$Cl (0.100 mM final). NH$_4^+$ was determined by the indophenol procedure (Chaney and Marbach, 1967) in supernatant solutions of samples collected at indicated times.

NH$_4^+$ excretion

Stagnant cultures of *A. brasilense* strains were grown in N-free semi-solid medium at 18–20 °C. NH$_4^+$ excreted into the medium during growth was determined as above.

Construction of a gene library of Herbaspirillum seropedicae *total DNA*

Total DNA of *H. seropedicae* strain HZ78 was purified, digested with *SalI* and fractionated in a sucrose gradient according to Maniatis *et al.* (1982). Fragments of 15–32 kb were ligated to the *SalI* site of cosmid pVK102 located in the Tc resistance gene, and the recombinant plasmids were *in vitro* packaged into bacteriophage Lambda heads. *Escherichia coli* HB101 was infected with the phage suspension and recombinant clones were isolated in LA medium containing Km. A gene bank composed of 460 recombinant clones (Km^r, Tc^s) was obtained. Clones were purified, pooled and stored in 50% glycerol at -20 °C. All recombinant DNA techniques were performed as described in Maniatis *et al.* (1982).

Identification of a clone carrying a H. seropedicae *nifA-like gene*

For this purpose, *A. brasilense* strain FP10 (*nifA$^-$*) was used as recipient for recombinant cosmids of *H. seropedicae*. The gene bank was conjugated 'en masse' with strain FP10 and the strain containing the pRK2013 helper plasmid. *Azospirillum brasilense* FP10 Nif$^+$ transconjugants were selected in N-free semi-solid medium under 0.5–1.0% O$_2$ at 30 °C. Three transconjugants (FP10.9, FP10.29 and FP10.31) were isolated and shown to carry recombinant plasmids. One such plasmid (pEMS1) carries a *H. seropedicae nifA*-like gene and was further studied.

Plasmid pEMS- curing from A. brasilense *FP10 transconjugants*

This was carried out using plasmid R68.45 of the same IncP group, as described before (Pedrosa and Yates, 1984).

Transformation of E. coli HB101 *with plasmid pEMS1*

This plasmid was purified from *A. brasilense* FP10.9 (pEMS1) transconjugants by alkaline lysis (Birnboim and Doly, 1979), and used to transform *E. coli* HB101.

158   *Pedrosa* et al.

Transfer of pEMS1 from E. coli *HB101 to* A. brasilense *FP10 NifA⁻*

Plasmid pEMS1 was transferred from *E. coli* HB101 to *A. brasilense* FP10 by triparental mating conjugation with pRK2013. *Azospirillum brasilense* FP10(pEMS1) transconjugants were isolated as above.

Total cell protein was determined by the Lowry procedure after alkaline lysis using bovine serum albumin as standard.

Results

Isolation of EDA resistant, Nif constitutive mutants of A. brasilense

Spontaneous mutants of *A. brasilense* FP2 resistant to EDA occurred at a frequency of 2.4×10^{-6}. All isolates tested (30) were capable of N_2-dependent growth and 15 of them expressed nitrogenase activity in the presence of high NH_4^+ concentrations (20 mM, Nif^c phenotype). The frequency of Nif^c was found to be 1.7×10^{-7}. Four isolates (HM14, HM26, HM53, HM210) were further characterized.

Growth characteristics of A. brasilense *FP2 Nif^c mutants*

The four isolates were prototrophs, showing no nutritional requirements for growth. All grew well

Table 1. Growth of Nif^c mutants of *A. brasilense* FP2 on different N-sources[b]

Strain	N_2^a	NH_4Cl^b		NO_3^{-b} 10 mM	Arg[b] 5 mM	Glu[b] 5 mM
		2 mM	20 mM			
FP2 (WT)	+	+	+	+	+	+
HM14	+	−^{c}	+	−	+	+
HM26	+	−^{c}	+	−	+	+
HM53	+	Poor	+	−	+	+
HM210	+	+	+	+	+	+

Arg, Arginine; Glu, Glutamate; WT, Wild Type.
[a] Growth in N-free semi-solid NFbHP medium. Diazotrophic growth of HM mutants was slower than that of FP2.
[b] Growth in solid NFbHP medium plus N-sources at 30 °C for 48 h.
[c] Very slow growth in liquid medium.

Table 2. Effect of NH_4^+ concentration of nitrogenase activity of Nif^c mutants of *A. brasilense* FP2

Strain	NH_4^+ concentration in growth medium[a] (mM)	Apparent nitrogenase activity (nmol C_2H_4 min^{-1} mg^{-1} protein)
FP2 (wild type)	Zero	9.1
	5	0.1
	20	0.0
HM14	Zero	8.3
	5	2.7
	20	0.7
HM26	Zero	5.2
	5	3.8
	20	2.0
HM53	Zero	7.5
	5	11.0
	20	6.4
HM210	Zero	5.6
	5	4.1
	20	1.7

[a] Cells grown in semi-solid NFbHP medium.

on arginine, glutamate or high NH_4^+ (20 mM) as sole N-sources (Table 1). N_2-dependent growth was slower than that of the wild type. All mutants, except HM210, failed to grow or grew poorly on low NH_4^+ or NO_3^-. Growth rates of the mutants in high NH_4^+ (doubling times, dt = 163–186 min) or in glutamate (dt = 204–227) were of the same order of magnitude as those of the wild type, 152 and 200 min, respectively.

Effect of NH_4^+ concentration on nitrogenase activity

Mutants HM14, HM26, HM53 and HM210 displayed nitrogenase activity in semi-solid medium containing 5 or 20 mM NH_4Cl, while the wild type strain did not under the same conditions (Table 2). At 20 mM NH_4^+ only mutant HM53 had nitrogenase activity levels similar to those in the absence of NH_4^+.

Effect of the N-source on glutamine synthetase activities

Azospirillum brasilense strains grown in liquid NFbHP medium containing low (2 mM) or high

Table 3. Glutamine synthetase activities in *A. brasilense* Nif^c mutants

Strain	N-source for growth (mM)		Glutamine synthetase activity (μmol glutamyl-hydroxamate min^{-1}·mg^{-1} protein)			
			Biosynth.	Transferase		
				Mn^{2+}	Mn^{2+} + Mg^{2+}	Unadenylylated (%)
FP2	NH$_4^+$	2	0.141	1.482	0.741	50
		20	0.041	0.907	0.202	22
	Glu	5	0.240	1.798	0.637	35
HM14	NH$_4^+$	2	0.160	1.158	0.093	8
		20	0.026	0.039	0.045	115
	Glu	5	0.033	0.023	0.016	70
HM26	NH$_4^+$	2	0.117	0.697	0.078	11
		20	0.091	0.060	0.016	27
	Glu	5	0.160	0.435	0.030	7
HM53	NH$_4^+$	2	0.090	0.210	0.047	22
		20	0.094	0.113	0.012	11
	Glu	5	0.109	0.072	0.072	100
HM210	NH$_4^+$	2	0.193	1.582	0.238	15
		20	0.152	0.300	0.065	22
	Glu	5	0.184	0.579	0.162	28

NH$_4^+$, NH$_4$Cl; Glu, Glutamate.
Cells grown in liquid NFbHP medium containing the indicated N-source.

(20 mM) NH$_4^+$, or sodium glutamate (5 mM), were harvested by centrifugation and assayed for biosynthetic and transferase GS activities. Under low NH$_4^+$ all Nif^c mutants had biosynthetic GS activities close to those of the wild type strain FP2 (Table 3), while total transferase activities were lower (< 50% and < 15%) in mutants HM26 and HM53. Unexpectedly, under these conditions, in all four mutants a high degree of GS adenylylation (> 80%) was observed. Under high NH$_4^+$ the levels of biosynthetic GS activity were lowered, as expected, except in mutants HM26, HM53 and HM210, while total transferase activities were very low in mutants HM14 and HM26. Growth on glutamate increased biosynthetic activities of the wild type FP2 and mutants HM26, HM53 and HM210, but had a strong depressing effect on the biosynthetic activity of mutant HM14. Mutants HM26 and HM53 appear to be defective in transferase activity. In addition, three mutants (HM26, HM53 and HM210) appear to be constitutive for biosynthetic activity, similar to what is observed in some *ntrB* mutants of *K. pneumoniae*. The high degree of adenylylation of GS from all mutants grown at low NH$_4^+$, also suggests a possible defect in *glnD* (Uridylyltransferase activity).

NH$_4^+$ uptake

Cells grown in glutamate, which allows for constitutive synthesis of the NH$_4^+$-transport system in *A. brasilense* (Hartmann and Kleiner, 1982) were harvested at mid log phase, resuspended in N-free medium and exposed to 0.100 mM NH$_4$Cl, and the residual ammonia present in culture fluids determined at specified time intervals. The time courses of NH$_4^+$ uptake by *A. brasilense* strains are presented in Fig. 1. NH$_4^+$ uptake activities were 32.4, 5.9, 11.1, 8.4, and 12.0 (nmol NH$_4^+$·min^{-1}·mg protein^{-1}) for strains FP2, HM14, HM53, and HM210, respectively. There was an apparent correlation between NH$_4^+$ uptake rates and biosynthetic GS activities. Plasmid pGE10 (*glnAntrBC*) increased NH$_4^+$-uptake activity of mutants HM26, HM14 and HM53 by 2.0, 3.1 and 4.0 times, respectively (Table 4) but had no effect on HM210. We have not yet determined which of the *K. pneumoniae* genes carried by pGE10 restored NH$_4^+$-uptake activity of these Nif^c mutants of *A. brasilense*. The *ntrC* gene is a possible candidate since it has been shown to be involved in ammonium transport in *Salmonella typhimurium* (Kustu *et al.*, 1979) and *Escherichia coli* (Jayaku-

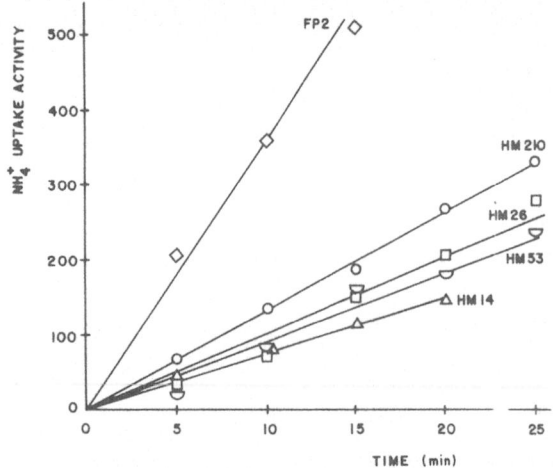

Fig. 1. NH$_4^+$ uptake (n moles NH$_4^+$ mg^{-1} protein) by Nif^c mutants of *A. brasilense*. Strains grown in liquid NFbHP glutamate medium and assayed for NH$_4^+$ uptake as described in Materials and methods.

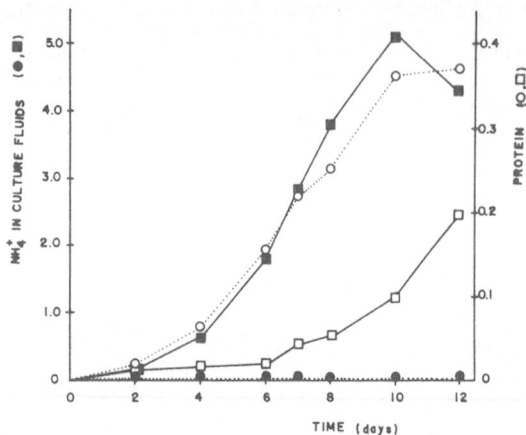

Fig. 2. NH$_4^+$ excretion by Nif^c mutants of *A. brasilense*. Strains FP2 and HM210 were grown in N-free semi-solid NFbHP medium at 18–20 °C. Samples were harvest by centrifugation and the supernatant fluid was assayed for NH$_4^+$ as described in Materials and methods. ○, ● FP2; □, ■ HM210. NH$_4^+$ in m*M*; protein in mg ml^{-1} culture.

mar *et al.*, 1986), although this seems improbable since *ntrC*-mutants of *A. brasilense* are Nif$^-$ (Pedrosa and Yates, 1984). The presence of pGE10 in *A. brasilense* transconjugants was confirmed by agarose gel electrophoresis following alkaline lysis.

NH$_4^+$ excretion

Nif constitutive mutants of *A. brasilense* FP2 excreted substantial amounts of NH$_4^+$ during diazotrophic growth in N-free semi-solid medium (Fig. 2 and Table 5). No trace of NH$_4^+$ was detected in the culture supernatant fractions of the wild type strain FP2 (Table 5). Excreted NH$_4^+$ was the product of nitrogen fixation and not of ammonifi-

cation of dead cells, since NH$_4^+$ excretion and total cell protein (Fig. 2) and cell number (data not shown) increased during growth. Reassimilation of NH$_4^+$ occurred slowly in older cultures, or more rapidly when cultures were grown at higher temperatures (data not shown).

Identification of a nifA-*like gene in* Herbaspirillum seropedicae

A recombinant plasmid, called pEMS1, capable of restoring Nif$^+$ phenotype to *A. brasilense* strain FP10 (*nifA$^-$*) was identified after conjugation of the gene bank of *H. seropedicae* with FP10, and selection for N$_2$-dependent growth and Km resistance. Three Nif$^+$ transconjugants of FP10, namely

Table 4. Rates of NH$_4^+$ uptake by *A. brasilense* Nif^c mutants and their pGE10 (*K. pneumoniae glnAntrBC*) transconjugants

Strain	NH$_4^+$ uptake activity (nmol NH$_4^+$ min^{-1}·mg^{-1} protein)	
	Plasmid	
	None	pGE10
FP2	32.4 (100)	44.4 (100)
HM14	5.9 (18)	18.5 (42)
HM26	11.1 (34)	22.7 (51)
HM53	8.4 (26)	33.7 (76)
HM210	12.0 (37)	15.0 (34)

[a] Percent wild type activity given in parenthesis.
Growth conditions and procedures as described for Fig. 1.

Table 5. Total NH$_4^+$ excretion by different Nif^c mutants of *A. brasilense*

Strain	Total excreted NH$_4^+$ (m*M*)	Final protein (mg^{-1} ml^{-1})
FP2 (wild type)	0	0.371
HM14	1.83	0.043
HM26	3.94	0.082
HM53	3.54	0.120
HM210	5.11	0.103

Cultures were grown in N-free semi-solid NFbHP medium at 18–20 °C for 10–12 days.

Table 6. Complementation of *A. brasilense* FP10 (*nif A⁻*) with cosmid containing *H. seropedicae* SalI DNA fragment

A. brasilense strains	Nitrogenase activity[a] (nmol $C_2H_4^{-1}$·culture^{-1})
FP2	307.2
FP10	0.0
FP10.9 (pEMS1)[b]	107.4
FP10.29 (pEMS1)	80.6
FP10.31 (pEMS1)	168.9
FP10.9 (R68.45)[c]	0.0
FP10.29 (R68.45)	0.0
FP10.31 (R68.45)	0.0
FP10 (pEMS1)[d]	193.4

[a] Nitrogenase activity of cultures grown in N-free semi-solid medium at 30 °C.
[b] pEMS1 = pVK102:*H. seropedicae* (DNA).
[c] After elimination of plasmid pEMS1 by plasmid R68.45.
[d] Plasmid pEMS1 was purified and used to transform *E. coli* HB101. Transformant *E. coli* HB101 (pEMS1) was used as donor in a conjugation with *A. brasilense* FP10.

FP10.9, FP10.29 and FP10.31 contained a recombinant plasmid of similar size (*ca.* 52 kb). This plasmid was purified from FP10.9 by alkaline lysis and used to transform *E. coli* HB101. Five transformants contained one plasmid of size identical to pEMS1. One HB101(pEMS1) transformant was conjugated with *A. brasilense* FP10 in a triparental mating with *E. coli* HB101(pRK2013), and *A. brasilense* FP10(pEMS1) Nif⁺ transconjugants were selected as above. Table 6 shows complementation of *A. brasilense* strain FP10 (*nifA⁻*) nitrogen fixation by plasmid pEMS1. Elimination of pEMS1 of FP10 by plasmid R68.45 of the same incompatibility group restored the Nif⁻ phenotype of FP10. The presence of pEMS1 or R68.45 in transconjugants and transformants was confirmed by agarose gel electrophoresis. These results strongly suggest that plasmid pEMS1 carries a *nifA*-like gene of *H. seropedicae*. The molecular characterization of this plasmid is in progress.

Discussion

We have isolated and partially characterized four Nif constitutive mutants of *A. brasilense* FP2. All mutants were prototrophs and grew well on several N-sources. N_2-dependent growth of the mutants was slower than that of the wild type strain FP2. NH_4^+ excretion and decreased NH_4^+-uptake rates was observed in all mutants suggesting deficiency in the NH_4^+ assimilatory enzymes. All Nif^c mutants isolated displayed GS activity, Nif^c phenotype, and had no glutamine requirement for growth. They are therefore different from those of Gauthier and Elmerich (1977) which were glutamine auxotrophs, GS⁻ and Nif⁻ or Nif^c. Based on the levels of biosynthetic GS activities, two classes of the Nif^c mutants were identified. The first class included mutant HM14 and had GS activities repressible by high NH_4^+ concentrations as did the wild type strain. The second class included mutants HM26, HM53 and HM210 and appeared to be constitutive for biosynthetic activity irrespective of the N-source or concentration. On the other hand, when analysed in terms of total transferase (Mn^{2+}) activity mutants HM14 and HM210 had wild type levels, while HM26 and HM53 had low levels of GS. All transferase activities were decreased by growth in high NH_4^+ concentrations. Although with one discrepancy, the GS data clearly indicate that in all mutants GS was highly adenylylated, irrespective of the NH_4^+ concentration present in the growth medium. This may explain Nif constitutivity and NH_4^+ excretion. A highly adenylylated, inactive enzyme may fail to generate the 'metabolic signal' required to trigger NH_4^+ repression of *nif* genes, and to assimilate NH_4^+ produced by nitrogenase. This phenotype resembles that of *E. coli* or *Klebsiella aerogenes glnD* mutants (Bloom *et al.*, 1978, Foor *et al.*, 1978), although the level of biosynthetic GS activity of *Azospirillum* mutants was high as compared to *K. aerogenes glnD* mutants. Mutation in *glnD* leads to a loss in uridylyltransferase and uridylyl-removing enzyme and a consequent inability to rapidly deadenylylate GS and a requirement for glutamine for normal growth (Foor *et al.*, 1978). Mutant HM53 apparently has a defect in GS transferase activity, which may explain NH_4^+ excretion, as suggested for *Nostoc muscorum* mutants (Singh *et al.*, 1983).

Although *ntr* genes have not been cloned from *A. brasilense*, genetic evidence (Pedrosa and Yates, 1984) suggests that *ntrC* and *ntrA* are functional in this organism. Assuming that the same two-tier regulatory circuitry for *nif/ntr* regulation described for *Klebsiella pneumoniae* (Dixon, 1984) is operational in *A. brasilense*, as suggested before (Pedrosa and Yates, 1984), the two Nif^c mutants with constitutive biosynthetic GS activity (HM53 and HM210) could as well be *ntrB*-mutants. Transport of NH$_4^+$ and amino acids in *Escherichia coli* and *Salmonella typhimurium* is under control of the general nitrogen regulatory system (*ntr* or *gln*; Kustu *et al.*, 1979, Jayakumar *et al.*, 1986). The *glnG* (*ntrC*) gene was found to be essential for the biosynthesis of the NH$_4^+$ transport system, and that GS activity had no role in NH$_4^+$ uptake in *E. coli* (Jayakumar *et al.* 1986). Stimulation of NH$_4^+$ uptake in *A. brasilense* Nif^c by a plasmid (pGE10) carrying active *glnAntrBC* gene of *K. pneumoniae* indicates the involvement of these genes in NH$_4^+$ - transport in *A. brasilense*. A potential role of these mutants as effective contributors to the nitrogen economy of associated plants is suggested.

No work on the genetics of nitrogen fixation by *Herbaspirillum seropedicae* is known. The present results strongly suggest that a *nif* specific activator gene *nifA* is present in *H. seropedicae* as in other diazotrophs such *K. pneumoniae* (Dixon, 1984) and *A. brasilense* (Pedrosa and Yates, 1984). Further characterization of pEMS1 (*H. seropedicae nifA*) is in progress.

Acknowledgement

The authors thank the Brazilian Research Council (CNPq) for financial support.

References

Baldani JI, Baldani VLD, Seldin L and Döbereiner J 1986 Characterization of *Herbaspirillum seropedicae* gen. nov. sp. nov., a root-associated nitrogen-fixing bacterium. Int. J. Syst. Bacteriol. 36, 86–93.

Bani D, Barberio C, Bazzicalupo M, Favilli F, Gallori E and Polsinelli M 1980 Isolation and characterization of glutamate synthase mutants of *Azospirillum brasilense*. J. Gen. Microbiol. 119, 239–244.

Birnboim H and Doly J 1979 A rapid alkaline extraction procedure for screening recombinant plasmid DNA. Nucleic Acids Res. 7, 1513–1525.

Bloom FR, Levin MS, Foor F and Tyler B 1978 Regulation of glutamine synthetase formation in *Escherichia coli*: characterization of mutants lacking the uridyl transferase. J. Bacteriol. 134, 569–577.

Bozouklian H, Fogher C and Elmerich C 1986 Cloning and characterization of the *glnA* gene of *Azospirillum brasilense* Sp7. Ann. Inst. Pasteur/Microbiol. 137 B, 3–18.

Chaney AL and Marbach EP 1967 Modified reagents for determination of urea and ammonia. Clin. Chem. 8, 130–132.

Ditta G, Stanfield S, Corbin D and Helinski D 1980 Broad host range DNA cloning system for Gram negative bacteria: Construction of a gene bank of *Rhizobium meliloti*. Proc. Nat. Acad. Sci. USA 77, 7347–7351.

Dixon RA 1984 The genetic complexity of nitrogen fixation. J. Gen. Microbiol. 130, 2745–2755.

Eady RR, Issack R, Kennedy C, Postgate JR and Ratcliffe HD 1978 Nitrogenase synthesis in *Klebsiella pneumoniae*: Comparison of ammonium and oxygen regulation. J. Gen. Microbiol. 104, 277–286.

Espin G, Alvarez-Morales A and Merrick M 1981 Complementation analysis of *gln*A-linked mutations which affect nitrogen fixation in *Klebsiella pneumoniae*. Mol. Gen. Genet. 184, 213–217.

Fischer M, Levy E and Geller T 1986. Regulatory mutation that controls nif expression and histidine transport in *Azospirillum brasilense*. J. Bacteriol. 167, 423–426.

Foor F, Cedergren RJ, Streicher SL, Rhee SG and Magasanik B 1978 Glutamine synthetase of *Klebsiella aerogenes*: Properties of *glnD* mutants lacking uridylyltransferase. J. Bacteriol. 134, 562–568.

Funayama S, Rigo LU and Pedrosa FO 1985 *NifA*-mutants of *Azospirillum brasilense* strain Sp245. *In* Nitrogen Fixation Research Progress. Eds. HJ Evans, PJ Bottomley and WE Newton. pp. 522. Martinus Nijhoff Publishers, Dordrecht, Boston, Lancaster.

Gauthier D and Elmerich C 1977 Relationship between glutamine synthetase and nitrogenase in *Spirillum lipoferum*. FEMS Microbiol. Lett. 2, 101–104.

Goldberg RB and Hanau R 1979 Relation between the adenylylation state of glutamine synthetase and the expression of other genes involved in nitrogen metabolism. J. Bacteriol. 137, 1282–1289.

Hartmann A and Kleiner D 1982 Ammonium (methylammonium) transport by *Azospirillum* spp. FEMS Microbiol. Lett. 15, 65–67.

Hartmann A 1982 Antimetabolite effects on nitrogen metabolism of *Azospirillum* and properties of resistant mutants. *In* *Azospirillum*, Genetics, Physiology, Ecology, Ed. W. Klingmüller. pp. 59–68. Experientia Suppl. 42. Birkhauser-Verlag, Basel.

Hartmann A, Fu H and Burris RH 1986 Regulation of nitrogenase activity by ammonium chloride in *Azospirillum* spp. J. Bacteriol. 165, 864–870.

Physiology, Ecology, Ed. W. Klingmüller. pp. 116–126. Springer-Verlag, Berlin, Heidelberg.

Hartmann A, Fu H and Burris RH 1986 Regulation of nitrogenase activity by ammonium chloride in *Azospirillum* spp. J. Bacteriol. 165, 864–870.

Hartmann A, Fusseder A and Klingmüller W 1983 Mutants of *Azospirillum* affected in nitrogen fixation and auxin production. *In* *Azospirillum* II, Genetics, Physiology, Ecology, Ed. W. Klingmüller. pp. 78–88. Experientia Suppl. 48. Birkhauser, Basel.

Jayakumar A, Schulman I, MacNeil D and Barnes EM Jr 1986 Role of the *Escherichia coli glnALG* operon in regulation of ammonium transport. J. Bacteriol. 166, 281–284.

Jouanneau Y, Meyer C and Vignais PM 1983 Regulation of nitrogenase activity through iron protein interconversion into an active and inactive form in *Rhodopseudomonas capsulata*. Biochim. Biophys. Acta 749, 318–328.

Kanemoto RH and Ludden PW 1984 Effect of ammonia, darkness, and phenazine methosulfate on whole-cell nitrogenase activity and Fe protein modification in *Rhodospirillum rubrum*. J. Bacteriol. 158, 713–720.

Kleiner D, Alef K and Hartmann A 1983 Uptake of methionine sulfoximine by some N_2 fixing bacteria, and its effect on ammonium transport. FEBS Lett. 164, 121–123.

Knauf VC and Nester EW 1982 Wide host range cloning vectors: A cosmid clone bank of an *Agrobacterium* Ti plasmid. Plasmid 8, 45–54.

Kustu S, McFarland N, Hui SP, Esmon B and Ferro-Luzzi AG 1979 Nitrogen control in *Salmonella*: Co-regulation of synthesis of glutamine synthetase and amino acid transport systems. J. Bacteriol. 138, 218–234.

Ludden PW, Okon Y and Burris RH 1978 The nitrogenase system of *Spirillum lipoferum*. Biochem. J. 173, 1001–1003.

Maniatis T, Fritsch EF and Sambrook J 1982 Molecular Cloning: A Laboratory Manual. Cold Spring Harbor Laboratory, Cold Spring Harbor, New York

Merrick MJ 1983 Nitrogen control of the *nif* regulon in *Klebsiella pneumoniae*: Involvement of the *ntrA* gene and analogies between *ntrC* and *nifA*. EMBO J. 2, 39–44.

Okon Y, Albrecht SL and Burris RH 1976 Carbon and ammonia metabolism of *Spirillum lipoferum*. J. Bacteriol. 128, 592–597.

Okon Y, Houchins JP, Albrecht SL and Burris RH 1977 Growth of *Spirillum lipoferum* at constant partial pressures of oxygen, and the properties of its nitrogenase in cell-free extracts. J. Gen. Microbiol. 98, 87–93.

Ow DW and Ausubel FM 1983 Regulation of nitrogen metabolism genes by *nifA* gene product in *Klebsiella pneumoniae*.

Nature (London) 301, 307–313.

Pedrosa FO and Yates MG 1983 Nif mutants of *Azospirillum brasilense*: Evidence for a *nifA*-type regulation. *In Azospirillum* II, Genetics, Physiology, Ecology. Ed. W. Klingmüller. pp. 67–77. Experientia Suppl. 48. Birkhauser-Verlag, Basel.

Pedrosa FO and Yates MG 1984 Regulation of nitrogen fixation (*nif*) genes of *Azospirillum brasilense* by *nifA* and *ntrC* (*gln*) type genes. FEMS Microbiol. Lett. 23, 95–101.

Postgate JR 1982 The Fundamentals of Nitrogen Fixation. Cambridge University Press, Cambridge, London, New York.

Singh HN, Singh RK and Sharma R 1983 An L-methionine-D, L-sulfoximine-resistant mutant of the cyanobacterium *Nostoc muscorum* showing inhibitor-resistant gama-glutamyl transferase defective glutamine synthetase and producing extracellular ammonia during N_2 fixation. FEBS Lett. 154, 5–9.

Singh M and Klingmüller W 1987 A Tn5 induced *nifA* like mutant of *Azospirillum brasilense*. *In Azospirillum* IV. Ed. W. Klingmüller. Birkhauser-Verlag, in press.

Tarrand JJ, Krieg NR and Döbereiner J 1978 A taxonomic study of the *Spirillum lipoferum* group, with descriptions of a new genus, *Azospirillum* gen. nov. and two species, *Azospirillum lipoferum* (Beijerinck) comb. nov. and *Azospirillum brasilense* sp. nov. Can. J. Microbiol. 24, 967–980.

Zumft WG 1985 Regulation of nitrogenase activity in anoxygenic phototrophic bacteria. *In* Nitrogen Fixation Research Progress. Eds. H.J. Evans, P.J. Bottomley and W.E. Newton. pp. 551–557. Martinus Nijhoff Publishers, Dordrecht, Boston, Lancaster.

F. A. Skinner et al. (Eds.), Nitrogen fixation with non-legumes, 165–172.
© 1989 by Kluwer Academic Publishers.

Homology between plasmids of *Azospirillum brasilense* and *Azospirillum lipoferum*

CLAIRE VIEILLE, I. ONYEOCHA, M. GALIMAND and CLAUDINE ELMERICH
Unité de Physiologie Cellulaire, Département des Biotechnologies, Institut Pasteur, 25 rue du Dr Roux, F-75724 Paris Cedex 15, France

Key words: *Azospirillum*, nodulation genes, plasmids, Tn5 site-directed mutagenesis

Abstract

Azospirillum strains examined so far contain large plasmids with which no phenotypic property has been associated. Homology was previously detected between total DNA from several *Azospirillum* strains and *Rhizobium meliloti* nodulation (*nod* and *hsn*) genes, which in *Rhizobium* are plasmid-borne. From *A. brasilense* Sp7, a 10 kb *Eco*RI fragment sharing homology with the *R. meliloti hsn* region was cloned in pUC18, to yield pAB502. The *hsn* homologous region was localized on the 90 Mda plasmid, p90, contained in strain Sp7. This was confirmed after mutagenesis of pAB502 by Tn5-Mob and recombination of the insertions into p90. The 10 kb *Eco*RI fragment was used as a probe to study homology between plasmids from eight *A. brasilense* strains and from five *A. lipoferum* strains. Hybridization with a plasmid of about 90 Mda was observed in two *A. lipoferum* strains and in all of the *A. brasilense* strains, except in strain R07 where homology was found with a 115 Mda plasmid. The three *A. lipoferum* strains, which did not hybridize with the probe, did not contain a 90 Mda plasmid.

Abbreviations. kb: kilobase pairs; Km^R: kanamycin resistance; Mda: megadalton; SDS: sodium dodecylsulphate.

Introduction

In general, as in many other soil bacteria, *Azospirillum* strains contain plasmids (for a review, see Elmerich, 1986). The number and size of the plasmids found in the two species *A. brasilense* and *A. lipoferum* are variable, but plasmids ranging from 90 to 120 Mda are often observed (Franche and Elmerich, 1981; Plazinsky *et al.*, 1983). Little is known about the molecular biology of *Azospirillum* plasmids which are difficult to cure, apparently non-conjugative and not easy to purify, due to their multiplicity and large size (Elmerich, 1986). The two taxonomic groups, *A. brasilense* and *A. lipoferum*, cannot be differentiated on the basis of their plasmid content. Moreover, there is no indication that plasmids with the same apparent molecular weight correspond to the same molecular species. Spontaneous loss of the 115 Mda plasmid of strain Sp7 was reported (Franche and Elmerich, 1981). Except for the AII prophage, which was shown to be maintained as a plasmid in *A. lipoferum* Br17 lysogenic strains (Elmerich *et al.*, 1982), no phenotype has been associated with the plasmids. In particular, *nif* genes are likely to be chromosomal (Plazinsky *et al.*, 1983).

In rhizobia, functions related to symbiotic nitrogen fixation were shown to be plasmid-borne (for a review, see Prakash and Atherly, 1986). Previously we reported homology between *Azospirillum* total DNA and the nodulation genes of *Rhizobium meliloti* 41 (Fogher *et al.*, 1985). In particular, homology was detected with a *R. meliloti* DNA fragment carrying *nodEFGH* (Kondorosi *et al.*, 1984), also named *hsnABCD* (Horvath *et al.*, 1986), and which is referred to as the '*hsn*' region.

When this fragment was used as a probe against DNA from strain Sp7, three *Eco*RI fragments of 12, 10 and 1.8 kb and three *Sal*I fragments of 5.6, 3.8 and 2.4 kb were detected (Fogher *et al.*, 1985). Gene banks, composed of *Azospirillum* Sp7 *Eco*RI or *Sal*I fragments, were constructed in pUC18. Colony hybridization was used to identify clones that carried homology to *hsn* and a recombinant plasmid, named pAB502, was isolated. Plasmid pAB502 contained a 10 kb *Eco*RI insert, which carried an internal 5.6 kb *Sal*I fragment (Elmerich *et al.*, 1987).

In this work, the DNA fragment cloned in pAB502 was shown to originate from the p90 plasmid carried by strain Sp7. In addition, when used as a probe against plasmids isolated from several *A. brasilense* and *A. lipoferum* strains, this fragment hybridized with a plasmid of molecular weight close to 90 Mda.

Materials and methods

Strains, plasmids, media and growth conditions

Azospirillum strains and plasmids are listed in Table 1. Strain 7030 is a mutant of Sp7, and strain 7800 was obtained after recombination of Tn5-Mob in p90 (see Results). Other strains are wild type except 131 which derives from Sp13. Complete medium for Sp7 was nutrient broth as described by Gauthier and Elmerich (1977). As nutrient broth was inadequate for optimal growth of several wild type strains, another rich medium, designated KYE, was employed. KYE was made of the salt base described by Franche and Elmerich (1981) and contained in addition (g. l^{-1}): sodium lactate, 5; NH$_4$Cl, 1; yeast extract, 2. *Escherichia coli* strains were JC5466 (Tn5-Mob): *trp*, *his*, *recA*, *rspE*, *rif*, Tn5-Mob (Km^R) (this work), and S17.1: *pro*, *thi*, *recA*, RP4 (Tc-Mu Km-Tn7) (Simon et al., 1983). *Escherichia coli* was grown in Luria broth (Miller, 1972). Plasmids used as DNA probes were pAB1, pAB502, and pSUP2021. Plasmid pAB1 derives from pACYC184 and contains the Sp7 *nifHDK* cluster (Quiviger *et al.*, 1982); pAB502 was isolated from a gene bank of Sp7 DNA in pUC18, it contains a 10 kb *Eco*RI fragment sharing homology with *hsn* (Elmerich *et al.*, 1987); pSUP2021 was used to prepare the Tn5 probe (Simon *et al.*, 1983).

Azospirillum *plasmid extraction and agarose gel electrophoresis*

Plasmid extraction was performed with 20 ml cultures, grown in nutrient broth or KYE, according to the technique of Kado and Liu (1981), with the following modifications. Cells were centrifuged and the pellet corresponding to 50 mg wet weight was suspended in 0.5 ml of Tris-acetate buffer 50 m*M*, pH 7.9. Lysis was achieved after addition of 1.5 ml of tris-acetate buffer, pH 12.6, containing 3% SDS and the suspension was incubated at 55 °C for 45 min. The lysate was treated with two volumes of phenol-chloroform, as described by Kado and Liu (1981) and centrifuged for 30 min at 15 000 g. The aqueous phase contained the plasmids and 150 μl volumes were routinely used for electrophoresis. Electrophoreses were performed with 0.7% agarose vertical slab gels as described by Franche and Elmerich (1981). Gels were run at 5 mA for 1 h, at 10 mA for 1 h and at 40 mA for 4 to 5 h.

Hybridization technique

Southern blotting, probe purification and labelling, and hybridization were performed as described by Fogher *et al.* (1985). Time of exposure to Kodak X-OMAT films varied depending on the intensity of the signal and are indicated in the Figures.

Tn5 mutagenesis

To mutagenize pAB502, it was first introduced by transformation into *E. coli* JC5466(Tn5-Mob). Selection of the transposition event was made by plating on a kanamycin gradient, according to Simon (1984). Tn5-Mob containing plasmids were isolated from the Km^R clones and introduced into *E. coli* S17.1 by transformation. Plasmids carrying Tn5-Mob in the 10 kb *Eco*RI fragment were subsequently introduced into *Azospirillum* strains by conjugation. *Azospirillum* recombinants were selected by plating on minimal medium containing kanamycin (20 μg/ml). The presence of Tn5-Mob in p90 was checked by hybridization with a Tn5 probe.

Results

Plasmid content of Azospirillum *strains*

Resolution of *Azospirillum* plasmids was previously obtained by the method of Casse *et al.* (1979). This method reveals the presence of megaplasmids over 300 Mda (Franche and Elmerich, 1981). The method of Kado and Liu (1981) employed here is rapid, but does not allow repeated demonstration of megaplasmids. This was also reported by Vanstockem *et al.* (1987). For each strain examined, several extractions and several gels were used and except for the presence of megaplasmids, the plasmid content was in agreement with our previous report (Franche and Elmerich, 1981). The plasmid content of strains not previously analysed is also reported. Table 1 summarizes the results. Above 50 Mda, there is no precision in the estimation of plasmid size. Without making any assumption on the identity of plasmids which, in the different strains, displayed an apparently identical molecular weight, it can be seen in Fig.

1A, that most of the strains examined contained a plasmid of about 90 Mda.

Localization in Azospirillum *genome of the 10 kb EcoRI fragment homologous to* R. meliloti *hsn region*

Preliminary experiments suggested that the cloned fragment in pAB502 originated from the Sp7 p90 plasmid. Since most of the *Azospirillum* strains contain a plasmid of this size range, it was of interest to use the cloned fragment to look for homology between plasmids carried by the various strains. Hybridization was performed with the 10 kb *Eco*RI fragment purified from pAB502. As a control, hybridization was also performed with a 5 kb *Bg*lII-*Xho*I fragment purified from pAB1, which carried the *nif*HDK operon of Sp7. If the probe utilized is homologous with one of the plasmids, a signal is expected with this plasmid and with linear DNA, which contained both chromosomal DNA and denatured plasmid. In contrast, if

Table 1. Plasmid content of *Azospirillum* strains and homology to the p90 plasmid of strain Sp7

Strain	Plasmid	Hybridization with pAB502	Strain reference
A. brasilense			
Sp7	megap; p115; p90	p90	Tarrand *et al.* 1978
7030	megap; p90	p90	Franche and Elmerich 1981
7800	megap; p115; p90-Tn5-Mob	p90-Tn5-Mob	This workrk
Sp13	megap; p85	p85	Tarrand *et al.* 1978
131	p85-Tn5-Mob	NT	This work
Sp107	p100; p40	p100	Baldani *et al.* 1983
Sp245[a]	megap; p90	p90	Baldani *et al.* 1983
Sp246	megap; p90; p80	p90 or p80	Döbereiner
SpCd (ATCC29710)	megap; p115; p90	p90	Eskew *et al.* 1977
SpCd (ATCC29729)	megap; p90	p90	Eskew *et al.* 1977
R07	megap; p115; p90; p4	p115	Franche and Elmerich 1981
A. lipoferum			
Br17	p150	ND	Tarrand *et al.* 1978
RG20a	p90	p90	Tarrand *et al.* 1978
4B	p150	ND	Bally *et al.* 1983
USA5a	p130; p90; p33; p24	p90	Tarrand *et al.* 1978
S28	p150; p115; p33; p29	ND	Franche and Elmerich 1981

A. brasilense strains were grown in nutrient broth, except Sp246 and Sp245 which were grown in KYE. *A. lipoferum* strains were grown in KYE except strain S28 which was grown in nutrient broth. Strain origin: R07 and S28, Senegal; SpCd and USA5a, USA; 4B, France; other strains, Brazil. Megaplasmids are indicated by megap; other plasmids are indicated by p followed by the approximate molecular weight in Mda; ND, not detected; NT, not tested.

[a] Plazinski *et al.* (1983) found three plasmids in strain Sp245. Therefore, plasmid extraction was performed on a new sample provided by Dr. Döbereiner. The same plasmids were detected, plus a plasmid of 75 Mda. It is thus assumed that strain Sp245, used in this work, was strain Sp245 spontaneously cured from the smaller plasmid.

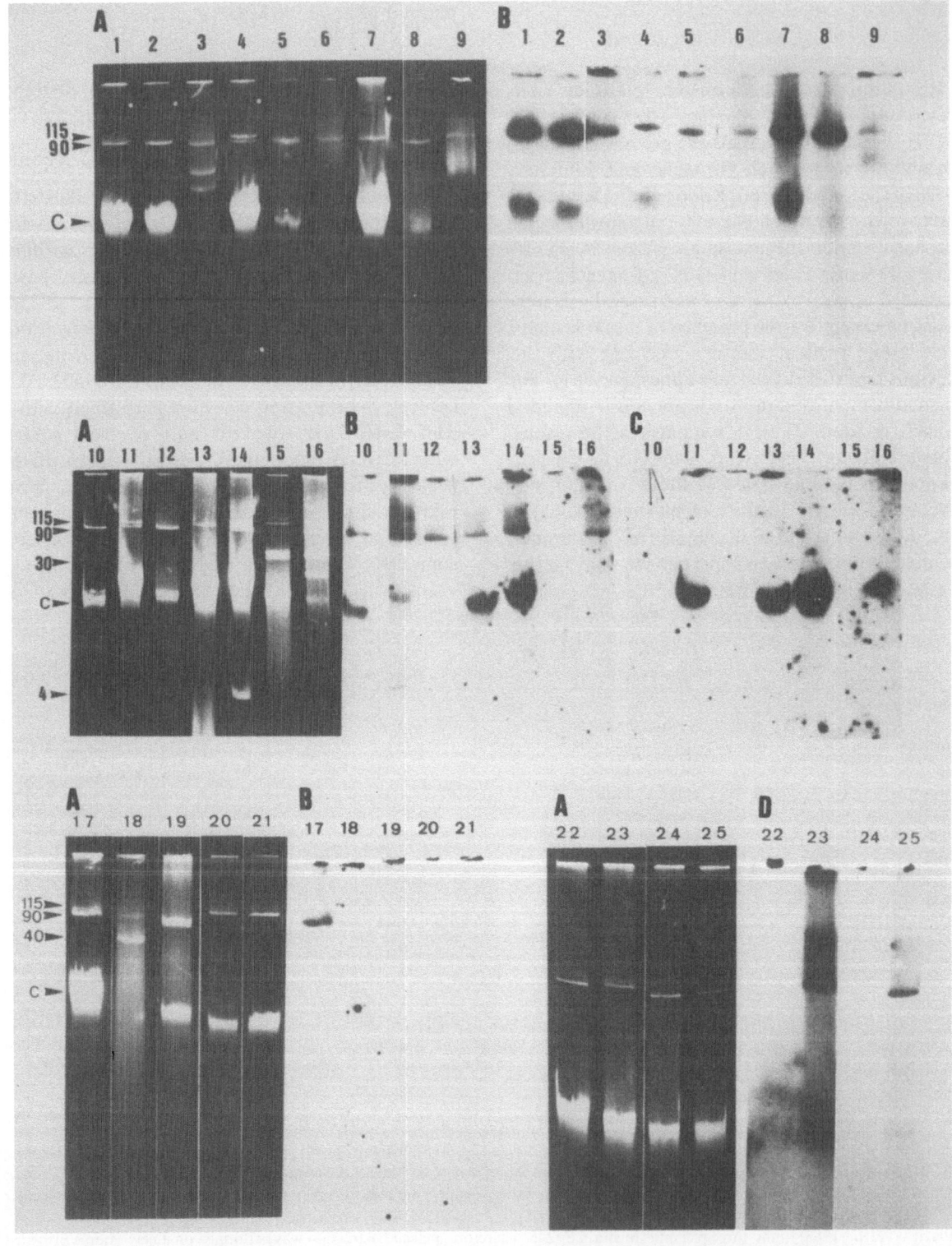

the probe is homologous with a chromosomal region, a signal is expected with linear DNA but not with plasmids.

The *nif* probe hybridized with linear DNA but not with plasmids (Fig. 1C, lanes 10 to 16). For example, in Fig. 1C, hybridization was observed with strains 7800, Sp245, R07 and USA5a (lanes 11, 13, 14, 16). Absence of hybridization in lanes 12 and 15 does not necessarily mean a lack of homology, but it is likely because the plasmid extraction method includes a heat treatment to eliminate most of the chromosomal DNA.

The pAB502 probe hybridized with p90 in strain Sp7 (Fig. 1B, lanes 1 and 17) and 7030 (Fig. 1B, lane 2) and revealed a plasmid larger than p90 in strain 7800 (Fig 1B, lane 11), the increased size of which is likely due to insertion of Tn5-Mob (see below). Hybridization of the pAB502 probe with a purified preparation of p90 from strain 7030 is shown in Fig. 1B, lane 10. The plasmid preparation was partially degraded as judged by the presence of linear DNA. Both plasmid and linear DNA hybridized with the pAB502 probe (Fig. 1B, lane 10) and not with the *nif* probe (Fig. 1C, lane 10).

All other *A. brasilense* strains and two *A. lipoferum* strains, RG20a and USA5a, contained a plasmid hybridizing with the pAB502 probe (Fig. 1B and Table 1). In most cases, the hybridizing plasmid had a molecular weight close to 90 Mda, except for strain R07 in which hybridization was found with p115 (Fig. 1B, lanes 4 and 14). For strain Sp246, it was not possible to distinguish which of the p90 and p80 plasmids hybridized with the probe (Fig. 1B, lanes 5 and 19).

Azospirillum lipoferum Br17, 4B, and S28, which did not contain a plasmid of about 90 Mda, did not hybridize with the pAB502 probe, though a faint signal was detected with p150 of strain 4B (not visible in Fig. 1B, lane 21).

Tn5-Mob site-directed mutagenesis of Azospirillum *plasmids*

Mutagenesis of pAB502 was performed in *E. coli* JC5466 containing Tn5-Mob inserted in the chromosome. Tn5-Mob contains a 3 kb *Bam*HI fragment carrying the *mob* region of RP4 (Simon, 1984). When introduced into the pUC18, the resulting plasmid can be transferred into *Azospirillum* and used for random or site-directed mutagenesis. Several insertions have been isolated in the 10 kb *Eco*RI fragment. Crosses between strain S17.1 (pAB502-Tn5-Mob) and a few *A. brasilense* strains were performed. Kanamycin resistant transconjugants were obtained with several recipients. To check whether the Tn5-Mob had recombined in the homologous plasmid, hybridization with a Tn5 probe was performed with a plasmid preparation from the wild type and from a Km^R clone. Correct recombination was observed with Sp7 and Sp13 Km^R clones. As shown in Fig. 1D, lanes 23 and 25, Km^R derivatives of strain Sp7 and Sp13 hybridized with the Tn5 probe and the hybridizing plasmid displayed an increased molecular weight as compared to p90 and -85. No hybridization was observed with Sp7 and Sp13 wild types, as expected (Fig. 1D, lanes 22 and 24).

Comparison of plasmid regions sharing homology with the 10 kb EcoRI fragment in A. brasilense *strains*

The total DNA of five *A. brasilense* strains, including Sp7, was restricted with *Pst*I and hybridized with the 10 kb *Eco*RI fragment cloned in PAB502, in order to compare regions of the plasmids that were homologous. As a control, hybridization with the *nif* probe was also performed.

Fig. 1. Plasmid content and hybridization with the 10 kb *Eco*RI fragment from pAB502. Part A: agarose gel electrophoresis of plasmids extracted from *Azospirillum* strains; Parts B, C and D, autoradiogram after hybridization with pAB502 (B), *nif*HDK (C), Tn5 (D). Lanes 1, 17 and 22, Sp7; 2, 7030; 3 and 16, USA5a; 4 and 14, R07; 5 and 19, Sp246; 6, RG20a; 7, SpCd (ATCC29710); 8 and 24, Sp13; 9 and 13, Sp245; 10, partially degraded preparation of 7030 p90; 11 and 23, 7800; 12, SpCd (ATCC29729); 15, S28; 18, Sp107; 20, Br17; 21, 4B; 25, 131. Time of exposure to Kodak X-OMAT films was: Part B: 1 day at room temperature, lanes 1, 2, 6 to 9; 1 day at −80 °C, 10 and 12; 3 days at −80 °C, 3 to 5, 11; 7 days at −80 °C, 13 to 16; Part C: 7 days at −80 °C; Part D: 2 days at −80 °C. The gel corresponding to Part C is not shown since it is a duplicate of the gel shown in Part A, lanes 10 to 16. The approximate size of plasmids is indicated in left margin in Mda, C in the left margin indicates migration of linear DNA, In Part A, 1 to 9, white dots indicate the migration of megaplasmids and of R07 p4.

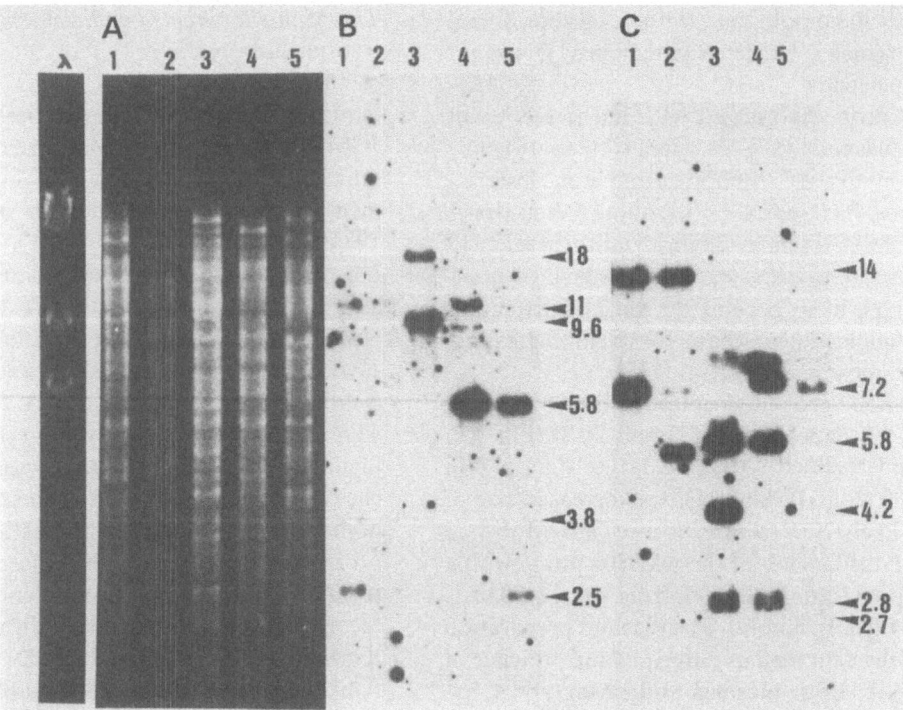

Fig. 2. Comparison of *Pst*I restriction fragments homologous to pAB502 and to the *nifHDK* operon from several *A. brasilense* strains. Part A: agarose gel electrophoresis of total DNA restricted with *Pst*I; Part B: autoradiogram after hybridization with *nifHDK*; Part C: autoradiogram after hybridization with pAB502. 1, Sp7; 2, SpCd (ATCC29710); 3, Sp107; 4, Sp245 and 5, Sp246. The gel corresponding to Part C is not shown, restriction samples were identical but the gel migration was slightly different from the gel shown in Part A. Size of the fragments is indicated in kb, λ: bacteriophage lambda *Hin*dIII restriction fragments (size in kb, 24; 9.5; 6.7; 4.3; 2.3 and 2). Time of exposure to Kodak X-OMAT films, 3 days at −80 °C.

With Sp7, the pAB502 probe hybridized with two *Pst*I fragments of 14 and 7.2 kb (Fig. 2C, lane 1). The 7.2 kb fragment was also observed in strains Sp245 and Sp246 (Fig. 2C, lanes 4 and 5), whereas the 14 kb fragment was observed in ATCC29710 (Fig. 2C, lane 2). This suggested that, in spite of carrying a homologous region, the plasmids were not identical. The same conclusion can be made with the *nif* region. With Sp7, the nif probe hybridized with three *Pst*I fragments of 2.5, 3.8 and 11 kb carrying *nifDK*, *nifH*, and *nifE* respectively (Elmerich *et al.*, 1987). ATCC 29710 displayed a pattern identical to Sp7; the other strains were different.

Discussion

The *Eco*RI fragment from pAB502 homologous to the *hsn* region from *R. meliloti*, and previously

cloned from Sp7 total DNA, appears to originate from the p90 plasmid. This conclusion is based on hybridization experiments between pAB502 and plasmid preparations from Sp7, 7030 and 7800. Strain 7030, an Sp7 mutant, does not contain p115, and thus hybridization with the probe is probably due to p90. In strain 7800, migration of p115 did not change, whereas that of p90 was slower suggesting an increase in size (Fig. 1A, lanes 11 and 23). This was in agreement with a recombination of Tn5-Mob in p90. In 7800, the modified plasmid hybridized both with the pAB502 and the Tn5 probes (Fig. 1B, lane 11 and Fig. 1D, lane 23). In addition, pAB502 hybridized with a preparation of p90, unfortunately partially degraded upon storage (Fig. 1B, lane 10).

It is not known precisely which part of the *hsn* region is conserved in p90, since pAB502 carries one of the three *Eco*RI fragments hybridizing with *hsn* (Fogher *et al.*, 1985). Data reported here showed that there was an homology between p90 of strain Sp7 and a part of *R. meliloti* pSym. Thus, pAB502 can be utilized as a probe to study relationship between plasmids in various *Azospirillum* strains. Physical mapping and cloning of plasmid p90 are in progress and should provide other probes to extend our analysis. At this stage, it is of interest to note that p90 might be related to a plasmid of similar size in most *A. brasilense* and some *A. lipoferum* strains, suggesting the possibility of the existence of a plasmid family in *Azospirillum*. It would be important to define whether these plasmids belong to the same incompatibility group. However, in strain R07, the homologous region is carried by p115 and not by p90. In addition, when the hybridizing *Pst*I restriction fragments from five *A. brasilense* strains were compared (Sp7, SpCd (ATCC29710), Sp107, Sp245 and Sp246) none showed a similar pattern. No homology was found with *A. lipoferum* Br17 and S28 but a faint signal was detected with p150 in strain 4B. This remains to be confirmed. The lack of homology observed in some strains may be due either to the existence of totally unrelated plasmids, to a reduced conservation of the probed fragment, to localization of the fragment on the chromosome, or to a poor transfer of plasmid to nitrocellulose.

Recombination of Tn5-Mob in p90 from strain Sp7 and p85 from strain Sp13 suggests that the DNA region in which Tn5-Mob is inserted is extremely conserved in the two plasmids. Further screening for recombination in other strains is in progress. This technique, when successful, would permit the introduction of a physical and genetic marker in related plasmids. Moreover, the introduction of Tn5-Mob could provide a tool for intra- or intergeneric mobilization of *Azospirillum* plasmids, as described for *Rhizobium* plasmids (Simon, 1984).

The analysis of homology between *Azospirillum* plasmids, and the possibility of performing transposon site-directed mutagenesis in heterologous hosts, should lead, in the near future, to a better understanding of the molecular biology of plasmids and should provide useful tools for genetic analysis of their possible role in the bacteria-plant association.

Acknowledgements

The authors wish to thank Drs J Döbereiner, Y Dommergues and J Balandreau for the gift of strains, Dr J-P Aubert for improving the typescript, and Ms A Paquelin and N Desnoues for skilful technical assistance. C.V. was the recipient of a fellowship from the Ministère de la Recherche et de l'Enseignement Supérieur and I.O. was the recipient of a fellowship from the French Government. This work was supported by research funds from the Université, Paris 7.

References

Baldani VLD, Baldani JI and Döbereiner J 1983 Effects of *Azospirillum* inoculation on root infection and nitrogen incorporation in wheat. Can. J. Microbiol. 29, 924–929.

Bally R, Thomas-Bauzon D, Heulin T, Balandreau J, Richard C and De Ley J 1983 Determination of the most frequent N$_2$-fixing bacteria in a rice rhizosphere. Can. J. Microbiol. 29, 881–887.

Casse F, Boucher C, Julliot JS, Michel M and Dénarié J 1979. Identification and characterization of large plasmids in *Rhizobium meliloti* using agarose gel electrophoresis. J. Gen. Microbiol. 113, 229–242.

Elmerich C 1986 *Azospirillum*. *In* Nitrogen Fixation Vol. IV: Molecular Biology. Eds. A. Pühler and W.J. Broughton. pp. 106–126. Clarendon Press, Oxford.

Elmerich C, Bozouklian H, Vieille C, Fogher C, Perroud B, Perrin A and Vanderleyden J 1987 *Azospirillum*: Genetics of nitrogen fixation and interaction with plants. Phil. Trans. R. Soc. Lond. B, 317, 183–192.

Elmerich C, Quiviger B, Rosenberg C, Franche C, Laurent P and Döbereiner J 1982 Characterization of a temperate bacteriophage for *Azospirillum*. Virology 122, 29–37.

Eskew DL, Focht DD and Ting IP 1977 Nitrogen fixation, denitrification and pleomorphic growth in a highly pigmented *Spirillum lipoferum*. Appl. Environm. Microbiol. 34, 582–585.

Fogher C, Dusha I, Barbot P and Elmerich C 1985 Heterologous hybridization of *Azospirillum* DNA to *Rhizobium nod* and *fix* genes. FEMS Microbiol. Lett. 30, 245–249.

Franche C and Elmerich C 1981 Physiological properties and plasmid content of several strains of *Azospirillum brasilense* and *Azospirillum lipoferum*. Ann. Microbiol. 132A, 3–18.

Gauthier D and Elmerich C 1977 Relationship between glutamine synthetase and nitrogenase in *Spirillum lipoferum*. FEMS Microbiol. Lett. 2, 101–104.

Horvath B, Kondorosi E, John M, Schmidt J, Török I, Györgypal Z, Barabas I, Wieneke U, Schell J and Kondorosi A 1986 Organization, structure and symbiotic function of *Rhizobium meliloti* nodulation genes determining host specificity for Alfalfa. Cell 46, 335–343.

Kado ST and Liu CI 1981 Rapid procedure for detection and isolation of large and small plasmids. J. Bacteriol. 145, 1365–1373.

Kondorosi E, Banflavi Z and Kondorosi A 1984 Physical and genetic analysis of a symbiotic region of *Rhizobium meliloti*: identification of nodulation genes. Mol. Gen. Genet. 193, 445–452.

Miller J 1972 Experiments in Molecular Genetics. Cold Spring Harbour Laboratory, Cold Spring Harbour, New York.

Prakash RK and Atherly A 1986 Plasmids of *Rhizobium* and their role in symbiotic nitrogen fixation. Int. Rev. Cytol. 104, 1–24.

Plazinski J, Dart PJ and Rolfe B 1983 Plasmid visualization and *nif* gene location in nitrogen fixing *Azospirillum* strains. J. Bacteriol. 155, 1429–1433.

Quiviger B, Franche C, Lutfalla G, Rice D, Haselkorn R and Elmerich C 1982 Cloning a nitrogen fixation (*nif*) gene cluster of *Azospirillum brasilense*. Biochimie 64, 495–502.

Simon R 1984 High frequency mobilization of Gram-negative bacterial replicons by the *in vitro* constructed Tn5-Mob transposon. Mol. Gen. Genet. 196, 413–420.

Simon R, Priefer U and Pühler A 1983 A broad host range mobilization system for *in vivo* genetic engineering: Transposon mutagenesis in Gram-negative bacteria. Bio/Technology 1, 784–791.

Tarrand J, Krieg NR and Döbereiner J 1978 A taxonomic study of the *Spirillum lipoferum* group, with description of a new genus, *Azospirillum* gen. nov. and two species *Azospirillum lipoferum* (Beijerinck) comb. nov. and *Azospirillum brasilense* sp. nov. Can. J. Microbiol. 24, 967–980.

Vanstockem M, Michiels K, Vanderleyden J and Van Gool A 1987 Transposon mutagenesis of *Azospirillum brasilense* and *Azospirillum lipoferum*: Physical analysis of Tn5 and Tn5-Mob insertion mutants. Appl. Environm. Microbiol. 43, 410–415.

F. A. Skinner et al. (Eds.), Nitrogen fixation with non-legumes, 173–178.

Self-transmissible *nif*-plasmids in *Enterobacter*

W. KLINGMÜLLER, SABINE HERTERICH and BYUNG-WHAN MIN
Department of Genetics, University of Bayreuth, D-8580 Bayreuth, FRG

Key words: *Enterobacter*, N_2-fixation, *nif*-genes, plasmids, transmissibility, transposon-tagging, wheat-rhizosphere

Abstract

Nitrogen fixing *Enterobacter agglomerans* strains, isolated from the innermost rhizosphere of wheat, contained all genes essential for nitrogen fixation on large (100–150 kb) indigenous plasmids. One of these plasmids, pEA9, was marked with either Tn1725 or Tn5, and studied with regard to its transmissibility by conjugation with other recipient strains. In a homologous system (cured parental strains as recipients) the rate of transfer was in the order of 10^{-5} per recipient cell. It could be drastically increased by introduction of a mob site (on Tn5) and offering plasmid RP4 as helper. However, the host range was narrow. To obtain *nif*-plasmids with a wide host range, the *nif*-genes have been cut out and ligated to a broad host range vector, pRK415, derivative of RP4. These studies are a prerequisite for later releasing experiments.

Introduction

In countries where nutrition is mainly based on cereals, e.g. maize, wheat and rice, non-symbiotic nitrogen-fixing bacteria have attracted considerable interest as potential biofertilizers. Among them, *Azospirillum brasilense* and *A. lipoferum* are those studied most frequently. The present state of investigations using these organisms was summed up in the IVth International Bayreuth Azospirillum workshop (Klingmüller, 1987). Although progress over the last 15 years has been considerable, much has to be done before practical applications of *Azospirillum* spp. do not form a substantial part of ics. Their applicability in temperate climates is doubtful, since here in contrast to tropical climates, *Azospirillum* spp. do not form a susbtantial part of the bacterial population in the rhizosphere of cereals and it is unlikely that inoculation could establish them permanently in sufficient numbers.

For temperate climates, we have identified the group of Enterobacter, as potentially useful biofertilizer candidates (Kleeberger *et al.*, 1983). These bacteria form up to 23% of the bacterial microflora in the innermost rhizosphere of wheat and barley; some of them fix N_2, and, in addition, they are related to *E. coli* and *K. pneumoniae*, so that genetic methods developed for those two species can be extrapolated to them.

Special attention was paid to *E. agglomerans* strains (14% of bacterial microflora). Of 50 isolates, 5 fixed nitrogen naturally. It could be shown that the nitrogenase structural genes KDH in these isolates are located on large plasmids (Singh *et al.*, 1983). One of the strains, representative of the others, *E. agglomerans* 333, contained a plasmid (111 kb), which we designated as pEA3, and analyzed further by genetechnological methods (Singh and Klingmüller, 1986; Singh *et al.*, 1988). A physical and genetic map of the plasmid was presented (Kreutzer, 1986). The plasmid contains a 23 kb cluster of *nif*-genes which shows homology (by Southern hybridization and heteroduplex analysis) not only to the KDH genes, but to the entire *nif*-gene cluster of *Klebsiella pneumoniae* M5A1. Apart from *nif* J, which is located on the left end of the pEA3 *nif*-gene cluster (near *nif* QB), all the other *nif*-genes on pEA3 are organized in the same manner as in *K. pneumoniae*. Preliminary experiments indicate that the same holds for the other 4 *nif*-plasmids, among them pEA9 from *E. agglomerans* 339. A BamHI restriction map of pEA3 and a detailed restriction map of the 23 kb *nif*-region on pEA3 has been presented recently (Singh et al. 1988).

Since the *nif*-genes are contiguous on these *nif*-plasmids, forming a package, these plasmids seem extremely well suited for transfer of the *nif* information to other, non-fixing soil bacteria. We have taken up this question, and first studied their self-transmissibility. The results are described in the following paragraphs.

Materials and methods

Bacterial strains

a) *Enterobacter cloacae* MF10 was isolated by Glatzle from, the rhizosphere of *Festuca heterophylla* (Kleeberger and Klingmüller, 1980). This strain is free of plasmids and is non-nitrogen-fixing.
b) *Enterobacter agglomerans* 333, 335, 339, all three nitrogen-fixing, were isolated from the rhizosphere of wheat and barley (Kleeberger et al., 1983).
c) *Escherichia coli* HB101 F⁻ hsd20 r_B^- m_B^- recA13 ara14 proA2 lacY galK2 rpsL20 xyl5 mtl1 supE44 Rif^r (Boyer and Rouillard-Dussioux, 1969).
Escherichia coli S17-1 pro r⁻ m⁺ Tpr^r Sm^r RP4-2-Tc::Mu-Kan::Tn7 (Simon *et al.*, 1983).

Plasmids and transposons

pMS175-2 Amp^r, Tet^r, Mob$^+$ (Singh and Klingmüller, 1986). PMS175-2::Tn1725 Amp^r, Tet^r, Cm^r, Mob$^+$ (Singh and Klingmüller, 1986).

pSUP5011, Amp^r, Cm^r, Tn5 Kan^r::Mob, col-replicon (Simon, 1984). pSUP202, Amp^r, Cm^r, Tet^r, col-replicon.

pPH1JI, Gent^r, Cm^r Str^r, Spec^r, IncP (Hirsch and Beringer, 1984). RP4, Tet^r, Kan^r, Cb^r, IncP (Datta et al., 1971).

pRK415, derivative of pRK404 (Ditta *et al.*, 1985, and N. Keen, personal communication).

Growth media

All bacterial strains were grown in LB medium (g per 1 distilled water): tryptone, 10; yeast extract, 5; NaCl, 5; *Escherichia coli* and *Enterobacter cloacae* were grown at 37 °C and *Enterobacter agglomerans* at 30 °C.

Antibiotics and amino acids

Carbenicillin, rifampicin and streptomycin were used at 100, chloroamphenicol, kanamycin and nalidixic acid at 50, and tetracycline at 15, $\mu g \cdot ml^{-1}$.

Acetylene reduction assay

Nitrogen fixation was assessed for bacterial suspensions grown in N-free minimal with glucose (NFDM, for *E. coli* strains) or sucrose (NFSM, for *E. agglomerans* strains), under anaerobic conditions, adding N_2 and using the C_2H_2 reduction assay. For details see Klingmüller (1985).

Bacterial matings

These, if not stated otherwise, were done as described earlier (Klingmüller, 1979). For triparental matings, fresh cell cultures were mixed as patches on solid complete medium, incubated overnight, and then restreaked on selective plates (G. Ditta, pers. comm.).

Preparation of plasmid DNA and electrophoresis

For rapid detection of plasmid DNA, crude lysates from a small volume of the bacterial culture (1.0 ml) were prepared following either the method described by Kado and Liu (1981) or abbreviated laboratory protocols. Whenever necessary, large quantities of plasmid DNA were purified by caesium chloride-ethidium bromide (CsCl-EtBr) density gradient centrifugation. Plasmids were separated on 0.7% agarose gels followed by staining with ethidium bromide. For separating restriction fragments, 0.8% agarose gels containing EtBr ($1 \mu g\, ml^{-1}$) were run overnight.

Southern transfer and hybridization

See Maniatis *et al.* (1982) and Singh *et al.* (1988).

Results

Transmissibility of nif-plasmid pEA9, homologous situation

Since most natural plasmids have a narrow host range, mating experiments with identical bacteria as recipients (homologous situation) were planned first. Plasmid curing (Singh *et al.*, 1983) had provided us with two Nif⁻ *E. agglomerans* strains, derivatives of Nif⁺ strains, that lacked their *nif*-plasmids, *viz. E. agglomerans* 335, Nif⁻ and 339⁻, Nif⁻. The eliminated plasmids were pEA5 and pEA9, respectively. The other three Nif-positive *E. agglomerans* strains could not be cured. As donor strain in the mating experiments therefore, the parental strain *E. agglomerans* 339, containing plasmid pEA9 was used.

To follow transmission of this plasmid (pEA9 = 150 kb) it was marked in *E. agglomerans* 339 Rif^r by *in vivo* transfer of Tn1725 (Cm^r). Rifampicin resistance was used to select against the donor (Herterich, 1986). Exconjugants of the desired type (Cm^r, Rif^r) were then superinfected with a closely related incompatible plasmid so that only exconjugants with Tn1725 integrated in their genome (either chromosome or plasmids) would remain Cm^r and become selected as such. This transposition was observed in one of 7000 exconjugants. The relevant class of exconjugants was then screened by Southern hybridization, using ^{32}P-labelled Tn1725 as probe. Three isolates were identified, harbouring Tn1725 in plasmid pEA9. One of them was designated 339 T1.

Enterobacter agglomerans cells containing the marked plasmid were then mated with cured, but Sm^r or Nal^r derivative strains of *E. agglomerans* 339 as recipients. Exconjugants were selected for Cm^r and Sm^r or Nal^r (Herterich, 1986), and again

checked further by screening for plasmid content, by Southern hybridizations against ^{32}P-labelled Tn1725, and by measuring acetylene reduction.

The rate of plasmid transfer thus obtained was in the order of 10^{-5} per recipient cell, depending on mating conditions, titre of donor and recipient cells, and type of donor and recipient strains used. The exconjugants contained the Tn-labelled plasmid pEA339, and reduced acetylene (were nif⁺) with rates similar to the wild type of the recipients. The highest rate of plasmid transfer ($\geqslant 10^{-3}$ per recipient cell) was obtained when cells from the logarithmic growth phase were incubated together for more than 20 h in a sediment (8000 rev.min^{-1}, Eppendorf centrifuge) with strain 339 T1 as donor and strain 339/22-1 Sm^r as recipient (Table 1).

It should be noted that the remaining two plasmid mutants mentioned before in addition to 339 T1, one of them designated as 339 T45, were ineffective in conjugation experiments. Their Tra⁻ phenotype could be due to integration of the transposon in a *tra* gene, or to deletions. One such deletion (of 14 kb) was osberved in plasmid 339 T45 upon HindIII digestion and gel electrophoresis. Using a recipient strain labelled with Nal^r in addition to Sm^r, transfer rates of 5×10^{-5} were obtained with 339 T1 as donor.

Transmissibility of plasmid pEA9, heterologous situation

In additional experiments, the host range of transfer was checked, offering other bacterial strains as recipients in matings with 339 T1. Closely related to *E. agglomerans* 339 is *E. agglomerans* 335 (Kleeberger *et al.*, 1983). Both strains had formerly been cured of their *nif* plasmid by heat treatment.

Table 1. Rates of plasmid transfer *versus* incubation time for *E. agglomerans* strain 339 T1 as donor and strain 339/22-1 Sm^r as recipient

Duration of incubation, (h)	Titre of recipient	Titre of donor	Titre of transconjugants	Conjugation rate per recipient
2	1.5×10^9	6.0×10^8	4.0×10^2	2.7×10^{-7}
4	1.1×10^9	7.0×10^8	4.4×10^3	4.0×10^{-6}
8	1.2×10^9	9.9×10^8	2.3×10^4	1.9×10^{-5}
16	8.0×10^8	4.0×10^8	3.2×10^5	4.0×10^{-4}
24	4.2×10^8	4.1×10^8	9.7×10^5	2.3×10^{-3}
32	1.4×10^8	2.0×10^8	5.0×10^5	3.6×10^{-3}
48	1.9×10^7	1.1×10^7	6.5×10^4	3.4×10^{-3}

Such cured 335⁻ cells were now used as recipients, again as Sm^r derivatives. Under the conditions found optimal for the former (homologous) case, transfer rates of 4×10^{-4} per recipient cell were obtained (Table 2, a). Again plasmid profile, Southern hybridization and acetylene reduction assay showed that the exconjugants had obtained the transposon-marked plasmid. Experiments with *E. cloacae* MF10 and *E. coli* HB101 as recipients were negative. This indicates that the host range for transfer of plasmid pEA9 is narrow. The potential ecological relevance of selftransmissibility of *nif*-plasmid pEA9 as such seems therefore rather small.

Increase of transmissibility by insertion of mob

To increase the transfer rate, insertion of a mob site into the plasmid pEA9 was envisaged (Min, 1986). This mob site could for example correspond to the transfer system of the broad host range plasmid RP4, to achieve good transfer in the laboratory, in combination with a suitable helper strain (Ditta *et al.*, 1980).

To introduce the mob site, a derivative of transposon Tn5 (Kan^r), harbouring this site was used (Simon *et al.*, 1983; Simon, 1984). It contained the mob region as the Sau3A fragment from the ori-T region of RP4. The transposon itself was delivered on plasmid pSUP5011, using *E. coli* S17-1 with RP4 in the chromosome as helper. Since plasmid pSUP5011 replicates in the recipient *E. agglomerans* 339, Rif^r, the exconjugants had to be

cured by superinfection with plasmid pSUP202 (Tet^r) of the same incompatibility group. Suitable drug combinations were used for selection. Those cells, now harbouring Tn5::mob on plasmid pEA9 itself, and not on the chromosome or on megaplasmids, were identified by secondary matings of the exconjugants to *E. agglomerans* 339/22-1, Nal^r.

A mutant of plasmid pEA9 was thus identified (designated as M13–19) which, if contained in *E. agglomerans* 339, Rif^r as donor, and mated to *E. agglomerans* 339/22-1 Nal^r as recipient (at about 2×10^9 recipient cells per ml final concentration, 30 °C, 24 h) gave about 5×10^{-6} exconjugants per recipient upon selection for Kan^r and Nal^r (Table 2, b). The identity of the exconjugants was verified by screening for their plasmids, and by running Southern hybridizations with the 4.6 kb HindIII fragment of plasmid pSUP5011 as probe, which contains the 2.7 kb central part of Tn5 with the inserted mob region (1.9 kb). Using a Sm^r-recipient, instead of Nal^r recipient, as in the experiments under a), rates of about 6×10^{-8} were obtained with uncontrolled pH. These rates could be increased, using controlled pH of 5, for matings to recipients with Nal^r (rate 4×10^{-5} per recipient) and Sm^r (rate 3×10^{-6} per recipient).

It was then investigated if the rate of transfer was increased by the introduced mob site, using two different suitable helper plasmids. The donor strain was *E. agglomerans* 339/22-1, Nal^r, containing the marked plasmid pEA9::Tn5::mob. In a first set of experiments, plasmid PH1JI was introduced into this strain by conjugation. The final recipient for

Table 2. Transfer of *nif*-plasmid pEA9

Donor *Enterobacter* *agglomerans* 339 (pEA9, *nif*)	Recipient	Rate of transfer per recipient
(a) Plasmid	*Enterobacter agglomerans* 339, Sm^r	$10^{-5} - 10^{-3}$
labelled with	*Enterobacter agglomerans* 339, Sm^r, Nal^r	5×10^{-5}
Tn1725, Cm^r	*Enterobacter agglomerans* 335, Sm^r	4×10^{-4}
	Enterobacter cloacae MF10, Sm^r	–*
	Escherichia coli HB101, Sm^r	–
(b) Plasmid labelled	*Enterobacter agglomerans* 339, Nal^r	$5 \times 10^{-6} - 4 \times 10^{-5}$
with Tn5-Mob$^+$, Kan^r	*Enterobacter agglomerans* 339, Sm^r	$6 \times 10^{-8} - 3 \times 10^{-6}$
+ helper plasmid: PH1JI	*Enterobacter agglomerans* 339, Sm^r	6×10^{-6}
RP4	*Enterobacter agglomerans* 339, Sm^r	10^{-2}
RP4	*Enterobacter cloacae* MF10, Sm^r	–
RP4	*Escherichia coli* HB101, Sm^r	–

* –, no transfer of plasmid pEA9.

the mating was *E. agglomerans* 339/22-1 Sm^r, as before. The rate of transfer thus obtained was 6×10^{-6} as compared to 6×10^{-8}, obtained under identical conditions without helper plasmid. In a second set of experiments, plasmid RP4 was introduced into the donor, again by conjugation. Using the same recipient as before, a transfer rate of 10^{-2} was found, as compared to the rate without helper plasmid of 6×10^{-8}, indicating an increase of more than 10^5.

Experiments with heterologous recipients, administering the pEA9 donor plasmid marked by Tn5::mob, and using either of both helper plasmids, gave no exconjugants, either with *E. cloacae* MF10 or *E. coli* HB101 as recipients, again indicating a narrow host range of pEA9. However, these negative results do not exclude transfer of pEA9 into the respective recipients. Transfer is likely to have occurred, since the RP4 system in combination with the mob site with its wide host range makes it possible, at least in principle. Rather, the fact that no exconjugants were found suggests that after transfer there is no replication of pEA9 in the heterologous recipients.

Broad host range constructs with the nif-*gene group from pEA3*

In our present work we are also using the complete *nif*-gene group of plasmid pEA3 (from *E. agglomerans* 333) alone, instead of whole indigenous *nif*-plasmids, as a means to spread the *nif* information to other non-fixing *Enterobacter* and related bacteria from the rhizosphere of wheat. In this context new plasmid constructs are being prepared. They contain the *nif*-gene group ligated into a safety vector plasmid with wide host range, but lacking transfer functions. Such vectors are available as smaller derivatives of plasmid RP4 (Ditta *et al.*, 1985). Conjugal transfer can then be achieved in the laboratory by adding a helper strain that provides the transfer functions needed (triparental matings).

Once a larger number of other strains have thus been infected with either the natural or the new hybrid type *nif*-plasmids, propagation of the recipients and release into the rhizosphere of wheat plants is being envisaged. The hybrid type system would be relatively safe, since no uncontrolled spread of the *nif*-plasmids could occur *in vivo*. Both

systems should be efficient since the recipient bacteria, originally taken from the rhizosphere, should easily re-establish there. Details of this approach are to be published elsewhere.

Discussion

In *E. agglomerans* strains from the rhizosphere of wheat and barley, all *nif*-genes essential for nitrogen fixation are located on large indigenous plasmids, in the form of a contiguous package. For one of these plasmids, pEA9, self-transmissibility is documented here, but with low efficiency, and only to closely related *Enterobater* strains. The rate of transfer could be increased by introducing the mob site into the plasmid, and adding a helper strain. Such studies serve as model experiments for elucidating the basic situation.

To obtain beneficial effects with cereals via inoculation, it seems important to make as many as possible of their rhizosphere bacteria Nif-positive, e.g. by *nif*-plasmid transfer as reported above. For this, high transfer rates, and a wide host range of the *nif*-plasmids is essential. To increase the transfer rate of *nif*-plasmid pEA9, mobilization was helpful. Also, derepression of the transfer functions by mutation would seem promising (Chandler and Krishnapillai, 1977). Additional transfer studies have to be carried out with the other *nif*-plasmids described here, and in particular with pEA3, since the latter plasmid is now the best characterized. To widen the host range, our new constructs with respective vector plasmids seem promising. Problems have arisen however, due to the instability of the vector plasmids. This is known to be an inherent property of all smaller derivatives of RP4. Therefore, in the long run it seems necessary, for stabilization, to achieve integration of the *nif*-genes into the chromosome of heterologous recipients. For this purpose, vector plasmids with special regions, favouring their random or even their site-specific integration (e.g. via homologous recombination) into the genome of the recipients, might be needed.

Acknowledgements

We thank Mrs. C Fentner for devoted technical assistance, Mrs. B Gubitz and Mrs. G Siebeker for typing and NM Stitt for reading the manuscript.

References

Boyer HW and Roulland-Dussoix D 1969 Complementation analysis of the restriction and modification of DNA in *Escherichia coli*. J. Mol. Biol. 41, 459–472.

Chandler PM and Krishnapillai V 1977. Characterization of *Pseudomonas aeruginosa* derepressed R-plasmids. J. Bacteriol. 130, 596–603.

Datta N, Hedges RW, Shaw EJ, Sykes RP, Richmond MH 1971 Properties of an R factor from *Pseudomonas aeruginosa*. J. Bacteriol. 108, 1244–1251.

Ditta G, Stanfield S, Corbin D and Helinski DR 1980 Broad host range cloning system for Gram-negative bacteria: Construction of a gene bank of *Rhizobium meliloti*. Proc. Natl. Acad. Sci. USA 77, 7347–7351.

Ditta G, Schmidhauser T, Yakobson E, Lu P, Liang X-W, Finlay DR, Guiney D and Helinski DR 1985 Plasmids related to the broad host range vector, pRK290, useful for gene cloning and for monitoring gene expression. Plasmid 13, 149–153.

Herterich S 1986 Untersuchung der Transfereigenschaft des Plasmids pEA9 aus *Enterobacter agglomerans* 339 und Entwicklung eines Transposon-Mutagenesesystems. Diplomarbeit, Universität Bayreuth.

Hirsch PR and Beringer JE 1984 A physical map of pPH1JI and pJBJI. Plasmid 12, 139–141.

Kado CI and Liu ST 1981 Rapid procedure for detection and isolation of large and small plasmids. J. Bacteriol. 145, 1365–1373.

Kleeberger A and Klingmüller W 1980 Plasmid-mediated transfer of nitrogen fixing capability to bacteria from the rhizosphere of grasses. Mol. Gen. Genet. 180, 621–627.

Kleeberger A, Castorph H and Klingmüller W 1983 The rhizosphere microflora of wheat and barley with special reference to gram-negative bacteria. Arch. Microbiol. 136, 306–311.

Klingmüller W 1979 Genetic engineering for practical application. Naturwissenschaften 66, 182–198.

Klingmüller W 1985 Studies on natural and artificial nitrogen-fixing soil bacteria. *In* Trends in Molecular Genetics. Eds. U Sinha and W Klingmüller pp. 37–61. Spectrum Publishing House, Patna, Delhi.

Klingmüller W (ed.) 1987 *Azospirillum* IV: Genetics, Physiology, Ecology. Springer-Verlag, Berlin, Heidelberg, New York, Tokyo.

Kreutzer R 1986 Restriktionskartierung eines *nif*-Plasmids von *Enterobacter agglomerans* und genetische Charakterisierung der von ihm getragenen *nif*-Gene. Diplomarbeit, Universität Bayreuth.

Maniatis T, Fritsch E and Sambrook J 1982 Molecular Cloning. Cold Spring Harbor Laboratory, Cold Spring Harbor, N.Y.

Min BW 1986 Erzeugung von Nif⁻-Mutanten des Stammes *Enterobacter agglomerans* 339 durch Transposonmutagenese und Untersuchung der Transfereigenschaft seines Plasmids pEA9. Diplomarbeit, Universität Bayreuth.

Simon R 1984 High frequency mobilization of gram-negative bacterial replicons by the in vitro constructed Tn5-Mob transposon. Mol. Gen. Genet. 196, 413–420.

Simon R, Priefer U and Pühler A 1983 Vectors for in vivo and in vitro manipulations of gram-negative bacteria. *In* Molecular Genetics of Bacteria-Plant Interaction. Ed. A Pühler. pp. 98–105. Springer-Verlag, Berlin, Heidelberg, New York, Tokyo.

Singh M, Kreutzer R, Acker G and Klingmüller W 1988 Localization and physical mapping of a plasmid-borne 23 kb *nif*-gene cluster from *Enterobacter agglomerans* showing homology to the entire *nif*-gene cluster of *Klebsiella pneumoniae* M5a1. Plasmid. (In press).

Singh M, Kleeberger A and Klingmüller W 1983 Location of nitrogen fixation (*nif*) genes on indigenous plasmids of *Enterobacter agglomerans*. Mol. Gen. Genet. 190, 373–378.

Singh M and Klingmüller W 1986 Cloning of pEA3, a large plasmid of *Enterobacter agglomerans* containing nitrogenase structural genes. Plant and Soil 90, 235–242.

F. A. Skinner et al. (Eds.), Nitrogen fixation with non-legumes, 179–187.
© 1989 by Kluwer Academic Publishers.

Identification of *Bacillus azotofixans* nitrogen fixation genes using heterologous *nif* probes

L. SELDIN, M.C.F. BASTOS and E.G.C. PENIDO
Instituto de Microbiologia, Universidade Federal do Rio de Janeiro, C.C.S., Bloco I, Ilha do Fundão, Rio de Janeiro, RJ — CEP: 21.944 — Brasil

Key words: *Bacillus azotofixans*, genes, *nif* genes, nitrogen fixation

Abstract

Twenty-two strains of *Bacillus azotofixans*, a new nitrogen-fixing species, were tested for the presence of DNA homology to the *Klebsiella pneumoniae nif* genes by the Southern hybridization technique. Total DNA from all *B. azotofixans* strains tested was shown to carry homologous sequences only to *K. pneumoniae* structural *nif* genes. When *B. azotofixans* DNA was digested with *Eco*RI, *Eco*RV, *Hind*III or *Sal*I and hybridized with pSA30 containing *nifHDK*, more than one fragment was obtained in all strains tested with the different restriction enzymes. The hybridization pattern also varied among the strains. When probes containing each of the *K. pneumoniae* structural *nif* genes were used, the homology detected between plasmid pSA30 and *B. azotofixans* DNA was limited to *nifD* and *nifH*. The results now reported also suggest that the *B. azotofixans nifDH* genes are contiguous although *nifH* probably underwent a rearrangement during evolution.

Introduction

Up to 1984, all *Bacillus* strains described as nitrogen-fixers belonged to the species: *B. polymyxa* (Grau and Wilson, 1962), *B. macerans* (Witz *et al.*, 1967) and *B. circulans* (Line and Loutit, 1971). Some other nitrogen-fixing *Bacillus* strains having an atypical pattern of classification for *B. polymyxa* were also isolated from different soils and plants (Hino and Wilson, 1958; Larson and Neal, 1978; Seldin *et al.*, 1983) and were tentatively identified as variants of this species. In 1984, a new nitrogen-fixing member of the family *Bacillaceae*, *Bacillus azotofixans*, was described (Seldin *et al.*, 1984). The 22 *B. azotofixans* strains were isolated from non-rhizosphere soil and from grass roots, obtained in five different locations in Brazil and also in Japan, suggesting that they may occur either as free-living bacteria or as root-associated bacteria. Additional biochemical studies were performed with *B. azotofixans* strains using API tests (Seldin and Penido, 1986) and the levels

of DNA homology among the strains showed 54 to 100% DNA relatedness (Seldin and Dubnau, 1985), indicating that these strains comprise a relatively homogeneous species.

The ability of many different species of prokaryotes to fix nitrogen has been shown to be due to the enzyme nitrogenase, composed of the MoFe protein and the Fe protein. The nitrogenase enzyme complex is highly conserved among evolutionarily distant organisms and the DNA that codes for nitrogenase from one nitrogen-fixing species has been shown to hybridize specifically with DNA from many other nitrogen-fixing species (Mazur *et al.*, 1980; Ruvkun and Ausubel, 1980).

The model organism for the study of genetics of nitrogen fixation is *Klebsiella pneumoniae*: the MoFe protein is an $\alpha_2 \beta_2$ tetramer and is encoded by *nifD* and *nifK*, while the Fe protein is a dimer of two identical subunits and is encoded by *nifH*. Besides these structural genes (*nifHDK*), 14 additional genes are necessary for nitrogen fixation in *K. pneumoniae* (Dixon, 1984). Genetic studies of *nif* were facilitated by the construction of small amplifiable plasmids carrying *nif* genes (Cannon *et al.*, 1977, 1979) and these plasmids have been used as probes in hybridization experiments (e.g. Fornari and Kaplan, 1983; Mathis *et al.*, 1986; Mazur *et al.*, 1980; Ruvkun and Ausubel, 1980; Sibold *et al.*, 1985). No information is available on the genetics of nitrogen fixation in *Bacillus*, as yet. Based on the idea that DNA coding for nitrogenase has been conserved among different organisms during evolution, it was decided to identify the nitrogenase genes in *Bacillus azotofixans* by hybridization with heterologous *nif* probes by the technique developed by Southern (1975). It was considered important to carry out these molecular analyses as a first step toward the subsequent cloning of the nitrogenase structural genes from this new *Bacillus* species. Recombinant plasmids containing all of the nitrogenase structural genes from *Klebsiella pneumoniae*, as well as these genes cloned separately, were used in this work. Results reported here show that *B. azotofixans* strains contain DNA sequences homologous to the nitrogenase structural genes of *K. pneumoniae* strain M5al, although the organization of these sequences appears to differ among the different *B. azotofixans* strains.

Materials and methods

Bacterial strains and plasmids

The properties of the *Bacillus azotofixans* and *Escherichia coli* strains used are summarized in Tables 1 and 2.

Propagation of cultures

For growth of *B. azotofixans* cells for DNA isolation, 1.5 ml from a 24 h culture grown in TBN liquid medium (Seldin and Penido, 1986), was inoculated into 150 ml of the same medium and incubated at 32 °C for 18 h. *Klebsiella pneumoniae* and *Escherichia coli* strains were grown in LB medium (Schleif and Wensink, 1981) at 37 °C. For plasmid isolation, *E. coli* strains were grown with appropriate antibiotics.

Isolation of chromosomal DNA

Bacillus azotofixans cells from a 150 ml culture were concentrated in 1 ml of lysis buffer (50 mM tris (hydroxymethyl) aminomethane (Tris), 50 mM ethylenediaminetetraacetate (EDTA), pH 7.8) and treated with 1 mg of lysozyme for 30 min at 37 °C. Sodium dodecyl sulphate was added to a final concentration of 1% and in order to purify the DNA, 1 mg of pronase (catalogue No. P5147, Sigma Chemical Co.) per ml was added, and the preparation incubated at 37 °C for 2 h. It was then shaken with an equal volume of phenol:chloroform :isoamyl alcohol (1:0.96:0.04) and incubated at 37 °C for 1 h. The aqueous phase containing the DNA was treated with an equal volume of chloroform:isoamyl alcohol (24:1), and the DNA was precipitated by adding 3 M sodium acetate (final concentation, 0.3 M) and 3 volumes of cold ethanol. DNA was pelleted by centrifugation at 12,100 **g** for 10 min at 4 °C, resuspended in TE (10 mM Tris, 1 mM EDTA) at pH 7.8 and then treated with pancreatic ribonuclease (50 μg ml^{-1}) and ribonuclease T$_1$ (1 U ml^{-1}) for 1 h at 37 °C. After two treatments with phenol and chloroform:isoamyl alcohol, the DNA was precipitated and dissolved in 1 ml of TE, pH 7.8.

Klebsiella pneumoniae DNA used as positive control and *Staphylococcus aureus* DNA used as negative control in some hybridization experiments were isolated using the same procedure as described for *B. azotofixans* with minor modifications: *K. pneumoniae* cells were treated with lysozyme (1 mg ml^{-1}) for 30 min on ice and *S. aureus* cells were treated with lysostaphin (0.5 mg ml^{-1}) for 30 min at 37 °C.

Isolation of plasmid DNA

The alkaline method of Birnboim and Doly (1979) was used for the isolation of plasmid DNA. All preparations were purified in CsCl-ethidium bromide gradients to eliminate chromosomal contamination in the hybridization experiments, and dialysed against 10 mM Tris, 0.5 mM EDTA, pH 7.8.

Restriction enzyme cleavage

Chromosomal DNA was routinely digested with 10 to 20 U of each restriction enzyme per μg of

Table 1. Bacterial strains used in this study

Strain	Relevant characteristics	Source or reference
Bacillus azotofixans groups:		
1. P3L-5, P3L-3, P3E-2, P3E-30, P5L-4	Nif⁺, isolated from wheat roots	Seldin *et al.* (1984)
2. P3E-20	Nif⁺, isolated from wheat roots	Seldin *et al.* (1984)
3. Te-10	Nif⁺, isolated from wheat roots	Seldin *et al.* (1984)
4. F-100, F-102, F-203, F-215, F-1532	Nif⁺, isolated from Fundão Garden soil	Seldin *et al.* (1984)
5. BE-1, BE-4, BE-5	Nif⁺, isolated from *Brachiaria* roots	Seldin *et al.* (1984)
6. I-9	Nif⁺, isolated from Itaguaí soil	Seldin *et al.* (1984)
7. Hino	Nif⁺, isolated from soil	Hino and Wilson (1958) Seldin *et al.* (1984)
8. C8R-1, C3L-4, C4L-3, CA-52, CA-72	Nif⁺, isolated from sugar cane roots	Seldin and Penido (1986)
Klebsiella pneumoniae		
M5al	Nif⁺, wild type	Mahl *et al.* (1965)
Escherichia coli K12		
AB2880	host of plasmid pSA30	F. Ausubel
GM4	host of plasmid pGR113	F. Ausubel
FMA185	host of plasmid pCRA37	F. Ausubel
MC1061	host of plasmid pPC1201, pPC1202 or pPC1203	L. Sibold

DNA, for at least 6 h at 37 °C. *Eco*RI, *Hind*III, *Eco*RV and *Sal*I were used according to protocols supplied by the manufacturer (Boehringer Mannheim Biochemicals).

Agarose gel electrophoresis

Agarose gel electrophoresis of restricted DNA samples was performed in 0.8% agarose in TEB buffer (89 m*M* Tris, 89 m*M* boric acid, 2.5 m*M* EDTA, pH 8.2) at 2 V cm⁻¹, for 16 h at room temperature. In all experiments, lambda bacteriophage DNA (Sigma Chemical Co.) digested with *Eco*RI or *Hind*III, was used as molecular weight marker.

Blotting and prehybridization conditions

DNA was blotted from gels to Millipore HAWP 345.05 nitrocellulose filters as described by Maniatis *et al.* (1982). Prehybridization of filters dried in a vacuum oven at 80 °C for 2 h was performed for 16 h at 65 °C in prehybridization solution (Barinaga *et al.*, 1981).

Preparation of plasmids DNA probes and hybridization experiments

All probes used in this study were nick translated as described by Davis *et al.* (1980). They were

Table 2. Plasmids used in this study

Plasmid	Vector	Insertion size (kb)	Phenotype or genotype	Reference
pSA30	pACY184	6.2	Tcᴿ *nif' EYKDH*	Cannon *et al.* (1979)
pGR113	pACY184	2.25	Tcᴿ *nif' NE'*	Riedel *et al.* (1983)
pCRA37	pMB9	16.95	Tcᴿ *his DGnifQBALF-MVSUXN'*	Cannon *et al.* (1977; 1979)
pPC1201	pGV822	0.9	Cbᴿ, Tcˢ, *nifH*	Sibold *et al.* (1985)
pPC1202	pBR322	1.5	Cbᴿ, Tcˢ, *nif'KD*	Sibold *et al.* (1985)
pPC1203	pBR322	1.4	Cbᴿ, Tcˢ, *nifK*	Sibold *et al.* (1985)

Tcᴿ, tetracycline resistance; Cmᴿ, chloramphenicol resistance; Cbᴿ, carbenicillin resistance. A prime before a gene indicates that the plasmid carries the 5′ end of the gene; a prime after a gene indicates that the plasmid carries the 3′ end of the gene.

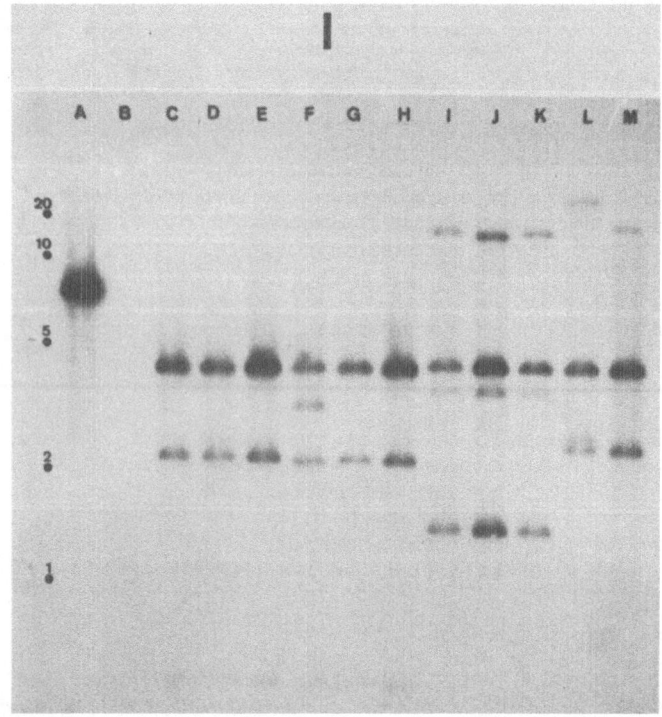

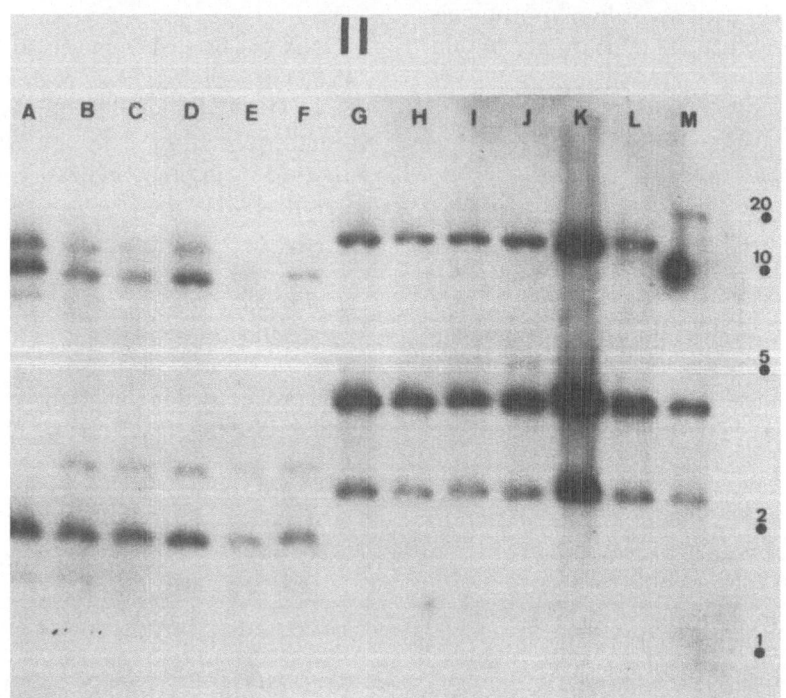

Fig. 1. (I, II). Southern blot hybridization of DNA from all *B. azotofixans* strains with *K. pneumoniae nifKDH* (pSA30 probe). All DNA was restricted by EcoRI endonuclease. IA: *K. pneumoniae* DNA digested with EcoRI (positive control), IB: *S. aureus* DNA (negative control), IC: P3L-3, ID: P3L-5, IE: P3E-2, IF: P3E-20, IG: P3E-30, IH: P5L-4, Il: BE-1, IJ: BE-4, IK: BE-5, IL: I-9, IM: TE-10, IIA: Hino, IIB: C8R-1, IIC: CA-52, IID: CA-72, IIE: C3L-4, IIF: C4L-3, IIG: F-100, IIH: F-102, IIl: F-203, IIJ: F-215, IIK: F-1532, IIL: TE-10, IIM: I-9. Size scale in kb.

Table 3. Size of fragments (originated from the digestion of *B. azotofixans* DNA with different restriction enzymes) hybridizing with *K. pneumoniae* nifHDK, nifH, nifD and nifK

B. azotofixans strains[a]	nifHDK (pSA30)				nifH (pPC1201)	nifD (pPC1202)	nifK (pPC1203)
	EcoRI	EcoRV	HindIII	SalI	EcoRI	EcoRI	EcoRI
P3L-5	4.1, 2.2, (0.6)*	11, (2.3)	6.8, 3.4, 1.5, (0.8)	(14), 6.6, 4.9	4.1, 2.2, (0.6)	4.1	ND
P3E-20	4.1, 3.1, 2.2, (0.6)	11, 8.8	6.8, 3.4, 1.5, (0.8)	17, 6.6, 4.9	4.1, 3.1, 2.2, (0.6)	4.1	ND
I-9	21, 4.1, 2.4	16.5, 9.8, 4.9	13, 3.4, 1.5, (0.8)	14, 6.6, 1.9	21, 4.1, 2.4	4.1	ND
BE-5	14, 4.1, (3.4), 1.4	14, 9.8, 4.9	19, 3.4, 1.5, (0.8)	14, 6.6, 1.9	14, 4.1, (3.4), 1.4	4.1	ND
F-100	15, (4.9), 4.1, 2.4	19, 9.8, 4.0	11, (7.8), 3.4, 1.5, (0.8)	14, 6.6, 1.9	15, (4.9), 4.1, 2.4	4.1	ND
TE-10	15, 4.1, 2.4	24, 9.8, 5.5	11, (7.8), 3.4, 1.5, (0.8)	17, 6.6, 1.9	15, 4.1, 2.4	4.1	ND
CA-52	13, 8.0, (6.5), (2.6), 1.7, 1.2	7.2, 1.7	4.0, 2.3		13, 8.0, (6.5), (2.6), 1.7	1.7, 1.2	ND
Hino	13, 8.0, (6.5), (4.7), 1.7, 1.2	21, 7.2, (4.2), 3.9, (3.2), 1.7	6.8, 2.3	26, 18, 14	13, 8.0, (6.5), (4.7), 1.7	1.7, 1.2	ND

Size of fragments in kb.

[a], One strain of each group of similar *EcoRI* fragments hybridizing with *K. pneumoniae* nifHDK (Fig. 1) was chosen.

ND, no band detected.

* Sizes listed in parentheses are for weakly hybridizing fragments.

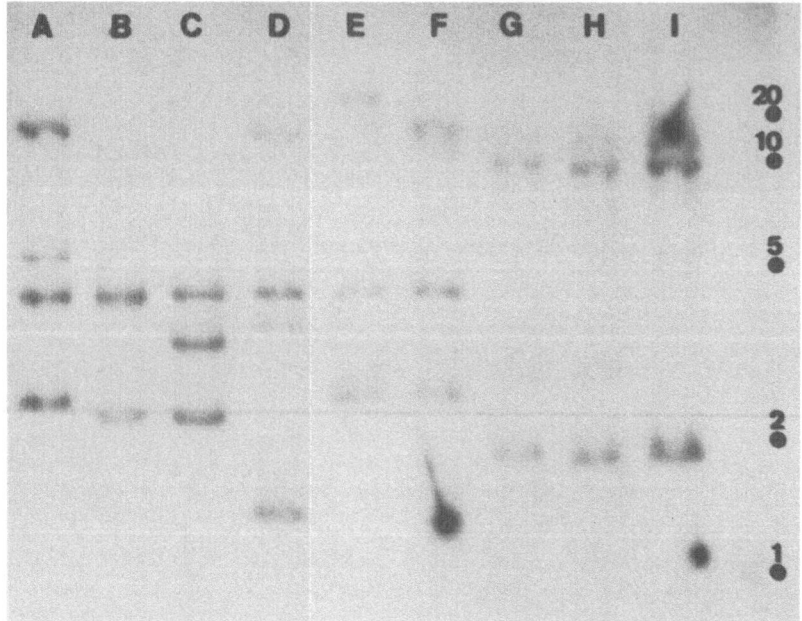

Fig. 2. Southern blot hybridization of DNA from *B. azotofixans* strains with *K. pneumoniae nifH* (pPC1201 probe). All DNA was restricted by EcoRI. A: F-100, B: P3L-5, C: P3E-20, D: BE-5, E: I-9, F: TE-10, G: C4L-3, H: CA-52, I: Hino. Size scale in kb.

radiolabelled with α^{32}P-deoxyadenosine triphosphate to a level of *ca.* 5×10^7 cpm μg^{-1} of DNA.

Hybridizations were carried out for 24 h at 65 °C in hybridization solution (Barinaga *et al.*, 1981) to which about 1×10^7 cpm of denatured probe had been added for each blot. The blots were then washed twice in a solution containing $2 \times$ SSC ($1 \times$ SSC is $0.15\,M$ NaCl plus $0.015\,M$ sodium citrate) and 0.1% sodium dodecyl sulphate for 30 min at room temperature, and twice in a solution containing $0.1 \times$ SSC and 0.1% sodium dodecyl sulphate for 30 min at 65 °C. The filters were dried at room temperature and then were exposed to X-ray film by using a Du Pont Cronex Lightning Plus intensifying screen at -70 °C for 3 to 10 days.

Results

The first goal of this study was to determine whether *nif* genes from *K. pneumoniae* share homology with DNA from *B. azotofixans.* For that, nick-translated *K. pneumoniae nif* DNA was hybridized to total DNA from *B. azotofixans* digested with EcoRI. When plasmid pSA30 containing the nitrogenase structural genes was the probe, DNA from all 22 *B. azotofixans* strains showed homology

with this plasmid (Fig. 1). The hybridization pattern from the 22 strains is presented in Table 3. All experiments were performed at least twice and *K. pneumoniae* chromosomal DNA homologous to the probe was used as a positive control. As a negative control, *Staphylococcus aureus* chromosomal DNA was included. Eight groups of different hybridization patterns were obtained among the 22 strains when *B. azotofixans* DNA was digested with EcoRI (Table 3). In all strains, excepting those isolated from sugar cane and strain Hino (see Table 1), a 4.1 kb fragment hybridized with *K. pneumoniae nifHDK* DNA. The strains Hino, C8R-1, C4L-3, C3L-4, CA-52 and CA-72 showed a similar hybridization pattern, varying in just one fragment with weak homology to pSA30.

When total DNA of *B. azotofixans* was hybridized with plasmids pGR113 (*nifNE*) and pCRA37 (*nifQBALFMVSUXN*) no bands were detected. As pSA30 and pGR113 are derived from the same vector (pACYC184 plasmid), hybridization with pSA30 cannot be due to the cloning vector but to the 6.2 kb fragment carrying the genes coding for the nitrogenase. As no homology was observed with the rest of the *nif* operons, the homology between *B. azotofixans* DNA and *K. pneumoniae* DNA must be limited to the nitrogenase structural genes.

To facilitate the cloning of the *B. azotofixans nif* structural genes, chromosomal DNA was digested with several restriction enzymes in order to find a single fragment containing all the nitrogenase structural genes. Considering the eight different hybridization patterns obtained when *B. azotofixans* DNA was digested with *Eco*RI, one strain of each group was chosen for the subsequent experiments. *Bacillus azotofixans* DNA was then treated with *Hae*III, *Bgl*II, *Hind*III, *Eco*RV, *Xho*I, *Hpa*II, *Sal*I and *Bam*HI. Very small fragments were obtained when *Hae*III and *Hpa*II were used, while large fragments were observed when *B. azotofixans* DNA was digested by *Xho*I and *Bam*HI hybridized to pSA30 (data not shown). Table 3 shows the results of the hybridization of pSA30 to *B. azotofixans* DNA cleaved with *Hind*III, *Eco*RV and *Sal*I. More than one fragment was obtained for all the strains tested with the different restriction enzymes. The previous classification of the strains into groups based on *Eco*RI restriction patterns was confirmed when the *B. azotofixans* DNA was digested with *Eco*RV. Again, eight different groups were observed, but the same did not occur with *Hind*III and *Sal*I. When pSA30 was used as a probe and the *B. azotofixans* DNA was digested with *Hind*III, strains P3L-5 and P3E-20 presented the same fragments, the same as occurred with strains F-100 and TE-10. With *Sal*I, strains I-9, BE-5 and F-100 showed the same hybridization pattern to pSA30. It should be noted that some fragments are common to almost all strains (3.4, 1.5 and 0.8 kb fragments in *Hind*III digestions and 6.6 kb fragment in *Sal*I digestions) except for strains CA-52 and Hino.

To determine which part of the pSA30 probe was involved in the homology, plasmids pPC1201, pPC1202 and pPC1203 were used as probes in other hybridization experiments with *B. azotofixans* DNA digested with *Eco*RI. These plasmids provide *nifH*, *nifD*, and *nifK* probes respectively (Sibold *et al.*, 1985). Figure 2 shows the hybridization pattern when pPC1201 (*nifH*) was used as probe. It can be observed that this pattern is the same as that obtained when pSA30 was used as a probe. The strains C4L-3, CA-52 and Hino represented the only exception, since the 1.2 kb fragment was not observed when *nifH* was the probe used. When pPC1202 (*nifD*) was used as a probe, the 4.1 kb fragment of *B. azotofixans* P3L-5, P3E-20, I-9, BE-5, F-100 and TE-10 also hy-

bridized to *nifD*. The 1.7 kb fragment of strains CA-52 and Hino hybridized to *nifH* and *D*, while the 1.2 kb fragment showed homology only with *nifD*. No fragments were detected when plasmid pPC1203 containing the *nifK* was used as a probe. The sizes of *Eco*RI fragments of *B. azotofixans* DNA hybridizing to *nifH* and *D* are summarized in Table 3. As pPC1202 and pPC1203 are derived from pBR322, a non-specific hybridization with the cloning vector is excluded since there was no homology between *nifK* and *B. azotofixans* DNA.

Discussion

Extensive studies of the genetics of the nitrogen fixation process have been developed in recent decades. Complementation analysis and physicochemical methods have shown that the structure of the nitrogenase complex is similar among the different nitrogen-fixing organisms studied to date (Mortenson and Thorneley, 1979). This similarity reflects the conservation of nucleotide sequences coding for nitrogenase, and recombinant DNA probes containing *Klebsiella nif* genes have been used to identify and clone DNA fragments with related sequences from several nitrogen-fixing organisms, such as *Anabaena* (Mazur *et al.*, 1980), *Azotobacter* (Ruvkun and Ausubel, 1980), *Bradyrhizobium japonicum* (Hennecke, 1981), *Rhizobium meliloti* (Ruvkun and Ausubel, 1980), *Rhodopseudomonas* (Fornari and Kaplan, 1983; Ruvkun and Ausubel, 1980), Archaebacteria (Possot *et al.*, 1986; Sibold *et al.*, 1985), and others (Ruvkun and Ausubel, 1980).

Concerning *Bacillus azotofixans*, extensive physiological studies had been performed (Seldin *et al.*, 1984; Seldin and Penido, 1986) but no genetic or molecular analysis had yet been introduced.

This paper reports the identification of *B. azotofixans* DNA fragments that contain nitrogen fixation genes by using heterologous probes (*Klebsiella nif* genes cloned in different plasmids). The 22 *B. azotofixans* strains showed homology to pSA30 which contains the *Klebsiella nif* structural genes, although the hybridization patterns differed among strains. When *B. azotofixans* DNA was digested with *Eco*RI and hybridized to pSA30, eight different hybridization patterns were observed. In almost all strains a 4.1 kb *Eco*RI fragment was

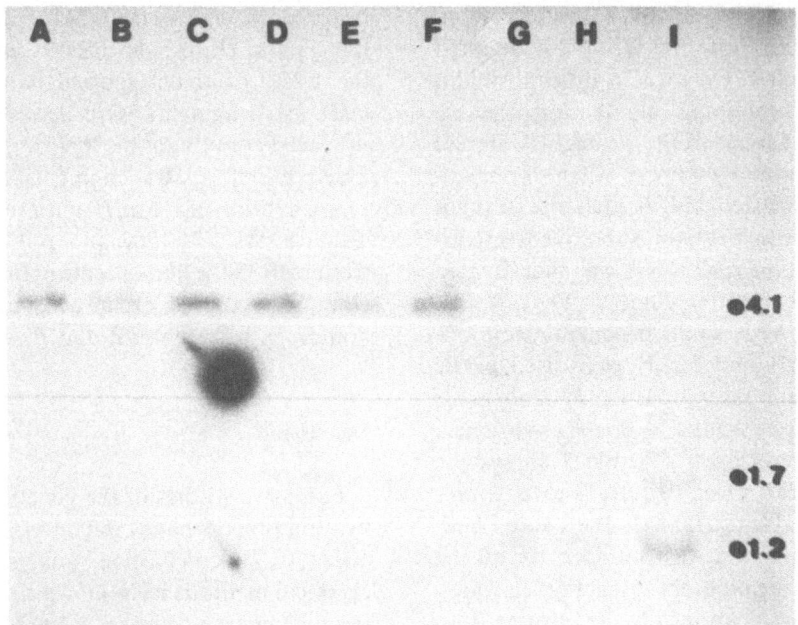

Fig. 3. Southern blot hybridization of DNA from *B. azotofixans* strains with *K. pneumoniae nifD* (pPC1202 probe). All DNA was restricted by EcoRI. A: F-100, B: P3L-5, C: P3E-20, D: BE-5, E: I-9, F: TE-10, G: C4L-3, H: CA-52, I: Hino. Size scale in kb.

present but the other fragments observed varied in length. Strains isolated from sugar cane, and strain Hino, showed a completely different hybridization pattern when they were compared to the other strains. The strains were similar to each other and it is possible they diverged earlier during evolution. This assumption is supported by the data obtained from the biochemical studies (Seldin and Penido, 1986) and from the determination of DNA relatedness among strains (Seldin and Dubnau, 1985). Strain Hino showed an intermediate level of homology to *B. azotofixans* P3L-5^T DNA (38%) indicating that it is not as closely related to the other *B. azotofixans* strains as those strains are one to another, but taxonomic data (Seldin *et al.*, 1984; Seldin and Penido, 1986) led us to consider it as a member of this species of *Bacillus*.

When *B. azotofixans* DNA was digested with *Eco*RV, *Sal*I and *Hind*III, the groups suggested by the *Eco*RI patterns could not be maintained, supporting the idea that *B. azotofixans* strains have diverged during evolution. The purpose of digesting *B. azotofixans* DNA with different restriction enzymes is to find one fragment containing all nitrogenase structural genes for a subsequent cloning, but this has not yet been achieved.

When the non-structural *nif* genes were used as probes, no homology was detected. Since homology was restricted to the *Klebsiella* nitrogenase structural genes, we determined which genes were involved. No homology was detected with *nifK*. Ruvkun and Ausubel (1980) also observed that *nifK* did not hybridize to DNA isolated from many different species. Hybridization with *Eco*RI fragment of 4.1 kb or 1.7 kb (see Table 3) was found when *nifH* and *nifD* were used as probes. This suggests that these two genes are contiguous in *B. azotofixans* DNA. The *nifH* probe consists of a 0.9 kb fragment cloned in pGV822 and the same *Eco*RI fragments were observed when pSA30 (*nif HDK*) or pPC1201 (*nifH*) were used as probes. In most strains three or more fragments larger than 0.9 kb were detected when *B. azotofixans* DNA was digested with *Eco*RI and hybridized to *nifH*. The numbers and sizes of these fragments indicate the possibility of a reiteration of *B. azotofixans nifH*. The possibility that multiple copies of this gene exist in the *B. azotofixans* genome is reasonable since reiteration of *nif* sequences has already been reported for some microorganisms (Chen *et al.*, 1986; Prakash and Atherly, 1984; Rice *et al.*, 1982). The organization and sequence of the nitrogenase structural genes of *B. azotofixans* remains to be elucidated.

Acknowledgements

We are indebted to Dr David Dubnau for critical reading of this manuscript. Thanks are due to all colleagues who were kind enough to provide strains, specially to F Ausubel and L Sibold for providing the plasmids containing the *nif* genes. We are also indebted to C Maia for supplying the a-^{32}P dATP. This work was supported by grants from the National Research Council of Brazil (CNPq) and from the Commission of the European Community grant no. CTl*/0062/00-BR(B).

References

Barinaga M, Franco R, Meinkoth J, Ong E and Wahl GM 1981 Methods for the Transfer of DNA, RNA and Protein to Nitrocellulose and Diazotized Paper Solid Supports. Schleicher & Schuell, Inc., Keene, N.H.

Birnboim HC and Doly J 1979 A rapid alkaline extraction procedure for screening recombinant plasmid DNA. Nucleic Acids Research 7, 1513–1523.

Cannon, FC Riedel GE and Ausubel FM 1977 Recombinant plasmid that carries part of the nitrogen fixation gene cluster of *Klebsiella pneumoniae.* Proc. Natl. Acad. Sci. 74, 2963–2967.

Cannon FC, Riedel GE and Ausubel FM 1979 Overlapping sequences of *Klebsiella pneumoniae nif* DNA cloned and characterized. Molec. Gen. Genet. 174, 59–66.

Chen KCK, Chen JS and Johnson JL 1986 Structural features of multiple *nifH*-like sequences and very biased codon usage in nitrogenase genes of *Clostridium pasteurianum.* J. Bacteriol. 166, 162–172.

Davis RW, Botstein D and Roth JR 1980 Advanced Bacterial Genetics. A Manual for Genetic Engineering. Cold Spring Harbor Laboratory, Cold Spring Harbor, N.Y.

Dixon RA 1984 The genetic complexity of nitrogen fixation. J. Gen. Microbiol. 130, 2745–2755.

Fornari CS and Kaplan S 1983 Identification of nitrogenase and carboxylase genes in the photosynthetic bacteria and cloning of a carboxylase gene from *Rhodopseudomonas sphaeroides.* Gene 25, 291–299.

Grau FH and Wilson PW 1962 Physiology of nitrogen fixation by *Bacillus polymyxa.* J. Bacteriol. 83, 490–496.

Hennecke H 1981 Recombinant plasmids carrying nitrogen fixation genes from *Rhizobium japonicum.* Nature (London) 291, 354–355.

Hino S and Wilson PW 1958 Nitrogen fixation by a facultative *Bacillus.* J. Bacteriol. 45, 403–408.

Larson RI and Neal JL Jr 1978 Selective colonization of the rhizosphere of wheat by nitrogen fixing bacteria. Ecol. Bull. (Stockholm) 26, 331–342.

Line MA and Loutit MW 1971 Non-symbiotic nitrogen fixing organisms from some New Zealand tussock-grassland soils. J. Gen. Microbiol. 66, 309–318.

Mahl MC, Wilson PW, Fife MA and Ewing WH 1965 Nitrogen fixation by members of the tribe Klebsiellae. J. Bacteriol. 89, 1482–1487.

Maniatis T, Fritsch EF and Sambrook J 1982 Molecular Cloning. A Laboratory Manual. Cold Spring Harbor Laboratory, Cold Spring Harbor, N.Y.

Mathis JN, Kuykendall LD and Elkan GH 1986 Restriction endonuclease and *nif* homology patterns of *Bradyrhizobium japonicum* USDA 110 derivatives with and without nitrogen fixation competence. Appl. Environ. Microbiol. 51, 477–480.

Mazur BJ, Rice D and Haselkorn R 1980 Identification of blue-green algae nitrogen fixation genes by using heterologous hybridization probes. Proc. Natl. Acad. Sci. 77, 186–190.

Mortenson LE and Thorneley RNF 1979 Structure and function of nitrogenase. Annu. Rev. Biochem. 48, 387–418.

Possot O, Henry M and Sibold L 1986 Distribution of DNA sequences homologous to *nifH* among archaebacteria. FEMS Microbiol. Lett. 34, 173–177.

Prakash RK and Atherly AG 1984 Reiteration of genes involved in symbiotic nitrogen fixation by fast-growing *Rhizobium japonicum.* J. Bacteriol. 160, 785–787.

Rice D, Mazur BJ and Haselkorn R 1982 Isolation and physical mapping of nitrogen fixation genes from the cyanobacterium *Anabaena* 7120. J. Biol. Chem. 257, 13157–13163.

Riedel GE, Brown SE and Ausubel FM 1983 Nitrogen fixation by *Klebsiella pneumoniae* is inhibited by certain multicopy hybrid *nif* plasmids. J. Bacteriol. 153, 45–56.

Ruvkun GB and Ausubel FM 1980 Interspecies homology of nitrogenase genes. Proc. Natl. Acad. Sci. 77, 191–195.

Schleif RF and Wensink PC 1981 Practical Methods in Molecular Biology. Springer-Verlag New York Inc.

Seldin L, van Elsas JD and Penido EGC 1983 *Bacillus* nitrogen fixers from Brazilian soils. Plant and Soil 70, 243–255.

Seldin L, van Elsas JD and Penido EGC 1984 *Bacillus azotofixans* sp. nov., a nitrogen-fixing species from Brazilian soils and grass roots. Int. J. Syst. Bacteriol. 34, 451–456.

Seldin L and Dubnau D 1985 Deoxyribonucleic acid homology among *Bacillus polymyxa, Bacillus macerans, Bacillus azotofixans,* and other nitrogen-fixing *Bacillus* strains. Int. J. Syst. Bacteriol. 35, 151–154.

Seldin L and Penido EGC 1986 Identification of *Bacillus azotofixans* using API tests. Antonie van Leeuwenhoek 52, 403–409.

Sibol L, Pariot D, Bhatnagar L, Henriquet M and Aubert JP 1985 Hybridization of DNA from methanogenic bacteria with nitrogenase structural gene (*nifHDK*). Mol. Gen. Genet. 200, 40–46.

Southern EM 1975 Detection of specific sequences among DNA fragments separated by gel electrophoresis. J. Mol. Biol. 98, 503–517.

Witz DF, Detroy RW and Wilson PW 1967 Nitrogen-fixation by growing cells and cell-free extracts of the *Bacillaceae.* Arch. Mikrobiol. 55, 369–381.

F. A. Skinner et al. (Eds.), Nitrogen fixation with non-legumes, 189–195.

Isolation and properties of *Azospirillum lipoferum* and *Azospirillum brasilense* surface polysaccharide mutants

K. MICHIELS[1], J. VANDERLEYDEN[1], A. VAN GOOL[1] and E.R. SIGNER[2]
[1]F.A. Janssens Memorial Laboratory for Genetics, B-3030 Heverlee, Belgium and [2]Department of Biology, Massachusetts Institute of Technology, Cambridge, MA 02139, USA

Key words: *Azospirillum*, calcofluor, complementation, EPS, *Rhizobium meliloti*, surface polysaccharide

Abstract

Polysaccharide production by *Azospirillum* strains was indicated by fluorescence on growth media containing calcofluor. Mutants showing decreased and increased levels of fluorescence were obtained from *A. lipoferum* strain Sp59b by chemical mutagenesis, and from *A. brasilense* strain 7030 by Tn5 mutagenesis. One class of calcofluor dark mutants from 7030 synthesized an exopolysaccharide (EPS) probably different in composition from the parent EPS. However, not EPS, but an unidentified component seemed to determine the fluorescence phenotype. Another class of 7030 mutants could be complemented to wild type fluorescence with a cosmid bank.

Introduction

Bacteria of the family Rhizobiaceae are classified into two genera: *Rhizobium* and *Agrobacterium*. *Rhizobium* species have the ability to live symbiotically in specialized nitrogen-fixing nodules which are formed on the roots of legumes (see Verma and Long, 1983). *Agrobacterium tumefaciens* and *Agrobacterium rhizogenes* respectively are responsible for crown gall and hairy root induction in infected sensitive plants (Nester *et al.*, 1984; Chilton *et al.*, 1982). Recently, genetic studies revealed similarities between both infection processes. In *A. tumefaciens*, chromosomal virulence genes (*chv*) are required for attachment to and tumour formation of plant cells (Douglas *et al.*, 1982; Douglas *et al.*, 1985). Through hybridization and complementation studies the occurrence of similar genes could be demonstrated in the genome of *R. meliloti*. In this *Rhizobium* species, these genes were found to be essential for nodule development and are therefore designated as *ndv* genes (Dylan *et al.*, 1986).

Tn5 mutants of *R. meliloti* that are defective in the synthesis of exopolysaccharides (EPS) and therefore impaired in effective nodulation, have been isolated (Finan *et al.*, 1985; Leigh *et al.*, 1985). Similarly EPS deficient (*exo*) mutants of *A. tumefaciens* have been isolated (Cangelosi *et al.*, 1987). EPS deficiencies of *R. meliloti* and *A. tumefaciens* can be complemented with cloned *A. tumefaciens* and *R. meliloti exo* loci, respectively. Recently, another locus, called *pscA*, was identified in *A. tumefaciens*, and which is necessary for EPS production and virulence (Thomashow *et al.*, 1987). Furthermore, it is suggested that the *pscA* locus is related to the *exoC* locus of *R. meliloti*. Finally, another class of Tn5-induced avirulent mutants of *A. tumefaciens* that fail to attach to plant cells was described by Mathysse (1987). This chromosomal locus, revealed by the genetic analysis of these mutants, appears not to be related to *chv* or *pscA*.

Unlike the above species, azospirilla interact with plant roots without the formation of clearly differentiated structures (Elmerich, 1984; Okon, 1985). Attachment to root hairs (Jain and Patriquin, 1984), penetration of roots (Baldani *et al.*, 1986) and effects on root hair morphology and development (Jain and Patriquin, 1984; Kapulnik *et al.*, 1985), have been reported, but the molecular mechanism underlying these processes is not well

Table 1. Bacterial strains and plasmids

Bacterial strain or plasmid	Relevant properties	Source or reference
A. brasilense		
Sp7		ATCC 29145
7030	Sm^r derivative of Sp7	Franche and Elmerich, 1981
7030TN5–1	mucoid; increased fluorescence with calcofluor	This study
7030TN5–11	Cal^d Km^r	This study
7030TN5–12	Cal^d Km^r	This study
7030TN5–22	Cal$^-$ Km^r	This study
7030TN5–23	Cal$^-$ Km^r	This study
R07		Y. Dommergues
Sp245		J. Döbereiner
Sp246		J. Döbereiner
A. lipoferum		
Sp59b		ATCC 29707
Sp59bNG11	Cal^{++}	This study
Sp59bNG12	Cal^{++}	This study
SpBr17		ATCC 29709
R. meliloti		
1021	Sm^r	F. Ausubel
7032	ExoA$^-$ Nm^r	Leigh *et al.*, 1985
7094	ExoB$^-$ Nm^r	Leigh *et al.*, 1985
7027	ExoC$^-$ Nm^r	Leigh *et al.*, 1985
7055	ExoF$^-$ Nm^r	Leigh *et al.*, 1985
Plasmids		
pLAFR1	Tc^R cosmid, broad host range	Friedman *et al.*, 1982
pRK600	pRK2013 *npt*::Tn9	T. Finan
pCal11–10	*Azospirillum* cosmid clone complementing 7030TN5–11	This study
pCal11–19	*Azospirillum* cosmid clone complementing 7030TN5–11	This study
pD2	*R. meliloti* cosmid clone complementing ExoB	Leigh *et al.*, 1985
pD34	*R. meliloti* cosmid clone complementing ExoA	Leigh *et al.*, 1985
pD56	*R. meliloti* cosmid clone complementing ExoB and ExoF	Leigh *et al.*, 1985

understood. The presence in *Azospirillum* of DNA sequences homologous to the *R. meliloti nod* and *hsn* genes (Fogher *et al.*, 1985), and to the *A. tumefaciens chv* genes (Waelkens *et al.*, 1987) has been demonstrated, but it is not yet known what their function is, if any, in *Azospirillum*-plant interactions. Given the important role of EPS in the *Rhizobium*- and *Agrobacterium*-plant interactions, we began a study on EPS synthesis by *Azospirillum*. We discuss here the isolation and characterization of *Azospirillum* mutants effective in exopolysaccharide synthesis.

Materials and methods

Plasmids, bacterial strains and culture conditions

Bacterial strains and plasmids used in this work are listed in Table 1. *Azospirillum* and *R. meliloti* strains were grown at 30 °C on LB agar or broth supplemented with 2.5 mM CaCl$_2$ and 2.5 mM MgCl$_2$. When appropriate, antibiotics were added in the following concentrations (µg ml^{-1}): tetracycline (Tc), 10; chloramphenicol (Cm), 20; streptomycin (Sm), 250; kanamycin (Km)

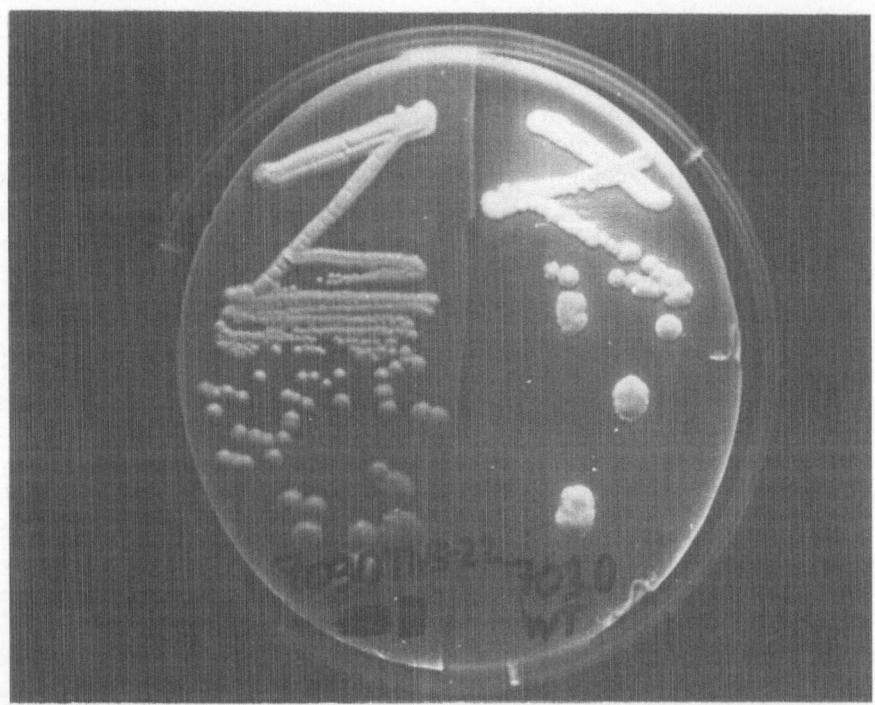

Fig. 1. Fluorescence on calcofluor-containing medium. Right, *A. brasilense* strain 7030, exhibiting wild-type fluorescence; left, 7030TN5–22, a Cal⁻ mutant.

25, for *A. brasilense* Tn5 mutants, and neomycin (Nm) 200 for *R. meliloti* Tn5 mutants. Calcofluor white (Sigma, St. Louis) was added at $200\,\mu g\,ml^{-1}$ to plates buffered with 25 mM HEPES, pH 7.2.

Isolation and analysis of EPS

Bacteria were grown to late log phase in supplemented LB, harvested and incubated another 24 h with shaking in NMF medium, which is MMAB (Vanstockem *et al.*, 1987) without N source and with 0.5% fructose as a sole carbon source. Bacteria were removed by centrifugation at 10 000 g for 20 min and the clear supernatant liquid was dialyzed extensively against water in dialysis tubing with a mol.wt cutoff of 2000 daltons. The solution was then concentrated by freezedrying, redissolved and eventually loaded on a 40 × 3 cm Sepharose 6B column and eluted with 10 mM phosphate buffer pH 7.0. Fractions were analyzed by the anthrone method for hexose content (Dische, 1962). Precipitation of EPS was performed by adding 0.3 part of a 1% solution of cetrimide to dialyzed NMF culture supernatant fractions. Bind-

ing affinities of EPS in solution for calcofluor were compared as follows: samples of EPS containing $200\,\mu g\,ml^{-1}$ calcofluor were applied on a Sepharose 6B column and the calcofluor content of the anthrone positive fractions was determined by measuring absorption at 360 nm (Wood and Fulcher, 1978). Capsule staining was done as described by Doetsch (1981).

Mutagenesis

Tn5 mutagenesis on *A. brasilense* 7030 was performed as described (Vanstockem *et al.*, 1987). Tn5 mutants were directly screened for fluorescence on calcofluor plates under long wave u.v. light. Nitrosoguanidine (NG) mutagenesis was performed as described by Miller (1972).

Complementation

Cosmid banks of *A. brasilense* Sp7 and Sp245 and of *R. meliloti* 1021 were in pLAFR1, in an *E. coli* HB101 background. Triparental matings were done overnight on solid medium with HB101 (pRK600) as a helper strain. After purification, plasmid DNA was isolated from brightly fluorescent exconjugants by an alkaline lysis method (Maniatis *et al.*, 1982), and transformed into *E. coli* HB101. Complementation was then confirmed by mating the plasmids back into the mutants. Hybridization was carried out as previously described (Vanstockem *et al.*, 1987).

Results

Isolation of Azospirillum *exopolysaccharides*

Azospirillum species, grown on agar medium containing the fluorescent stain calcofluor, have a blue-purple fluorescent colony phenotype under u.v. light (Fig. 1.). Calcofluor is a stilbene dye that binds predominantly to β-1,4 and β-1,3 linked glycans (Wood and Fulcher, 1978). Among the strains tested (see Table 1), it was observed that *A. lipoferum* strains Sp59b and spBr17 show a brighter fluorescence under u.v. than *A. brasilense* strains. Among the *A. brasilense* strains tested, strain 7030 was examined more carefully for fluorescence in the presence of calcofluor and production of exopolysaccharides, since Tn5 transposon mutagenesis can only efficiently be carried out in *A. brasilense* strains (Vanstockem *et al.*, 1987; Singh and Klingmüller, 1986), and in strain 7030 in particular (Vanstockem *et al.*, 1988). In the presence of 0.02% calcofluor white, this strain showed a brightly fluroescent colony phenotype under u.v. on LB agar.

The exopolysaccharides of strain 7030 were isolated from culture supernatant fractions as described in Materials and methods. The yield was about $0.1 \, \mathrm{mg \, ml^{-1}}$ of supernatant liquid, as measured by the anthrone method. In contrast to the acidic exopolysaccharides of *R. meliloti* strain 1021, the exopolysaccharides of *A. brasilense* strain 7030 cannot be precipitated with cetrimide, a polycationic detergent. The exopolysaccharides could be separated by gel filtration chromatography into two major fractions: 95% of the EPS could be eluted in a peak corresponding to a $MW_r \geqslant 4.10^6$ daltons; a minor fraction could be eluted in a second peak corresponding to a MW_r of $\sim 2.10^4$ daltons.

Isolation and properties of EPS mutants

Azospirillum lipoferum strain Sp59b was treated with nitrosoguanidine, and mutants with an altered fluorescence colony phenotype (in the presence of 0.02% calcofluor white) under u.v. were selected. Of these, only prototrophic clones with a growth rate similar to the growth rate of the corresponding wild-type strains were retained for further study. Two mutants, selected on the basis of their non-fluorescent colony phenotype (Cal⁻), had lost the ability to form floc in liquid cultures. Floc formation of wild-type *Azospirillum* strains has been described (Sadasivan and Neyra, 1985). From strain Sp59b, two mutants that showed an increased fluorescence under u.v. (Cal⁺⁺) could be isolated as well (Sp59bNG11 and Sp59bNG12). Although their visual appearance was unaltered, these Cal⁺⁺ mutants formed colonies with a rubbery consistency (with a transfer loop, the whole colony, rather than a portion of it, comes off the agar surface). Their growth in broth culture was poor and, unlike Sp59b, cells did not aggregate as flocs, but as a very viscous sediment at the bottom of the culture tube. As compared to the wild type, the viscosity of culture supernatant fractions was lower for Cal⁻ and higher for Cal⁺⁺ mutants of Sp59b.

Of *ca.* 4000 Tn5-induced mutants of *Azospirillum brasilense* strain 7030, five mutants with an altered fluorescence pattern in the presence of calcofluor were isolated. They grew well on all media tested and could be divided into three classes on the basis of fluorescence level, mucoidy and floc formation in NMF medium, as summarized in Table 2. 7030TN5-22 and 7030TN5-23 were completely non-fluorescent (Cal⁻, see Fig. 1), 7030TN5-11 and

Table 2. Characteristics of *A. brasilense* 7030 Tn5 mutants

Strain	Cal phenotype[a]	Mucoidy	Floc formation in NMF medium
7030	Cal⁺	+	+
7030TN5–1	Cal⁺⁺	+	+
7030TN5–11	Cal^d	+	+
7030TN5–12	Cal^d	+	+
7030TN5–22	Cal⁻	+	−
7030TN5–23	Cal⁻	+	−

[a] refers to intensity of fluorescence on calcofluor containing medium. Cal⁻, no fluorescence; Cal^d, dim fluorescence; Cal⁺, fluorescence of parental strain 7030; Cal⁺⁺, fluorescence than 7030.

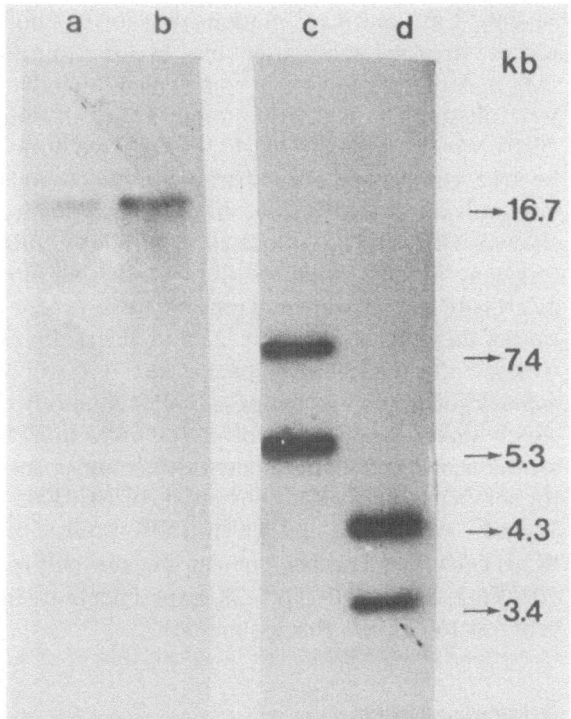

Fig. 2. Analysis of Tn*5* mutants by hybridization with a Tn*5* specific probe. Lane a, *Eco*RI digest of 7030TN5–11 DNA. Lane b, *Eco*RI digest of 7030TN5–12. Tn*5* has no recognition site for *Eco*RI, and both mutants have Tn*5* inserted in an *Eco*RI fragment of the same size. Lane c, *Sal*I digest of 7030TN5–11 DNA. Lane d, *Sal*I digest of 7030TN5–12 DNA. Tn*5* is cleaved once by *Sal*I. The *Sal*I fragment containing Tn*5* in 7030TN5–11 is 7.0 Kb in size (Tn*5* itself measures 5.7 Kb), but the *Sal*I fragment containing Tn*5* in 7030TN5–12 is only 2.0 Kb. This indicates that the location of Tn*5* in both mutants is different, but probably in the same *Eco*RI fragment.

7030TN5-12 showed a dim fluorescence (Cal^d) and 7030TN5-1 fluoresced slightly brighter than wild type and had a more mucoid phenotype. *Azospirillum brasilense* 7030 does not normally form flocs in a broth culture, but when log phase cells are transferred from a broth culture into NMF medium, they start overproducing EPS and flocculate drastically. The Cal$^-$ mutants 7030TN5-22 and 7030TN5-23 never show flocculation. Surprisingly, EPS could be isolated from both Cal$^-$ mutants, in slightly higher amounts (0.15 mg ml^{-1}) than from 7030, and with the same fractionation pattern on a Sepharose 6B column and the same binding affinity for calcofluor as the 7030 EPS.

However, unlike 7030 EPS, at least part of the EPS from Cal$^-$ mutants could be precipitated with cetrimide. This might be explained by assuming that the mutant EPS carries more acidic groups (e.g. uronic acids or succinyl, pyruvyl, acetyl substituents), whose negative charges can interact with the cetrimide polycation. In addition, the presence of similarly charged molecules on the cell surface will cause the cells to repel one another, which might also explain the absence of floc formation in conditions of EPS overproduction with these mutants.

Mutant 7030 and the Cal$^-$ mutant 7030TN5-22 were compared for attachment to plant cells in a quantitative assay (Douglas *et al.*, 1982; Eyers *et al.*, 1988). The mutant appears not to be affected in plant cell attachment. Microscopic analysis revealed the presence of a capsule of about cell-thickness in 7030 and in all five Tn*5* mutants. All mutants were shown to be equally motile as the parent strain.

Complementation of 7030 Cal^d mutants

A cosmid bank of *A. brasilense* Sp7 was mobilized into 7030TN5-11, 7030TN5-22 and 7030TN5-23. Cosmids from normally fluorescing colonies were isolated, transformed into *E. coli* HB101 and mated back into the mutant to confirm complementation. Two different clones complementing 7030TN5-11 for fluorescence could be isolated (pCal11–10 and pCal11–19), but none complementing 7030TN5-22 or 7030TN5-23 for fluorescence could be obtained. pCal11–10 and pCal11–19 could not restore fluorescence in the Cal$^-$ mutants, but the presence of pCal11–19 in either Cal$^-$ mutant caused increased mucoidy. 7030TN5-12 however, with the same phenotype as 7030TN5-11, could be complemented by both pCal11–10 and pCal11–19. In addition, the hybridization data in Fig. 2 show that both Cal^d mutants have a Tn*5* insertion in an *Eco*RI fragment of the same size (11 kb). These data strongly suggest that Tn*5* is in the same locus in 7030TN5-11 and 7030TN5-12.

Attempts to complement Tn*5*-induced Cal$^-$ mutants of 7030 with the cosmids pD2, pD34 and pD56 which carry cloned *R. meliloti exo* genes, or with a *R. meliloti* gene bank, were unsuccessful.

Discussion

We were able to isolate *Azospirillum* mutants with altered surface polysaccharide properties. As cell surface components are shown to be key determinants in bacteria-plant interactions, these mutants will be very valuable in assessing the role, if any, of such components in the *Azospirillum*-plant root association.

It follows from our NG mutagenesis on an *A. lipoferum* strain, that is very easy to obtain a variety of mutants with altered Cal phenotype. For Sp59b, the level of fluorescence of a mutant seems to be correlated with the viscosity of a broth culture of that mutant. This could be due to the level of EPS production, as in *R. meliloti* and *A. tumefaciens*, where Cal⁻ mutants are EPS deficient. However, this seems not to be the case for *A. brasilense* 7030. Our Cal⁻ Tn*5* mutants from this strain produced only slightly higher amounts of EPS. In addition, this EPS was shown to bind calcofluor to the same extent as the parent EPS. This means that fluorescence of 7030 is caused by another polysaccharide than EPS. This could be cellulose. From *A. tumefaciens*, Cal⁻ mutants defective in cellulose synthesis have been isolated (Matthyse, 1983), but so far there is no evidence for cellulose synthesis by *Azospirillum*.

Two observations support the idea that EPS from the 7030 Cal⁻ mutants is more acidic than the parent EPS. First, in conditions of C-source excess and N-source limitation where EPS is over-produced, 7030 forms flocs, probably by apolar interactions between neutral EPS (and capsular polysaccharide) molecules. If the mutant EPS carries negatively charged groups, bacteria will be electrostatically repulsed from each other and remain in suspension. Second, the presence of such charged groups is more directly evidenced by the precipitability of the mutant EPS with cetrimide. We can conclude that the 7030 Cal⁻ mutants are probably altered in more than one surface component. This was also found for some *R. meliloti* and *A. tumefaciens* Cal⁻ mutants (Leigh *et al.*, 1985; Cangelosi *et al.*, 1987).

Complementation analysis of 7030 Cal⁻ and Cal^d mutants was severely hindered by a high background of partial, unspecific restoration of fluorescence. Cosmids isolated from such colonies had no common insert, increased fluorescence partially in both Cal⁻ and Cal^d mutants, and often fluorescence dropped again after some cycles of purification. Although consistent complementation data were obtained with pCal11-10 and pCal11-19 for 7030TN5-11 and 7030TN5-12, this too might not be true complementation because both cosmids have only a 6 kb *Eco*RI insert in common (data not shown) whereas Tn*5* is located in an 11 kb *Eco*RI fragment in both mutants. pCal11-19, but not pCal11-10, causes increased mucoidy (overproduction of an EPS) in 7030TN5-22 and 7030TN5-23 without affecting the Cal phenotype. A possible explanation for this is that pCal11-19 contains two closely linked functions involved somehow in EPS synthesis or its regulation: one which complements (or compensates) the Cal^d phenotype of 7030TN5-11 and 7030TN5-12 and which is also present on pCal11-10, and another one causing mucoidy of 7030TN5-22 and 7030TN5-23. Experiments are in progress to confirm this assumption.

Acknowledgements

This work was supported by a grant from the Fonds voor Kollektief Fundamenteel Onderzoek (F.K.F.O. 2.0013.85). Part of it was carried out at the Massachusetts Institute of Technology, Cambridge, U.S.A., with financial support of the N.F.W.O. and the Ministerie van de Vlaamse Gemeenschap. K.M. is a recipient of a fellowship from the N.F.W.O.

References

Baldani VLD, Alvarez MA de B, Baldani JI and Döbereiner J 1986 Establishment of inoculated *Azospirillum* spp. in the rhizosphere and in roots of field grown wheat and sorghum. Plant and Soil 90, 35–46.

Cangelosi GA, Hung L, Puvanesarajah V, Stacey G, Orga DA, Leigh JA and Nester EW 1987 Common loci for exopolysaccharide synthesis in *Agrobacterium tumefaciens* and *Rhizobium meliloti*, and their role in plant interactions. J. Bacteriol. 169, 2086–2091.

Chilton M-D, Tepfer DA, Petit A, David C, Casse-Delbart F and Tempé J 1982 *Agrobacterium rhizogenes* inserts T-DNA into the genomes of the host plant root cells. Nature 295, 432–434.

Dische Z 1962 *In* Methods in Carbohydrate Chemistry. Eds. RL Whistler and ML Wolfrom. 1, 477–479.

Doetsch RN 1981 Determinative methods of light microscopy. *In* Manual of Methods for General Bacteriology. Eds. P

Gerhardt *et al.* American Society for Microbiology, Washington, D.C.

Douglas CJ, Halperin W and Nester EW 1982 *Agrobacterium tumefaciens* mutants affected in attachment to plant cells. J. Bacteriol. 152, 1265–1275.

Douglas CJ, Staneloni RJ, Rubin RA and Nester EW 1985 Identification and genetic analysis of an *Agrobacterium tumefaciens* chromosomal virulence region. J. Bacteriol. 161, 850–860.

Dylan T, Ielpi L, Stanfield S, Kashyap L, Douglas C, Yanofsky M, Nester E, Helinski DR and Ditta G 1986 *Rhizobium meliloti* genes required for nodule development are related to chromosomal virulence genes in *Agrobacterium tumefaciens*. Proc. Natl. Acad. Sci. USA 83, 4403–4407.

Elmerich C 1984 Molecular biology and ecology of diazotrophs associated with non-leguminous plants. Bio/Technology 2, 967–978.

Eyers M, Waelkens F, Vanderleyden J and Van Gool A 1987 Quantitative measurement of *Azospirillum* plant cell attachment. *In Azospirillum* IV: Genetics, Physiology, Ecology. Ed. W. Klingmüller. pp 174–180. Springer-Verlag, Heidelberg.

Finan TM, Hirsch AM, Leigh JA, Johansen E, Kuldau GA, Deegan S, Walker GC and Signer ER 1985 Symbiotic mutants of *Rhizobium meliloti* that uncouple plant from bacterial differentiation. Cell 40, 869–877.

Franche C and Elmerich C 1981 Physiological properties and plasmid content of several strains of *Azospirillum brasilense* and *Azospirillum lipoferum*. Ann. Inst. Pasteur 132A, 3–18.

Fogher C, Dusha I, Barbot P and Elmerich C 1985 Heterologous hybridization of *Azospirillum* DNA to *Rhizobium nod* and *fix* genes. FEMS Microbiol. Letters 30, 245–249.

Friedman AM, Long SR, Brown SE, Buikema WJ and Ausubel FM 1982 Construction of a broad host range cloning vector and its use in the genetic analysis of *Rhizobium* mutants. Gene 18, 289–296.

Jain DK and Patriquin DG 1984 Root hair deformation, bacterial attachment and plant growth in wheat-*Azospirillum* associations. Appl. Environm. Microbiol. 48, 1208–1213.

Kapulnik Y, Okon Y and Henis Y 1985 Changes in root morphology of wheat caused by *Azospirillum* inoculation. Can. J. Microbiol. 31, 881–887.

Leigh JA, Signer ER and Walker GC 1985 Exopolysaccharide-deficient mutants of *Rhizobium meliloti* that form ineffective nodules. Proc. Nat. Acad. Sci USA 82, 6231–8730.

Maniatis T, Fritsch EF and Sambrook J 1982 Molecular Cloning. A Laboratory Manual. Cold Spring Harbor Laboratory, Cold Spring Harbor, New York.

Matthyse AG 1983 Role of bacterial cellulose fibrils in *Agrobacterium tumefaciens* infection. J. Bacteriol. 154, 906–915.

Matthysse AG 1987 Characterization of nonattaching mutants of *Agrobacterium tumefaciens*. J. Bacteriol. 169, 313–323.

Miller JH 1972 Experiments in Molecular Genetics. Cold Spring Harbor Laboratory, Cold Spring Harbor, New York.

Nester EW, Gordon MP, Amasino RM and Yanofsky MF 1984 Crown gall: A molecular and physiological analysis. Annu. Rev. Plant Physiol. 35, 387–413.

Okon Y 1985 *Azospirillum* as a potential inoculant for agriculture. Trends in Biotechnology 3, 223–228.

Sadasivan L and Neyra CA 1985 Flocculation in *Azospirillum brasilense* and *Azospirillum lipoferum*: exopolysaccharides and cystformation. J. Bacteriol. 163, 716–723.

Singh M and Klingmüller W 1986 Transposon mutagenesis in *Azospirillum brasilense*: isolation of auxotrophic and *nif* mutants and molecular cloning of the mutagenized nif DNA. Mol. Gen. Genet. 202, 136–142.

Thomashow MF, Karlinsey JE, Marks JR and Hurlbert RE 1987 Identification of a new virulence locus in *Agrobacterium tumefaciens* that affects polysaccharide composition and plant cell attachment. J. Bacteriol. 169, 3209–3216.

Vanstockem M, Michiels K, Vanderleyden J and Van Gool A 1987 Transposon mutagenesis of *Azospirillum brasilense* and *Azospirillum lipoferum*: Physical analysis of Tn5 and Tn5-*Mob* insertion mutants. Appl. Environm. Microbiol. 53, 410–415.

Vanstockem M, Milcamps A, Michiels K and Vanderleyden J 1988 Tn5 mutagenesis in *Azospirillum brasilense*. *In Azospirillum* IV: Genetics, Physiology, Ecology. Ed. W. Klingmüller. pp 32–39. Springer-Verlag, Heidelberg.

Verma DPS and Long S 1983 The molecular biology of *Rhizobium* legume symbiosis. Int. Rev. Cytol. 14 (Suppl), 211–245.

Waelkens F, Maris M, Verreth C, Vanderleyden J and Van Gool A 1987 FEMS Microbiol. Letters 43, 241–246.

Wood PJ and Fulcher RG 1978 Interaction of some dyes with cereal β-glucans. Cereal Chem. 55, 952–966.

Session 7

Mechanisms of association of N$_2$-fixing bacteria with grasses and cereals

F. A. Skinner et al. (Eds.), Nitrogen fixation with non-legumes, 199–207.

Plant-bacteria interactions with special emphasis on the kallar grass association

BARBARA REINHOLD, T. HUREK and I. FENDRIK
Institute of Biophysics, University of Hannover, Herrenhäuser Str. 2, D-3000 Hannover 21, FRG

Key words: *Azospirillum*, kallar grass, N_2-fixing bacteria, plant-bacteria interactions, root zone specificity

Abstract

Kallar grass is a highly salt-tolerant grass grown as a pioneer plant on alkaline, salt-affected soils in Pakistan. Nitrogen-fixing bacteria and kallar grass were found to be in close association, which was even root-zone specific: rhizoplane and endorhizosphere were colonized by two different populations. Among the *Azospirillum* isolates originating from the root surface, some were of a new species, now named *A. halopraeferens*. To study plant-bacterium interactions, this natural kallar grass association was chosen. The possible role of bacterial chemotaxis and oxygen tolerance are discussed.

Introduction

Various N_2-fixing microorganisms have been found to be present in the rhizosphere of grasses (Döbereiner, 1961; Döbereiner and Day, 1976; Haahtela *et al.*, 1983; McClung *et al.*, 1983a) but opinions concerning the contribution of fixed nitrogen to the plant are controversial (Giller and Day, 1985; Lima *et al.*, 1987). Among the diazotrophs associated with grass roots, bacteria of the genus *Azospirillum* have been the most intensively studied. Although positive responses of plants to inoculation have been obtained, they were mainly attributed to factors other than nitrogen fixation (Kapulnik and Okon, 1983; Okon, 1985; Okon *et al.*, 1983). For a directed manipulation of crop-bacterium associations towards a higher yield, a better understanding of the mechanisms of interactions between both partners would be helpful. To study these interactions, we wanted to focus attention on a natural association, *i.e.* a plant which had not been bred by man and which grew in a habitat where there was a high selection pressure for development of an association with diazotrophs: kallar grass became the object of our study. In this communication, we summarize some of the results which were obtained for this association.

Kallar grass

Leptochloa fusca (L.) Kunth, formerly *Diplachne fusca*, is widely distributed in tropical and subtropical regions. Its distribution ranges from Africa to Asia (*e.g.* Arabia, China, Korea) and to Australia. In Pakistan it is commonly called 'kallar grass'. As it is highly tolerant of soil salinity, alkalinity, and waterlogged conditions (Khan, 1966) it was introduced as a pioneer plant on salt-contaminated, infertile soils about a decade ago (Sandhu and Malik, 1975). It grows well there without addition of nitrogenous fertilizer and produces 20–40 t of green fodder per ha per year (Sandhu *et al.*, 1981). Detection of acetylene reduction activity in the rhizosphere of this grass (Malik *et al.*, 1980) gave further indication for the occurrence of an association with nitrogen-fixing bacteria. First, the partners of the association had to be defined.

Microbiology of the association

Several different N_2-fixing microorganisms have been isolated from root pieces or MPN dilutions of kallar grass. Some of the isolates have been described as being related to *Klebsiella* and *Beijerinc-*

Table 1. Organic acids exuded by sterile plants of *Leptochloa fusca* (L.) Kunth[a]

Substance	Amount in root exudates (μg)	Exudate per dry weight of root ($\mu g \cdot mg^{-1}$)
Succinic acid	87.4	3.2
Fumaric acid	104.2	3.8
Malic acid	5514.9	202.8
Citric acid	345.8	12.7
Total	6052.4	222.5

[a] Data from Kloss et al. (1984).

kia (Zafar *et al.*, 1987), to *Zoogloea* (Bilal and Malik, 1987) and to *A. brasilense* (Bors *et al.*, 1982); the last-mentioned has been lost during subculture. Also, numbers of diazotrophs in the rhizosphere of kallar grass were found to vary throughout the year (Zafar *et al.*, 1986).

Since only those diazotrophs occurring in highest numbers can be relevant to the association, our aim was to count and isolate the organisms which were most abundant in distinct root zones. Because of the diversity of diazotrophs that had been isolated previously, we were interested to see if it would be possible to specify a reproducibly occurring population of diazotrophs at a given kallar grass site at a given time of the year. The media for enumeration and isolation were adapted to the natural conditions. In a previous study (Kloss *et al.*, 1984), malic acid had been found to be the dominant organic acid exuded by kallar grass (Table 1). Therefore, we used DL-malate as a C-source in semisolid media, one of them similar to the soil in salt composition (Reinhold *et al.*, 1986).

The dominating N_2-fixing bacteria from MPN-counts were differentiated by morphological, serological and physiological criteria (Table 2). High numbers of diazotrophs could be found on and in the roots. Bacteria isolated from the endorhizosphere fraction can be regarded as colonizers of the root interior, because the success of surface sterilization had been checked by an appropriate control. The distribution of diazotrophs was reproducible throughout the experiments carried out in October and November: the rhizoplane population consisted of *Azospirillum lipoferum* and *A. halopraeferens* in almost equal numbers, whereas in the endorhizosphere, motile rods producing a yellow cell-bound pigment predominated. Such a root-zone specific association has neither been described for rice (Watanabe and Barraquio, 1979) nor for *Spartina alterniflora* (McClung *et al.*, 1983b) during studies in which appropriate controls had also been included.

A consideration of the factors that might play a role in interactions of bacteria and hosts has to be extended to the mechanisms likely to be involved in a stable colonization of distinct microhabitats of the rhizosphere. The association between this grass and diazotrophs seems to be rather close, since there is a high preferential enrichment of diazotrophs over non-diazotrophic microorganisms on and in roots compared with non-rhizosphere soil (Reinhold *et al.*, 1986).

For those isolates which could not be assigned to a known species, the ability to fix nitrogen has been demonstrated unequivocally by the incorporation of ^{15}N (Reinhold *et al.*, 1986). During enumeration

Table 2. Recovery of N_2-fixing bacteria from different root zones of *in situ* grown kallar grass and from non-rhizosphere soil[a]

Fraction	Bacterial number[b] (per g root dry weight/soil dry weight)	
	SMV-medium	SSM-medium
Rhizoplane	$2.0 \times 10^7 \pm 1.3$[c]	$1.0 \times 10^7 \pm 0.5$[d]
Rhizoplane after surface sterilization	$2.2 \times 10^4 \pm 1.8$	$7.7 \times 10^4 \pm 10.1$
Endorhizosphere	$2.4 \times 10^4 \pm 1.5$	$7.3 \times 10^7 \pm 7.0$[e]
Soil	$1.2 \times 10^4 \pm 0.4$	$3.1 \times 10^4 \pm 1.2$

[a] Table from Reinhold et al. (1986).
[b] Bacterial number given with standard deviation, derived from 5 (rhizoplane), 3 (rhizoplane after 'surface sterilization'), 3 (endorhizosphere), or 3 (soil) samplings, respectively.
[c] Differentiated into *A. lipoferum* $6.4 \times 10^6 \pm 5.7$ and *A. halopraeferens* $7.4 \times 10^6 \pm 5.1$.
[d] Determined to be *A. halopraeferens*.
[e] Unidentified straight rods.

Table 3. Results from DNA:rRNA and DNA:DNA hybridizations[a]

Strain used	mol % G + C	[14C]rRNA from *Azospirillum brasilense* ATCC 29145[T]		% DNA binding with DNA from	
		$T_{m(e)}$ (°C)	rRNA binding (%)	*Azospirillum halopraeferens* Au 4[T]	*Azospirillum amazonense* Y1[T] (Gent)
Azosprillum brasilense DSM 1690[T]	67.5	81.6	0.18	< 25[b]	< 25[b]
Azospirillum lipoferum SpBr 17	69.0	ND	ND	< 25[b]	ND
Azospirillum amazonense Y1 (Gent)	66.6	73.4	0.03	< 25[b]	100
Azospirillum amazonense Y13	66.4	75.2	0.07	< 25[b]	63
Azospirillum amazonense Y9	66.7	75.3	0.06	< 25[b]	64
Azospirillum halopraeferens Au 4[T]	70.4	74.8	0.12	100	< 25
Azospirillum halopraeferens Au 5	69.6	74.0	0.07	106	< 25
Herbaspirillum seropedicae Z67[T]	64.5	61.0	0.03	ND	ND
Herbaspirillum seropedicae Z78	64.7	59.2	0.03	ND	ND

[a] Table from Reinhold *et al.* (1987b).
[b] 25% is the lower reliability border of the method used.

and isolation of rhizoplane bacteria, we observed cells resembling *Azospirillum* with unusual morphological and serological characters. Differences from known species were also found in physiological tests (Reinhold *et al.*, 1986). As a new genus of spirillum-like microaerophilic diazotrophs has been described, *Herbaspirillum* (Baldani *et al.*, 1986), the generic status of our isolates had to be confirmed. DNA:rRNA hybridizations were carried out by a method in which the degree of similarity can be estimated from the $T_{m(e)}$ value (DeSmedt and DeLey, 1977), which is the temperature at which 50% of the DNA-rRNA hybrid is denatured. High thermal stability of the hybrid indicates a high degree of relatedness. Like *Azospirillum* our isolates belonged in rRNA super-

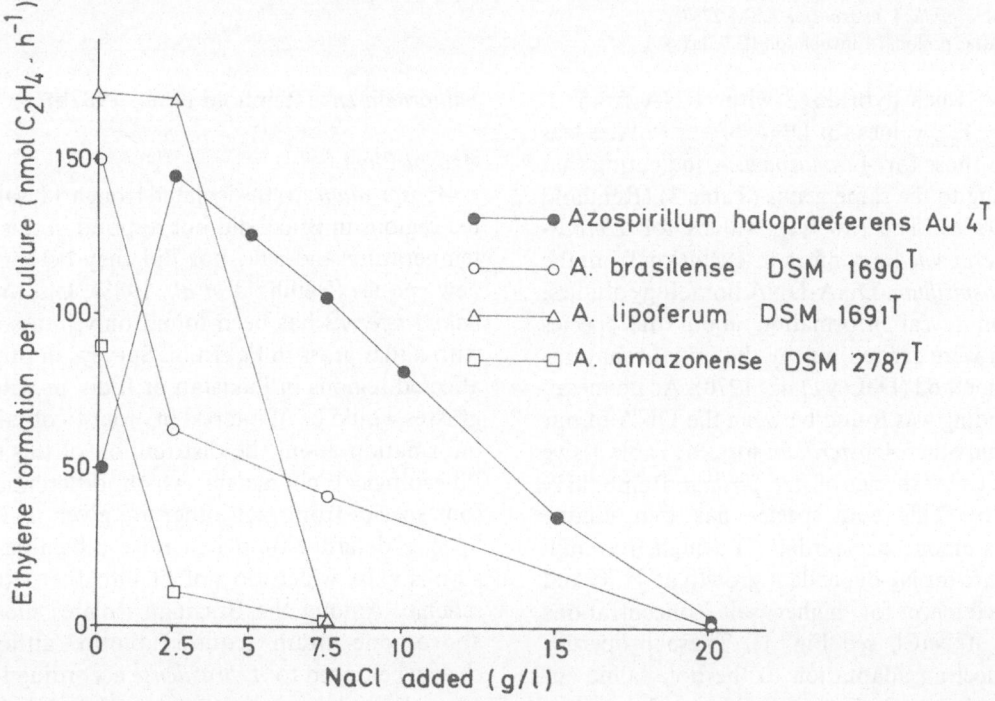

Fig. 1. Effect of NaCl concentration on N_2-fixation and growth of *Azospirillum halopraeferens* Au 4 (●), *Azospirillum lipoferum* DSM 1691 (△), *Azospirillum brasilense* DSM 1690 (○) and *Azospirillum amazonense* DSM 2787 (□). NaCl was added to N-free semi-solid SM medium containing 0.01% NaCl. From Reinhold *et al.*, (1987b).

Table 4. Characteristics differentiating *Azospirillum* species from each other[a]

Characteristic	A. lipoferum	A. brasilense	A. amazonense	A. halopraeferens
Cell width (μm)	1.0 − 1.7	1.0 − 1.2	0.9 − 1.0	0.7 − 1.4
Cell length exceeding 5 μm dominating in alkaline N-free semisolid medium	+[b]	−	−	−
Optimum temperature for growth (°C)	37	37	35	41
Growth at pH 6.0	+	+ or w	+	−
> 6.8	+	+	w	+
Growth in presence of 3% NaCl	− (−)[c]	d (−)	− (−)	+
Biotin requirement	+ (+)	−(−)	−(−)	+
sole carbon sources for growth in N-free semisolid medium:				
D-Glucose	+ (+)	− (−)	+ (+)	−
D-Mannitol	+ (+)	− (−)	(−)	w[d]
Sucrose	− (−)	− (−)	+ (+)	−
Acidification of peptone-based glucose broth	+ (+)	− (−)	(−)	−
Acid from glucose or fructose broth anaerobically	+ or v (+)	− (−)	(−)	−
Nitrate reductase	+ (+)	+ (+)	d	+
Denitrification	d (+)	d (+)	−	+
DNA base composition (mol% G + C)	69 − 70	70 − 71	67 − 68	69 − 70

[a] Table from Reinhold *et al.* (1987b).

[b] +, Positive in more than 90% of the strains; d, positive in 11 to 89% of the strains; −, negative in more than 90% of the strains; w, scant growth; v, strain instability.

[c] Data in parenthesis were determined by us for reference strains: *Azospirillum lipoferum* DSM 1691[T], *Azospirillum brasilense* DSM 1690[T] and *Azospirillum amazonense* DSM 2787[T].

[d] Growth after prolonged incubation (5–7 days).

family IV; when hybridized with rRNA from *A. brasilense*, $T_{m(e)}$ values for DNA of our isolates was similar to those for *A. amazonense*, indicating that they belong to the same genus (Table 3) (Reinhold *et al.*, 1987b). The low $T_{m(e)}$ values for *Herbaspirillum seropedicase* confirm its exclusion from the genus *Azospirillum*. DNA-DNA homology studies, which can reveal information about the species affiliation were carried out by the initial renaturation rate method (DeLey *et al.*, 1970). As no meaningful binding was found between the DNA of our isolates and other *Azospirillum* species (Table 3), we proposed a new species of *Azospirillum* Reinhold *et al.*, 1987b). This new species has two unique properties among azospirilla: (i) a high optimum temperature for N_2-dependent growth at 41 °C and a (ii) preference for higher salt concentrations ($2.5 \, g \, l^{-1}$ of NaCl, see Fig. 1). These properties might reflect an adaptation to the hot, saline environment in the Punjab of Pakistan. Due to the response of NaCl, we named the new species *A.*

halopraeferens (Reinhold *et al.*, 1987b).

Azospirillum

Azospirillum strains isolated from arid, salt-affected regions in Brazil did not respond in this way to temperature and salt, nor did they belong to the new species (Reinhold *et al.*, 1988). Until now, *A. halopraeferens* has been found only in association with kallar grass in Pakistan. Surveys in other salt-affected regions in Pakistan or India or with other grasses would be of interest in order to obtain more information about the distribution of this species. Physiological characters which differentiate the four species from each other are given in Table 4. Species determination can raise difficulties, since strains exist which do not fit into the taxonomic scheme. Among the Brazilian isolates mentioned above, one strain required biotin, although it clearly belonged to *A. brasilense* according to other physiological characters and DNA-DNA hybridizations (Reinhold *et al.*, 1988).

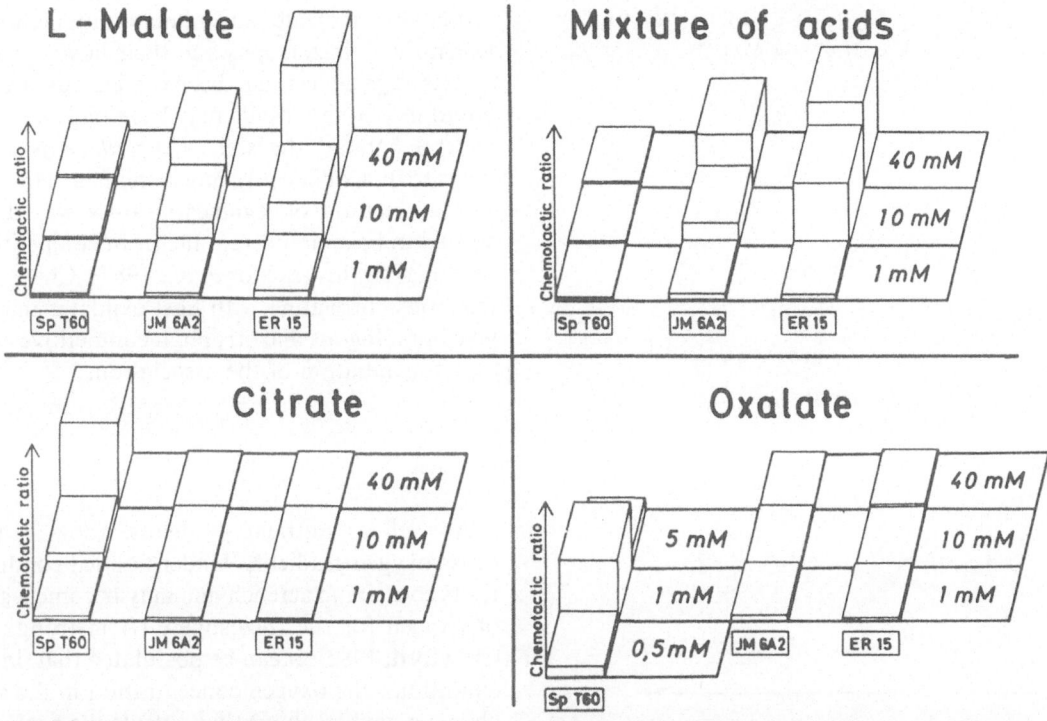

Fig. 2. Patterns of chemotactic response of *A. brasilense* SpT60, *A. brasilense* JM6A2, and *A. lipoferum* ER15 determined by the channelled-chamber technique. A mixture of acids contained the following organic acids in the same composition and molar ratio as found in the root exudates of kallar grass: L-malate-citrate-fumarate-succinate (55:2.4:1.2:1) with L-malate at the given molarity. From Reinhold *et al.*, (1985).

Chemotaxis

Chemotactic reactions enable bacteria to carry out directed movements in a chemical gradient. In nutrient gradients, bacteria capable of chemotaxis have a competitive advantage (Pilgram and Williams, 1976). In the rhizosphere, roots can serve as a source of nutrients and so create gradients. Thus, chemotaxis may be important in assisting motile bacteria to approach their host. We compared chemotactic responses of three *Azospirillum* strains in an *in vitro* assay, using the channelled chamber technique which allows the measurement of a chemotactic response without interference of aerotaxis (Barak *et al.*, 1982, 1983). The strains originated from a C_3-plant, wheat, and C_4-plants, maize and kallar grass.

Among the amino acids, sugars and organic acids tested, inter-strain differences were most pronounced for the organic acids (Reinhold *et al.*, 1985). Reactions of the strains isolated from C_4-plants were similar and differed from those of the

strain originating from wheat (Fig. 2). The chemotactic response partially correlated with the exudation patterns of the host plants. Malate, the strongest attractant for the C_4-plant isolates, is by far the dominant organic acid exuded by kallar grass (Kloss *et al.*, 1984) and occurs in large amounts in maize roots (Neyra and Hageman, 1976) in contrast to oxalate, which was the major organic acid in exudates of rice (Boureau, 1977) and wheat (Vancura, 1964). Thus, strain specificity of the chemotactic response of *Azospirillum* might reflect an adaptation of the micro-environment created by the host plant.

An even more specific response, which was not related to a group of bacteria but to the homologous isolate, was found towards a high-molecular-weight, heat-labile attractant found in the exudates of kallar grass (Reinhold *et al.*, 1985). Maize mucilage has also been found to attract only the homologous *Azospirillum* and *Enterobacter* strains (Mandimba *et al.*, 1986). In birdsfoot-trefoil root exudates, a high-molecular-weight attractant ident-

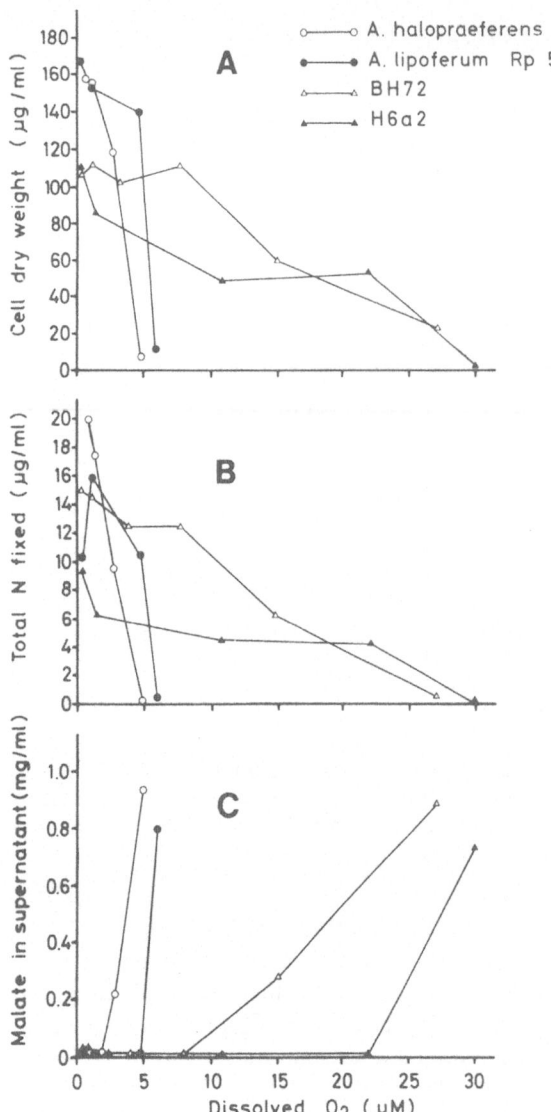

Fig. 3. Effect of O_2 on cell dry weight (A), total nitrogen fixed (B), and malate in supernatant fractions (C) of steady-state cultures of diazotrophs predominating on the rhizoplane (*A. halopraeferens* Au 4, ●; *A. lipoferum* Rp5 ●) and in the endorhizosphere (BH72, △; H6a2, ▲) of kallar grass. Continuous cultures of the strains were grown at a dilution rate of 0.083 h^{-1}. Data represent means of each steady-state culture (maximum standard deviation of cell dry weight, 9%; of total nitrogen fixed, 8%; and of malate in supernatant fraction, 5%). All values close to the axis of dissolved O_2 in panel C correspond to a malate concentration of $\leqslant 0.3\,\mu g/ml$. From Hurek *et al.*, (1987a).

ified as a glycoprotein has been found to attract rhizobia specifically (Currier and Strobel, 1977, 1981) without having the agglutinating properties of a lectin (Currier, 1985). Occurrence of such

attractants indicate an intimate interaction between these diazotrophs and their host.

Although all tests had been carried out in vitro in liquid media, the results may have relevance in the natural habitat, the soil. *Azospirillum* is able to migrate in a sufficiently moist soil (Bashan, 1986). For nodulation of legumes motility was demonstrated to be an important factor in competition of rhizobial strains (Mellor *et al.*, 1987). Chemotactic reactions correlating with host exudates may also give homologous diazotrophs a competitive advantage for initiation of the association.

Oxygen

As well as nutrient gradients, roots can also cause oxygen gradients. Under flooded conditions, roots containing aerenchyma may become a source of oxygen for the rhizosphere (Armstrong, 1979; Crawford, 1982). It can be postulated that, in these conditions, the oxygen concentration in the aerenchyma would be above that outside the root and in the rhizosphere soil. The kallar grass site which was under study for isolation of diazotrophs is often flooded. Kallar grass, which is highly tolerant of waterlogged conditions, contains aerenchymatic tissue with large air spaces (Reinhold *et al.*, 1987a). As we found different populations of diazotrophs in and on the root (Reinhold *et al.*, 1986), we were interested in their oxygen tolerance.

To use conditions as controlled and reproducible as possible bacteria were grown in an oxygen-controlled chemostat (Hurek *et al.*, 1987a, b) under malate limitation. As representative strains, two diazotrophic rods isolated from the highest positive endorhizosphere dilutions, together with *A. lipoferum* and *A. halopraeferens*, originating from the highest rhizoplane dilutions, were compared. The effect of O_2 on steady-state cultures of N_2-grown cells can be seen in Fig. 3. There was a clear difference between the two groups. The *Azospirillum* strains showed a sharp decrease in nitrogen fixation at increasing O_2-concentrations, with a concomitant shift from malate-limitation to an internal nitrogen-limitation (Hurek *et al.*, 1987a). In contrast, the endorhizosphere isolates were still able to fix nitrogen, even while maintaining the malate limitation, at significantly higher oxygen concentrations.

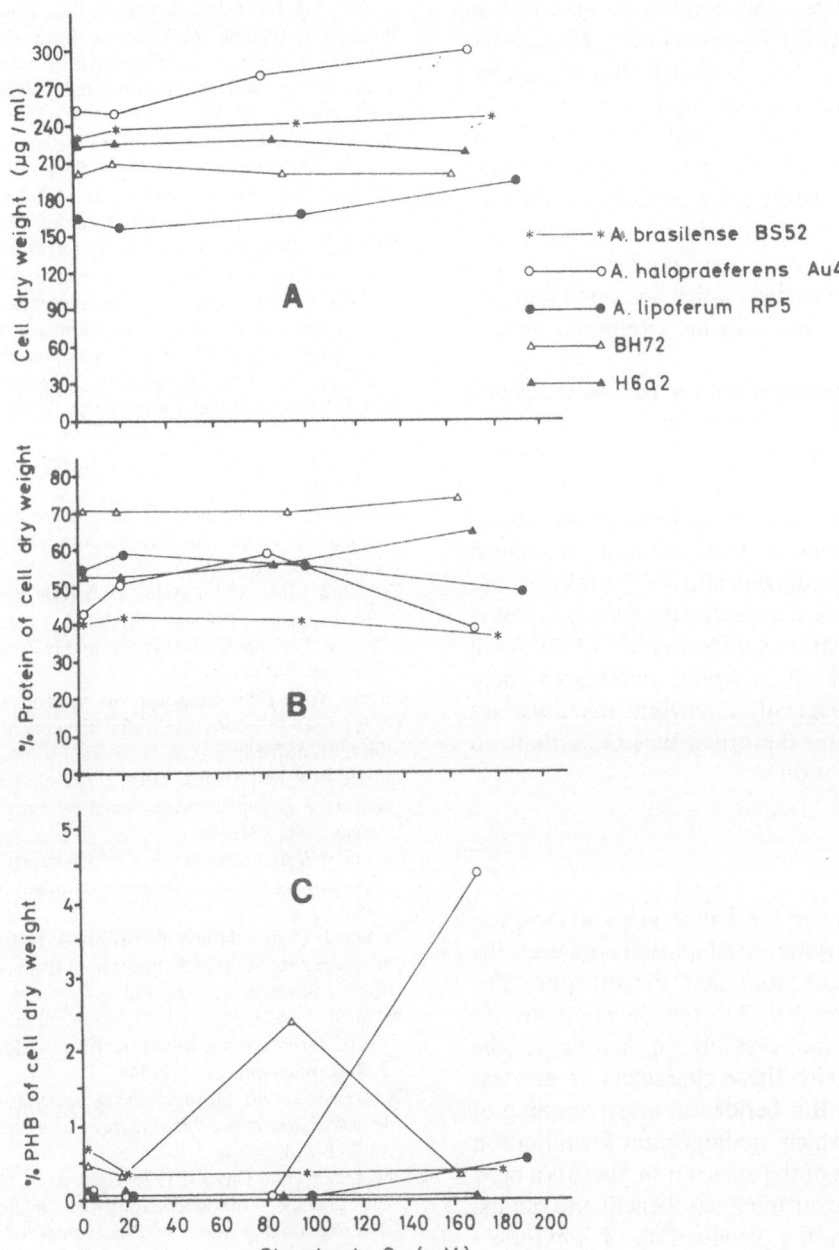

Fig. 4. Effect of O_2 on cell dry weight (A), percent protein of cell dry weight (B), and percent poly-β-hydroxybutyrate (PHB) of cell dry weight (C) in NH_4^+-grown steady-state cultures of diazotrophs either associated (root interior, △, ▲; root surface, O, ●) or unassociated (*) with kallar grass. Data represent means of each steady-state culture (maximum standard deviation of cell dry weight, 7%; of percent protein of cell dry weight, 8%; and of percent PHB of cell dry weight, 6%). From Hurek *et al.*, (1987b).

Such root-zone specific differences were not obtained when the isolates were grown on ammonium (Fig. 4). Growth of all strains, including an *A. brasilense* strain isolated from non-rhizosphere soil at the kallar grass site (Hurek *et al.*, 1987b), was not microaerophilic. Increasing the oxygen concentration had no pronounced effect on the biomass concentration, with all steady state cultures malate limited.

Thus there was a correlation between the oxygen

tolerance of isolates and conditions provided in their microhabitat for N_2-grown cells, but not for NH_4^+-grown ones. This indicates that N_2-dependent growth is likely to be important for the differential colonization of the distinct root zones of kallar grass. Therefore oxygen tolerance may also be one feature of relevance for establishment of the association. Colonization of the root interior by the diazotrophic rods has been confirmed by immunofluorescence studies, which demonstrated colonization of the aerenchyma (Reinhold *et al.*, 1987a).

Interstrain differences within the same species were found to occur in oxygen tolerance and affinity. The type strain of *A. lipoferum* (Sp 59) was significantly more oxygen tolerant than the kallar grass isolate of the same species when grown on N_2. On NH_4^+, *A. brasilense* BS52 showed no oxygen limitation at O_2-concentrations (Hurek *et al.*, 1987b) below those which already resulted in oxygen limitation of strain Cd (Nur *et al.*, 1982). As in chemotaxis, such inter-strain differences may indicate that ecologically important reactions are more dependent on the origin of strains than on their species affiliation.

Conclusions

Data obtained for the kallar grass association indicate a high degree of adaptation between the host and the most abundant diazotrophs. The microhabitat provided by the host seems to generate a selection pressure in favour of the microorganisms with those characters of greatest value to the host. But besides an understanding of the mechanisms which are important for initiation and establishment of the association, there is a need to know how diazotrophs may benefit the plants. Investigations on the production of phytohormones by bacteria (Morgenstern and Okon, 1987) or on the action of siderophores, which have been found to be produced by *Azospirillum* (Saxena *et al.*, 1986), might help to understand this aspect of interactions.

References

Armstrong W 1979 Aeration in higher plants. *In* Advances in Botanical Research, vol. 7. Ed. HW Woolhouse. pp. 226–332.

Academic Press, Inc. (London), Ltd., London.

Baldani JI, Baldani VLD, Seldin L and Döbereiner J 1986 Characterization of *Herbaspirillum seropedicae* gen. nov., sp. nov., a root-associated nitrogen-fixing bacterium. Int. J. Syst. Bacteriol. 36, 86–93.

Barak R, Nur I and Okon Y 1983 Detection of chemotaxis in *Azospirillum brasilense*. J. Appl. Bacteriol. 54, 399–403.

Barak R, Nur I, Okon Y and Henis Y 1982 Aerotactic response in *Azospirillum brasilense*. J. Bacteriol. 152, 643–649.

Bashan Y 1986 Migration of the rhizosphere bacteria *Azospirillum brasilense* and *Pseudomonas fluorescens* towards wheat roots in the soil. J. Gen. Microbiol. 132, 3407–3414.

Bilal R and Malik KA 1987 Isolation and identification of a N_2-fixing zoogloea-forming bacterium from Kallar grass histoplane. J. Appl. Bacteriol. 62, 289–294.

Bors J, Kloss M, Zelles I and Fenrik I 1982 Nitrogen fixation and nitrogen-fixing microorganisms from the rhizosphere of *Diplachne fusca* Linn. (Beauv.) J. Gen. Appl. Microbiol. 28, 111–118.

Boureau M 1977 Application de la chromatographie en phase gazeuse à l'etude de l'exsudation recinaire du riz. Cah. ORSTOM Ser. Biol. 12, 75–81.

Crawford MM 1982 Physiological responses to flooding. *In* Encyclopedia of Plant Physiology, vol. 12 B. Eds. OL Lange, PS Nobel, CB Osmond and H Ziegler. pp. 453–477. Springer Verlag, Berlin.

Currier WW 1985 Response of birdsfoot-trefoil nodulating *Rhizobium* to lectins and trefoil chemotactin. Can. J. Microbiol. 31, 587–589.

Currier WW and Strobel GA 1977 Chemotaxis of *Rhizobium* spp. to a glycoprotein produced by birdsfoot trefoil roots. Science 196, 434–436.

Currier WW and Strobel GA 1981 Characterization and biological activity of trefoil chemotactin. Plant Sci. Lett. 21, 159–165.

De Ley J, Cattoir H and Reynaerts A 1970 The quantitative measurement of DNA hybridization from renaturation rates. Eur. J. Biochem. 12, 133–142.

De Smedt J and De Ley J 1977 Intra- and intergeneric similarities of *Agrobacterium* ribosomal ribonucleic acid cistrons. Int. J. Syst. Bacteriol. 27, 222–240.

Döbereiner J 1961 Nitrogen-fixing bacteria of the genus *Beijerinckia* Derx in the rhizosphere of sugar cane. Plant and Soil 15, 211–216.

Döbereiner J and Day JM 1976. Associative symbioses in tropical grasses: Characterization of microorganisms and dinitrogen-fixing sites. *In* Proceedings of the First International Symposium on N_2 Fixation. Eds. WE Newton and CJ Nymann. pp. 518–537. Washington State University Press, Pullman.

Giller KE and Day JM 1985 Nitrogen fixation in the rhizosphere: significance in natural and agricultural systems. *In* Biological Interactions in Soil. Ed. AH Fitter. pp. 127–147. Blackwell Scientific Publications, Oxford.

Haahtela K, Helander I, Numiaho-Lassila E-L and Sundmann V 1983 Morphological and physiological characteristics and lipopolysaccharide composition of N_2-fixing (C_2H_2-reducing) root-associated *Pseudomonas* sp. Can. J. Microbiol. 29, 874–880.

Hurek T, Reinhold B, Fendrik I and Niemann E-G 1987a

Root-zone-specific oxygen tolerance of *Azospirillum* spp. and diazotrophic rods closely associated with Kallar grass. Appl. Environ. Microbiol. 53, 163–169.

Hurek T, Reinhold B and Niemann E-G 1987b Effect of oxygen on NH$_4^+$-grown continuous cultures of *Azospirillum* spp. and diazotrophic rods closely associated with Kallar grass. Can. J. Microbiol. 33, 919–922.

Kapulnik Y and Okon Y 1983 Benefits of *Azospirillum* inoculation on wheat: Effects on root development, mineral uptake, nitrogen fixation, and crop yield. *In Azospirillum II*: Genetics, Physiology, Ecology. Ed. W Klingmüller. pp. 163–170. Birkhäuser Verlag, Basel.

Khan MD 1966 'Kallar grass', a suitable grass for saline lands. Agric. Pak. 17, 375.

Kloss M, Iwannek K-H, Fendrik I and Niemann E-G 1984 Organic acids in the root exudates of *Diplachne fusca* (Linn.) Beauv. Environ. Exp. Bot. 24, 179–188.

Lima E, Boddey RM and Döbereiner J 1987 Quantification of biological nitrogen fixation associated with sugar cane using a 15-N aided nitrogen balance. Soil Biol. Biochem. 19, 165–170.

Malik KA, Zafar Y and Hussain A 1980 Nitrogenase activity in the rhizosphere of Kallar grass (*Diplachne fusca* (Linn.) Beauv.) Biologia 26, 107–112.

Mandimba G, Heulin T, Bally R, Guckert A and Balandreau J 1986 Chemotaxis of free-living nitrogen-fixing bacteria towards maize mucilage. Plant and Soil 90, 129–139.

McClung CR, Patriquin DG and Davis RE 1983a *Campylobacter nitrofigilis* sp. nov., a nitrogen-fixing bacterium associated with roots of *Spartina alterniflora* Loisel. Int. J. Syst. Bacteriol. 33, 605–612.

McClung CR, VanBerkum P, Davis RE and Sloger C 1983b Enumeration and localization of N$_2$-fixing bacteria associated with roots of *Spartina alterniflora* Loisel. Appl. Environ. Microbiol. 45, 1914–1920.

Mellor HY, Genn AR, Arwas R and Dilworth MJ 1987 Symbiotic and competitive properties of motility mutants of *Rhizobium trifolii* TA1. Arch. Microbiol. 148, 34–39.

Morgenstern E and Okon Y 1987 The effect of *Azospirillum brasilense* and auxin on root morphology in seedlings of *Sorghum bicolor × Sorghum sudanense*. Arid Soil Res. Rehab. 1, 115–127.

Neyra CA and Hageman RG 1976 Relationship between carbon dioxide, malate, and nitrate accumulation and reduction in corn (*Zea mays* L.) seedlings. Plant Physiol. 58, 726–730.

Nur I, Okon Y and Henis Y 1982 Effect of dissolved oxygen tension on production of carotenoids, poly-β-hydroxybutyrate, succinate oxidase and superoxide dismutase by *Azospirillum brasilense* Cd grown in continuous culture. J. Gen. Microbiol. 128, 2937–2943.

Okon Y 1985 *Azospirillum* as a potenial inoculant for agriculture. Trends Biotechnol. 3, 223–228.

Okon Y, Heytler PG and Hardy RWF 1983 N$_2$-fixation by *Azospirillum brasilense* and its incorporation into host *Setaria italica*. Appl. Environ. Microbiol. 46, 694–697.

Pilgram WK and Williams FD 1976 Survival value of chemotaxis in mixed cultures. Can. J. Microbiol. 22, 1771–1773.

Reinhold B, Hurek T, Baldani I and Döbereiner J 1988 Temperature and salt tolerance of *Azospirillum* spp. from salt-affected soils in brazil. *In Azospirillum IV*: Genetics, Physiology, Ecology. Ed. W Klingmüller. pp. 234–241. Springer Verlag, Berlin.

Reinhold B, Hurek T and Fendrik I 1985 Strain-specific chemotaxis of *Azospirillum* spp. J. Bacteriol. 162, 190–195.

Reinhold B, Hurek T and Fendrik I 1987a Cross reaction of predominant nitrogen-fixing bacteria with enveloped, round bodies in the root interior of Kallar grass. Appl. Environ. Microbiol. 53, 889–891.

Reinhold B, Hurek T, Fendrik I, Pot B, Gillis M, Kersters K, Thielemans S and De Ley J 1987b *Azospirillum halopraeferens* sp. nov., a nitrogen-fixing organism associated with roots of Kallar grass (*Leptochloa fusca* (L.) Kunth). Int. J. Syst. Bacteriol. 37, 43–51.

Reinhold B, Hurek T, Niemann E-G and Fendrik I 1986 Close association of *Azospirillum* and diazotrophic rods with different root zones of Kallar grass. Appl. Environ. Microbiol. 52, 520–526.

Sandhu GR, Aslam Z, Salim M, Sattar A, Qureshi RH, Ahmad N and Wyn Jones RG 1981 The effect of salinity on the yield and composition of *Diplachne fusca* (Kallar grass). Plant Cell Environ. 4, 177–181.

Sandhu GR and Malik KA 1975 Plant succession – a key to the utilization of saline soils. Nucleus 12, 35–38.

Saxena B, Modi M and Modi VV 1986 Isolation and characterization of siderophores from *Azospirillum lipoferum* D-2. J. Gen. Microbiol, 132, 2219–2224.

Vancura V 1964 Root exudates of plants. I. Analysis of root exudates of barley and wheat in their initial phases of growth. Plant and Soil 21, 231–248.

Watanabe I and Barraquio WL 1979 Low levels of fixed nitrogen required for isolation of free-living N$_2$-fixing organisms from rice roots. Nature (London) 277, 565–566.

Zafar Y, Ashraf M and Malik KA 1986 Nitrogen fixation associated with roots of Kallar grass (*Leptochloa fusca* L. Kunth). Plant Soil 90, 93–105.

Zafar Y, Malik KA and Niemann E-G 1987 Studies on N$_2$-fixing bacteria associated with the salt-tolerant grass, *Leptochloa fusca* (L.) Kunth. MIRCEN J. 3, 45–56.

F. A. Skinner et al. (Eds.), Nitrogen fixation with non-legumes, 209–218.

Location of diazotrophs in the root interior with special attention to the kallar grass association

BARBARA REINHOLD and T. HUREK

Institute of Biophysics, University of Hannover, Herrenhäuser Str. 2, D-3000 Hannover 21, FRG

Key words: colonization, light microscopy, kallar grass, N_2-fixation, protein A-gold, root interior

Abstract

There is increasing evidence that nitrogen-fixing bacteria are able to colonize the interior of grass roots. Several techniques have been used to demonstrate the sites of colonization and are discussed. Among the techniques useful for specific labelling, gold-labelled reagents have been frequently successfully applied in other fields. We propose the use of the protein A-gold technique coupled with silver amplification and light microscopy to render diazotrophs visible in semi-thin sections of roots, and we have applied this technique to gnotobiotically grown kallar grass. LR white resin soft grade was a suitable resin for embedding. Diazotrophic rods predominating in the endorhizosphere of naturally growing kallar grass had the potential to colonize the aerenchyma of gnotobiotically-grown plants. Penetration by the bacteria probably occurred at epidermal cell junctions and at points of emergence of lateral roots, as also proposed for *Azospirillum*. Larger cell aggregates described by us for naturally occurring kallar grass plants were not detected in the gnotobiotic system. Physiological consequences of colonization of the aerenchyma are discussed.

Introduction

Various nitrogen-fixing bacteria are known to have the potential to colonize plant roots. Among the genera of interest, such as *Beijerinckia* (Döbereiner, 1961), *Pseudomonas* (Barraquio *et al.*, 1983; Haahtela *et al.*, 1983), *Campylobacter* (McClung and Patriquin, 1980; McClung *et al.*, 1983b), *Enterobacter* and *Klebsiella* (Haahtela *et al.*, 1981; Ladha *et al.*, 1983), *Bacillus* (Seldin *et al.*, 1984), and *Herbaspirillum* (Baldani *et al.*, 1986b), *Azospirillum* has been studied most intensively.

In several studies, isolation procedures including appropriate controls gave evidence for the occurrence of diazotrophs in the root interior (Baldani *et al.*, 1986a; Reinhold *et al.*, 1986; McClung *et al.*, 1983a; Watanabe and Barraquio, 1979); in *Spartina alterniflora* and kallar grass (*Leptochloa fusca* (L.) Kunth), the same types of bacteria were found in the endorhizosphere repeatedly (McClung and Patriquin, 1980; Reinhold *et al.*, 1986).

The apparent occurrence of diazotrophs in the root is of considerable interest, as it results in a very close contact of both partners of the association. But isolation procedures can give only indirect evidence for the colonization of inner tissues of the root. Knowledge about these sites is of importance for a better understanding of plant-bacteria interactions.

Since numerous different bacteria may colonize plants growing in their natural surroundings, and since the bacteria have to be clearly distinguished from plant material, especially when intracellular infection is taken into account, the methodology for detection of diazotrophs is crucial. In this communication we propose protein A-gold labelling followed by silver amplification as an immunological technique to localize bacteria specifically in plant roots.

We have studied kallar grass, which is grown as a pioneer plant on salt-affected, low-fertility soils in Pakistan (Sandhu and Malik, 1975). This plant has been found to be in close, root-zone-specific association with nitrogen-fixing bacteria (Reinhold *et al.*, 1986; see also Reinhold *et al.*, 1988).

Common techniques to locate diazotrophs

A simple way to observe bacterial colonization is to examine roots by light microscopy. In many studies, 2,3,5-triphenyl-tetrazolium dichloride (TTC) has been used as a vital staining agent, and its conversion into water-insoluble, red-coloured. formazans upon its reduction has been regarded as evidence for the presence of nitrogenase and thus N_2-fixing microorganisms (Malik and Zafar, 1985; O'Hara *et al.*, 1983; Patriquin and Döbereiner, 1978; Patriquin *et al.*, 1980). However, reduction of tetrazolium has also been considered to be a measure of respiratory electron transport (Zimmermann *et al.*, 1978), and TTC is reduced by a variety of obligately anaerobic fermentative bacteria (Oren, 1987), too. Therefore, reduction of TTC should be interpreted with caution. Light microscopic examination of plants without application of a specific bacterial stain may make it difficult to prove unequivocally the presence of bacteria in root tissues, especially if an intracellular colonization is to be demonstrated. For more detailed studies, electron microscopy is an important tool. Distinguishing bacteria from plant material is possible by a skilled electron microscopist. One disadvantage of transmission electron microscopy is the very limited area of the root which can be covered by one section. This makes a screening procedure for structures which do not occur frequently, rather time consuming. Another problem which arises in light microscopy is to differentiate any diazotrophs of interest from other microorganisms, especially when field-grown plants are used. 'Gnotobiotic' systems incubated for a long period may present the same problem, since contamination of the system often cannot be avoided. In these cases especially, a more definite determination of the diazotroph under study would be helpful.

A more specific labelling of bacteria can be achieved by immunological techniques. Polyclonal specific antibodies can be produced against cellular components of the bacteria, such as fimbriae (Korhonen *et al.*, 1986) or whole bacterial cells, by injecting them into *e.g.* rabbits. Fluorochromes like fluorescein isothiocyanate which emit light in the visible region of the spectrum when excited by ultraviolet light, can be coupled to the antibodies without loss of immunological specificity (Coons *et al.*, 1942). For the 'direct' fluorescent antibody technique, the antibody against the bacterium of interest is conjugated itself, whereas in the 'indirect' or 'sandwich' technique, a fluorescent-labelled antibody specific for the globulins of the animal in which the bacterium-specific antibody has been produced, is used (for an extensive review see Bohlool and Schmidt, 1980).

For localization of diazotrophs in the root interior by immunofluorescence, sections have been obtained from non-embedded roots by freezing microtome (Schank *et al.*, 1979) or from paraffin-embedded roots (Reinhold *et al.*, 1987a, Fig. 1). Although non-specific staining can to some extent be avoided by the use of gelatin-rhodamine conjugates (Bohlool and Schmidt, 1968), problems with autofluorescence of roots have been described (Diem *et al.*, 1978); this is especially strong in the stele (Schank *et al.*, 1979; own observations). Additionally, the use of cryosections or deparaffinized sections raise the problem that single bacteria which are not firmly attached to the plant tissue may be washed out of or into the section during the staining procedure. This can be overcome by embedding roots in resins, which results in a fixed position of particles even after sectioning.

The protein A-gold technique

In electron microscopy, electron-dense stains are employed to provide a sufficiently high contrast. The use of ferritin, an iron-containing protein, was introduced as early as 1959 (Singer, 1959). An alternative is the application of colloidal suspensions of particles of metal, *e.g.* gold as a marker in immuno-electron microscopy (Faulk and Taylor, 1971). Colloidal gold was considered to be of advantage because it is simple to prepare, it gives high resolution, and it can be used to quantify immunocyto-staining (Roth, 1982).

Colloidal gold can be coated with different macromolecules. Common applications are the coating of the particles with enzymes to detect specific substrates (Mace *et al.*, 1977) or with lectins to detect sugar sequences (Horisberger and Vonlanthen, 1977), or immunogold staining, including coating with immunoglobulins or staphylococcal protein A (Geoghegan *et al.*, 1978; Horisberger and Vonlanthen, 1977; Romano and Romano, 1977; Roth *et al.*, 1978). Staphylococcal protein A binds with a non-antibody type reaction predominantly to the F_c region and to a lesser extent to the F_{AB} region of immunoglobulins (Forsgren and Sjöquist, 1966, 1967; Endresen, 1979). An advantage of protein A is that binding to immunoglobulins is not species-specific (Forsgren and Sjöquist, 1967; Goudswaard *et al.*, 1978). Thus, antibodies against different antigens originating from different mammals can be labelled with the same second marker.

The use of gold-labelled reagents is widespread in electron microscopy. They have also been applied in the examination of plant tissues in the field of N_2-fixation, *e.g.* for locating leghaemoglobin in pea nodules (Robertson *et al.*, 1984) and of the enzyme uricase in soybean nodules (Nguyen *et al.*, 1985). The immunogold technique has also made its entry into light microscopy (for extensive review, the reader is referred to Luqcoq and Roth, 1985 and Roth, 1986).

Applications of gold-labelled reagents for light microscopy

Structures labelled with colloidal gold appear light to dark red by bright-field illumination. If contrast is regarded as being too weak, heavier staining can be achieved by an additional photochemical silver reaction, called silver amplification, which has recently been introduced by Danscher (1981). The principle of the action of the physical developer is the following: the developer contains silver ions (as silver lactate) and a reducing substance (hydroquinone); colloidal gold, and later, metallic silver, catalyze the reduction of adhering silver ions to an enlarging shell of silver around the gold particles (Danscher and Nörgaard, 1983); after this reaction, gold particles become visible as black spots by bright field illumination. Additional silver amplification allows the application of low concentrations of protein A-gold, thereby reducing the possibility of non-specific staining. Such low concentrations would often result in only a faint staining in light microscopy without silver amplification (Taatjes *et al.*, 1987). Due to diffusion

problems it would be difficult to carry out an immunological pre-embedding labelling for locating diazotrophs in roots. Therefore, embedded tissue has to be stained after sectioning (post-embedding labelling).

Embedding into resins in order to retain bacteria in a well fixed position raises the problem that fixation, dehydration and embedding procedures may lower antigenicity. Embedding procedures may affect the strength of the antigenic reaction, depending on the antigen under study (see *e.g.* Roth, 1986). The choice of a resin will often be a compromise dictated by the need to retain antigenicity and preserve ultrastructural details, especially when the embedded material is also used for electron microscopic studies. The use of a low temperature dehydration and embedding technique gave excellent retention of antigenicity (Roth *et al.*, 1981). Lowicryl K4M, a methacrylic resin developed for this technique (Carlemalm *et al.*, 1982), has already been applied successfully for embedding legume nodules (Robertson *et al.*, 1984). However, long infiltration times (Titus and Becker, 1985) and compression during sectioning (Craig and Miller, 1984) have caused difficulties when plant tissues were embedded in this resin.

To locate diazotrophs in roots of kallar grass, we applied post-embedding immunogold labelling

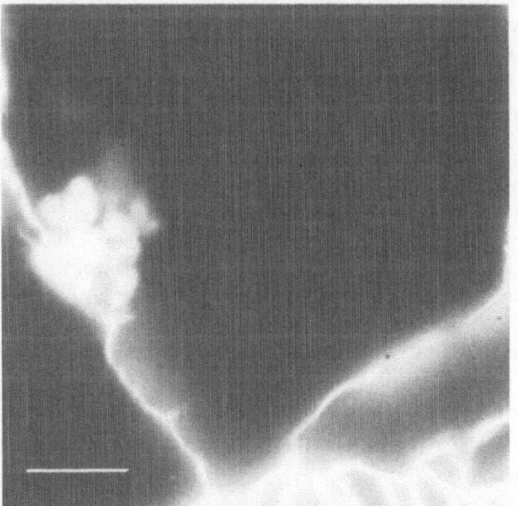

Fig. 1. Round bodies specifically stained by indirect immunofluorescence with serum against diazotrophic rods BH72 and H6a2 in aerenchymatic tissue of kallar grass roots. Plants originated from a natural site in Punjab, Pakistan. Bar represents 20 μm. Photo taken from Reinhold *et al.* (1987b).

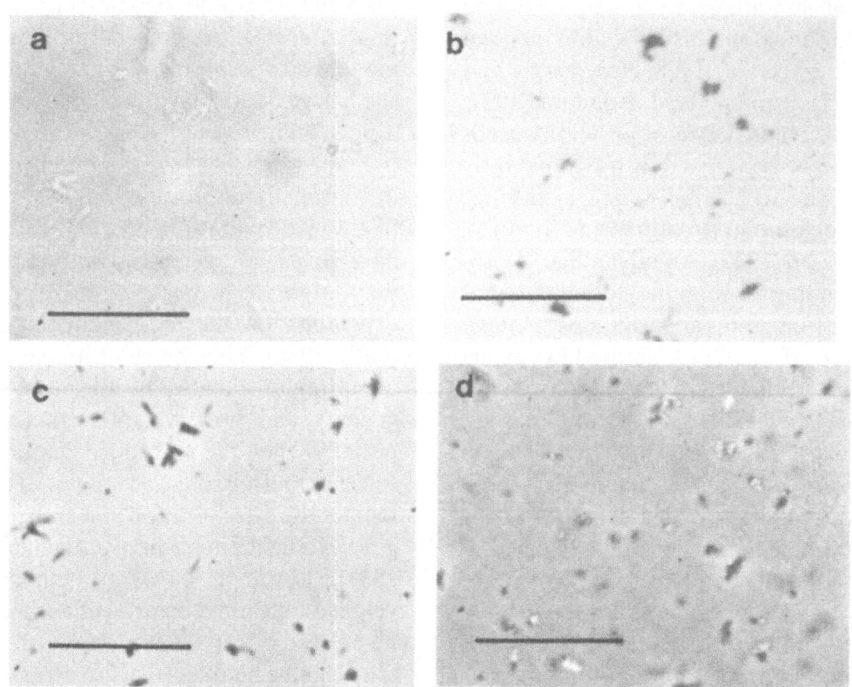

Fig. 2. Light microscopic appearance of diazotrophic rods in LR white soft grade sections without (a, c) and with (b, d) colloid-gold labelling using bright field (a, b) or phase contrast (c, d) illumination. Bars represent 20 μm. Bacteria were grown in liquid SM-medium (Reinhold *et al.*, 1985) and suspended in solidifying agar (2%); agar blocks of 1 mm were fixed for 1 h with 1% glutaraldehyde in 50 mM Na-cacodylate buffer pH 7.2, immersed in 50 mM cacodylate buffer supplemented with 50 mM NH$_4$Cl, dehydrated in an ethanol series, and infiltrated with LR white resin soft grade (Agar Aids Ltd., Essex, U.K.) overnight. Polymerization was done at 54°C for 4 d. Sections were cut with glass knives and a Sorvall 'Porter Blum' ultra-microtome MT, Newton, USA, and mounted on gelatine-coated glass slides. Immunostaining was carried out similarly as described by Roth (1986): serum and the protein A-gold complex (Auro Probe EM pA G15, Janssen Life Sci. Prod., Nettetal, FRG) were diluted 1:25 and 1:10, respectively, in PBS containing 0.1% Tween 20, 0.1% Triton X-100 and 1% bovine serum albumin. Sections were covered for 1 h with serum, washed 2 × 3 min in PBS, incubated for 1 h with the protein A-gold complex and washed again. Silver amplification followed a protocol given by Taatjes *et al.* (1987), with slight modification (all chemicals were dissolved in double-distilled water). Two volumes of citrate buffer (0.5 M, pH 3.9) were mixed with three volumes of hydroquinone (0.85 g in 15 ml) and 15 volumes of double-distilled water; sections were covered with this solution for 5 min after which it was replaced by the same solution containing 3 volumes of silver lactate solution (0.11 g in 15 ml, Ventron-Alfa Products, Karlsruhe, FRG) instead of 3 volumes of water. After 10 min, sections were washed for 1 min in double-distilled water, placed for 3 min in a photographic fixative, and washed again twice with double-distilled water. After drying, sections were mounted with LR white resin. Photographs were taken on Ektachrome 160 tungsten film through a Zeiss microscope. Sections shown in (b) and (c) were incubated with a mixture of serum prepared against strains H6a2 and BH72 (Reinhold *et al.*, 1987a), whereas sections in (a) and (c) were incubated with serum prepared against strain W1 (Reinhold *et al.*, 1986) as a control.

with protein A-gold and subsequent silver amplification. We decided to use an acrylic resin, LR white, soft grade (Agar aids Ltd., Essex, U.K.), which is easy to handle and has been stated to give superior labelling of uricase in soybean nodules but worse preservation of ultrastructure, than Spurr's epoxy resin (Vanden Bosch, 1986). The suitability of LR white resin for our studies was checked by embedding two representative strains of the diazotrophic rods which had been found to be predominant in the endorhizosphere of naturally occurring

kallar grass (Reinhold *et al.*, 1986), and roots of axenically grown kallar grass plants.

Post-embedding staining and immunolabelling of LR white-embedded strains H6a2 (Fig. 1) and BH72 resulted in a specific staining. When sections were treated with serum, which had previously been described to be sufficiently specific against these isolates (Reinhold *et al.*, 1987a), those parts of the bacteria which were obviously located at the surface of the section and were thus accessible to the reagents, appeared black under bright field illumination (Fig. 2b). The same spots appeared strongly refractile (Fig. 2d) with a golden to red-brownish colour in phase contrast illumination. The experimenter may use both illumination systems to assess if a structure has been stained. No visible staining was obtained when an antiserum cross-reacting with pseudomonad-like bacteria occurring on the root surface of kallar grass (Reinhold *et al.*, 1986) was applied (Fig. 2a, c), or the serum was omitted. Axenically grown kallar grass plants showing large aerenchymatic air spaces were not stained when sera against BH72 and H6a2 were used (Fig. 3a), and the structure was reasonably preserved. Therefore we regarded the method as suitable for locating the diazotrophic rods in kallar grass roots. In contrast to immunofluorescence, this method results in a permanent stain which does not fade. A simple phase contrast microscope without u.v.-illumination device is sufficient to examine the object.

Colonization of the root interior

Kallar grass has been described as being in close root-zone-specific association with N_2-fixing bacteria (Reinhold *et al.*, 1986). Diazotrophic, motile rods producing a yellow pigment were found in the 'root interior' in high numbers (about 10^8 per g root dry weight) during several experiments, whereas *Azospirillum halopraeferens* (Reinhold *et al.*, 1987b) and *Azospirillum lipoferum* were the predominant diazotrophs on the rhizoplane. Evidence for the occurrence of diazotrophs in the histosphere of kallar grass has also been reported by others (Zafar *et al.*, 1986).

To confirm the presence of diazotrophic rods in the root interior and to detect sites of colonization, immunofluorescence studies were carried out with roots of field-grown kallar grass (Reinhold *et al.*, 1987a). Roots had been sampled at the same plot and at the same time as for the studies on population mentioned above. In cross-sections of paraffin-embedded roots, large round bodies were detected in the aerenchyma, each corresponding in size to about 10^4 tightly packed bacteria (Fig. 1). These structures were apparently harbouring the diazotrophic rods, as they were specifically stained with serum against these bacteria. The structures appeared to be surrounded by a no cross-reacting

envelope (Reinhold *et al.*, 1987a). Probably owing to the method used, single bacterial cells or small aggregates were difficult to find in the sections (Reinhold *et al.*, 1987a).

We next applied the protein A-gold technique because of its advantages outlined above. The plants we used were grown for 4 weeks in a (more or less) gnotobiotic system: modified Leonard jars with quartz sand and Hoagland nutrient solution supplemented with $1 \, g \, l^{-1}$ of NaCl and $5 \, mg \, N$ as KNO_3 per jar, were each inoculated with 5×10^8 cells of strains H6a2 and BH72. Bacteria could be specifically stained on the sections. They colonized the root surface (Fig. 3b), but could also be found in the root interior in older parts of the roots which had developed large air spaces in the cortex region. Fig. 3c and Fig. 3d show small aggregates of bacteria attached to plant cell walls in the aerenchymatic tissue, sometimes close to the endodermis. Roots of the respective sections still had intact layers of epidermal cells as well as root hairs. No large structures as found in the field-grown plants could be detected, which may be due to the age of the plants or culture conditions. The cell layers within the endodermis appeared to be free of bacteria.

The colonization of aerenchymatic tissue of healthy-looking plants raises the question of how the diazotrophic rods penetrate the root. *Beijerinckia* cells have been shown to colonize monoaxenically-grown rice (Diem *et al.*, 1978). Electron micrographs have also shown bacteria in the aerenchyma of field-grown *Spartina alterniflora* associated with *Campylobacter* (McClung *et al.*, 1983a, b). Colonization of stem tissues and of the cortex and xylem of roots of sugar cane by diazotrophs was proposed after tetrazolium staining (Patriquin *et al.*, 1980). Tetrazolium staining was also observed in the aerenchymatic tissue of kallar grass (Malik and Zafar, 1985). Among diazotrophs associated with grasses, colonization of the root interior has been most intensively studied for *Azospirillum*. Tetrazolium staining has been reported to indicate presence of these bacteria in root hairs, epidermal and outer cortex cells of monoaxenic *Sorghum* systems (Patriquin and Döbereiner, 1978), between epidermal cells and outer cortex cells in monoaxenic wheat and *Sorghum* systems (Patriquin and Döbereiner, 1978), in epidermal and cortical tissues in several sand:vermiculite grown monocotyledons

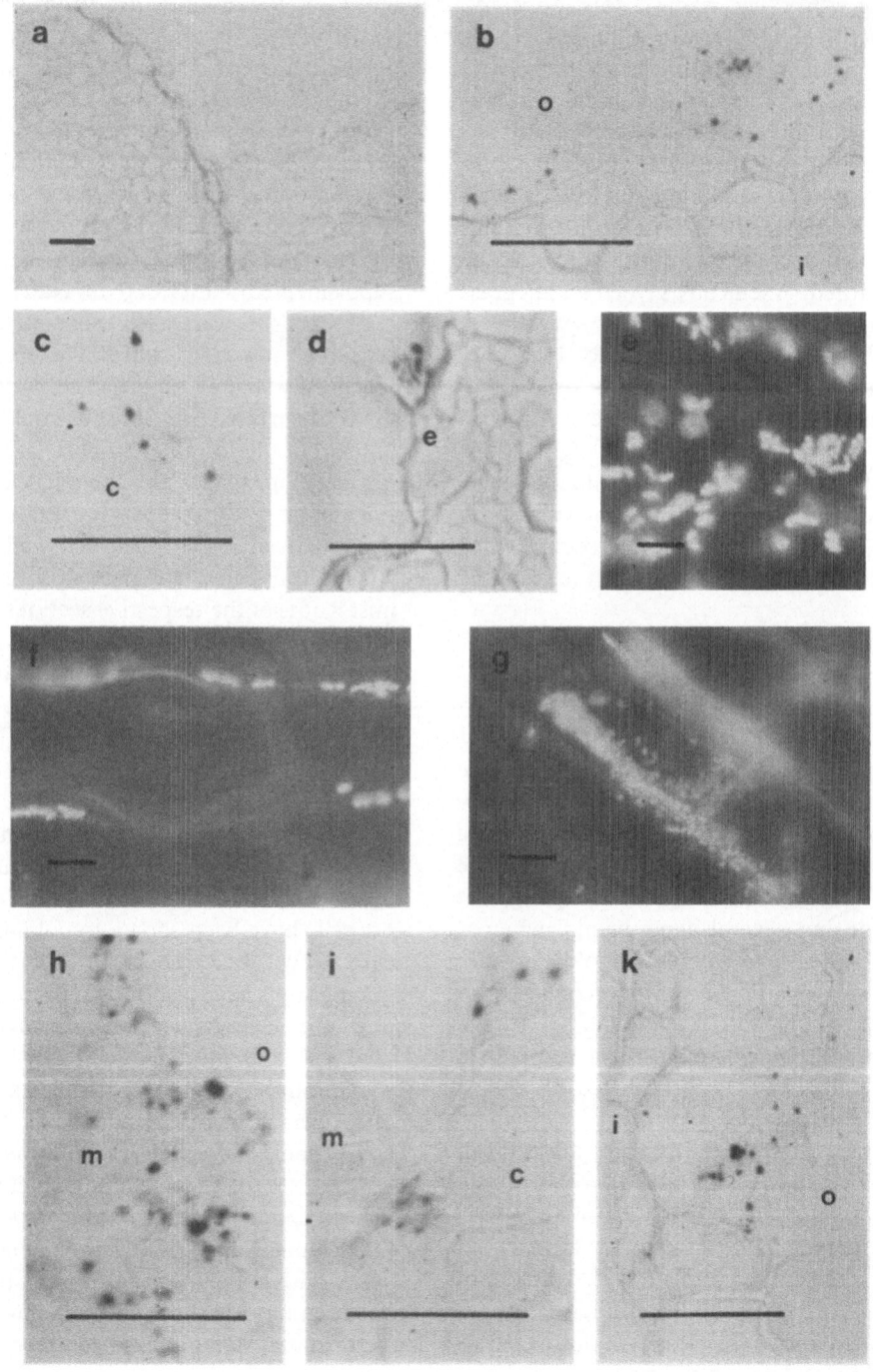

and *Phaseolus vulgaris* (O'Hara *et al.*, 1983), and in the inner cortex and stele of field-grown maize (Magalhaes *et al.*, 1979; Patriquin and Döbereiner, 1978). Light microscopic observation of *Azospirillum* cells in the xylem of monoaxenically-grown rice seedlings (Lakshmi *et al.*, 1977) has not been accompanied by specific staining. Immunofluorescence studies have demonstrated the occurrence of *Azospirillum* in the cortex of field-grown *Cynodon dactylon* roots (Schank *et al.*, 1979). In monoaxenic systems, *Azospirillum* has also been shown, by electron microscopy, to penetrate the middle lamella of cortical cells of *Panicum maximum* (Umali-Garcia *et al.*, 1978). Detection of pectinolytic activity in *Azospirillum* (Plazinski and Rolfe, 1985; Tien *et al.*, 1981; Umali-Garcia *et al.*, 1980) revealed additional evidence that an active process may contribute to the penetration of grass roots. It has also been proposed that *Azospirillum* enters roots at points of emergence of lateral roots (Patriquin and Döbereiner, 1978; Umali-Garcia *et al.*, 1978).

When we applied a non-specific fluorescent stain, europium chelate (Anderson and Westmoreland, 1971) to reveal the presence of bacteria on the root surface of inoculated kallar grass plants, we found young parts of the root about 1 cm from the root tip only weakly colonized. In contrast to some Enterobacteriaceae which were stated to adhere preferentially to root hairs of grasses (Korhonen *et al.*, 1986), almost no bacteria could be seen on root hairs. Further from the tip, bacteria occurred singly or in patches on the epidermis (Fig. 3e) as well as being aligned with cell junctions (Fig. 3f). In older parts of the roots, heavy colonization of cell junctions or intercellular spaces between epidermal

cells could be seen (Fig. 3g), as well as heavy colonization of areas of emergence of lateral roots.

The latter could be confirmed in immunogold-stained sections of the roots too. Numerous bacteria were stained on the root surface at points of emergence of young growing laterals (Fig. 3h) as well as inside the root between emerging lateral roots and epidermal or cortical cells (Fig. 3i). Bacteria also seemed to attack and invade intact epidermal cells (Fig. 3k). This observation and the colonization of junctions may indicate the involvement of an active process of penetration although the diazotrophic rods have not yet been assayed for presence of pectinolytic or cellulolytic enzymes. Entering of points of emergence of laterals would also be consistent with the finding that *Azospirillum* could be isolated mainly from the root interior of older parts of the root (Baldani *et al.*, 1986a).

Physiological consequences

Colonization of the root interior results in a very close contact between the partners of the association. The diazotrophs will have less competition with other rhizosphere bacteria for the substrate. This effect will be pronounced especially inside kallar grass roots, as diazotrophs have been found to be highly enriched over other bacteria in this region (Reinhold *et al.*, 1986). Exchange of metabolites will also be faciliated in the direction from the bacteria to the plant. Therefore, endorhizosphere bacteria may be more important for the nitrogen supply to the plant than rhizoplane or rhizosphere soil bacteria. Even in the cortex of older parts of the roots, substrate for the bacteria

Fig. 3. Colonization of kallar grass roots in gnotobiotic culture as revealed by gold-labelled reagents on semi-thin LR white sections or europium-chelate staining. Plants were grown after surface sterilization and germination (Reinhold *et al.*, 1985) in modified Leonard jars with quartz sand and Hoagland solution, supplemented with 1 g per litre of NaCl and 5 mg N as KNO_3 per jar for 4 weeks in a growth chamber. Roots of the same plants were used either for embedding and immunostaining as described in Fig. 2, or for the europiumchelate staining procedure. In the latter procedure Europium (III) Thenoyltrifluoroacetonate (Eastman Kodak Company, Rochester, USA) was dissolved in 50% ethanol. Roots were incubated in this solution for 1.5 h, washed for 2 × 2 min in 50% ethanol, mounted in glycerol, and then observed with a Zeiss fluorescence microscope equipped with a Zeiss Plan Neofluar 40 objective. (a) to (d), (h) to (k) bright field images of immunostained sections, (e) to (g) fluorescent images of europiumchelate-stained roots. (a) Roots of uninoculated kallar grass exhibiting large aerenchymatic air spaces, showing no cross reaction with diazotrophic rods after application of gold-labelled reagents. (b) Epidermis colonized with diazotrophic rods; bacteria occur near cell walls of aerenchymatic tissue singly (c) or in small aggregates (d) near endodermis. (e) Bacteria colonizing the epidermis of older parts of the roots singly and in patches; bacteria aligned with cell junctions (f) and colonizing them heavily (g). Heavy colonization by diazotrophic rods at the point of emergence of young, growing lateral roots (h); cell aggregates between emerging lateral roots and cortical cells (i); diazotrophic rods penetrating an epidermal cell (k). o, outside and i, inside the root; c, cortex cells, e, endodermis; m, meristematic tissue. Bars represent 20 μm.

may be sufficiently available: in maize, bare roots with a highly lignified cortex were surprisingly as metabolically active as sheathed roots (McCully and Canny, 1985).

When soil is flooded, aerenchymatic tissue may become a source of oxygen for the rhizosphere (Armstrong, 1979; Crawford, 1982). In cortical spaces of wetland plants, oxygen concentrations of at least 2% have been suggested (Armstrong and Gaynard, 1976). Our diazotrophic rods are able to fix N_2 in up to $25\,\mu M$ dissolved O_2 (Hurek *et al.*, 1987). Hence, during N_2-dependent colonization as single cells, they may have a competitive advantage over the more oxygen-sensitive *Azospirillum* strains occurring on the rhizoplane.

In larger cell aggregates described for field-grown plants (Reinhold *et al.*, 1987a), O_2 concentration will probably be reduced. For an aggregate of 30 000 soybean bacteroids, an external supply concentration of $10\,\mu M$ O_2 has been calculated to maintain respirable O_2 concentrations inside the aggregate without presence of leghaemoglobin (Bergersen, 1984). This would make the large aggregates ideal units for N_2-fixation when situated in the aerenchyma. The tight packing of cells and the apparent occurrence of an envelope may have an impact on the physiological state of these diazotrophs, resulting probably in an immobilization of the bacteria in these structures.

Conclusions

The protein A-gold technique coupled with silver amplification proved to be a simple and suitable technique to locate bacteria inside roots specifically. Diazotrophic rods predominating in the endorhizosphere of kallar grass had the potential to colonize the root interior of their host in gnotobiotic culture. They apparently share, with *Azospirillum*, the ability to penetrate the roots via cell junctions and points of emergence of lateral roots. The aerenchyma is likely to be an important site of colonization resulting in physiological advantages for both partners of the association.

Acknowledgements

We would like to thank Dr. I Fendrik and Prof. E-G Niemann for continuous interest and support, and Prof. F Schönbeck, Institute of Phytopathology, Hannover, for making the microtome available for us.

References

Anderson J R and Westmoreland D 1971 Direct counts of soil organisms using a fluorescent brightner and a europium chelate. Soil Biol. Biochem. 3, 85–87.
Armstrong W 1979 Aeration in higher plants. *In* Advances in Botanical Research, vol. 7. Ed. H W Woolhouse. pp. 226–332. Academic Press, Inc. (London), Ltd., London.
Armstrong W and Gaynard T J 1976 The critical oxygen pressures for respiration in intact plants. Physiol. Plant. 37, 200–206.
Baldani V L D, de B Alvarez M A, Baldani J I and Döbereiner J 1986a Establishment of inoculated *Azospirillum* spp. in the rhizosphere and in roots of field grown wheat and sorghum. Plant and Soil 90, 35–46.
Baldani J I, Baldani V L D, Seldin L and Döbereiner J 1986b Characterization of *Herbaspirillum seropedicae* gen. nov. sp. nov., a root-associated nitrogen-fixing bacterium. Int. J. Syst. Bacteriol. 36, 86–93.
Barraquio W L, Ladha J K and Watanabe I 1983 Isolation and identification of N_2-fixing *Pseudomonas* associated with wetland rice. Can. J. Microbiol. 29, 867–873.
Bergersen F J 1984 Oxygen and the physiology of diazotrophic microorganisms. *In* Advances in Nitrogen Fixation Research. Eds. C Veeger and W E Newton. pp. 171–180. Martinus Nijhoff, Dr. W. Junk Publishers, The Hague.
Bohlool B B and Schmidt E L 1968 Nonspecific staining: Its control in immunofluorescence examination of soil. Science 162, 1012–1014.
Bohlool B B and Schmidt E L 1980 The immunofluorescence approach in microbial ecology. *In* Advances in Microbial Ecology. Ed. M Alexander. pp. 203–241. Plenum Publishing Co., New York.
Carlemalm E, Garavito M and Villiger W 1982 Resin development for electron microscopy and an analysis of embedding at low temperature. J. Microsc. 126, 123–143.
Coons A H, Creech H J, Jones R N and Berliner E 1942 The demonstration of pneumococcal antigen in tissues by the use of fluorescent antibody. J. Immunol. 45, 159–170.
Craig S and Miller C 1984 LR White resin and improved on-grid immunogold detection of vicilin, storage protein. Cell Biol. International Reports 8, 879–886.
Crawford M M 1982 Physiological responses to flooding. *In* Encyclopedia of Plant Physiology, vol. 12B. Eds. O L Lange, P S Nobel, C B Osmond and H Ziegler. pp. 453–477. Springer Verlag, Berlin.
Danscher G 1981 Localization of gold in biological tissue: A photochemical method for light and electron microscopy. Histochemistry 71, 81–88.
Danscher G and Nörgaard J O R 1983 Light microscopic visualization of colloidal gold on resin-embedded tissue. J. Histochem. Cytochem. 31, 1394–1398.

Diem H G, Schmidt E L and Dommergues Y R 1978 The use of the fluorescent-antibody technique to study the behaviour of *Beijerinckia* isolate in the rhizosphere and spermosphere of rice. Ecol. Bull. (Stockholm) 26, 312–318.

Döbereiner J 1961 Nitrogen-fixing bacteria of the genus *Beijerinckia* Derx in the rhizosphere of sugar cane. Plant and Soil 15, 211–216.

Endresen C 1979 The binding to protein A of immunoglobulin G and of Fab and Fc fragments. Acta Pathol. Microbiol. Scand. Sect. C 87, 185–190.

Faulk W P and Taylor G M 1971 An immunocolloid method for the electron microscope. Immunochemistry 8, 1081–1083.

Forsgren A and Sjöquist J 1966 "Protein A" from *S. aureus*. I. Pseudoimmune reaction with human γ-globulin. J. Immunol. 97, 822–827.

Forsgren A and Sjöquist J 1967 "Protein A" from *S. aereus*. III. Reaction with rabbit γ-globulin. J. Immunol. 99, 19–24.

Geoghegan W D, Scillian J J and Ackerman G A 1978 The detection of human B-lymphocytes by both light and electron microscopy utilizing colloidal gold labeled anti-immunoglobulin. Immunol. Commun. 7, 1–12.

Goudswaard J, van der Donk J A, Noordzi J A, van Dam R H and Vaerman J P 1978 Protein A reactivity of various mammalian immunoglobulins. Scand. J. Immunol. 8, 21–28.

Haahtela K, Helander I, Nurmiaho-Lassila E-L and Sundman V 1983 Morphological and physiological characteristics and lipopolysaccharide composition of N_2-fixing (C_2H_2-reducing) root-associated *Pseudomonas* sp. Can. J. Microbiol. 29, 874–880.

Haahtela K, Wartioraara T, Sundmann V and Skujins J 1981 Root-associated N_2 fixation (acetylene reduction) by *Enterobacteriaceae* and *Azospirillum* strains in cold-climate spodosols. Appl. Environ. Microbiol. 41, 203–206.

Horisberger M and Vonlanthen M 1977 Location of mannan and chitin on thin sections of budding yeasts with gold markers. Arch. Microbiol. 115, 1–8.

Hurek T, Reinhold B, Fendrik I and Niemann E-G 1987 Root-zone-specific oxygen tolerance of *Azospirillum* spp. and diazotrophic rods closely associated with Kallar grass. Appl. Environ. Microbiol. 53, 163–169.

Korhonen T K, Nurmiaho-Lassila E-L, Laakso T and Haahtela K 1986 Adhesion of fimbriated nitrogen-fixing enteric bacteria to roots of grasses and cereals. Plant and Soil 90, 59–69.

Lakshmi V, Rao A S, Vijayalakshmi K, Lakshmi-Kumari M, Tilak K V B R and Subba Rao N S 1977 Establishment and survival of *Spirillum lipoferum*. Proc. Indian Acad. Sci. Sect. B 86, 397–404.

Ladha J K, Barraquio W L and Watanabe I 1983 Isolation and identification of nitrogen-fixing *Enterobacter cloacae* and *Klebsiella planticola* associated with rice plants. Can. J. Microbiol. 29, 1301–1308.

Lucocq J M and Roth J 1985 Colloidal gold and colloidal silver-metallic markers for light microscopic histochemistry. *In* Techniques in Immunocytochemistry, Vol. 3. Eds. G R Bullock and P Petrusz. pp. 203–236. Academic Press, London.

Mace M L, Van N T and Conn P M 1977 Electron microscopic localization of DNA-dependent RNA polymerase binding sites on DNA using enzyme immobilized on colloidal gold. Cell Biol. Intern. Rep. 1, 527–538.

Magalhaes L M S, Patriquin D and Döbereiner J 1979 Infection of field-grown maize with *Azospirillum* spp. Rev. Bras. Biol. 39, 587–596.

Malik K A and Zafar Y 1985 Quantification of root associated nitrogen fixation in Kallar grass as estimated by ^{15}N isotope dilution. *In* Nitrogen and the Environment. Eds. K Malik, S H M Naqvi and M I H Aleem. pp. 161–171. Nuclear Institute for Agriculture and Biology, Faisalabad, Pakistan.

McClung C R, Van Berkum P, Davis R E and Sloger C 1983a Enumeration and localization of N_2-fixing bacteria associated with roots of *Spartina alterniflora* Loisel. Appl. Environ. Microbiol. 45, 1914–1920.

McClung C R and Patriquin D G 1980 Isolation of a nitrogen-fixing *Campylobacter* species from the roots of *Spartina alterniflora* Loisel. Can. J. Microbiol. 26, 881–886.

McClung C R, Patriquin D G and Davis R E 1983b *Campylobacter nitrofigilis* sp. nov., a nitrogen-fixing bacterium associated with roots of *Spartina alterniflora* Loisel. Int. J. Syst. Bacteriol. 33, 605–612.

McCully M E and Canny M J 1985 Localisation of translocated ^{14}C in roots and root exudates of field-grown maize. Physiol. Plant 65, 380–392.

Nguyen T, Zlechowsla M, Foster V, Bergamn H and Verma D P S 1985 Primary structure of the soybean nodulin-35 gene encoding uricase II localized in the peroxysomes of uninfected cells of nodules. Proc. Natl. Ac. Sci. USA 82, 5040–5044.

O'Hara G W, Davey M R and Lucas J A 1983 Association between the nitrogen fixing bacterium *Azospirillum brasilense* and exised plant roots. Z. Pflanzenphysiol. 113, 1–13.

Oren A 1987 On the use of tetrazolium salts for the measurement of microbial activity in sediments. FEMS Microbiol. Ecol. 45, 127–133.

Patriquin D G and Döbereiner J 1978 Light microscopy observations of tetrazolium-reducing bacteria in the endorhizosphere of maize and other grasses in Brazil. Can. J. Microbiol. 28, 734–742.

Patriquin D G, Gracioli L A and Ruschel A P 1980 Nitrogenase activity of sugar cane propagated from stem cuttings in sterile vermiculite. Soil Biol. Biochem. 12, 413–417.

Plazinski J and Rolfe B C 1985 Analysis of the pectolytic activity of *Rhizobium* and *Azospirillum* strains isolated from *Trifolium repens*. J. Plant Physiol. 120, 181–187.

Reinhold B, Hurek T and Fendrik I 1985 Strain-specific chemotaxis of *Azospirillum* spp. J. Bacteriol. 162, 190–195.

Reinhold B, Hurek T and Fendrik I 1987a Cross reaction of predominant nitrogen-fixing bacteria with enveloped, round bodies in the root interior of Kallar grass. Appl. Environ. Microbiol. 53, 889–891.

Reinhold B, Hurek T, Fendrik I, Pot B, Gillis M, Kersters K, Thielemans S and De Ley 1987b *Azospirillum halopraeferens* sp. nov., a nitrogen-fixing organism associated with roots of Kallar grass (*Leptochloa fusca* (L.) Kunth). Int. J. Syst. Bacteriol. 37, 43–51.

Reinhold B, Hurek T, Niemann E-G and Fendrik I 1986 Close association of *Azospirillum* and diazotrophic rods with different root zones of Kallar grass. Appl. Environ. Microbiol. 52, 520–526.

Robertson J G, Wells B, Bisseling T, Farnden K J F and Johnston A W B 1984 Immuno-gold localization of leghaemoglobin in cytoplasm in nitrogen-fixing root nodules of pea.

Nature 311, 254–256.

Romano E L and Romano M 1977 Staphylococcal protein A bound to colloidal gold: A useful reagent to label antigen-antibody sites in electron microscopy. Immunochemistry 14, 711–715.

Roth J 1982 Applications of immunocolloids in light microscopy: Preparation of protein A-silver and protein A-gold complexes and their application for localization of single and multiple antigens in paraffin sections. J. Histochem. Cytochem. 30, 691–696.

Roth J 1986 Post-embedding cytochemistry with gold-labelled reagents: A review. J. Microsc. 143, 125–137.

Roth J, Bendayan M, Carlemalm E, Villiger W and Garavito M 1981 Enhancement of structural preservation and immunocytochemical staining in low temperature-embedded pancreatic tissue. J. Histochem. Cytochem. 29, 663–671.

Roth J, Bendayan M and Orci L 1978 Ultrastructural localization of intracellular antigens by the use of protein A-gold complex. J. Histochem. Cytochem. 26, 1074–1081.

Sandhu G R and Malik K A 1975 Plant succession — a key to the utilization of saline soils. Nucleus 12, 35–38.

Schank S C, Smith R L, Weiser G C, Zuberer D A, Bouton J H, Quesenberry K H, Tyler M E, Milam J R and Littell R C 1979 Fluorescent antibody technique to identify *Azospirillum brasilense* associated with roots of grasses. Soil Biol. Biochem. 11, 287–295.

Seldin L, Elsas J D and van Penido E G C 1984 *Bacillus azotofixans* sp. nov., a nitrogen-fixing species from Brazilian soils and grass roots. Int. J. Syst. Bacteriol. 34, 451–456.

Singer S J 1959 Preparation of an electron-dense antibody conjugate. Nature 183, 1523–1524.

Taatjes D J, Schaub U and Roth J 1987 Light microscopical detection of antigens and lectin binding sites with gold-labelled reagents on semithin Lowicryl K4M sections: Usefulness of the photochemical silver reaction for signal amplification. Histochem. J. 19, 235–245.

Tien T M, Diem H G, Gaskins M H and Hubbell D H 1981 Polygalacturonic acid transeliminase production by *Azospirillum* species. Can. J. Microbiol. 27, 426–431.

Titus D E and Becker W M 1985 Investigation of the glyoxysome-peroxysome transition in germinating cucumber cotyledons using double-label immunoelectron microscopy. J. Cell Biol. 101, 1288–1299.

Umali-Garcia M, Hubbell D H and Gaskins M H 1978 Process of infection of *Panicum maximum* by *Spirillum lipoferum*. Ecol. Bull. (Stockholm) 26, 373–379.

Umali-Garcia M, Hubbell D H, Gaskins M H and Dazzo F B 1980 Association of *Azospirillum* with grass roots. Appl. Environ. Microbiol. 39, 219–226.

Vanden Bosch K A 1986 Light and electron microscopic visualization of uricase by immunogold labelling of sections of resin-embedded soybean nodules. J. Microsc. 143, 187–196.

Watanabe I and Barraquio W L 1979 Low levels of fixed nitrogen required for isolation of free-living N_2-fixing organisms from rice roots. Nature (London) 277, 565–566.

Zafar Y, Ashraf M and Malik K A 1986 Nitrogen-fixation associated with roots of Kallar grass (*Leptochloa fusca* L. Kunth). Plant and Soil 90, 93–105.

Zimmermann R, Iturriaga R and Becker-Birck J 1978 Simultaneous determination of the total number of aquatic bacteria and the number thereof involved in respiration. Appl. Environ. Microbiol. 36, 926–935.

F. A. Skinner et al. (Eds.), Nitrogen fixation with non-legumes, 219–224.

Field inoculation of sorghum and rice with *Azospirillum* spp. and *Herbaspirillum seropedicae*

J. A. R. PEREIRA, V. A. CAVALCANTE, J. I. BALDANI and JOHANNA DÖBEREINER
EMBRAPA-UAPNPBS, Km 47, Seropédica, Rio de Janeiro 23851, Brazil

Key words: antibiotic resistance, *Azospirillum* inoculation, establishment of *Azospirillum*, establishment of *Herbaspirillum*, *Herbaspirillum* inoculation, rice, sorghum

Abstract

Two field experiments were carried out at the UAPNPBS experimental station, Seropédica, with two sorghum and one rice cultivars. The establishment, and inoculation effects, of *Azospirillum* spp. and *Herbaspirillum* strains marked with antibiotic resistance were investigated. One grain sorghum (BR 300) and one sugar sorghum (Br 505) cultivar were used. *Azospirillum lipoferum* strain S82 (isolated from surface sterilized roots of sorghum) established in both cultivars and comprised 40 to 80% of the *Azospirillum* spp. population in roots and stems 60 days after plant emergence (DAE). *Azospirillum amazonense* strain AmS91 (isolated from surface-sterilized roots of sorghum) reached only 50%. At 90 DAE, S82 almost disappeared (less than 30% of establishment) while the establishment of AmS91 remained constant in roots and stems. No establishment of *H. seropedicae* strain H25 (isolated from surface-sterilized roots of sorghum) or *A. lipoferum* strain S65 (isolated from the root surface of sorghum) could be observed on inoculated roots. Inoculation with S82, AmS91 or S65 but not with *H. seropedicae* H25, increased plant dry weight of both cultivars and total N in grain of the grain sorghum. In rice, *A. lipoferum* Al 121 and *A. brasilense* Sp 245 (isolated from surface sterilized rice and wheat roots respectively) established in the roots but there was no increase in *Azospirillum* spp. numbers due to inoculation. None of the strains affected plant growth or rice grain yield. *Azospirillum amazonense* A82 and *H. seropedicae* Z95, which did not establish in roots, significantly enhanced seed germination.

Introduction

Forage grasses and cereal grains represent the major source of food in the modern world. Although they cannot support *Rhizobium*-like symbiotic associations they may stimulate nitrogen-fixing bacteria on and in roots.

Enhancement of crop yields of cereals by inoculation with N_2-fixing bacteria has been observed in many field experiments (Baldani *et al.*, 1983; Baldani *et al.*, 1987; Boddey *et al.*, 1986; Kapulnik *et al.*, 1981a, b; Kapulnik *et al.*, 1982; Kapulnik *et al.*, 1987; Sarig *et al.*, 1984). At least 10 new species of root-associated microaerobic diazotrophs have been isolated and described during the past 15 years (Döbereiner and Pedrosa, 1987).

Yield increases obtained in inoculated plants have been attributed to the production of plant growth substances by the root-colonizing bacteria. Promotion of root growth resulted in enhanced nutrient and water uptake from the soil (Lin *et al.*, 1983; Okon, 1985) and in increased tillering (Kapulnik *et al.*, 1982). There is, however, no clear evidence from field experiments that inoculation with diazotrophs increased yields by biological nitrogen fixation (Okon, 1985). In most of these experiments, one strain of *Azospirillum brasilense* (Sp 7) or its red-pigmented derivative, strain Cd, was used. The strain Sp 7 was one of the first *Azospirillum* spp. isolates, obtained from rhizosphere soil under *Digitaria decumbens* (Döbereiner, 1983). Increases in yield by inoculation with strain Cd have been consistent in field experiments mainly in Israel and other semi-arid regions (Kapulnik *et al.*, 1981a, b; Kapulnik *et al.*, 1982; Kapulnik *et al.*, 1987; Smith *et al.*, 1984; Okon, 1985). In Brazil no significant increase in yield has been observed with strain Sp 7-Cd except for one experiment where the soil was fumigated before inoculation (Boddey *et al.*, 1986). The favourable effect of strain Cd in Israeli soils was probably obtained following optimal colonization of roots, due to the absence of competing *Azospirillum* spp and other rhizosphere bacteria for colonization sites on the roots (Y. Okon pers. com.). Tropical soils of Brazil usually contain 10^4–10^6 cells of *Azospirillum* per g of soil. Inoculation of wheat with two *A. brasilense* strains (245 and 107), isolated from surface sterilized wheat roots has resulted in repeatable yield increases (Baldani *et al.*, 1983; Baldani *et al.*, 1987 and Boddey *et al.*, 1986). The favourable effect of strains Sp 245 and Sp 107 on wheat in contrast to strain Sp 7 could be due to the establishment of the inoculant in the internal parts of wheat roots (Baldani *et al.*, 1986).

In the present paper we describe our attempts to establish selected *Azospirillum* spp. and *Herbaspirillum* strains in roots of sorghum and rice.

Materials and methods

Field experimental design

Two field experiments were carried out at the UAPNPBS experimental fields in Planossol (Itaguaí series) soil. A split plot experimental design was used for *Sorghum bicolor* with two cultivars (Br 300, grain sorghum and Br 505, sugar sorghum), with six inoculation treatments or N applications and six replicates. The seeds were obtained from the Centro Nacional de Pesquisa de Milho e Sorgo, Sete Lagoas-MG, Brazil. The plot size used was 3 × 2 m with 0.5 m spacing between rows and 0.2 m between plants. The experimental design for rice (*Oryza sativa* cv. IAC-25, the upland rice cultivar most used in Brazil) was a complete randomized block with 8 inoculation treatments or N applications and 6 replicates. The plot size was 10 m^2 with 0.4 m spacing between rows and 50 seeds per linear metre.

Fertilization for the two experiments consisted of

120 kg.ha^{-1} P$_2$O$_5$, 80 kg.ha^{-1} K$_2$O, 1 ton.ha^{-1} CaCO$_3$ and micronutrients (40 kg.ha^{-1} FTE) added before sowing. In the sorghum experiment all treatments received 10 kg N.ha^{-1} as ammonium sulphate 20 days after planting. The nitrogen treatment received in addition 75 kg N.ha^{-1} applied in two doses, at 45 and 75 days after sowing. In the rice experiment, 20 kg N.ha^{-1} were applied 20 days after sowing in the nitrogen treatment only and at 43 days after sowing in all treatments.

Selection of antibiotic resistant mutants

In order to identify the inoculated strains (Table 1) spontaneous antibiotic resistant mutants were selected. Suspensions (0.1 ml) with 10^9 cells, were spread on potato agar (Baldani and Döbereiner, 1980) containing increasing concentrations of antibiotics. Individual mutant colonies appearing on the highest concentration of antibiotic were streaked out on potato agar containing the same antibiotic concentration, and on plates with higher concentrations. The antibiotic resistance levels of all strains were then rechecked three times by observing growth on potato agar plates with antibiotic. N$_2$ fixing capability of the selected strains was checked in semi-solid NFb or LGI medium (Baldani, 1984; Baldani and Döbereiner, 1980) in the presence of the same concentration of antibiotic. Once the resistance level to one antibiotic was established, some of the strains were treated with a second antibiotic in the same way for double marking. A list of the levels of resistance of the strains to the various antibiotics is presented in Table 1.

Inoculation

The individual strains were grown in 0.5 litre of NFb or LGI medium (Baldani, 1984; Baldani and Döbereiner, 1980) at 30°C for 48 h and the bacterial suspension was mixed with 2.5 kg of granular peat. Counts for bacteria in the granular inoculant revealed 10^7 cells.g^{-1} after 36 h at 30°C. The inoculant (1 g/seed) was applied at sowing. Control plots were treated similarly with peat granules mixed with autoclaved cells.

Table 1. Origin and characteristics of *Azospirillum* spp. and *Herbaspirillum* sp. used

Strain	Characteristic	Origin
Sp 245	*A. brasilense*, resistant to 40 mg.l^{-1} of streptomycin	Sterilized roots of wheat (1% Chloramine-T-15 min)
Al 121	*A. lipoferum*, resistant to 20 mg.l^{-1} of spectinomycin	Rhizosphere soil of rice
S 65	*A. lipoferum*, resistant to 40 mg.l^{-1} of streptomycin	Surface of washed roots of sorghum
S 82	*A. lipoferum*, resistant to 5 mg.l^{-1} of streptomycin and 5 mg.l^{-1} of spectinomycin	Sterilized roots of sorghum (1% Chloramine-T-15 min)
Am 73	*A. amazonense*, resistant to 40 mg.l^{-1} of streptomycin	Sterilized roots of rice (1% Chloramine-T-15 min)
Am 82	*A. amazonense*, resistant to 40 mg.l^{-1} of streptomycin	Sterilized roots of rice (1% Chloramine-T-5 min)
Am S91	*A. amazonense*, resistant to 50 mg.l^{-1} of streptomycin	Sterilized roots of sorghum (1% Chloramine-T-15 min)
H 25	*H. seropedicae*, resistant to 80 mg.l^{-1} of streptomycin	Sterilized roots of sorghum (1% Chloramine-T-15 min)
Z 89	*H. seropedicae*, resistant to 240 mg.l^{-1} of streptomycin	Sterilized roots of rice (1% Chloramine-T-5 min)
Z 95	*H. seropedicae*, resistant to 5 mg.l^{-1} of spectinomycin	Washed roots of rice

Estimates of Azospirillum *spp. and* Herbaspirillum *numbers and establishment of inoculated bacteria*

Establishment of the inoculated strain was verified 60 and 90 days after planting. Composite samples of 4 plants per plot were taken and 10 g subsamples of washed roots, sterilized roots (1% Chloramine-T for 15 min) or washed stems were each macerated with 90 ml of 4% sucrose solution. Serial dilutions were made to 10^{-5} for stems, 10^{-6} for sterilized roots and to 10^{-7} for washed roots. Three 0.1 ml portions of each individual dilution, were placed in nitrogen-free semi-solid NFb or LGI medium (Baldani, 1984; Baldani and Döbereiner, 1980). The numbers of bacteria were estimated according to McGrady (MPN) tables for three replicates. Cultures from the three highest positive dilution vials were replicated into vials of semi-solid NFb or LGI medium containing the various antibiotics. The percentage of establishment was calculated according to the proportion of vials showing growth with the specific antibiotic for which each strain was marked (Baldani *et al.*, 1986). From the control plots, cultures were replicated into media with all the antibiotics used for the inoculated strains.

Plant material analysis

Plant samples were dried at 65°C for 3, 5 and 8 days (for leaves, roots and panicles, and stems respectively), ground to pass through a 0.5-mm sieve and analysed for total N according to the method described by Boddey *et al.* (1987).

Results and discussion

There were differences between diazotroph species and between strains within species, in their ability to establish on the root surface and in sorghum roots, as tested after root sterilization with Chloramine-T.

Sixty days after planting, *A. lipoferum* S82, predominated among the *Azospirillum* spp. population of both sorghum cultivars on the root surface, within roots and also in the stems (Table 2). At the grain filling stage, strain S82 almost disappeared while total bacterial numbers remained steady. When strain S82 was inoculated in a mixed inoculant with all other strains (Table 2) it also could be established in roots and it predominated at the flowering stage. This strain had been shown previously to establish well in sorghum roots. Due

Table 2. Establishment of inoculated *Azospirillum* spp. and *Herbaspirillum* sp. in field grown sorghum (means of 5 replicates)[a]

Treatment	Washed roots		Surface sterilized roots		Stems	
	FL	GRF	FL	GRF	FL	GRF
Grain Sorghum						
Control[b]	0	0	0	0	11	0
Control[c]	12	0	50	0	22	0
A. lipoferum S 82	80	8	75	0	75	28
A. lipoferum S 65	0	0	0	0	0	0
A. amazonense S 91	50	50	29	50	30	6
H. seropedicae H 25	0	0	0	0	0	0
Mixture[b]	60	0	50	0	93	0
Mixture[c]	40	50	45	25	50	33
Sugar Sorghum						
Control[b]	40	0	33	0	25	0
Control[c]	67	0	0	0	60	0
A. lipoferum S 82	50	29	75	21	100	11
A. lipoferum S 65	0	0	0	0	0	0
A. amazonense S 91	67	50	50	44	18	45
H. seropedicae H 25	0	0	0	0	0	0
Mixture[b]	40	0	40	0	70	0
Mixture[c]	60	17	60	75	35	40

[a] percent of the 3 highest positive MPN vials containing the inoculated strain; [b] tested for *A. lipoferum* S 82; [c] tested for *A. amazonense* Am S91. FL, flowering stage (60 days after planting). GRF, grain-filling stage (90 days after planting)

to its continuous presence along single roots it was suggested that the organism may multiply along the xylem stream (Baldani *et al.*, 1986). For this reason, in the present experiment, stem samples were also examined. The results showed that bacteria inoculated into the soil could move upwards into the stems (samples were 5 cm from the ground) and predominate in the *Azospirillum* population there (Table 2).

Azospirillum lipoferum strain S65 could not be recovered from roots or stems in this experiment although in a previous similar experiment with sorghum (J. A. Ramos Pereira, unpublished data) S65 was recovered from 65% of MPN dilution vials from surface-sterilized roots. This discrepancy may possibly be explained by the fact that in the present experiment, the antibiotic resistance label of this strain was not stable. For that reason the strain was not recognised on analysis even though the bacteria might have been present.

The evaluation of establishment of *A. amazonense* strain AmS91 was difficult because, in the non-inoculated controls there was a high percentage of *A. amazonense* with a similar antibiotic resistance (Table 2). At the second harvest, however, only *A. amazonense* AmS91 could be isolated from surface sterilized roots, but not from the unin-

oculated controls, indicating a definite establishment in the roots and also in the stems. Similar results were obtained when strain AmS91 was inoculated in the mixture with other diazotrophs.

The results obtained in this experiment confirmed a similar observation in a previous experiment where the *A. amazonense* strain used was present in 36 and 70% of MPN dilution vials of washed and surface-sterilized roots, respectively (J. A. Ramos Pereira, unpublished data).

The *Herbaspirillum seropedicae* strain H25 could not be recovered from any of the samples.

The effects of *Azospirillum* and *Herbaspirillum* inoculation on plant growth seems to be in accordance with the establishment data (if we assume that strain S65 established but was not recognised) (Table 3). However, yield increases were significant only for the grain sorghum. Inoculation with *A. lipoferum* S82 increased the plant dry weight of grain sorghum significantly (37% above controls). Panicle weight and panicle total N were increased by 54 and 66%, respectively (Table 3). These increases were equivalent to those obtained by the application of 75 kg N ha^{-1}. Unfortunately, no final grain yields at harvest are available because of bird damage.

Inoculation effects of *A. amazonense* strain

Table 3. Effect of inoculation with *Azospirillum* spp. and *Herbaspirillum seropedicae* on sorghum growth in the field (means of 5 replicates)

Treatment	Total plant dry weight[a] (g/plant)		Panicles (90 days)[b]	
	60 Days	90 Days	Dry weight (g/plant)	Total N (mg/plant)
Grain Sorghum				
Control	15.7[b]	41.6	4.17[B]	41[B]
75 kg N ha^{-1}	19.6[ab]	53.1	7.85[AB]	98[A]
A. lipoferum S 82	21.5[a]	51.2	6.44[AB]	68[AB]
A. lipoferum S 65	23.1[a]	58.2	8.94[A]	89[A]
A. amazonense S 91	19.6[ab]	59.4	7.46[AB]	78[AB]
H. seropedicae H 25	15.6[b]	44.0	5.43[AB]	57[AB]
Sugar Sorghum				
Control	29.1	63.7	3.34	36
75 kg N ha^{-1}	26.2	71.4	3.42	42
A. lipoferum S 82	31.3	69.3	3.50	39
A. lipoferum S 65	24.7	68.7	3.51	39
A. amazonense S 91	20.6	71.6	3.10	36
H. seropedicae H 25	24.2	70.0	2.84	29
	n.s	n.s	n.s	n.s

n.s., not significant
[a] Values in the same column followed by the same small letter are not significantly different at $P = 0.05$ (Duncan test on individual cultivars).
[b] Values in the same column followed by the same capital letter are not significantly different at $P = 0.05$ (Tukey test on data from both cultivars).

AmS91 were similar to those obtained with *A. lipoferum* strains S82 and S65.

In a similar field experiment with rice, where several other strains of the three species were compared, establishment of the inoculated strains on roots was observed only at flowering (Table 4). *Azospirillum lipoferum* strain Al 121 isolated from rhizosphere soil of rice completely dominated the azospirilla population on washed roots (no observation on surface-sterilized roots was made). The only other organism which was able to dominate the (washed) root population of *Azospirillum* was strain Sp 245 isolated from surface-sterilized wheat roots and which in many experiments has been shown to increase wheat yields (Baldani *et al.*, 1986; Baldani *et al.*, 1987; Boddey *et al.*, 1986). *Herbaspirillum seropedicae* did not establish itself well. Only one strain was recovered from 36% of the dilution vials. Independently of their establishment on roots, there was an unexpected significant effect of two strains, one of *A. amazonense* and one of *H. seropedicae* on seed germination (Table 4). This observation became apparent by the number of plants per plot. Neither the effect on germination

nor the establishment of the strains in roots had any effect on rice growth or grain yield. The yields, if extrapolated for Brazilian dry rice conditions, are very good (around 3500 kg ha^{-1}). Moreover, nitrogen fertilizer did not increase yields or plant growth.

The data in this paper support the need for extensive strain and species selection of the many available diazotrophs before reproducible inoculation results can be expected. Although the use of homologous strains isolated from surface sterilized roots seems to indicate a satisfactory procedure the fact that the best strain for wheat encountered so far was able to establish itself on rice roots may indicate that there are some exceptions to this rule.

The establishment of the inoculated strains in the roots or in the rhizosphere must be confirmed. The selection of beneficial strains and their effect on plant growth must be followed by different tests at various growth stages of the plant. Spermosphere models as suggested by the Balandreau group (Thomas-Banzon *et al.*, 1982) may speed up the selection process. Effects on root growth, root surface area at early stages of plant development may

Table 4. Effects of inoculation of field grown rice with *Azospirillum* spp. and *Herbaspirillum seropedicae* and establishment[A] of inoculated strains in washed roots (means of 6 replicates)

Treatment	*Seed germination (No. plants/plot)	Percentage establishment[A]		Plant dry weight ($kg\,4\,m^{-2}$)		Grain ($kg\,4\,m^{-2}$)
		FL	GRF	FL	GRF	
Control	350[c]	0	0	2.42	3.44	1.40
40 kg N ha^{-1}	366[c]	–	–	2.70	3.96	1.43
Inoculation with:						
A. brasilense 245	406[bc]	82	0	2.32	3.57	1.38
A. lipoferum Al 121	387[bc]	100	0	2.33	3.70	1.43
A. amazonense Am 82	489[a]	0	0	2.49	4.00	1.46
A. amazonense Am 73	332[c]	0	0	2.33	4.05	1.57
H. seropedicae Z 95	433[ab]	38	0	1.81	3.49	1.28
H. seropedicae Z 89	402[bc]	0	0	2.40	3.13	1.37
				n.s	n.s	n.s

* Same letters in the same column have been calculated at the Duncan 5% level and were not significant.
A, percent of the 3 highest positive MPN vials contain the inoculated strain. FL, flowering stage; GRF, grain-filling stage.

also help to select the best strains (Okon, 1985). Finally N_2-fixation must be evaluated by the ^{15}N dilution technique.

There are now more than 10 species of root-associated diazotrophs, of which perhaps 15 strains of four of them have been tested. Much remains to be done in this respect.

References

Baldani J I 1984 Ocorrência e caracterização de *Azospirillum amazonense* em comparação com as outras espécies deste gênero em raïzes de milho, sorgo e arroz. Rio de Janeiro, Universidade Federal Rural, MSc. Thesis.

Baldani V L D and Döbereiner J 1980 Host plant specificity in the infection of cereals with *Azospirillum* spp. Soil Biol. Biochem. 12, 433–439.

Baldani V L D, Baldani J I and Döbereiner J 1983 Effect of *Azospirillum* inoculation on root infection and nitrogen incorporation in wheat. Can. J. Microbiol. 29, 924–929.

Baldani V L D, Alvarez M A B, Baldani J I and Döbereiner J 1986 Establishment of inoculated *Azospirillum* spp. in the rhizosphere and in roots of field grown wheat and sorghum. Plant and Soil 90, 35–46.

Baldani V L D, Baldani J I and Döbereiner J 1987 Inoculation of field grown wheat with *Azospirillum* spp in Brasil. Biol. Fertil. Soil 4, 37–40.

Boddey R M, Baldani V L D, Baldani J I and Döbereiner J 1986 Effect of inoculation of *Azospirillum* spp. on nitrogen accumulation by field-grown wheat. Plant and Soil 95, 109–121.

Boddey R M *et al.*, 1987 Methods for the study of nitrogen assimilation and transport in grain legumes. MIRCEN J. 3, 3–22.

Döbereiner J 1983 Ten year *Azospirillum. In Azospirillum*:

Genetics, Physiology, Ecology. Ed. W Klingmüller. pp. 9–23. Experientia Supplementum 48. Basel.

Döbereiner J and Pedrosa F O 1987 Nitrogen Fixing Bacteria in Nonleguminous Crop Plants. 155 pp. Brock/Springer Series in Comtemporary Biosciences. Science Tech Publishers, Madison.

Kapulnik Y, Kiegel J, Okon Y, Nur I and Henis Y 1981a Effect of *Azospirillum* inoculation of some growth parameters on N-content of wheat, sorghum and *Panicum*. Plant and Soil 61, 65–70.

Kapulnik Y, Sarig S, Nur I, Okon Y, Kigel J and Henis Y 1981b Yield increases in summer cereal crops in Israeli fields inoculated with *Azospirillum*. Expl. Agric. 17, 179–187.

Kapulnik Y, Okon Y and Henis Y 1987 Yield response of spring wheat cultivars (*Triticum aestivum* and *T. turgidum*) to inoculation with *Azospirillum brasilense* under field conditions. Biol. Fertil. Soils 4, 27–35.

Kapulnik Y *et al.*, 1982 The effect of *Azospirillum* inoculation on growth and yield of corn. Isr. J. Bot. 3, 247–256.

Lin W, Okon Y and Hardy R W F 1983 Enhanced mineral uptake by *Zea mays* and *Sorghum bicolor* roots inoculated with *Azospirillum brasilense*. Appl. Environ. Microbiol. 45, 1775–1779.

Okon Y 1982 *Azospirillum*: Physiological properties, mode of association with roots and its application for the benefit of cereal and forage grass crops. Isr. J. Bot. 31, 214–220.

Okon Y 1985 *Azospirillum* as a potential inoculant for agriculture. Trends Biotecn. 3, 223–228.

Sarig S, Kapulnik Y, Nur I and Okon Y 1984 Response of non-irrigated *Sorghum bicolor* to *Azospirillum* inoculation. Expl. Agric. 20, 59–66.

Smith R L, Schank S C, Milan J R and Baltensperger A A 1984 Response of sorghum and pennisetum species to the N_2-fixing bacteria *Azospirillum brasilense*. Appl. Environ. Microbiol. 47, 1331–1336.

Thomas Banzon D, Weinhard P, Villecourt P and Balandreau J 1982 The spermosphere model. I. Its use in growing, counting and isolating N_2-fixing bacteria from the rhizosphere of rice. Can. J. Microbiol 28, 922–928.

Session 8

Mechanisms of response of cereal
crops to *Azospirillum* inoculation

F. A. Skinner et al. (Eds.), Nitrogen fixation with non-legumes, 227–234.

Involvement of IAA in the interaction between *Azospirillum brasilense* and *Panicum miliaceum* roots

AMALIA HARARI, J. KIGEL and Y. OKON
Faculty of Agriculture, The Hebrew University of Jerusalem, P.O. Box 12, Rehovot 76100, Israel

Key words: *Azospirillum*, IAA, *Panicum*, roots

Abstract

The possible involvement of IAA in the effect that *Azospirillum brasilense* has on the elongation and morphology of *Panicum miliaceum* roots was examined by comparing in a Petri dish system the effects of inoculation with a wild strain (Cd) with those of an IAA-overproducing mutant (FT-326). Both bacterial strains produced IAA in culture in the absence of tryptophan. At the stationary growth phase, production of IAA by FT-326 was *ca.* 12 times greater than that of Cd. When inoculation was made with bacterial concentrations higher than 10^6 colony forming units ml^{-1} ($CFU\,ml^{-1}$), both strains inhibited root elongation to the same extent. At lower concentrations Cd enhanced elongation by 15–20%, while FT-326 was ineffective. Both strains promoted root-hair development, and root-hairs were produced nearer the root tip the higher the bacterial concentration (*e.g.* root elongation region was reduced). Effects of FT-326 on root-hair development were greater than those of Cd. Acidified ether extracts of Cd and FT-326 cultures had inhibitory or promoting effects on root elongation depending on the dilution applied. At low dilutions, extracts from FT-326 were more inhibitory for elongation than those from Cd. At higher dilutions root elongation was promoted, but FT-326 extracts had to be more diluted than those from Cd. Dilutions that promoted root elongation contained supra-optimal concentrations of IAA, 1–3 orders of magnitude higher than those required for optimal enhancement by synthetic IAA. It is suggested that the bacteria produce in culture an IAA-antagonist or growth inhibitor that decreases the effectiveness of IAA action. The large variability reported for the effects of *Azospirillum* on root elongation could be the result of the opposite effects on root elongation of IAA and other compounds produced by the bacteria.

Introduction

Extensive work during the last few years has shown that inoculation of cereals and forage grasses with *Azospirillum brasilense* improved plant growth and productivity in many cases (Boddey and Döbereiner, 1982; Okon, 1985, Patriquin *et al.*, 1983). Enhancement of plant growth by *A. brasilense* seems not to be limited to its role of N_2 fixation by the *Azospirillum*/plant association (Giller and Day, 1985). Therefore, other mechanisms have been proposed to explain its beneficial effects. It has frequently been observed that inocu-

lation with *Azospirillum* enhances root development (Hartmann *et al.*, 1983a; Okon and Kapulnik, 1986; Umali-Garcia *et al.*, 1980), and improves mineral (Barton *et al.*, 1986; Kapulnik *et al.*, 1984; Lin *et al.*, 1983) and water uptake by the root system (Sarig *et al.*, 1985). Several morphological parameters that determine the characteristics of the root system are affected by *Azospirillum* — *e.g.* root elongation, initiation of adventitious roots, root branching, root-hair differentiation, etc. (Kapulnik *et al.*, 1985; Morgenstern and Okon, 1987; Patriquin *et al.*, 1983; Umali-Garcia *et al.*, 1980). However, the assessment of the separate effects of the bacteria on each one of the above growth processes is difficult due to the correlative processes and compensatory responses of the root system (*e.g.* root elongation *vs* root branching). As a result, contradictory results are frequently found in the relevant literature. Since these developmental and morphogenetic processes involve the action of endogenous growth regulators (Scott, 1972; Torrey, 1976; Wightman *et al.*, 1980), and since *A. brasilense* produces phytohormones, such as auxins, gibberellins and cytokinins (Hartmann *et al.*, 1983b; Reynders and Vlassak, 1979; Tien *et al.*, 1979), it has been proposed that the effects of the bacteria on the morphology of the root system are due to the production of such growth regulators by the colonizing bacteria, or by the plant as a reaction to colonization (Inbal and Feldman, 1982; Okon, 1985; Tien *et al.*, 1979).

Auxins comprise one of the main groups of phytohormones involved in the regulation of root growth, and they may act in both promotive or inhibitory capacities (Scott, 1972). Therefore, experimental use of *Azospirillum* mutants differing in their capability for auxin production (Hartmann *et al.*, 1983b) may contribute to the elucidation of the mechanism by which the bacteria influence the development of the root system.

The main objective of the present work was to study the effects on root growth of an IAA-overproducing mutant of *A. brasilense* Cd strain (FT-326) (Hartmann *et al.*, 1983b) compared to those of the wild Cd strain. The effects of inoculation and of IAA-containing extracts obtained from the liquid culture medium in which the bacteria were grown, were examined on *Panicum miliaceum* seedlings in a Petri dish bioassay system.

Materials and methods

The following *Azospirillum brasilense* strains were used: Cd (ATCC-27929) isolated in California from roots of *Cynodon dactylon*, and FT-326, an IAA-overproducing mutant isolated by Hartmann *et al.* (1983b) from *A. brasilense* Cd (ATCC-29710).

Bacteria were grown without tryptophan in a malate synthetic liquid medium, supplemented with 0.05% (w/v) NH_4Cl in a shaking bath at 30°C for 24 h. Before use, each batch was washed three times by centrifugation (10 min at 3000 g) in sterile 0.06 M phosphate buffer pH 6.8, and bacteria were finally resuspended in the same buffer to the cell concentration required for inoculation. Bacteria were counted by the plate dilution method, and their concentration expressed in colony-forming units (CFU) ml^{-1}.

Seeds of *Panicum miliaceum* L. were obtained from the Hazera Co., Israel. The seeds were surface sterilized in 1% sodium hypochlorite for 30 min under partial vacuum, washed with sterile water and then allowed to germinate on moist sterile filter paper at 25°C in darkness for 24 h. Only germinated seeds (root protrusion stage) were subsequently used.

Chemicals and solvents

Main chemicals used were: indole-3-acetic acid (IAA, Sigma); (^{14}C)IAA (radiochemical Centre, Amersham); benzyladenine (BA, Sigma); butylated hydroxytoluene (BHT, Sigma) and polyvinylpyrrolidone (PVP, insoluble, Sigma). Main organic solvents used were: methanol (AR, BDH); ethyl acetate (AR, Frutarom), diethyl ether (AR, BDH). Solvents were distilled before use.

Experiments in Petri dishes

Five germinated seeds were placed on a single layer of Whatman No. 1 filter paper in each 9 cm diameter Petri dish with 4 ml of sterile tap water. For inoculation, 1 ml of the required bacterial suspension was added to each Petri dish. In experiments in which the effect of IAA was examined, 1 ml of the required solution was added. Germinated seeds were incubated at 25°C in darkness and root elongation was measured after 5 days.

Scanning electron microscope (SEM)

Inoculated roots grown for 48 h in Petri dishes were fixed in 5% aqueous gluteraldehyde solution for 5 h at 4°C. Roots were dehydrated in a series of acetone-water solutions of 50, 70 and 100% for 30 min each, and finally for 1 h in 100% acetone (Echline, 1971). Root segments were dried in a critical point dryer (DRUVA) in a CO_2 atmos-

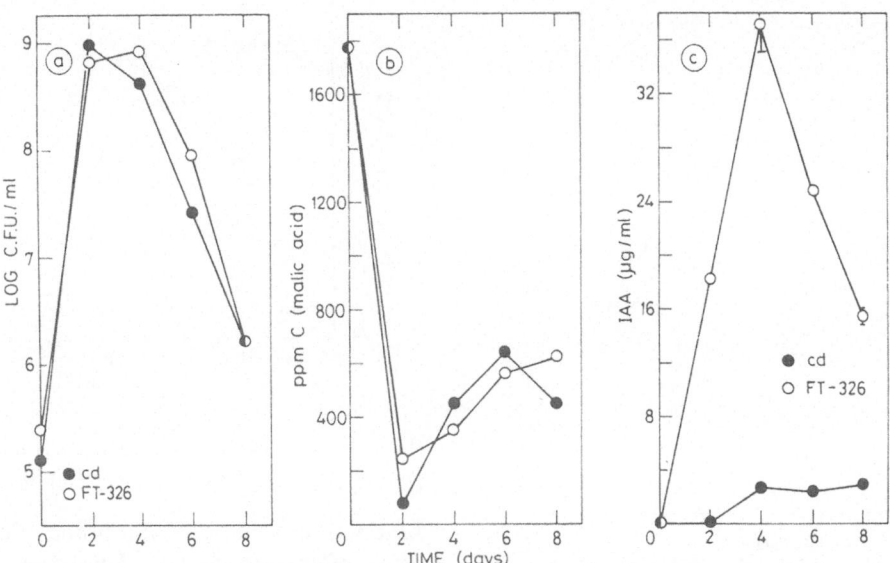

Fig. 1. Time course of growth (a), C utilization (b) and IAA excretion (c) in *A. brasilense* wild type Cd (●) and FT-326 (○).

phere. Dry segments were coated with gold and examined in a Joel ISM-35 SEM, at a working distance of 39 mm.

IAA extraction and determination

The methodology used was adapted from Iino *et al.* (1980). The supernatant fraction obtained after culture centrifugation at 3000 g for 10 min, was filtered through 0.45 μm pore size Millipore filter and brought to pH 8. After filtration the supernatant liquid was applied to a PVP column (3 cm diameter and 20 cm length, 6 g PVP) that had been previously equilibrated with $0.1\,M$ K_2HPO_4 at pH 8.0. The column was eluted with 3×10 ml volumes of $0.1\,N$ K_2HPO_4. The eluate was adjusted to pH 3 with $2.5\,N$ HCl and partitioned 5 times with water-washed ether containing $100\,mg\,l^{-1}$ of the antioxidant BHT. The ether phase was partitioned 3 times with $0.05\,M$ K_2HPO_4 pH 8.0, and the resulting aqueous phase divided into two aliquots. One was adjusted to pH 7.3 and used to study its effect on root elongation of *Panicum miliaceum* germinated seeds in Petri dishes. The other aliquot was used for the quantitative determination of IAA. The aqueous phase was adjusted to pH 3, partitioned 3 times against ether, and the combined organic fractions dried under vacuum at 35°C. The residue was dissolved in 1 ml of methanol and ap-

plied to a polyamid 11 F_{254} plate (0.15 mm, 20×20 cm) for TLC. Plates were developed with benzene-ethyl acetate -acetic acid (70:25:5) containing $100\,mg\,ml^{-1}$ BHT. Elution of IAA from the scaped solid support was carried out with H_2O-saturated ethyl acetate containing $10\,mg\,ml^{-1}$ BHT. Eluates were dried under vacuum and the IAA in the residue was determined by the specific indolo-α-pyrone fluorescence method (Stoessl and Venis, 1970) using the modifications suggested by Iino *et al.* (1980).

Statistical analysis

Petri dish experiments were repeated at least three times, with five replicates for each treatment. Data were subjected to analysis of variance. Significance at 5% level was considered to demonstrate true differences by using Duncan's multiple range test.

Results

Growth and IAA production by Cd and FT-326

The time-course of bacterial growth and C-utilization in culture were similar for wild Cd and the IAA-overproducing FT-326 (Fig. 1). Both

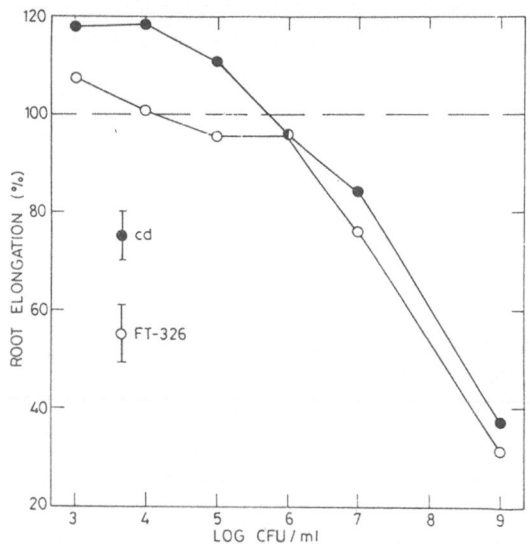

Fig. 2. Effects of *Azospirillum* inoculum concentration on root elongation of *Panicum* seedlings. (●) Cd, wild type; (○) FT-326. Vertical bars represent SE of the mean.

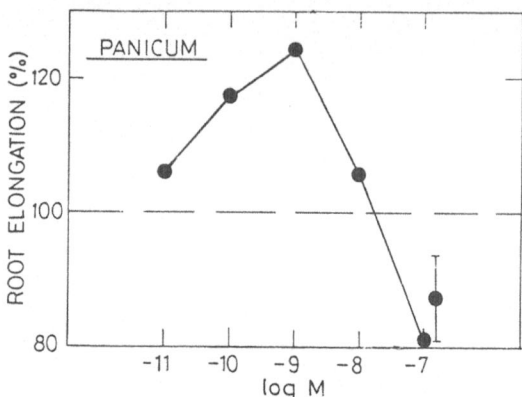

Fig. 3. Effects of IAA (indoleacetic acid) on root elongation of *Panicum* seedlings. Vertical bars represent SE of the mean.

strains reached the stationary growth phase two days after the onset of the culture, and it lasted for a further two days. During this phase both strains maintained the same concentration of bacteria and most of the available C-sources (85–95%) were consumed. With the onset of the decline phase, however, the concentration of C-sources increased gradually, probably due to the death and decomposition of the bacteria. Both strains produced IAA in the absence of tryptophan, and the hormone was excreted to the growth medium. IAA production by FT-326 was much higher compared to Cd, *e.g.* 36.6 *vs* 2.9 $\mu g \, ml^{-1}$, on the 4th day, respectively. In FT-326 IAA was mainly excreted during the stationary phase and its concentration decreased gradually after the onset of the decline phase.

Effects of Azospirillum *strains and IAA on root elongation*

The effects of inoculation with Cd and FT-326 on root elongation of germinated *Panicum* seeds was studied in Petri dishes. When grown in darkness a *Panicum* seedling produces only one (em-

bryonic) root that does not branch during the first week after germination. Therefore, confounding compensatory responses by the growing root are avoided, and the action of the bacteria on root elongation can be directly evaluated.

In both strains, bacterial numbers higher than $10^6 \, CFU \, ml^{-1}$ inhibited root elongation to the same extent (Fig. 2). At lower concentrations Cd induced a relatively small (15–20%) but significant ($P = 0.05$) promotion of elongation, while FT-326 did not affect elongation.

Application of IAA to germinated *Panicum* seeds in the same system of Petri dishes resulted in typical optimal dose- and response curves (Fig. 3). Optimal concentrations were in the range 10^{-9}–$10^{-10} \, M$ and maximal promotion of root elongation was similar to that observed after inoculation with Cd. Higher concentration inhibited root elongation.

Effects on root morphology

From visual examination of the roots it was evident that inoculation with Cd and FT-326 strongly promoted root-hair development compared to the non-inoculated controls. Therefore, 2 cm apical segments were taken from the roots 48 h after inoculation and examined in more detail by SEM (Fig. 4). In both strains, increasing concentrations of bacteria increased the length of mature root-hairs and shortened the distance between the

Fig. 4. Effects of *Azospirillum* inoculum concentration on the morphology of the apical 2 cm of *Panicum* roots. Seedlings were grown in Petri dishes in darkness at 25°C. Roots were taken 5 days after inoculation. Controls were similar to Cd roots inoculated with 10^3 $CFU \, ml^{-1}$ bacteria concentration.

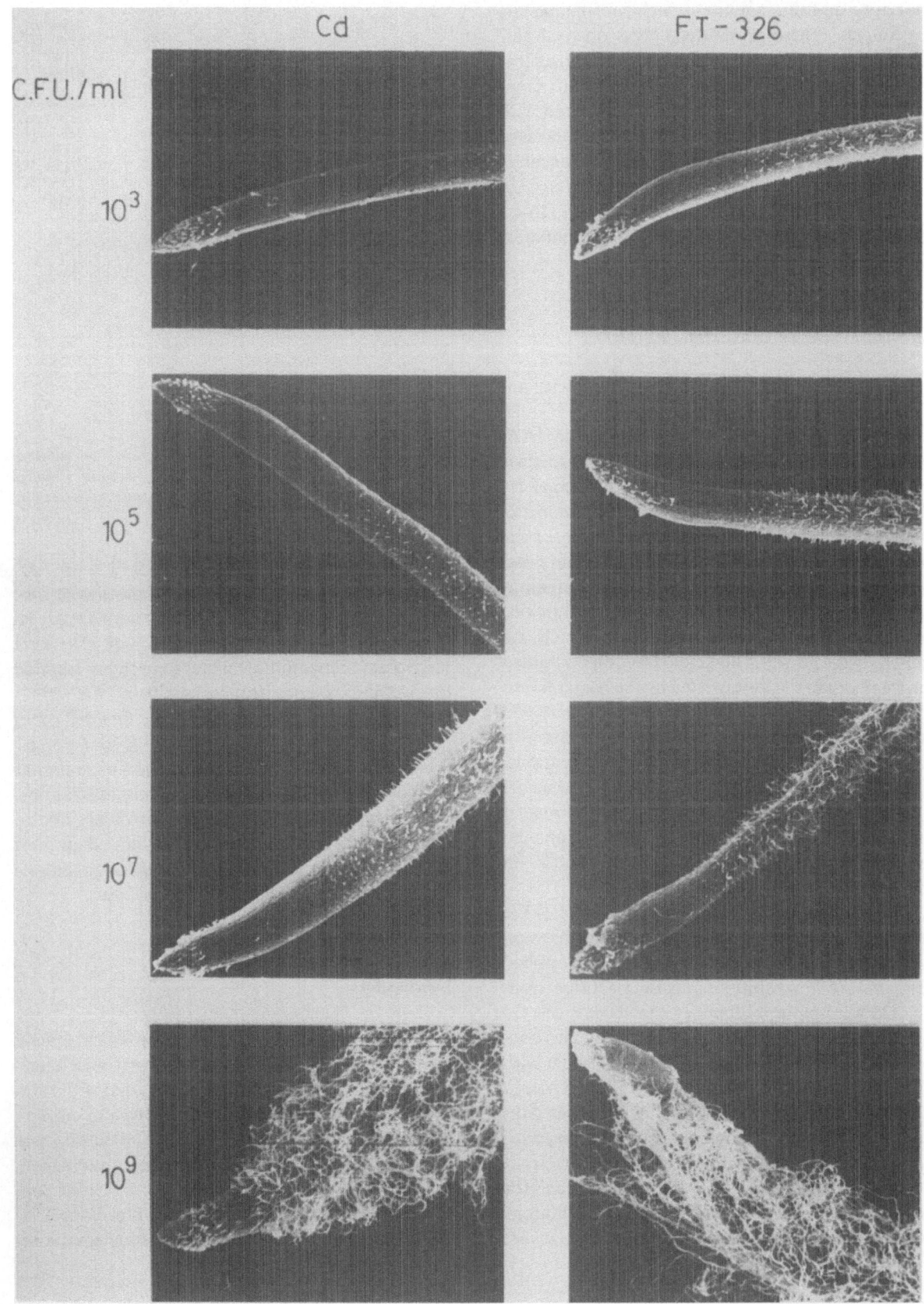

root apex and the region at which root-hairs start to elongate. These promotive effects on root-hair elongation were stronger the higher the concentration of bacteria. Promotion of root-hair elongation, however, is not necessarily correlated with inhibition of root elongation. Root-hair development was promoted even at bacterial concentrations that did not affect root elongation (*e.g.* 10^5 CFU ml^{-1}, Figs. 2 and 4). Effects of FT-326 on root-hair development were stronger compared to Cd (*e.g.* 10^3 and 10^5 CFU ml^{-1}, Fig. 4).

Effects of A. brasilense *culture extracts*

As shown in Fig. 2, root elongation was not enhanced by the IAA-overproducing FT-326 strain. This fact may indicate that beside IAA, other factors are involved in the effects of *Azospirillum* inoculation of root growth. It is conceivable that, in addition to IAA, the bacterium excretes other compounds having promoting or inhibiting effects on root growth. This possibility was explored by studying the action on root elongation of a crude acidic ether fraction (acidic fraction) extracted from the liquid medium in which the bacteria were grown. This acidic fraction contained the weak organic acids excreted by the bacteria into the growth medium. Thus, liquid cultures of Cd and FT-326 at the end of the stationary phase (4th day) and with similar bacterial concentrations, were extracted and the amount of IAA in the acidic fractions was measured. A series of dilutions of these fractions were applied to germinated panicum seeds in the Petri dish system, and their action on root elongation was examined.

The acidic fractions from both Cd and FT-326 cultures showed inhibitory or promoting effects on root elongation, depending on the dilution applied (Fig. 5). At low dilutions, acidic fractions from FT-326 were more inhibitory than those from Cd cultures. A two-fold dilution of the Cd acidic fraction inhibited root elongation by 50%, while with FT-326 this level of inhibition was obtained after a 100-fold dilution. On the other hand, higher dilutions promoted root elongation. For example, dilutions of acidic fractions in the range 10^1–10^3 from Cd and 10^4–10^5 from FT-326 were optimal for promotion of root elongation. At these dilutions IAA concentrations measured in the extracts were

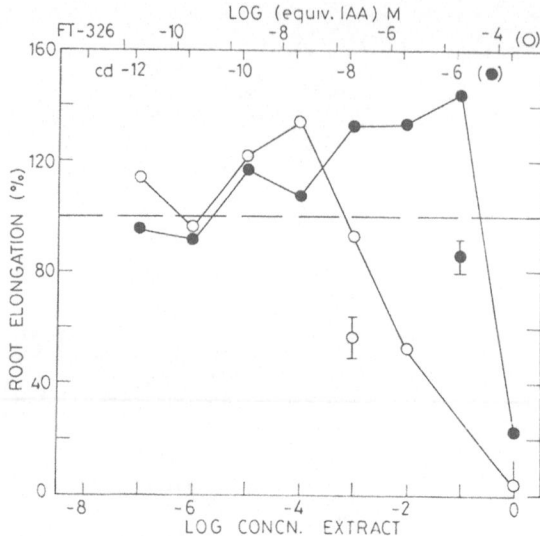

Fig. 5. Effects of dilutions of acidic ether fraction extracted from the liquid media of Cd (●) and FT-326 (○) cultures. Cultures were 4 days old and had similar bacteria concentrations (see Fig. 1).

in the ranges of 10^{-6}–$10^{-8}\,M$ and 10^{-8}–$10^{-9}\,M$, respectively (Fig. 5). Therefore, promotion of root elongation was reached in dilutions with apparently supra-optimal concentrations of IAA, that were 1–3 orders of magnitude higher than those required for optimal enhancement by synthetic IAA (10^{-9}–$10^{-10}\,M$) in the Petri dish system. On the other hand, levels of maximal promotion of root elongation (*ca* 25–35%) by the acidic fraction were similar to those found for optimal IAA concentrations. This upper level of response was probably due to growth restrictions imposed by the limited amount of storage in the seeds, and by the dark conditions under which the assays were carried out.

Discussion

In *Panicum* (Fig. 2) as well as in other grasses (Kapulnik *et al.*, 1985) inhibition of root elongation by *A. brasilense* is usually associated with root-hair proliferation. In sorghum increasing concentrations of *A. brasilense* increased the density and the length of root-hairs (Morgenstern and Okon, 1987). At higher ($\geq 10^8$ CFU ml^{-1}) bacterial concentrations root-hairs covered all the elongation region of the root, including the meristematic region near the root cap (Fig. 4). This enhancement of root-hair production near the root apex may result from the inhibition of root elongation, and not from a direct and specific effect of *Azospirillum* on root-hair differentiation itself.

In *Panicum*, however, threshold bacterial concentrations for the enhancement of root-hair differentiation were lower than those required for the inhibition of root elongation (10^5 *vs* 10^7 CFU ml^{-1}, respectively). Thus a differential effect of *Azospirillum* on root elongation and root-hair differentiation remains a possibility that has to be studied.

Inoculation of C_3 and C_4 species of grasses with *Azospirillum* promoted root elongation in many instances. This response was observed after the inoculation of plants grown in hydroponic culture (wheat, Kapulnik *et al.*, 1984, 1985; sorghum, unpublished observations), or in pots with soil (*Setaria*, Kapulnik *et al.*, 1981), vermiculite (*Setaria*, Kapulnik *et al.*, 1981; wheat, Kapulnik *et al.*, 1985; sorghum, unpublished observations), or quartz sand (maize, Cohen *et al.*, 1980, Hartmann *et al.*, 1983a; *Setaria*, Cohen *et al.*, 1980), as well as after inoculation of germinated seeds in Petri dishes (*Panicum*, Fig. 2; wheat, Kapulnik *et al.*, 1985; sorghum, Morgenstern and Okon, 1987; barley, unpublished observations). However, inoculation with higher concentrations of *Azospirillum* always results in an inhibition of root elongation. Thus, an optimum relationship between root elongation and bacterial concentration is of common occurrence. However, concentrations optimal for promotion as well as threshold concentrations for inhibition vary for different plant species and *Azospirillum* strains. Furthermore, the promotion of root elongation by *Azospirillum* has not always been observed (Morgenstern and Okon, 1987).

The effect of *A. brasilense* on root growth and morphology can be mimicked by applying IAA (Morgenstern and Okon, 1987), or mixtures of IAA, GA_3 and kinetin (Tien *et al.*, 1979) to the roots. These results, and the fact that *A. brasilense* in liquid culture produces IAA, cytokinins and gibberellins (Hartmann *et al.*, 1983b) support the proposition that *Azospirillum* effects on root growth are mediated by phytohormones.

In this work we studied the possible involvement of IAA in the effects of *A. brasilense* on root elongation. This study was based on a comparison of the action of the IAA-overproducing FT-326 mutant and Cd wild type on *Panicum* root elongation. Despite the fact that FT-326 produces in culture more IAA than does Cd (Fig. 1; Hartmann *et al.*, 1983b) it did not promote root elongation in Petri dishes (Fig. 2). Furthermore, the same degree of inhibition of root elongation was found for both types. This lack of correlation between the capacity for IAA production and the effects on root elongation has been reported by others. In maize and wheat (Kapulnik *et al.*, 1985), Cd and FT-326 enhanced elongation to the same extent, while in sorghum (Morgenstern and Okon, 1987) both strains did not promote elongation at all. On the other hand, FT-326 did not inhibit root growth of wheat seedlings, while in sorghum it inhibited root elongation at lower bacterial concentrations than did Cd (10^7 *vs* 10^5 CFU ml^{-1}). Only in this last case the strongest inhibition by FT-326 may be attributed to a larger production of supra-optimal amounts of IAA.

The general lack of correlation between the potential of *Azospirillum* strains for IAA production in liquid culture and their action on root growth can be explained assuming that IAA synthesis by the bacteria in the rhizosphere is different from that in the culture. For example, since IAA synthesis by *A. brasilense* is increased in the presence of tryptophan (Hartmann *et al.*, 1983b) IAA production by the bacteria in the rhizosphere can be constrained if tryptophan secreted by the roots becomes a limiting factor.

Analysis of the effects of the acidic fraction on root elongation of *Panicum* (fig. 5) indicates a second possible explanation. Root elongation was promoted by dilutions of the acidic fraction containing supra-optimal IAA inhibitor concentrations. On the other hand, dilutions containing optimal concentrations of IAA did not affect root elongation. Thus, in the acidic fraction the dose-response curve from IAA has been shifted toward higher concentrations. We suggest that this shift is due to the production by the bacteria of an IAA-antagonist or growth inhibitor that decreases the effectiveness of IAA action. Furthermore, the fact that the FT-326 acidic fraction has to be diluted 100–1000 times more than the fraction from Cd to promote root elongation (Fig. 5), probably indicates that FT-326 produces larger amounts of this IAA-antagonist or inhibitor than does Cd. Thus, the large variability and sometimes contradictory results reported in the literature for the promotive effects of *Azospirillum* on root elongation, can be the result of the balanced action of auxin and auxin antagonists or growth inhibitors produced by the bacteria.

References

Barton L L, Johnson G V and Orbok Miller S 1986 The effect of *Azospirillum brasilense* on iron absorption and translocation by sorghum. J. Plant Nutr. 9, 557–565.

Boddey R M and Döbereiner J 1982 Association of *Azospirillum* and other diazotrophs with tropical graminae. *In* 12th International Congress of Soil Science, New Delhi, International Society of Soil Science, FAO, Rome. Vol. 1, 28–47.

Cohen E, Okon Y, Kigel J, Nur I and Henis Y 1980 Increase in dry weight and total nitrogen content in *Zea mays* and *Setaria italica* associated with nitrogen-fixing *Azospirillum* spp. Plant Physiol, 66, 746–749.

Echline P 1971 Preparation of labile biological material for examination in the scanning electron microscope. *In* Scanning Electron Microscopy. Ed. V H Haywood. pp 307–315. Academic Press, NY.

Giller K E and Day J M 1985 Nitrogen fixation in the rhizosphere: Significance in natural and agricultural systems. *In* Biological Interaction in Soil. Ed. A H Fitter. pp 137–147. Blackwell Scientific Publications, Oxford.

Hartmann A, Fuseder A and Klingmüller W 1983a Mutants of *Azospirillum* affected in nitrogen fixation and auxin production. *In Azospirillum* II. Genetics, Physiology, Ecology. Ed. W Klingmüller. pp 78–87. Birkhauser Verlag, Basel.

Hartmann A, Singh M and Klingmüller W 1983b Isolation and characterization of *Azospirillum* mutants excreting high amounts of indoleacetic acid. Can. J. Microbiol. 29, 916–923.

Iino M, Yu R S T, and Carr D J 1980 Improved procedure for the estimation of nanogram quantities of indole-3-acetic acid in plant extracts using the indolo-pyrone fluorescence method. Plant Physiol. 66, 1099–1105.

Inbal E and Feldman M 1982 The response of a hormonal mutant of common wheat to bacteria of the genus *Azospirillum*. Isr. J. Bot. 31, 257–263.

Kapulnik Y, Okon Y, Kigel J, Nur I and Henis Y 1981 Effects of temperature, nitrogen fertilization and plant age on nitrogen fixation by *Setaria italica* inoculated with *Azospirillum brasilense* (strain Cd) Plant Physiol. 68, 340–343.

Kapulnik Y, Gafny R and Okon Y 1984 Effect of *Azospirillum* spp. inoculation on root development and NO₃ uptake in wheat (*Triticum aestivum* cv. Miriam) in hydroponic systems. Can. J. Bot. 63, 627–631.

Kapulnik Y, Okon Y and Henis Y 1985 Changes in root morphology of wheat caused by *Azospirillum* inoculation. Can. J. Microbiol. 31, 881–887.

Lin W, Okon Y and Hardy R W F 1983 Enhanced mineral uptake by *Zea mays* and *Sorghum bicolor* roots inoculated with *Azospirillum brasilense*. Appl. and Environ. Microbiol. 45, 1775–1779.

Morgenstern E and Okon Y 1987 The effect of *Azospirillum brasilense* and auxin on root morphology in seedlings of *Sorghum bicolor* x *Sorghum sudanense*. Arid Soil Res. Rehabil. 1, 115–127.

Okon Y 1985 *Azospirillum* as a potential inoculant for agriculture. Trends in Biotechnology 3, 223–228.

Okon Y and Kapulnik Y 1986 Development and function of *Azospirillum*-inoculated roots. Plant and Soil 90, 3–16.

Patriquin D G, Döbereiner J and Jain D K 1983 Sites and processes of association between diazotrophs and grasses. Can. J. Microbiol. 29, 900–915.

Reynders L and Vlassak M 1979 Conversion of tryptophan to indoleacetic acid by *Azospirillum brasilense*. Soil Biol. Biochem. 11, 547–548.

Sarig S, Okon Y and Blum A 1985 Improvement of growth and yield of non-irrigated *Sorghum bicolor* by *Azospirillum* inoculation. *In* Nitrogen Fixation Research Progress. Eds. H J Evans, P J Bottomley and W E Newton. pp 707. Martinus Nijhoff, Dordrecht.

Scott T K 1972 Auxin and roots. Annu. Rev. Plant Physiol. 23, 235–258.

Stoessl A and Venis M A 1970 Determination of submicrogram levels of indole-3-acetic acid: a new, highly specific method. Anal. Biochem. 34, 344–351.

Tien T M, Gaskins M H and Hubbell D H 1979 Plant growth substances produced by *Azospirillum brasilense* and their effect on the growth of pearl millet (*Pennisetum americanum* L.) Appl. Environ. Microbiol. 37, 1016–1024.

Torrey J G 1976 Root hormones and plant growth. Annu. Rev. Plant Physiol. 27, 435–459.

Umali-Garcia M, Hubbell D H, Gaskins M H and Dazzo F B 1980 Association of *Azospirillum* with grass roots. Appl. Environ. Microbiol. 39, 219–226.

Wightman F, Schneider E A and Thimann K V 1980 Hormonal factors controlling the initiation and development of lateral roots. II. Effects of exogenous growth factors on lateral root formation in pea roots. Physiol. Plant. 49, 304–314.

F. A. Skinner et al. (Eds.), Nitrogen fixation with non-legumes, 235–239.

Influence of inoculation with *Azospirillum brasilense* and *Glomus fasciculatum* on sorghum nutrition*

R. S. PACOVSKY

U.S. Department of Agriculture, Agricultural Research Service, Albany, CA 94710, USA

Key words: amino acids, carbohydrates, fatty acids, micronutrients, vesicular-arbuscular mycorrhiza

Abstract

Sorghum (*Sorghum bicolor* (L.) Moench cv. Bok 8) plants were inoculated with either the vesicular-arbuscular mycorrhizal (VAM) fungus *Glomus fasciculatum*, with a strain of *Azospirillum brasilense*, or with both endophytes together. Non-inoculated plants were fertilized with quantities of N and P that had been found to compensate for the input of nutrients following azospirillum or glomus colonization. Total plant dry weight in all treatments was statistically indistinguishable at harvest (10 weeks). In general, plants colonized by Glomus contained less P, Mn, starch and sucrose, but more Cu, Zn and proline than P-fertilized plants. Azospirillum-inoculated sorghum contained less N, glucose, threonine and glutamine, but more Fe and glutamate than N-amended plants. Mycorrhizal roots contained five specific fatty acids not found in non-VAM plants. Inoculated plants displayed altered nutrient requirements, membrane composition and metabolite levels, indicating that colonization by these endophytes influenced host physiology, even under conditions where N or P input was negligible.

Introduction

Crop productivity is usually limited by low N and P availability, especially in tropical lateritic soils. The growth of cereals can be improved substantially by inoculating with both the rhizosphere bacteria *Azospirillum* spp. (Okon and Kapulnik, 1986) and a vesicular-arbuscular mycorrhizal (VAM) fungus (Subba Rao *et al.*, 1985). Although these dramatic biomass increases indicate the potential of these endophytes, comparisons are limited when the 'controls' are nutrient-deficient plants. In this study, plants inoculated with *Azospirillum brasilense* or the VAM fungus *Glomus fasciculatum* were compared to plants fertilized with N or P, respectively. The soil used was a high-fertility growth medium that was edaphically similar to an intensively cultivated soil (Pacovsky *et al.*, 1985a). Under these conditions, the nutrient input provided by the two endophytes did not have

such a profound effect on host growth. Other, more subtle changes in the host following inoculation could be studied.

The purpose of this study was to examine the effect of inoculation with azospirillum or glomus on the mineral, micronutrient, carbohydrate, amino acid and fatty acid content in sorghum.

Materials and methods

Sorghum (*Sorghum bicolor* (L.) Moench cv. Bok 8) seeds were sterilized with 0.1% $HgCl_2$ and were planted in a steam-sterilized high-fertility potting mix (U.C. mix, Pacovsky *et al.*, 1985a) that had been inoculated with either *Azospirillum brasilense* (strain 29729, 10^9 cells in 5 ml peptone-succinate-salts broth) or *Glomus fasciculatum* (300 spores/pot). Other plants received both endophytes, while some plants were not inoculated (controls). All plants were furnished with a basal (N- and P-free) nutrient solution. Sorghum not inoculated with

* Contribution from the Western Regional Research Center, USDA-ARS (CRIS No. 5325-41000-008).

azospirillum received *ca.* 300 ml of 0.4 mM NH_4NO_3 once a week, and plants not inoculated with glomus were fertilized with *ca.* 300 ml of 0.2 mM KH_2PO_4 once a week so that plants of all treatments were of similar size and developmental stage (Pacovsky *et al.*, 1985b, 1986). Plants were grown in a glasshouse at 24–33°C with 55–95% relative humidity and a 16 h photoperiod: the plants were exposed to a range of light intensities (photosynthetic active, photon flux density; PPFD) of 900–1500 μmol m^{-2} s^{-1}. Plants were harvested after 10 weeks (7 replicates per treatment). Leaf areas were determined, plant parts were harvested separately, frozen in liquid N_2 and lyophilized.

The N, P and micronutrient contents of the various plant parts and the soil were determined by standard methods (Chapman and Pratt, 1982). Root segments were fixed, stained and examined microscopically for evidence of azospirillum colonization or were scored for VAM-fungal infection. Leaf and root samples were extracted with methanol:H_2O (9:1, v/v), resuspended in citrate buffer (pH 2.2) and analysed for free amino acids using HPLC (Lee, 1974). Additional samples were extracted with hexane:isopropanol (3:2, v/v) using 200 μg of margaric acid as an internal standard,

and the lipids were derivatized using BF_3/methanol (Morrison and Smith, 1964). Fatty acid (FA) methyl esters were quantified using capillary gas chromatography (Murphy and Stumpf, 1979), and the identity of the FAs was verified using a GC-mass spectrometer operating in the selective ion monitoring mode (Pacovsky and Fuller, 1988).

Percent VAM-fungal colonization was determined histologically (Pacovsky *et al.*, 1985a), and none of the non-inoculated plants showed infection by VAM fungi. The characteristic root morphology of azospirillum-colonized roots (Lin *et al.*, 1983) was present in plants inoculated with the bacteria but not in non-inoculated plants. Plant, soil and endophyte data were subjected to analysis of variance and Student's *t* test.

Results

Total dry weight was statistically indistinguishable ($P > 0.05$) between the fertilized and inoculated plants (Table 1). Mycorrhizal plants had relatively more leaf growth and relatively less root growth than P-fertilized plants.

Sorghum inoculated with azospirillum had a

Table 1. Biomass (g dry wt), and macronutrient (%) and micronutrient content (μg g^{-1}) for sorghum inoculated with microbial endophytes or fertilized with N and P

	Phosphorus + Nitrogen	*Glomus* + Nitrogen	Phosphorus + *Azospirillum*	*Glomus* + *Azospirillum*
Total Dry Wt (g)	35.98 a	36.34 a	36.15 a	37.14 a
Shoot/Root Ratio	1.52 b	1.71 a	1.56 b	1.68 a
Leaf				
Area (cm^2)	2251 c	2674 a	2466 b	2692 a
N content (%)	0.84 a	0.76 b	0.70 c	0.73 bc
P content (%)	0.16 a	0.13 b	0.15 a	0.12 b
Fe (μg g^{-1})	33.6 a	38.3 b	40.7 ab	43.5 a
Mn (μg g^{-1})	58.4 ab	43.2 c	62.9 a	55.7 b
Zn (μg g^{-1})	17.3 c	22.7 a	19.0 b	24.2 a
Cu (μg g^{-1})	4.5 d	5.6 a	4.8 bc	5.1 b
Root				
N content (%)	0.83 a	0.80 a	0.68 b	0.62 b
P content (%)	0.14 ab	0.12 b	0.15 a	0.13 ab
Fe (μg g^{-1})	328 b	351 b	412 a	427 a
Mn (μg g^{-1})	63.5 a	49.2 b	58.6 a	46.4 b
Zn (μg g^{-1})	29.1 b	36.3 a	22.7 c	30.7 b
Cu (μg g^{-1})	7.8 a	6.4 b	8.2 a	6.0 b

Values represent the mean of four replicates selected at random, and values sharing a common letter within a row are not significantly different ($P < 0.05$).

lower N content in both leaf (11% difference) and root (20% difference) than N-fertilized plants (Table 1). The leaf P content of VAM plants was 19% lower than P-fertilized sorghum, and there was no statistically significant difference in root P content between treatments. The Fe content of azospirillum-inoculated sorghum plants was higher than that of N-amended plants (Table 1). VAM plants contained less Mn, but more Cu, than sorghum that received P.

Leaves of plants inoculated with *A. brasilense* contained 23% less glucose and 32% less sucrose than leaves from N-fertilized sorghum (Table 2). Similarly, roots of azospirillum-inoculated plants also contained 27% less glucose. The starch content of leaves and roots from VAM plants was significantly ($P < 0.05$) lower than from P-amended sorghum. The sucrose content of VAM roots was similarly decreased compared to P-fertilized plants.

Inoculation with glomus apparently did not influence leaf or root amino acid content, with the exception of the elevated proline level in VAM roots (Table 2). Leaves from azospirillum-inoculated plants contained less threonine, proline, glycine, cysteine, leucine and glutamine, but these leaves contained more glutamate and γ-aminobutyric acid (GABA) than leaves from sorghum fertilized with N. Roots colonized with azospirillum contained less threonine, glycine, leucine and glutamine, but more glutamate and GABA than roots of N-fertilized plants (Table 2).

The chloroplast-specific FA 16:1(3 *t*) was relatively lower in VAM plants compared to P-amended sorghum (Table 3). Leaves from plants colonized by both endophytes had the highest FA content of any treatment. Mycorrhizal roots contained five lipids not found in P-fertilized contol roots (Table 3). These unusual FAs were 16:1(11 *c*); 18:3(6,9,12); 20:3(8,11,14); 20:4(5,8,11,14) and

Table 2. Carbohydrate content (mg g^{-1}) and content of selected amino acids (μmol g^{-1}) for sorghum inoculated with microbial endophytes*

	Phosphorus + Nitrogen	*Glomus* + Nitrogen	Phosphorus + *Azospirillum*	*Glomus* + *Azospirillum*
Leaf				
Glucose	16.3 a	15.6 a	11.7 b	12.8 b
Fructose	10.2 c	14.3 a	9.8 c	12.1 b
Sucrose	48.4 a	40.7 b	33.4 c	26.5 c
Starch	35.1 b	26.5 c	44.6 a	24.9 c
Threonine	2.14 b	2.86 a	1.85 c	1.90 c
Glutamate	4.18 b	3.91 b	6.02 a	6.34 a
Proline	0.08 a	0.06 a	0.03 b	0.03 b
Glycine	0.60 a	0.48 a	0.25 b	0.30 b
Leucine	1.82 a	1.69 a	1.34 b	1.26 b
Glutamine	1.47 a	1.28 b	1.14 c	1.08 c
GABA	0.16 b	0.22 b	0.43 a	0.50 a
Root				
Glucose	26.5 a	28.5 a	18.8 b	21.6 b
Fructose	17.8 c	32.0 a	21.7 bc	29.1 ab
Sucrose	132.1 a	110.7 b	137.5 a	98.4 c
Starch	14.9 a	9.0 b	15.3 a	7.8 b
Threonine	1.24 a	1.33 a	0.85 b	0.97 b
Glutamate	2.73 c	3.09 c	4.06 b	5.14 a
Proline	0.04 b	0.08 a	0.03 b	0.07 a
Glycine	0.51 a	0.47 a	0.26 b	0.30 b
Leucine	2.73 a	2.58 a	1.84 b	1.93 b
Glutamine	1.20 a	1.32 a	0.88 b	0.85 b
GABA	0.19 c	0.36 b	0.53 a	0.58 a

* Values represent the mean of four replicates, and values sharing a common letter within a row are not significantly different ($P < 0.05$). GABA = γ-aminobutyric acid.

Table 3. Content of selected fatty acids (mg g^{-1}) for sorghum inoculated with microbial endophytes

	Phosphorus + Nitrogen	*Glomus* + Nitrogen	Phosphorus + *Azospirillum*	*Glomus* + *Azospirillum*
Leaf				
16:1 (3 *t*)	0.23 a	0.17 b	0.25 a	0.18 b
Total F.A.	6.91 c	7.96 b	7.67 b	8.86 a
Root				
16:0	0.75 b	1.50 a	0.82 b	1.37 a
16:1(11c)	0.06 c	2.14 b	0.05 c	2.81 a
18:3(6,9,12)	0.0 b	0.29 a	0.0 b	0.33 a
20:3(8,11,14)	0.0 b	0.14 a	0.0 b	0.15 a
20:4(5,8,11,14)	0.0 b	0.08 a	0.0 b	0.10 a
20:5(5,8,11,14,17)	0.0 b	0.15 a	0.0 b	0.14 a
Total F.A.	2.50 c	6.59 b	2.42 c	7.30 a

Values represent the mean of four replicates, and values sharing a common letter within a row are not significantly different ($P < 0.05$).

20:5(5,8,11,14,17), and they were absent from non-infected roots. These VAM-derived FAs have previously been found in citrus roots infected with VAM fungi (Nagy *et al.*, 1980). Δ11-*c*-Hexadecenoic acid comprised between 32 and 38% of all lipid-derived FAs in VAM roots. These VAM-specific FAs were not found in the leaves of VAM or non-VAM plants, and they were absent from non-inoculated roots. In previous work with soybean, FA 16:1(11 *c*) content and vesicle number were highly correlated (Pacovsky and Fuller, 1988). Quantifying these non-host FAs may be a means of determining fungal infection, physiology or effectiveness with time.

Discussion

A number of physiological and biochemical effects of colonization by azospirillum (Okon and Kapulnik, 1986) or mycorrhizal fungi (Nemec and Guy, 1982) may have been overlooked in previous studies examining N and P nutritional effects in these beneficial symbioses. More attention should be paid to increases in plant hormone levels, chlorophyll content, resistance to water stress, carbon allocation and substrate metabolism in these two- and three-way symbioses.

Sorghum plants inoculated with *A. brasilense* contained less glutamine but more glutamate than N-fertilized plants. This suggests that N assimilation and allocation in azospirillum-inoculated

plants was altered compared to sorghum that received fertilizer N.

Sorghum roots infected with *G. fasciculatum* were shown to contain five FAs that were not of plant origin. These FAs have been observed in VAM fungi previously (Beilby, 1980; Nagy *et al.*, 1980). FA 16:1(11 *c*) constituted over 32% of total root FA indicating that the fungus is capable of efficiently converting host carbon to storage lipid. Endomycorrhizal fungi are phycomycetes that characteristically contain γ-linolenic rather than α-linolenic acid (Weete, 1974). Phycomycetes also contain greater percentages of C$_{20}$ and C$_{22}$ FAs as compared to higher fungi or higher plants. Chemotaxonomically, the presence of 18:3(6,9,12), 20:3, 20:4 and 20:5 clearly indicates the presence of a VA mycorrhizal symbiont.

The response of the host plant to the bacterial or fungal endophyte must be further understood if optimum yields from glomus- or azospirillum-inoculated sorghum plants are to be obtained under field conditions.

Conclusions

The response of sorghum to these endophytes cannot be accounted for solely as an N or P effect. Even in high fertility substrates where nutrient input by azospirillum or glomus is not important, inoculation with these endophytes alters host physiology (Boddey *et al.*, 1986). Changes in metabolite levels and lipid composition may play a role in the establishment or maintenance of the host-endophyte association.

In future studies of nutrient interaction in cereals it will be essential to inoculate plants with rhizosphere bacteria and mycorrhizal fungi. More work is needed to elucidate the molecular mechanisms that regulate these complex symbioses.

Acknowledgements

This work was supported by the U.S. Department of Agriculture, Agricultural Research Service (CRIS No. 5325-41000-008). The author would like to thank G Fuller, A Noma, J Milam, A Stafford and E Grey for their support and technical assistance. Reference to a company and/or product named by the Department is only for purposes of information and does not imply recommendation of the product to the exclusion of others that may also be suitable.

References

Beilby J P 1980 Fatty acid and sterol composition of ungerminated spores of the vesicular-arbuscular mycorrhizal fungus *Acaulospora levis*. Lipids 15, 949–952.

Boddey R M, Baldani V L D, Baldani J I and Döbereiner J 1986 Effect of inoculation of *Azospirillum* spp. on nitrogen accumulation by field-grown wheat. Plant and Soil 95, 109–121.

Chapman H D and Pratt P F 1982 Methods of Analysis of Soils, Plants and Waters, Second Edn., 310 pp. Univ. California Press, Berkeley, CA.

Lee P L Y 1974 Three hour single column sodium citrate physiological analysis. Durrum Application Notes No. 10, Dionex Corporation, Sunnyvale, California.

Lin W, Okon Y and Hardy R W F 1983 Enhanced mineral uptake by *Zea mays* and *Sorghum bicolor* roots inoculated with *Azospirillum brasilense*. Appl. Environ. Microbiol. 45, 1775–1779.

Morrison W R and Smith L M 1964 Preparation of fatty acid methylesters and dimethylacetals from lipids with boron fluoride and methanol. J. Lipid Res. 5, 600–608.

Murphy D J and Stumpf P K 1979 Light-dependent induction of polyunsaturated fatty acid biosynthesis in greening cucumber cotyledons. Plant Physiol. 63, 328–335.

Nagy S, Nordby H E and Nemec S 1980 Composition of lipids in roots of six cultivars infected with the vesicular-arbuscular mycorrhizal fungus, *Glomus mosseae*. New Phytol. 85, 377–384.

Nemec S and Guy G 1982 Carbohydrate status of mycorrhizal and non-mycorrhizal citrus root stocks. J. Am. Soc. Hort. Sci. 107, 177–180.

Okon Y and Kapulnik Y 1986 Development and function of *Azospirillum*-inoculated roots. Plant and Soil 90, 3–16.

Pacovsky R S, Fuller G and Paul E A 1985a Influence of soil on the interactions between endomycorrhizae and *Azospirillum* in sorghum. Soil Biol. Biochem. 17, 525–531.

Pacovsky R S, Paul E A and Bethlenfalvay G J 1985b Nutrition of sorghum plants fertilized with nitrogen or inoculated with *Azospirillum brasilense*. Plant and Soil 85, 145–148.

Pacovsky R S, Bethlenfalvay G J and Paul E A 1986 Comparison between phosphorus-fertilized and mycorrhizal plants. Crop Sci. 26, 151–156.

Pacovsky R S and Fuller G 1988 Mineral and lipid composition of *Glycine-Glomus-Bradyrhizobium* symbioses. Physiol. Plant. 72, 733–746.

Subba Rao N S, Tilak K V B R and Singh C S 1985 Synergistic effect of vesicular-arbuscular mycorrhizas and *Azospirillum brasilense* on the growth of barley in pots. Soil Biol. Biochem. 17, 119–121.

Weete J D 1974 *In* Fungal Lipid Biochemistry. Ed. J D Weete. pp. 74–79, 158–161. New York: Plenum Press.

F. A. Skinner et al. (Eds.), Nitrogen fixation with non-legumes, 241–246.
© 1989 by Kluwer Academic Publishers.

Inoculation with associative N₂-fixers in Egypt

Y. Z. ISHAC

Unit of Biofertilizers, Faculty of Agriculture, Ain Shams University, Cairo, Egypt

Key words: associative N₂-fixers, *Azotobacter* carriers, endomycorrhizae, inoculation trials, organic matter

Abstract

This paper reviews our work on biofertilization in Egyptian soils which are rich in asymbiotic N₂-fixers. Several investigators have reported a positive effect of inoculation with *Azotobacter* and *Azospirillum* on cereals and other crops: this could be attributable to nitrogen fixation and formation of growth-promoting substances (indoles and gibberellins). Seed bacterization proved to be better than soil or phyllosphere inoculation techniques, and organic matter addition enhanced *Azotobacter* and *Azospirillum* proliferation. In pot experiments with only half the normal field rate of inorganic N-fertilizer, inoculation with asymbiotic N₂-fixers and endomycorrhizae in the presence of rock phosphate gave a wheat crop of high straw and grain yield. Moreover, this beneficial effect extended to onion or millet cultivated after the wheat in the same pots. Inoculation with associative N₂-fixers enhanced the rate of infection by indigenous endomycorrhizae and earlier panicle initiation. Several inoculant carriers were employed. Field application of our work might save about 50% of the cost of nitrogen and phosphorus fertilizers.

Introduction

Until 1962 *Azotobacter* was thought to be of purely laboratory interest but Gibson (1964) predicted its future role as a nitrogen fixer. Azospirilla also have a beneficial effect with grasses (Döbereiner and Boddey, 1980). The plant response to inoculation with associative N₂-fixers is due to nitrogen fixation and production of growth-promoting substances (van Berkum and Bohlool, 1980). The importance of interactions between mycorrhizae and asymbiotic N₂-fixers has also been reported (Amijee *et al.*, 1986; Bagyaraj and Menge, 1978; Barea *et al.*, 1983; Daft *et al.*, 1985; Gianinazzi-Pearson and Diem, 1982; Hayman, 1983).

The practical application of biofertilizers is important economically by reducing the cost of fertilizers, and ecologically, by reducing pollution of the environment caused by using nitrogen fertilizer.

N₂-fixing bacteria in Egyptian soils

Associative nitrogen fixers are ubiquitous soil inhabitants in Egypt. Dense population(s) of *Azotobacter* (Abd-El-Malek and Ishac, 1962; Abd-El-Malek, 1971) and *Azospirillum* (Hegazi *et al.*, 1980; Hegazi, 1983; Ishac *et al.*, 1986b) normally prevail in all types of Egyptian soil and in the phyllosphere (Ishac *et al.*, 1981). Application of organic matter has had beneficial effects on the proliferation of asymbiotic N₂-fixers in general (Abd-El-Malek and Ishac, 1977; El-Borollosy *et al.*, 1986; Ishac and Abd-El-Malek, 1981; Ishac *et al.*, 1980; Ishac *et al.*, 1983; Ishac *et al.*, 1986a; Saleh *et al.*, 1986b) and on the rate of nitrogen fixation in particular (Abd-El-Malek and Ishac, 1980; Ishac *et al.*, 1980; Ishac and Abd-El-Malek, 1981). It is observed, however, that application of organic matter with wide C/N ratio favours higher N₂-fixation (Ishac *et al.*, 1980). Organic matter application causes a greater evolution of CO₂

(Ishac *et al.*, 1983), which proved to have a beneficial effect on N_2 fixation (Becking, 1971): this has been confirmed by laboratory experiments (Ishac *et al.*, 1984).

Generally, the populations of N_2-fixing clostridia were reported to be as dense as, or even denser, than those of *Azotobacter* (Abd-El-Malek, 1971). Moreover, other genera of diazotrophic bacteria *e.g. Klebsiella* and *Bacillus* were recorded in soil and rhizosphere of desert plants (Saleh, unpublished data) as well as in Nile Delta soils (Hegazi and Monib, 1983).

Occurrence of mycorrhizae in Egyptian soils

All the tested soils contained endomycorrhizal symbiotic endophytes and some had comparatively high concentrations of spores (Ramadan *et al.*, 1983; Mahmoud *et al.*, 1985a, b). Biological fertilizers are potentially of value to large areas where intrinsic fertility is low Mahmoud *et al.*, 1985b).

Inoculation trials with N_2-fixing bacteria

The first trial in studying the effect of inoculation of plant growth was made with the cooperation of Bonn University. In greenhouse experiments, inoculation with *Azotobacter vinelandii* and *A. paspali* increased the total microbial count in rhizosphere of wheat and the weight of dry matter. Inoculation of *Azotobacter* in soil amended with compost significantly increased the nitrogen gain in Egyptian soil more than in German soils. However, with wheat straw, nitrogen gain was greater in German soil (Ishac *et al.*, 1982, 1983). The positive effect of inoculation could be attributed to nitrogen fixation and production of growth-promoting substances as indicated by culture filtrates and water culture experiments (Ishac *et al.*, 1982; Mahmoud *et al.*, 1982; Ishac *et al.*, 1983).

Monib *et al.* (1979a) studied the association between a selected strain of *Azotobacter chroococcum* and seven plants in water cultures under sterile conditions. *Azotobacter* population progressively increased in the nutrient solution and on the rhizoplane. Microbial proliferation depends on type of plant, being much higher in the presence of wheat,

followed by barley, maize, broad bean, and cotton, while in the presence of fenugreek and lentil, lower rates of multiplication were recorded. Inoculation increased the dry weight of plants by 5–12% and their length by 3–18%, in addition to increasing the nitrogen content of plants and nutrient solution. Nitrogen balance showed no significant change in systems devoid of *Azotobacter*, but association between plants and the microorganism invariably showed positive results. The extent of N_2-fixation depends on the type of plant; higher gains were recorded in presence of non-leguminous plants.

Monib *et al.* (1979b) studied the effects of seed bacterization with *Azotobacter chroococcum* in soils with different populations of naturally occurring azotobacters. Inoculation of barley grains had no effect on counts of total microflora, neither in rhizosphere nor in root-free soil, but significantly increased the *Azotobacter* population, especially in the rhizosphere. The rate of colonization in the root region was much higher when soil initially harboured small *Azotobacter* populations. Bacterization improved plant growth and increased soil nitrogen through nitrogen fixation. Nitrogen balance in soils showed higher gains in the inoculated treatments over the uninoculated controls of 30–98 mg. kg^{-1}. The effect of bacterization of barley grain with a selected strain of *Azotobacter chroococcum* was studied by Monib *et al.* (1979c). In nitrogen-deficient sand, seed inoculation increased plant length, dry weight, and nitrogen content in addition to a significant increase in soil nitrogen. Irrigation with KNO_3-containing nutrient solution suppressed nitrogen fixation, but inoculation still had a pronounced stimulating effect on plants. In presence of mixed soil microflora the beneficial effect of bacterization was less than in monobacterial cultures. Azotobacters naturally present in soil also colonized in heavy densities on the rhizoplane, but their effect on plant growth and soil nitrogen were less if compared with that resulting from bacterization.

El-Haddad *et al.*, (1986a) studied the interaction between *Azospirillum lipoferum* and maize, wheat, barley and broad bean grown in water cultures. Numbers of azospirilla, amounts of plant growth and dry matter, and nitrogen content significantly increased in inoculated treatments where indoles and gibberellin were detected.

It is worth noting that seed inoculation is a

promising method of bacterization as compared with soil or foliar application (El-Haddad *et al.*, 1986b). Therefore, consequent upon the experimental results and feasibility of application, seed inoculation is recommended in Egypt (El-Haddad *et al.*, 1986b).

Seed bacterization with N_2-fixing bacteria resulted in higher yield and/or decreased (by up to 50%) costs of nitrogen fertilizers (Emam *et al.*, 1985; El-Haddad *et al.*, 1986b; Fayez *et al.*, 1986; Gohar *et al.*, 1986; Ishac *et al.*, 1983, 1985, 1986; Yousef *et al.*, 1986; Saleh *et al.*, 1987). The beneficial effect and economic advantage were reported for wheat (Emam *et al.*, 1985; Fayez *et al.*, 1985; El-Haddad *et al.*, 1986b; Gohar *et al.*, 1986; Ishac *et al.*, 1983, 1985, 1986a, c; Saleh *et al.*, 1987; Yousef *et al.*, 1986) maize (Hegazi *et al.*, 1979; Hegazi and Monib, 1983; El-Borollosy *et al.*, 1986; Saleh *et al.*, 1986; El-Demerdash *et al.*, 1987), barley (Helemish *et al.*, 1986), rice (Omar, 1987), sorghum (El-Mokadem *et al.*, 1986) as well as Egyptian henbane (*Hyoscyamus muticus*) (El-Sawy *et al.*, 1986; Saleh *et al.*, 1986). Preliminary results showed that this is also true with tomatoes and potatoes.

Dual inoculation trials with endomycorrhizae and asymbiotic N₂-fixers

Ishac *et al.* (1985) studied the combined effect of inoculation with *Azotobacter* and/or *Azospirillum* on the growth of wheat plants and N_2-ase activity in a pot experiment with soil amended with pulverized maize stalks and 0.1% of either superphosphate or rock phosphate. Results obtained indicate that inoculation with a mixture of *Azotobacter*, *Azospirillum* and endomycorrhizal fungi in the presence of rock phosphate gave the best growth of wheat plants and highest yields of grain and straw. Nitrogenase activity in the rhizosphere of wheat plants was also higher in this case at nearly all growth stages under investigation. Thus, application of organic manure and inoculation with asymbiotic N_2-fixers and endomycorrhizal fungi in the presence of rock phosphate and only half the normal field rate of inorganic N-fertilizer gave a wheat crop of high straw and grain yields. Such application can save a high proportion of agricultural costs by reducing the amounts of

inorganic nitrogen and phosphorus fertilizers used (Ishac *et al.*, 1985). The same pots were used in another study to find out any possible residual effect of the above treatments. Thus, millet and onion were grown as a second host, in a greenhouse. Results revealed that the effects of organic amendment and phosphate fertilization combined with the inoculation of the wheat plants with diazotroph and endomycorrhizal fungi subsequently enhanced the vegetative growth as well as the nitrogen and phosphorus contents of both millet and onion plants as second hosts. Mixed inoculation with endomycorrhizal fungi, *Azotobacter* and *Azospirillum* proved to be the most effective treatment particularly with rock phosphate application (The author, unpublished data).

It is of interest to note that inoculation with associative N_2-fixers stimulated infection with indigenous endomycorrhizae (Ishac *et al.*, 1986a; Ishac *et al.*, 1986c; El-Demerdash *et al.*, 1987; and Saleh *et al.*, 1987).

Dual inoculations of associative N₂-fixers and rhizobia

Nodulation of soybean and N_2-fixation was enhanced by inoculation with associative N_2-fixers (Yanni and Mohamed, 1984 and Salem *et al.*, 1986).

Azotobacter carriers

Saleh *et al.* (1987) studied five raw materials namely, Irish and Nitragin peat (imported materials), soil-berseem straw, Nile silt, soil and charcoal (local materials), were selected as carrier materials for *A. chroococcum* strains. Sterilization of carrier materials either by autoclaving or gamma radiation was enough to eliminate all contaminating bacteria and fungi from the carriers. Nitragin peat and soil-berseem straw were highly efficient in permitting survival of *Azotobacter* cells for 60 days at all temperatures used (4, 15 and 30°C). Seed inoculation of wheat plants significantly increased the plant height, number of tillers and number and length of spikes. Soil-berseem straw or Nitragin peat carriers could be safely used as carrier materials for *Azotobacter* inoculants which can save half the recommended field rate of inorganic N-fertilizer (Zaki *et al.*, 1987).

Acknowledgement

Thanks are due to Dr M A El-Borollosy, Associate Professor of Agricultural Microbiology for his help during the preparation of this manuscript.

References

Abd-El-Malek Y 1971 Free-living nitrogen-fixing bacteria in Egyptian soils and their possible contribution to soil fertility. Plant and Soil, Special Volume 2, 423–442.

Abd-El-Malek Y and Ishac Y Z 1962 Abundance of *Azotobacter* in Egyptian soils. Eighth Internat. Cong. Microbiol., Montreal, Canada. Abstract p. 57.

Abd-El-Malek Y and Ishac Y Z 1977 Effect of crop refuse addition to soil on certain aspects of soil fertility. Fifth International Conference on Global Impacts of Applied Microbiology. 21–26 November, 1977, Bangkok, Thailand. Abstract p. 19.

Abd-El-Malek Y and Ishac Y Z 1980 Quantitative changes in soil nitrogen and in free-living nitrogen-fixing bacteria induced by organic matter decomposition. Sixth International Conference on Global Impacts of Applied Microbiology. 30 August–6 September, 1980. Lagos, Nigeria. Abstract p. 12.

Amijee F, Stribley D P and Tinker P B 1986 Relations between growth, colonization by a vesicular-arbuscular mycorrhizal fungus and soluble carbohydrates in roots of leek (*Allium porrum* L.) as affected by soil phosphorus. Transactions XIII. Congress of the International Society of Soil Science. Hamburg, 13–20 August, 1986. pp. 544–545.

Bagyaraj D J and Menge J A 1978 Interaction between VA mycorrhiza and *Azotobacter* and their effects on rhizosphere microflora and plant growth. New Phytol. 80, 567–573.

Barea J M, Bonis A F and Olivares J 1983 Interaction between *Azospirillum* and VA mycorrhiza and their effects on growth and nutrition of maize and ryegrass. Soil Biol. Biochem. 15, 705–709.

Becking J H 1971 Biological Nitrogen Fixation and its Economic Significance. pp. 189–222. International Atomic Energy Agency, Vienna.

van Berkum P and Bohlool B B 1980 Evaluation of nitrogen fixation by bacteria in association with roots of tropical grasses. Microbiological Reviews 44, 491–517.

Daft M J, Clelland D M and Gardner I C 1985 Symbiosis with endomycorrhizas and nitrogen fixing organisms. Proceedings of the Royal Society of Edinburgh 85 B, 283–298.

Döbereiner J and Boddey R M 1980 Nitrogen-fixation in association with Graminae. *In* Proceedings of Fourth International Symposium on Nitrogen Fixation, Canberra, December 1980. Ed. A H Gibson. Australian Academy of Sciences.

El-Borollosy M A, Ishac Y Z, El-Haddad M E, Saleh E A and

El-Demerdash M E 1986 Response of maize plants to inoculation with asymbiotic N$_2$-fixers in the presence of different nitrogen levels and organic amendment with maize stalks. Second Conference of the African Association for Biological Nitrogen Fixation. 15–19 December, 1986, Cairo, Egypt. Abstract p. 33.

El-Demerdash M E, Abd-El-Hafez A E, Saleh E A and El-Borollosy M A 1987 Maize growth as influenced by inoculation with diazotrophs and application of different phosphate forms under Egyptian field conditions. 12th International Congress for Statistics, Computer Science, Social and Demographic Research, Cairo, Egypt, 28 March–2 April, 1987. pp. 245–258.

El-Haddad M E, Ishac Y Z, El-Borollosy M A, Wedad E E and Girgis M G Z 1986a Studies on *Azospirillum* in Egypt. 2-Associative symbiosis with higher plants. XIV International Congress of Microbiology, 7–13 September, 1986, Manchester, England. Abstract p. 286.

El-Haddad M E, Ishac Y Z, Saleh E A, El-Borollosy M A, Refaat A A and El-Demerdash M A 1980b Comparison of different methods of inoculation with asymbiotic N$_2$-fixers on wheat plants. Second Conference of the African Association for Biological Nitrogen Fixation, 15–19 December, 1986, Cairo, Egypt. Abstract p. 57.

El-Mokadem M T, Zeinab Y M, Abou-Bakr and Helemish F A 1986 Effect of *Azospirillum* inoculation on nitrogen content and growth parameters of sorghum and maize in sterilized and non-sterilized soil. Second Conference of the African Association for Biological Nitrogen Fixation, 15–19 December, 1986, Cairo, Egypt. Abstract p. 59.

El-Sawy M, Saleh E A, Abd-el-Fattah M K, El-Borollosy M A and Shafar M S 1986 Hyoscyamine content of *Hyoscyamus muticus* L. as affected by inoculation with *Azotobacter* and/or *Azospirillum* strains. Second Conference of the African Association for Biological Nitrogen Fixation, 15–19 December, 1986, Cairo, Egypt. Abstract p. 57.

Emam N F, Fayez M and Makboul H E 1985 Wheat growth as affected with *Azotobacter* isolated from different soils. Zbl. Bakt. II Abt. 141, 17–23.

Fayez M, Emam N F and Makboul H E 1985 The possible use of nitrogen fixing *Azospirillum* as biofertilizer for wheat plants. Egypt. J. Microbiol. 20, 199–206.

Fayez M, Enayat H G, Emam N F, Khalafalla G and El-Sayed F F 1986 Biofertilization with non-symbiotic N$_2$-fixers for economizing N-fertilization in wheat cropping. Second Conference of the African Association for Biological Nitrogen Fixation, 15–19 December, 1986, Cairo, Egypt. Abstract p. 61.

Gianinazzi-Pearson V and H G Diem 1982 Endomycorrhizae in the tropics. *In* Microbiology of Tropical Soils and Plant Productivity. Eds Y R Dommergues and H. G Diem. pp. 210–251. Martinus Nijhoff/Dr. W. Junk Publishers, The Hague/Boston/London.

Gibson T 1964 Progress in agricultural microbiology. J. Appl. Bact. 27, 1–3.

Gohar M R, Khalil O S, Higazy A G and El-Hadidy M 1986 Locally-isolated asymbiotic diazotrophs as inoculant for wheat under field conditions. Second Conference of the African Association for Biological Nitrogen Fixation, 15–19 December, 1986, Cairo, Egypt. Abstract p. 60.

Hayman D S 1983 Practical aspects of vesicular-arbuscular mycorrhiza. *In* Advances in Agricultural Microbiology. Ed. N.S. Subba Rao. pp. 325-373. Oxford and IBH Publishing Co.

Hegazi N A 1983 Contribution of *Azospirillum* spp. to asymbiotic N₂-fixation in soils and roots of plants grown in Egypt. Experientia supplementum 48, *Azospirillum* II, Ed. W. Klingmüller. pp. 171-189. Birkhauser Verlag, Basel.

Hegazi N A, Amer H A and Monib M 1980 Studies on N₂-fixing spirilla (*Azospirillum* spp.) in Egyptian soils. Rev. Ecol. Biol. Soil 17, 491-499.

Hegazi N A and Monib M 1983 Response of maize plants to inoculation with azospirilla and (or) straw amendment in Egypt. Can. J. Microbiol. 29, 888—894.

Hegazi N A, Monib M and Vlassak K 1979 Effect of inoculation with N₂-fixing spirilla and *Azotobacter* on nitrogenase activity on roots of maize grown under sub-tropical conditions. Appl. Environ. Microbiol. 38, 621-625.

Helemish F A, Zeinab Y M, Abou-Bakr and El-Mokadem M T 1986 Influence of *Azospirillum* inoculation on growth and nitrogen content of barley, sunflower and peanut. Second Conference of the African Association for Biological Nitrogen-Fixation, 15-19 December, 1986, Cairo, Egypt. Abstract p. 62.

Ishac Y Z and Abd-El-Malek Y 1981 Effect of ploughing under crop residues on soil nitrogen under wheat and maize. Workshop on Nitrogen Cycling in Ecosystems of Latin America and the Caribbean, 16-21 March, CIAT, Cali, Colombia. p. 58.

Ishac Y Z, Abd-El-Malek Y and El-Sawy M 1980 Changes in soil nitrogen and in asymbiotic nitrogen fixers induced by organic matter addition. Annals Agric., Sci., Fac. Agric., Ain Shams Univ., Cairo, Egypt. 25, 119-140.

Ishac Y Z, El-Borollosy M A and Refaat A A 1981 Nitrogen-fixing microorganisms in the phyllosphere of some crop plants in Egypt. *In* Soil Biology and Conservation of the Biosphere. Ed. J Szegi. pp. 541-555. Godollo, Hungary.

Ishac Y Z, El-Haddad M E, Daft M J, Ramadan E M and El-Demerdash M E 1986a Effect of seed inoculation, mycorrhizal infection and organic amendment on wheat growth. Plant and Soil 90, 373-383.

Ishac Y Z, El-Haddad M E, El-Borollosy M A and Ismail M 1983 Effect of organic amendments on nitrogen fixation by asymbiotic bacteria associated with wheat and maize plants. Proceedings of the 5th International Symposium on Nitrogen-Fixation, Noordwijkerhout, The Netherlands, August 28-September 3, 1983. Eds. C Veeger and W E Newton. p. 343. Martinus Nijhoff/Dr W Junk Publishers, The Hague, Boston, Lancaster.

Ishac Y Z, El-Haddad M E, El-Borollosy M A, Rahal A G and Mostafa M I 1984 Effect of organic acids and carbon dioxide on asymbiotic nitrogen-fixation. Egyptian J. Microbiol. Special Issue, pp. 1-12.

Ishac Y Z, El-Haddad M E, El-Borollosy M A, Wedad E E and Girgis M G Z 1986b Studies on *Azospirillum* in Egypt. Densities as affected by some ecological factors. XIV International Congress of Microbiology, 7-13 September, 1986, Manchester, England. Abstract p. 286.

Ishac Y Z, El-Haddad M E, El-Kharbawy M I, Saleh E A, El-Borollosy M A and El-Demerdash M E 1986c Effect of

seed bacterization and phosphate supplementation on wheat yield and mycorrhizal development. Transactions of the XIII. Congress of International Society of Soil Science. Hamburg, 13-20 August, 1986. p. 588.

Ishac Y Z, El-Haddad M E, Saleh E A, El-Borollosy M A, El-Demerdash M E and Refaat A A 1985 Effects of inoculation with endomycorrhizas and asymbiotic N₂-fixers on growth of wheat plants. Proceedings of the 6th International symposium on Nitrogen Fixation, Corvallis, OR 97331, August 4—10, 1985 (p 704). Eds. H J Evans, P J Bottomley and W E Newton. p. 704. Martinus Nijhoff Publishers, Dordrecht, The Netherlands.

Ishac Y Z, Mahmoud S A Z, Kramer J, El-Demerdash M E and Eweda W 1983 Factors affecting wheat inoculation with *Azotobacter*. Proceedings of the 5th International Symposium on Nitrogen Fixation, Noordwijkerhout, The Netherlands, August 28- September 3, 1983. Martinus Nijhoff/Dr W Junk Publishers, The Hague, Boston, Lancaster.

Ishac Y Z, Mahmoud S A Z, Kramer J, Hazem A and Eweda W 1982 *Azotobacter* inoculation and wheat growth. Second International Symposium on N₂-fixation with Non-Legumes, 5-10 September, Banff, Canada. Abstract p. 26.

Mahmoud S A Z, Ishac Y Z, Kramer J and Wedad E E 1982 Associative symbiosis of *Azotobacter* and some higher plants. Second International Symposium on N₂-fixation with Non-Legumes, 5-10 September, Banff, Canada. Abstract p. 26.

Mahmoud S A Z, Ishac Y Z, Ramadan E M and Daft M J 1985 Occurrence and infectivity of endomycorrhizas in Egyptian soil. Egyptian J. Microbiol. Special Issue, pp. 47-55.

Mahmoud S A Z, Ishac Y Z, Ramadan E M and Daft M J 1985 Some effects of soil amendments and fertilizer treatments on mycorrhizal plants. Egyptian J. Microbiol. Special Issue, pp. 37-46.

Monib M, Abd-El-Malek Y, Hosny I and Fayze M 1979a Associative symbiosis of *Azotobacter chroococcum* and higher plants. Zbl. Bakt. II Abt. 134, 132-139.

Monib M, Abd-El-Malek Y, Hosny I and Fayez M 1979b Effect of *Azotobacter* inoculation on plant growth and soil nitrogen. Zbl. Bakt. II Abt. 134, 140-148.

Monib M, Abd-El-Malek Y, Hosny I and Fayez M 1979c Seed inoculation with *Az. chroococcum* in sand cultures and its effect on nitrogen balance. Zbl. Bak. II Abt. 139, 243-248.

Omar M N A 1987 Selection of nitrogen fixing bacteria from Egyptian Soils and their inoculation for improving rice yield. Ph.D. Thesis, Nancy Univ., France.

Ramadan E M, Ishac Y Z and Fares C N 1983 Occurrence of vesicular arbuscular mycorrhizae in the rhizosphere and root systems of some plants. Proc. V. Conf. Microbiol., Cairo, Egypt, May 1983, Vol. II. Soil and Water Microbiol. pp. 325-337.

Saleh E A, El-Borollosy M A, El-Hafez A E and El-Demerdash M E 1987 Application of rock phosphate and diazotrophs in cereals. 1- Effects on wheat growth. 12th International Congress for Statistics, Computer Science, Social and Demographic Research, Cairo, Egypt, 28 March-2 April, 1987. pp. 227-244.

Saleh E A, El-Haddad M E, Ishac Y Z, Eid M, El-Borollosy M A and El-Demerdash M E 1986a Effect of seed bacterization and amendment with garbage compost on the growth of maize plants. Second Conference of the African Association for Biological Nitrogen Fixation, 15–19 December, 1986, Cairo, Egypt. Abstract p. 32.

Saleh E A, El-Sawy M, Refaat A A, Abdel-Fattah M K and Sharaf M S 1986b Effect of seed inoculation with some asymbiotic N_2-fixers on the growth of Egyptian henbane plant (*Hyoscyamus muticus* L.). Second Conference of the African Association for Biological Nitrogen Fixation, 15–19 December, 1986, Cairo, Egypt. Abstract p. 30.

Saleh E A, Zaki M M and Selim S M 1987 Preparation of *Azotobacter* inoculants. Ann. Agric. Sci. Fac. Agric. Ain Shams Univ., Cairo, Egypt 32, 987–1003.

Salem S H, Gawaily E M, Tahia Khater and El-Zamek F I 1986 Effect of inoculation with mixed cultures of *Rhizobium japonicum* and other soil bacteria on soybean nodulation and symbiotic nitrogen fixation. Second Conference of the African Association for Biological Nitrogen Fixation, 15–19 December, 1986, Cairo, Egypt. Abstract p. 118.

Yanni Y G and Mohamed Y Z H 1984 Enhancement of nodulation and symbiotic nitrogen fixation by *Rhizobium japonicum* in soybean plant as contribution of the asymbiotic nitrogen-fixer *Azotobacter chroococcum*. Beitrag zu Biologie der Pflanzen 60, 207–221.

Yousef A N, Al-Azawi S K, Al-Nassiri A S and Munaam B H 1986 Response of wheat to inoculation with *Azotobacter chroococcum*. Second Conference of the African Association for Biological Nitrogen Fixation, 15–19 December, 1986, Cairo, Egypt. Abstract p. 58.

Zaki M M, Saleh E A, Abdel A E and Selim S M 1987 Use of *Azotobacter* inoculant as a biofertilizer in wheat planting. Mansaura Univ. Conf. Agric. Sci. On Food Deficiency Overcoming through Autonomous Efforts in Egypt, 22 June, 1987. pp. 1025–1038.

Session 9

Quantification of plant-associated biological nitrogen fixation

F. A. Skinner et al. (Eds.), Nitrogen fixation with non-legumes, 249–262.

Association between N$_2$-fixing bacteria and pearl millet plants: Responses, mechanisms and persistence*

S. P. WANI, S. CHANDRAPALAIH, M. A. ZAMBRE and K. K. LEE
International Crops Research Institute for the Semi-Arid Tropics (ICRISAT) Patancheru P.O., Andhra Pradesh 502 324, India

Key words: ELISA, grain yield, inoculation, millet, N$_2$-fixing bacteria, NO$_3$ reductase

Abstract

Responses to inoculation with N$_2$-fixing bacteria were studied in relation to genotypic differences in pearl millet, effect of nitrogen levels, and FYM additions in India. In some experiments, inoculation increased mean grain yield up to 33% over the uninoculated control, whereas in the remaining 11 experiments there was no significant increase. Increased grain yields, > 10% over the uninoculated controls were observed in 46% of the experiments with *Azospirillum lipoferum* (18.7% average increase) and with *Azotobacter chroococcum* (13.6% average increase). Yield increases were nil or reduced in three experiments with *Azos. lipoferum* and four experiments with *Aztb. chroococcum*. In two experiments continued inoculation for two or three years resulted in increased grain, plant biomass yield, and N uptake. Interactions of bacterial cultures with cultivars or years were not observed. The counts of the inoculated strains increased two to three-fold when inoculation was continued for three years. Repeated inoculations increased the mean cumulative N uptake from season 1 to season 3 by 19 kg ha^{-1}. Repeated inoculations with *Aztb. chroococcum* and *Azos. lipoferum* increased mean grain yield of a succeeding crop by 14.4% and 9.8%, respectively, over the uninoculated control. Inoculation increased the efficiency of N-assimilation by pearl millet. Marginal increase in nitrogenase activity, associated with the inoculated plants was observed during later stages of plant growth. Increased leaf nitrate reductase activity (NRA) was observed after inoculation with these bacteria. The responses to inoculation are mainly attributable to increased plant N assimilation which could be the effect of growth promoting substances secreted by the bacteria; and thus the contribution from BNF may be small.

Introduction

Pearl millet (*Pennisetum americanum* (L.) Leeke) is grown on nutritionally poor soils in the semi-arid tropics, often without the addition of fertilizer nitrogen. In many cases increased plant yields and/or increased N accumulation by plants have been observed from inoculations with *Azospirillum* spp. (Boddey and Döbereiner, 1982) *Azotobacter chroococcum* (Fedorov, 1952). Similar responses from inoculation with azospirilla and azotobacters have been reported in cereals (Avivi and Feldman, 1982; Reynders and Vlassak, 1982; Kapulnik *et al.*,

1983; Sarig *et al.*, 1984; Smith *et al.*, 1984; Wani *et al.*, 1985; Boddey *et al.*, 1986; Wani, 1986; Tilak and Subba Rao, 1987; Baldani *et al.*, 1987; Hussain *et al.*, 1987). The positive benefits from inoculation have been attributed to several mechanisms such as biological nitrogen fixation (BNF) (Cohen *et al.*, 1980; Kapulnik *et al.*, 1981a; Sarig *et al.*, 1984) and increased root uptake capacity because of enhanced root development and root hair formation in response to secretion of plant growth hormones (Tien *et al.* 1979, Umali-Garcia *et al.* 1980; Vlassak and Reynders, 1981; Okon, 1985). Other mechanisms such as enhanced uptake of nitrate, phosphate and potassium (Okon, 1982; Kapulnik *et al.*, 1985) and stimulation of NO$_3$ assimilation due to inocu-

* ICRISAT, journal article 732.

lation (Villas Boas and Döbereiner, 1981; Boddey et al., 1986), are also believed to increase yields.

There are several reports on positive benefits of inoculation, but information has been scanty on the benefits of continued inoculation on the yields of the main and the succeeding crops and on the survival of the inoculated bacteria. This paper summarizes the results of 25 field inoculation trials with pearl millet at different locations in India. The effects of continued inoculations on yields of main and succeeding millet crops, and also the persistence of inoculated bacteria under field conditions are discussed.

Materials and methods

Bacterial cultures

The *Azospirillum brasilense* (SL 33) culture was obtained from Dr. F. V. MacHardy, University of Alberta, Edmonton, Canada. *Azospirillum lipoferum* (ICM 1001) and *Azotobacter chroococcum* (ICM 2001) were isolated at the ICRISAT Center from the rhizosphere of sorghum cv. CSH 1 and *Cenchrus ciliaris* as reported earlier (Wani et al., 1985).

Preparation of inoculants

Peat inoculants of azospirilla and *Azotobacter* were prepared by injecting 30 ml of culture broth into a packet containing 40 g γ-irradiated peat, (Agriculture Laboratories Pty. Ltd., Australia: Wani et al., 1985). At the time of field inoculation all the inoculants had 10^8 bacterial cells g^{-1} of peat and were free from contamination at the 10^{-5} dilution. For uninoculated control treatments the peat packets were inoculated with sterile N-free sucrose medium.

Experiment details

The detailed experiments were conducted during the rainy season on alfisols (Table 1) at ICRISAT Center, Patancheru, India ($17°36'N$, $78°16'E$, 545 m altitude).

Liquid inoculum was prepared by thoroughly mixing the peat inoculum in unchlorinated tap water ($1 g l^{-1}$).

All the experiments were sown on ridges spaced 75 cm apart. Plant-to-plant spacing of 10 cm was maintained by thinning the plants 12–16 days after sowing (DAS). Top dressing with N fertilizer was done 18–20 DAS as required. Weeding and inter-row cultivations were carried out as and when required. Plant parts above ground level were harvested from the net plot area. The ears were separated and threshed. The plant matter was then chopped. The chopped plant matter and grain were dried at 70°C in an oven for 3 days and their dry weight recorded. Total nitrogen contents of powdered grain and plant dry matter subsamples from each treatment were estimated by a micro-Kjeldahl digestion method using a Technicon Autoanalyser (for details refer to Industrial method No. 218-72A, II, Technicon, Industrial Systems, Tarry Town, NY 10591, U.S.A.).

Multilocational trials

During 1982–86, 25 field experiments were conducted at the ICRISAT Center and other locations in India, using different millet cultivars, N doses,

Table 1. Details of pearl millet inoculation trials on Alfisols at ICRISAT Center, rainy seasons[a]

Experiment	Year	Soil pH	EC (m.mhos cm^{-1})	Organic Carbon (g kg^{-1})	Total N (mg kg^{-1})	NH$_4$ + NO$_3$ − N (mg kg^{-1})	Gross plot size (m^2)	Harvest area (m^2)
I	1983	ZF[b] 6.8	0.16	3.45	635	7.0		
		LF 6.8	0.18	3.60	728	13.5	18	9
		HF 7.3	0.20	4.20	755	15.2		
III	1985	7.9	0.35	4.10	585	33.1	27	12

[a] Presowing soil samples to 60 cm depth were collected only from Expts I and III before imposing the treatments.
[b] ZF, Zero fertility (no N or P added); LF, 20 kg N and 16 kg P$_2$O$_5$ ha^{-1}; HF, 56 kg P$_2$O$_5$ and 100 kg N ha^{-1}.

and FYM additions to study the responses to inoculation with N_2-fixing bacteria. Results are summarised, but full details are available from the authors on request.

Experiment I. The experiment was laid out in a split-split plot design. Nitrogen levels (0, 20 and 100 kg N ha^{-1}) were in the main plot, cultures *Azospirillum lipoferum* (ICM 1001), *A. brasilense* (SL 33), *Azotobacter chroococcum* (ICM 2001) were in sub-plots, with an uninoculated sub-plot as control, and cultivars in sub-sub plots.

For the 100 kg N ha^{-1} treatment a basal dose of 56 kg N and 56 kg P_2O_5 ha^{-1} treatment was applied as mixed fertilizer (28:28:0), and for the 20 kg N ha^{-1} treatment, 16 kg P_2O_5 ha^{-1} was applied as single superphosphate. The crop was machine sown on 28 June 1983, 16 June 1984, and 22 June 1985. Each treatment was replicated four times. First inoculation was done at 7 DAS. For inoculation, a small furrow was opened by the side of the seedlings, inoculum added at 100 ml m^{-1} row length, and the furrow then closed. Similarly, a second inoculation was done 20 DAS. The remaining dose of N as urea was applied after thinning.

Experiment II. This experiment was conducted in the plot of Experiment I to study the effects of continued inoculations with N_2-fixing bacteria for 3 years on the yield of a cover crop during the 4th year. A basal dressing with N and P and a top dressing with N treatments was given as mentioned above for Experiment I. A uniform cover crop of millet cv. ICMV 1 was machine sown on 17 June 1986. This experiment did not include inoculation with N_2-fixing bacteria.

Experiment III. This experiment was laid out in a split-split plot design. Farmyard manure (FYM) levels (0 and 5 t ha^{-1}) served as the main plot, N levels (0 and 20 kg ha^{-1}) were the sub-plots, and cultures *Azospirillum lipoferum* (ICM 1001), *Azotobacter chroococcum* (ICM 2001) and the uninoculated control served as sub-sub plots.

The treatment plots received FYM, containing 10% moisture, organic carbon (86 g kg^{-1}) total N (12.4 g kg^{-1}) NO_3-N (51 mg kg^{-1}), NH_4-N, (22 mg kg^{-1}) and available P (575 mg kg^{-1}) (pH 7.1), 15 days before sowing. The manure was mixed in the soil with a rotovator. Before sowing,

20 kg P_2O_5 ha^{-1} as single superphosphate was applied as a basal dose and 20 kg N ha^{-1} was applied as urea in the plots receiving N. Each treatment was replicated six times. Millet cv. BJ 104 was dibbled manually on 26 June 1985 and 17 June 1986. At the time of sowing, a furrow was opened on the ridge and 100–120 ml peat suspension m^{-1} row length (1 g peat inoculant L^{-1}) was applied; a second inoculation was done 20 DAS by opening a small furrow by the side of the seedlings.

Nitrogenase activity

Nitrogenase activity associated with the roots was measured 50 and 75 DAS at the flowering and grain filling stages in Experiment I in the 1984 and 1985 seasons following the improved soil-core assay technique (Wani *et al.*, 1983). Three-ml gas samples were taken from each container 1 and 6 h after incubation and were analyzed for C_2H_2 and C_2H_4 concentrations by gas chromatography (Wani *et al.*, 1984).

Most probable number (MPN) of N_2-fixing bacteria

Counts for MPN of N_2-fixing bacteria associated with the rhizosphere soil and roots of millet plants were done from each plot at 74 DAS in Experiment II and at harvest in Experiment III. Four plants from each plot were randomly selected and pulled out by hand or dug out. Roots and rhizosphere soil from the same plot were pooled, and the fresh weight of roots and soil recorded and subsampled. The same samples were also used for ELISA studies. The subsamples of roots and rhizosphere soil were kept at 80°C for moisture content determinations. Tenfold serial dilutions were prepared from each sample, and 0.1 ml from each dilution was added to semi-solid N-free medium in tubes. For the azospirilla and *Azotobacter* treatments, N-free malate medium and N-free sucrose medium enriched with 100 ml L^{-1} yeast extract, respectively, were used. The MPN counts from the uninoculated control plots were made with N-free sucrose medium. Soon after inoculation, the cotton plugs of tubes were replaced with sterilized Subaseals and 1% C_2H_2 injected (Balandreau, 1983; Wani, 1986).

The inoculated tubes were incubated at 33°C under 1% C_2H_2 for 48 h. Gas samples from the tubes were collected in 1 ml syringes and analyzed for C_2H_2 and C_2H_4.

Enzyme-linked immunosorbant assay (ELISA) counts

The counts of *Azos. lipoferum* (ICM 1001) and *Aztb. chroococcum* (ICM 2001) from the rhizosphere soil and roots were made from the plots inoculated with *Azospirillum lipoferum* (ICM 1001) and *Azotobacter chroococcum* (ICM 2001), and also from the uninoculated plots in Experiments II and III. Antisera for *A. lipoferum* (ICM 1001) and *A. chroococcum* were prepared by injecting live cells (10^8–10^9 cells ml^{-1} physiological saline) into New Zealand white rabbits. Gamma globulins (antibodies) were collected from the antisera with 1:1600 titre by sodium sulphate precipitation (Van Weeman and Schuurs, 1971). The purified γ-globulins were conjugated with the enzyme alkaline phosphatase, as described by Kishinevsky and Bar-Joseph (1978).

The subsamples from pooled rhizosphere soil and roots from each plot, obtained for MPN counts, were used for counting N_2-fixing bacteria, using ELISA. Ten g of rhizosphere soil was added to 10 ml extraction buffer (Phosphate buffered saline containing: 0.02 $mol\,l^{-1}$ phosphate, 0.15 $mol\,L^{-1}$ NaCl, 0.003 $mol\,l^{-1}$ KCl, pH 7.4, plus 0.05% Tween 20 and 2% polyvinylpyrrolidone, PVP-40T (PBS)) and mixed well. Tenfold serial dilutions were prepared in extraction buffer. Similarly, serial dilutions from 10 g roots macerated in 15 ml extraction buffer in a sterilized mortar were prepared.

For estimating the concentration of antigen in a given sample, the procedure for the direct (double-antibody sandwich) ELISA (Kisihnevsky and Bar-Joseph, 1978) was used wherein alkaline phosphatase enzyme was used to conjugate with γ-globulin, and *p*-nitrophenyl phosphate was used as a substrate. Reactions were stopped at the end of 30 min incubation with 50 μl (per well) of 3 M NaOH and the O.D. of *p*-nitrophenol produced in individual wells was read at 410 nm in a Dynatech MR 590 reader. Suitable standards with different concentrations of the standard antigen were included in each experiment along with suitable blanks. The counts of *A. lipoferum* (ICM 1001) and *A. chroococcum* (ICM 2001) were calculated from the standard curves obtained by using varying concentrations of the standard antigen.

Leaf nitrate reductase activity (NRA)

NRA in leaves of millet plants from Experiment I (at 43 and 58 DAS in 1984 and at 45 DAS in 1985) and Experiment III (at 40 DAS in 1986) was estimated. At each sampling, four plants randomly selected from each plot were cut at ground level, transported to the laboratory in polythene bags, and stored in a cold room (4°C). The top four leaves from each plant were separated and from each leaf three discs of 8 mm diameter were cut, their weight recorded, and NRA measured by the method of Jaworski (1971). The discs from the plants from each treatment were incubated in 15 ml sodium phosphate buffer (0.1 M sodium phosphate, pH 7.5, 5% propanol, and 0.02 M KNO_3) in 30 ml glass bottles. The discs were subjected to vacuum infiltration for 2 min at 1×10^3 pascals, and incubated at 30°C for 30 min. Ten ml of the incubation mixture was pipetted to a test tube and the nitrite content estimated using Szechrome NIT as described by Hunter *et al.* (1982). The total area of all the green leaves was measured, the leaves were then oven-dried at 60°C for 48–72 h and the total weight was recorded for calculating NRA per plant.

Results

Responses in multilocational trials

Mean grain yields increased significantly (up to 33%) due to inoculation with N_2-fixing bacteria over the respective uninoculated controls in 14 of the 25 experiments. Of the 24 experiments with *Azos. lipoferum* (ICM 1001), in 11 experiments increases in grain yields (average 18.7%) were significantly ($P < 0.05$) high; in 10 experiments the increases in grain yields (9.3%) were not statistically significant; in one experiment no response was observed and in two of them grain yields decreased (2.7%) after inoculation. Similarly, of the 24 experiments with *Aztb. chroococcum* (ICM 2001), in 8 trials mean grain yields across the cultivars/ treatments increased significantly ($P < 0.05$) (average increase 13.6%); in 12 experiments grain yield increases (with an average increase of 8.3%) were not statistically significant; in two experiments no response was observed, and in two other experiments grain yields decreased (by 4.5%) after inoculation. *Azospirillum brasilence* (SP 7) caused a reduction in grain yield in the two experiments where this strain was used. In a few other experiments, inoculation with other strains of *Azos. brasilense* resulted in higher grain yields by an average of 8% over the uninoculated control.

Results of continued inoculation experiments

Grain yield

Experiment I. A pooled analysis of three years' data from the experiment revealed that the mean grain yield of pearl millet varied significantly between seasons and cultivars. The results of the pooled analysis in Table 2 show that the mean grain yield of pearl millet cultivars increased significantly after the addition of nitrogen fertilizer, with a maximum grain yield of 2.73 t ha⁻¹ with 100 kg N ha⁻¹ application. Mean grain yield of cultivars across the years also increased significantly over the uninoculated treatment after inoculation with N₂-fixing bacteria (Table 2). The three inoculants were equally effective in terms of increased grain yield. The interaction between N levels and inocula was not significant.

Experiment II. The results of a uniform crop of millet cv ICMV1, grown in the plots which were inoculated and treated with nitrogen for three consecutive years previously showed increased grain yield in comparison with the respective control plots (Table 3). A maximum mean grain yield of 2.45 t ha⁻¹ (14.4% increase) was observed from the plots inoculated previously with *Aztb. chroococcum*.

Experiment III. The mean grain yield of cv. BJ 104 was significantly ($P < 0.01$) greater in the 1985 season (2.4 t ha⁻¹) than in the 1986 season (1.10 t ha⁻¹). A pooled analysis of the data from

both the years showed that mean grain yield increased significantly with FYM addition. Similarly, addition of 20 kg N ha⁻¹ increased yield significantly (1.89 t ha⁻¹) over the control (1.59 t ha⁻¹). A higher grain yield was observed with *A. chroococcum* inoculation (1.82 t ha⁻¹), followed by the *A. lipoferum* inoculation (1.77 t ha⁻¹) and the uninoculated control treatments (1.64 t ha⁻¹).

Total plant biomass

Experiment I. Total plant biomass yield of millet cultivars varied significantly across seasons and cultivars. Mean total plant biomass increased significantly with addition of N and also from inoculation with N₂-fixing bacteria (Table 2).

The interaction between millet cultivars and inoculations with N₂-fixing bacteria for plant biomass was significant, however, significant increases were observed only with cv. ICMV1 inoculated with *Azos. brasilense* (5.9%) and with cv. BJ 104 inoculated with *Aztb. chroococcum* (by 9.4%) and *Azos. lipoferum* (by 8.6%). No significant interactions were observed between N levels and inoculations, cultivars and N levels, and years and inoculations.

Experiment II. Cultivar ICMV1 yielded similar amounts of total plant biomass from each of the plots that had carried different cultivars over a 3-year period. The mean plant biomass yields of cv. ICMV1 were significantly higher from the plots supplied with 20 kg N ha⁻¹ (6.69 t ha⁻¹) and 100 kg N ha⁻¹ (7.48 t ha⁻¹) than from those where no N was applied (5.81 t ha⁻¹). Earlier inoculations with *Azos. lipoferum* and *Aztb. chroococcum* during experiment I resulted in increased plant biomass (6.77 and 6.94 t ha⁻¹, respectively) over that 6.40 t ha⁻¹ obtained from the uninoculated control plots (Table 3).

Experiment III. A larger total plant biomass yield was observed in 1985 (5.50 t ha⁻¹) than in 1986 (3.50 t ha⁻¹). Addition of FYM increased plant yield (4.63 t ha⁻¹) over the control yield (4.32 t ha⁻¹). Inoculation with *Azos. lipoferum* and *Aztb. chroococcum* significantly increased plant biomass yield over the uninoculated control (Table 4). A significant interaction between N levels and

Table 2. Mean grain and total plant biomass yield (t ha^{-1}), mean total plant N uptake (kg ha^{-1}) and plant dry matter nitrogen percentage of pearl millet cultivars inoculated with N$_2$-fixing bacteria at three N levels across three years in Experiments I[a]

N applied (kg ha^{-1})	Nitrogen fixing bacteria			Uninoculated control	Mean	SE ±
	Azos. lipoferum (ICM 1001)	*Azos. brasilense* (SL 33)	*Aztb. chroococcum* (ICM 2001)			
	Grain yield					
0	1.97	1.91	1.92	1.79	1.90	
20	2.50	2.48	2.58	2.43	2.50	0.047*
100	2.66	2.79	2.84	2.62	2.73	
Mean	2.38	2.40	2.45	2.28		0.033*
CV (%)			13.2			
	Total plant biomass					
0	5.68	5.56	5.51	5.42	5.54	
20	6.82	6.81	6.96	6.51	6.78	0.092*
100	7.62	7.75	7.83	7.44	7.66	
Mean	6.71	6.71	6.77	6.46		0.077*
CV (%)			11.4			
	Total plant N uptake					
0	37.6	36.4	36.5	32.8	35.8	
20	56.3	54.9	59.1	52.9	55.8	3.05*
100	92.1	90.3	89.7	83.5	88.9	
Mean	62.0	60.6	61.8	56.3		1.18*
CV (%)			19.9			
	Plant dry matter nitrogen (%)					
0	0.31	0.33	0.30	0.26	0.30	
20	0.39	0.36	0.42	0.37	0.39	0.031*
100	0.70	0.63	0.65	0.62	0.65	
Mean	0.47	0.44	0.45	0.42		0.009*
CV (%)			27.2			

[a] Average of 48 replications.
* $P = <0.05$.

bacterial cultures was observed for mean biomass. Increased ($P < 0.05$) yields of 4.24 and 4.96 t ha^{-1} were observed with *Azos. lipoferum* inoculations at zero and 20 kg N ha^{-1} addition, respectively.

Total plant N uptake

Experiment I. The mean total N uptake of cultivars varied significantly from year to year. A maximum N uptake (68 kg ha^{-1}) was observed in 1983 followed by 65 kg ha^{-1} in 1985 and 48 kg ha^{-1} in 1984. The mean nitrogen uptake increased ($P < 0.05$) following inoculation and addition of N (Table 2). There was no interaction between N levels and bacterial cultures for plant N uptake, although there was a significant variety × bacterial culture interaction for total plant N uptake. Inoculation of

cv. BJ 104 with *Aztb. chroococcum* and *Azos. lipoferum* increased plant N uptake to 67.6 kg ha^{-1} and 65.9 kg ha^{-1} respectively, compared with the N uptake of 56.1 kg N ha^{-1} in the uninoculated control. With any other combinations of variety × bacterial cultures, increase in plant N uptake was not significant.

Experiment II. Previous inoculations in experiment I with N$_2$-fixing bacteria resulted in increased ($P < 0.05$) N uptake of cv. ICMV 1 (Table 3).

Experiment III. The mean plant N uptake was greater in 1985 (37.3 kg ha^{-1}) than in 1986 (30.0 kg ha^{-1}). Increased plant N uptake (30 kg ha^{-1}) was observed with FYM addition, compared to the zero FYM treatment (27 kg ha^{-1}).

Nitrogen uptake also increased after application of N and inoculations with N_2-fixing bacteria (Table 4).

Nitrogen content in grain and plant

Experiment I. Nitrogen content in millet grains increased after the addition of 20 and 100 kg N ha^{-1} (1.5 and 2.07%, respectively) over that of the zero N treatment (1.27%).

The mean N content in shoot increased ($P < 0.05$) due to inoculation and the application of N (Table 2).

Cumulative plant nitrogen uptake

Data on cumulative nitrogen uptake in the above-ground plant biomass in Experiment I during the three seasons showed significant increases ($P < 0.001$) after the addition of 20 and 100 kg N ha^{-1}a^{-1}. In the zero applied N treatments, a mean cumulative N uptake of 107 kg ha^{-1}

was recorded; with 20 kg N ha^{-1}a^{-1} it increased to 167 kg N ha^{-1}. A maximum N uptake of 262 kg ha^{-1} was recorded in the 100 kg N ha^{-1}a^{-1} treatment. Similarly, inoculation with N_2-fixing bacteria increased ($P < 0.05$) mean cumulative N uptake. A maximum cumulative plant N uptake of 185 kg ha^{-1} was observed in cultivars inoculated with *Azos. lipoferum* (ICM 1001), followed by 182 kg N ha^{-1} with *Azos. brasilense* (SL 33) and *Aztb. chroococcum* (ICM 2001) inoculated treatments, compared to 166 kg N ha^{-1} in the uninoculated millet cultivars.

Nitrogenase (C_2H_2 reduction) activity

In Experiment I during the 1984 rainy season nitrogenase activity associated with millet cultivars inoculated with N_2-fixing bacteria was greater at 75 DAS than at 50 DAS (Table 5). Inoculation with *Azos. lipoferum*, *Azos. brasilense*, and *Aztb. chroococcum* tended to increase nitrogenase activity although there were some anomalous results (Table 5). Significantly reduced mean activity was obser-

Table 3. Mean grain and total plant biomass yield (t ha^{-1}) and plant nitrogen uptake (kg ha^{-1}) of millet cv. ICMV 1 grown in the plots inoculated earlier with N_2-fixing bacteria, Experiment II during 1986 rainy season

Nitrogen applied (kg ha^{-1})	Nitrogen fixing bacteria			Unino-culated control	Mean	SE ±
	Azos. lipoferum (ICM 1001)	*Azos. brasilense* (SL 33)	*Aztb. chroococcum* (ICM 2001)			
	Grain yield					
0	2.13	2.01	1.98	1.67	1.95	
20	2.33	1.99	2.55	2.01	2.22	0.076**
100	2.60	2.82	2.83	2.74	2.75	
Mean	2.35	2.27	2.45	2.14		0.070**
CV (%)			15.1			
	Total plant biomass yield					
0	6.01	5.78	6.01	5.42	5.81	
20	7.00	6.24	7.20	6.31	6.69	0.193**
100	7.28	7.56	7.60	7.48	7.48	
Mean	6.77	6.53	6.94	6.40		0.115**
CV (%)			9.2			
	Total plant nitrogen uptake					
0	41.0	43.1	41.5	34.1	39.9	
20	51.4	45.2	56.7	43.0	49.0	5.43**
100	86.3	87.5	86.0	81.9	85.4	
Mean	59.6	58.6	61.4	53.0		1.68**
CV (%)			13.6			

**$P = < 0.01$

Table 4. Mean grain and total plant biomass yield (t ha^{-1}) and total plant N uptake (kg ha^{-1}) of millet cv. BJ 104 inoculated with N$_2$-fixing bacteria in experiment III

Nitrogen applied (kg ha^{-1})	Nitrogen fixing bacteria			Mean	SE ±
	Azos. lipoferum (ICM 1001)	*Aztb. chroococcum* (ICM 2001)			
Grain yield					
0	1.65	1.58	1.54	1.59	
20	1.89	2.06	1.73	1.89	0.042**
Mean	1.77	1.82	1.64		0.034**
CV (%)		13.6			
Plant biomass yield					
0	4.24	3.94	3.85	4.01	
20	4.96	5.28	4.56	4.93	0.092**
Mean	4.60	4.61	4.20		0.075**
CV (%)		11.6			
Plant N uptake					
0	26.6	25.5	24.7	25.6	
20	31.5	33.8	29.1	31.5	0.71**
Mean	29.1	29.6	26.9		0.64**
CV (%)		15.6			

[1] A pooled analysis of data from 1985 and 1986 rainy seasons.
**$P = < 0.01$.

Table 5. Mean nitrogenase (C$_2$H$_2$ reduction) activity (nmol C$_2$H$_4$ plant^{-1} h^{-1}) associated with millet cultivars inoculated with N$_2$-fixing bacteria in experiment I during 1984 rainy season

Nitrogen applied (kg ha^{-1})	Nitrogen fixing bacteria			Uninoculated control	Mean	SE ±
	Azos. lipoferum (ICM 1001)	*Azos. brasilense* (SL 33)	*Aztb. chroococcum* (ICM 2001)			
50 DAS						
0	71	71	100	40	70	
20	57	22	38	43	40	4.49**
100	41	40	49	11	34	
Mean	56	44	62	31		4.75**
CV %			78			
75 DAS						
0	257	212	194	190	213	
20	249	191	177	168	196	11.19**
100	115	96	151	131	123	
Mean	207	166	174	163		9.04**
CV %			44			

**$P = < 0.01$.

ved at 100 kg N ha^{-1} in comparison to the zero N treatment activity at both growth stages.

Leaf nitrate reductase activity (NRA)

Experiment I. The specific mean leaf NRA of cultivars in experiment I during the 1984 rainy season was higher ($P < 0.01$) at 43 DAS (4.1 μmol NO$_2$ g^{-1} fresh leaf tissue) than at 58 DAS (1.5 μmol NO$_2$ g^{-1} fresh leaf tissue). Addition of 100 kg N ha^{-1} increased NRA in leaves (3.54 μmol NO$_2$ g^{-1} fresh leaf) over the zero and 20 kg N ha^{-1} treatments (2.3 and 2.5 μmol NO$_2$ g^{-1} fresh leaf, respectively). Mean NRA of cultivars varied significantly with cultivars with a maximum specific

NRA (3.3 μmol g^{-1} fresh leaf) in cv. BJ 104 and also with inoculation with N$_2$-fixing bacteria (Table 6). There was significant interaction between cultivars and bacterial cultures.

The specific leaf NRAs at 43 and 58 DAS were positively correlated with grain yield, total plant biomass yield, grain and plant N uptake, and grain N content of cultivars, and the relationship was stronger at 58 DAS. At 58 DAS, NRA was positively correlated with grain yield (r = 0.46), total plant biomass (r = 0.54), grain N uptake (r = 0.74), plant N uptake (r = 0.77), and grain N percentage (r = 0.75).

During 1985 the results for mean leaf NRA at 45 DAS were similar with the leaf NRA results observed in 1984 season (Table 6).

Experiment III. The mean leaf NRA of millet cv. BJ 104 during 1986 marginally increased with *Azos. lipoferum* and *Aztb. chroococcum* inoculation up to 3.5 and 3.9 μmol NO$_2$ g^{-1} fresh leaf, respectively, as against the uninoculated treatment activity of 3.2 μmol NO$_2$ g^{-1} fresh leaf. There was no appreciable effect of N and FYM addition on NRA of cv. BJ 104. The mean leaf NRA on a per plant basis increased significantly (P < 0.05) with *Aztb.*

chroococcum inoculation (658 μmol NO$_2$ plant^{-1}) as against to NRA with the uninoculated control (478 μmol NO$_2^{-1}$ plant^{-1}). Inoculation with *Azos. lipoferum* increased leaf NRA of cv. BJ 104 to 535 μmol NO$_2$ plant^{-1}; this increase was however marginal over the uninoculated control.

Counts of N$_2$-fixing bacteria

Experiment II. Earlier inoculations with *Azos. lipoferum* and *Aztb. chroococcum* resulted in increased MPN counts over the uninoculated controls; increases, however, were statistically not significant. The mean MPN counts of N$_2$-fixers in the rhizosphere soil increased to 9.3 $\times$ 10^4 g^{-1} dry soil with *A. chroococcum* inoculation and 6.4 $\times$ 10^4 g^{-1} dry soil with *A. lipoferum*, as against to 4.7 $\times$ 10^4 g^{-1} dry soil from the uninoculated control. The MPN counts in the rhizosphere soil did not change with millet cultivars. Similar results were found for MPN counts on a 'per plant' basis. The mean MPN count of N$_2$-fixers from macerated roots increased significantly (P < 0.05) in plots fertilized with 100 kg N ha^{-1} (9.8 $\times$ 10^5 g^{-1} dry roots) compared to MPN counts from 20 kg N ha^{-1} and zero N

Table 6. Mean specific leaf nitrate reductase activity (μ mole NO$_2$ g^{-1} fresh leaf) of millet cultivars inoculated with N$_2$-fixing bacteria in experiments during the 1984 and 1985 rainy seasons

Cultivar	Nitrogen fixing bacteria				Mean	SE $\pm$
	Azos. lipoferum (ICM 1001)	*Azos. brasilense* (SL 33)	*Aztb. chroococcum* (ICM 2001)	Uninoculated control		
	1984					
ICMV 1	2.9	3.0	2.9	2.2	2.7	
ICMV 4	3.2	2.6	2.5	2.1	2.6	
BJ 104	4.4	3.2	3.2	2.5	3.3	0.07**
Ex-Bornu	2.7	3.5	2.6	2.0	2.5	
Mean	3.3	2.8	2.8	2.2		0.08**
CV (%)			33			
	1985					
ICMV 1	2.5	3.0	2.4	2.4	2.6	
ICMV 4	3.3	3.2	2.7	2.7	3.0	
BJ 104	3.3	3.1	3.6	2.7	3.1	0.01**
Ex-Bornu	3.3	3.2	3.1	3.1	3.2	
Mean	3.1	3.1	2.8	2.7		0.10**
CV (%)			25			

Four 1984 season each value is mean of two samplings at 43 and 58 DAS; three N levels, 0, 20, and 100 kg N ha^{-1} and four replications at each sampling. For the 1985 season, sampling was done at 45 DAS and other details are the same as for the 1984 season.
**P = < 0.01.

treatments (4.0×10^5 and $3.8 \times 10^5 g^{-1}$ dry roots, respectively). Previous inoculations with *Azos. lipoferum* and *Aztb. chroococcum* increased the MPN counts from the roots of cv. ICMV 1 up to 6.7×10^7 and $6.0 \times 10^7 g^{-1}$ dry roots, respectively, as against $5 \times 10^7 g^{-1}$ dry roots in the uninoculated treatment. Similar results for MPN counts from the roots were observed on a per plant basis also.

The counts of *Azos. lipoferum* in the rhizosphere soil and macerated roots of cv. ICMV 1 grown in the plots inoculated earlier in Experiment I increased significantly (Table 7). The *Azos. lipoferum* counts in the rhizosphere soil of cv. ICMV 1 on a per plant basis increased ($P < 0.05$) to $3.2 \times 10^6 plant^{-1}$ where inoculations had been done in Experiment I, as against $2.2 \times 10^6 plant^{-1}$ in the uninoculated plant rhizosphere soil. Similarly, with addition of 20 and $100 \, kg \, N \, ha^{-1}$, *Azos. lipoferum* counts increased to 2.9×10^6 and $3.4 \times 10^6 plant^{-1}$, respectively, compared to $1.8 \times 10^6 plant^{-1}$ with zero N treatment.

The ELISA counts of *Aztb. chroococcum* in the rhizosphere soil and macerated roots increased significantly up to $3.9 \times 10^3 g^{-1}$ dry soil and $9.88 \times 10^5 g^{-1}$ dry roots, compared to the uninoculated control counts of $2.0 \times 10^3 g^{-1}$ dry soil and $3.61 \times 10^5 g^{-1}$ dry roots (Table 7).

Experiment III. The mean MPN counts of N_2-fixing bacteria in the rhizosphere soil of cv. BJ 104 increased significantly in 1986 after inoculation with N_2-fixing bacteria (Table 8). Similarly, MPN counts of N_2-fixing bacteria in the rhizosphere soil on a per plant basis also increased ($P < 0.01$) with *Azos. lipoferum* and *Aztb. chroococcum* inoculation (5.8×10^6 and $5.6 \times 10^6 plant^{-1}$, respectively), compared to the uninoculated control counts ($1.5 \times 10^6 plant^{-1}$). The MPN counts from the macerated roots on a per plant basis varied significantly with addition of $20 \, kg \, N \, ha^{-1}$, and also after inoculation with N_2-fixing bacteria (Table 8).

The mean ELISA counts of *Azos. lipoferum* in the rhizosphere soil of cv. BJ 104 increased significantly ($P < 0.01$) with inoculation ($9.6 \times 10^3 g^{-1}$ dry soil, compared to $5.8 \times 10^3 g^{-1}$ dry soil with the uninoculated control plants) (Table 9). Similarly, ELISA counts of *Azos. lipoferum* with roots increased twofold over the uninoculated control after inoculation. Similarly, increased counts of *Azos. lipoferum* in the rhizosphere soil and from the plant roots were observed on a per plant basis.

The ELISA counts of *Aztb. chroococcum* associated with the rhizosphere soil and roots of cv. BJ 104 increased significantly ($P < 0.01$) after inoculation (Table 10). Addition of FYM had no effect on the population of *Aztb. chroococcum*, and 20 kg N addition reduced *Aztb. chroococcum* compared to the zero N treatment (Table 10). Similar results were observed for *Aztb. chroococ-*

Table 7. Number of *A. lipoferum* and *A. chroococcum* in experiment II using ELISA ($\times 10^3 g^{-1}$ dry rhizospheric soil/dry root) associated with millet cv. ICMV 1 grown in the plots which were inoculated earlier in experiment I[a]

Nitrogen applied (kg ha^{-1})	Rhizosphere soil			Root macerate		
	Azos. lipoferum	Control	Mean	*Azos. lipoferum*	Control	Mean
0	7.3	5.1	6.2	31.6	20.5	26.1
20	9.0	7.4	8.2	36.2	29.2	32.7
100	8.3	6.5	7.4	49.6	40.1	44.8
Mean	8.2[a]	6.3[b]		39.2[a]	29.9[b]	
CV (%)	2			3		

Nitrogen applied (kg ha^{-1})	Rhizosphere soil			Root macerate		
	Aztb. chroococcum	Control	Mean	*Aztb. chroococcum*	Control	Mean
0	2.9	0.6	1.8	712	452	378
20	4.5	1.5	3.0	1050	416	733
100	4.4	4.0	4.2	1202	622	912
Mean	3.9[a]	2.0[b]		988[a]	361[b]	
CV (%)	10			12		

[a] Average of eight replications, mean across the cultivars. Log transformations of data used for analysis and figures with different letters vary significantly ($P = <0.05$) from each other.

Table 8. Most probable number of N$_2$-fixing bacteria associated with the rhizospheric soil and roots of millet cv. BJ 104 in experiment III during the 1986 rainy season

Nitrogen applied (kg ha^{-1})	Nitrogen fixing bacteria		Uninoculated control	Mean	SE $\pm$
	Azos. lipoferum (ICM 1001)	*Aztb. chroococcum* (ICM 2001)			
Rhizosphere soil					
		($\times 10^4$ g^{-1} dry soil)			
0	2.4	2.2	0.6	1.7	
20	2.5	2.0	0.9	1.8	0.381
Mean	2.5	2.1	0.7		0.315***
		($\times 10^6$ plant^{-1})			
0	5.5	5.6	1.1	4.1	
20	6.2	5.6	1.9	4.6	0.943
Mean	5.8	5.6	1.5		0.792
Macerated roots					
		($\times 10^4$ g^{-1} dry roots)			
0	9.9	4.8	1.3	5.4	
20	9.9	6.3	2.3	6.2	1.48
Mean	9.9	5.6	1.8		1.58**
		($\times 10^5$ plant^{-1})			
0	7.0	4.0	0.9	4.0	
20	11.5	8.3	2.6	7.5	1.06*
Mean	9.3	6.2	1.7		1.33**

*$P = <0.05$; **$P = <0.01$; ***$P = <0.001$.

cum counts in the rhizosphere soil and with roots on a per plant basis.

Discussion

The results from multilocation experiments conducted in fields where millet had been grown several times before under different environmental and soil conditions, indicated a higher success rate and more increases with *Azos. lipoferum* than with *Aztb. chroococcum*. In the USSR, from a comprehensive survey of the data obtained with *Aztb. chroococcum* inoculation experiments, increased yields of cereal and vegetable crops were obtained in 890 out of 1095 trials and the increase in yield

Table 9. Mean population of *Azos. lipoferum* associated with the rhizospheric soil ($\times 10^3$ g^{-1} dry soil) and roots ($\times 10^4$ g^{-1} dry roots) of millet cv. BJ 104 in experiment III using ELISA, during the 1986 rainy season

Nitrogen applied (kg ha^{-1})	*Azos. lipoferum* (ICM 1001)	Uninoculated control	Mean	SE $\pm$
Rhizosphere soil				
0	9.4	5.0	7.2	
20	9.9	6.6	8.3	0.53
Mean	9.6	5.8		0.72**
Macerated roots				
0	2.5	1.5	2.0	
20	3.5	1.6	2.5	0.15*
Mean	3.0	1.5		0.18**

* $P = <0.05$; **$P = <0.01$.

Table 10. Population of *Aztb. chroococcum* associated with the rhizospheric soil ($\times 10^2$ g^{-1} dry soil) and roots ($\times 10^3$ g^{-1} dry roots) of millet cv. BJ 104 in experiment III using ELISA, during the 1986 rainy season

Nitrogen applied (kg ha^{-1})	*Aztb. chroococcum*	Uninoculated control	Mean	SE $\pm$
Rhizosphere soil				
0	2.2	1.5	1.8	
20	2.4	0.8	1.6	0.13**
Mean	2.3	1.2		0.17**
Macerated roots				
0	1.9	1.0	1.4	
20	1.2	0.4	0.8	0.13**
Mean	1.5	0.7		0.12**

**$P = <0.01$.

amounted to $> 10\%$ in 514 experiments (47%) (Fedorov, 1952). In a few experiments, no increases or small reductions in yields were observed in our studies. Such non-significant effects and a small reduction in plant yield and total N uptake were observed in earlier studies also (see Boddey and Döbereiner, 1982; Fedorov, 1952; Bouton et al., 1979; Ruschel et al., 1982).

In Experiments I (Table 2) and III (Table 4) inoculation caused increases in mean grain and total plant biomass yields and the increases with different cultures were similar, indicating no specific affinity between the cultures and millet cultivars tested. Strains isolated from the roots of the same crop into which they were subsequently inoculated, have been termed 'homologous' (Boddey and Döbereiner, 1982). The strains used in the present studies were not homologous and except for *Azos. brasilense* (SP 7), in general, inoculations with all the strains increased the yields. The MPN counts of N_2-fixers in the pre-sowing soil samples from Experiments I and III were 10^2 and $10^3\,g^{-1}$ dry soil, respectively. Boddey et al. (1986) suggested that when azospirilla populations are low, *Azospirillum* strains of diverse origin may cause significant response, but in the areas where these bacteria are abundant, 'homologous' strains are more likely to stimulate yield increases. The lack of interaction between inoculations and years in experiments I and III suggest that subsequent inoculations in the same plot increased the yields. The MPN and ELISA counts in Experiments II (Table 7) and III (Tables 8–10) revealed that when the same plots were inoculated thrice and twice, respectively, the counts of the inoculated strains showed only a 1.8–3.0 fold increase over the uninoculated control. In other studies inoculating once resulted in a 2-3-fold increase in the MPN of N_2-fixers (Rao and Venkateswarlu, 1987). Smith et al. (1984) reported a continued decline in the population of *Azos. brasilense* to less than 10^2 by the 5th week after inoculation. These results reveal the inability of these bacteria to establish in the rhizosphere in large numbers and it may be a reason for lack of interaction between cultures and seasons.

Application of combined N significantly increased grain, plant biomass, and N content in grains and plant tissues. Inoculation did not increase grain N content in any experiment. However, plant N content increased in Experiment I

after inoculation (Table 2). Inoculation of *Azospirillum* often causes increases in plant dry matter with decreases or no increases in N concentrations (Avivi and Feldman, 1982; de Freitas et al., 1982; Millet and Feldman, 1984), and these responses have, therefore, been attributed to effects of plant growth substances. In other experiments, increased plant N concentration with *Azospirillum* inoculation indicated effects of inoculation on N_2 fixation or more nitrogen assimilation by plants (Kapulnik et al., 1981a, b; Baldani et al., 1983; Hegazi et al., 1983; Pacovsky et al., 1985; Wani et al., 1985).

Azospirillum lipoferum inoculation increased mean cumulative plant N uptake ($185\,kg\,ha^{-1}$) by $19\,kg\,N\,ha^{-1}$ more than the uninoculated control plant N uptake during three seasons. The mean cumulative total N uptake by three millet crops in Experiment I with $20\,kg\,N\,ha^{-1}$ treatment was $167\,kg\,N\,ha^{-1}$, as against $107\,kg\,N\,ha^{-1}$ from the zero N plots. These figures showed 100% recovery of added combined N during three years at $20\,kg\,N\,ha^{-1}$ which is a remarkably high value for N recovery studies. With $100\,kg\,N\,ha^{-1}$, however, the N recovery value was just 52%. These results indicated efficient N assimilation by the inoculated plants over the uninoculated control at a low level ($20\,kg\,N\,ha^{-1}$) of combined N. In all, three years of continued inoculation enabled the crops (3 main crops and one succeeding crop) to assimilate $25.6\,kg$ extra $N\,ha^{-1}$ over the uninoculated control plots. These increases were observed along with a 2–3 fold increase in the MPN and ELISA counts of *Azos. lipoferum* and *Aztb. chroococcum* (Table 7) associated with the succeeding crop. The lack of significant interaction between the cultures and seasons in both the experiments and only a 2–3 fold increase in the number of inoculated bacteria after three years of repeated inoculations suggest that these bacteria do not establish well in the soil and continued inoculation may be necessary for obtaining increased yields.

Such positive benefits in terms of increased grain, plant biomass, and N uptake could be attributed to a small increase in N input from BNF (Cohen et al., 1980), development and branching of roots (Umali-Garcia et al., 1980; Tilak and Subba Rao, 1987), production of plant growth hormones (Tien et al., 1979; Vlassak and Reynders, 1981; Brown, 1974); and increased uptake of NO_3^-, K^+, and H_2PO_4

(Lin *et al.*, 1983; Boddey *et al.*, 1986). In the present studies nitrogenase (C_2H_2 reduction) activity associated with the inoculated plants was increased (Table 5), but such increased activity was observed only during later stages of plant growth (Table 5, Wani *et al.*, 1983; Wani, 1988) for a shorter period. As most of the N required for plant growth in millet is taken up before flowering (45–50 DAS) (S.P. Wani, unpublished data) and increased nitrogenase activity was observed after flowering for a short period, the nitrogenase activity may not account for the increased N uptake observed in these studies. Inoculation always increased leaf NRA, suggesting a greater supply of NO_3 to the leaves over the uninoculated control (Table 6) and the increased NO_3 uptake may relate to increased root development in response to the production of hormones by these bacteria (Tien *et al.*, 1979; Umali-Garcia *et al.*, 1980; Brown, 1974; Tilak and Subba Rao, 1987). Further physiological, morphological, and biochemical studies on the plant-bacteria interaction should provide a better understanding of the increased N uptake mechanism.

Acknowledgements

We acknowledge the help of Dr K L Sahrawat for soil and plant chemical analysis and Dr V Subramanian for grain N content analysis.

References

Avivi Y and Feldman M 1982 The response of wheat to bacteria of the genus *Azospirillum*. Israel J. Bot. 31, 237–245.

Balandreau J 1983 Microbiology of the association. Can. J. Microbiol. 29, 851–859.

Baldani V L D, Baldani J I and Döbereiner J 1983 Effects of *Azospirillum* inoculation on root infection and nitrogen incorporation in wheat. Can. J. Microbiol. 29, 924–929.

Baldani V L D, Baldani J I and Döbereiner J 1987 Inoculation of field-grown wheat (*Triticum aestivum*) with *Azospirillum* spp. in Brazil. Biol. Fertil. Soils 4, 37–40.

Boddey R M and Döbereiner J 1982 Association of *Azospirillum* and other diazotrophs with tropical Graminae. *In* Non-symbiotic Nitrogen Fixation and Organic Matter in the Tropics. Transactions of the 12th International Congress of Soil Science, 8–16 Feb 1982, New Delhi, India, pp. 28–47. Indian Agricultural Research Institute, New Delhi, India.

Boddey R M, Baldani V L D, Baldani J I and Döbereiner J 1986 Effect of inoculation of *Azospirillum* spp. on nitrogen

accumulation by field-grown wheat. Plant and Soil 95, 109–121.

Bouton J H, Smith R L, Schank S C, Burton G W, Tyler M E, Littell R C, Gallaher N R and Queensberry K H 1979 Response of pearl millet inbreds and hybrids to inoculation with *Azospirillum brasilense*. Crop Sci. 19, 12–16.

Brown M E 1974 Seed and root bacterization. Annu. Rev. Phytopath. 12, 181–197.

Cohen E, Okon Y, Kigel J, Nur I and Henis Y 1980 Increases in dry weight and total nitrogen in *Zea mays* and *Setaria italica* associated with nitrogen-fixing *Azospirillum* spp. Plant Physiol. 66, 746–749.

Fedorov M V 1952 Biologicheskaya Fiksatsiya Azotaatmosfery. Sel'khozgiz, Moskva.

Freitas J L M de, Rocha R E M da, Pereira P A A and Döbereiner J 1982 Materia organica e inoculacao com *Azospirillum* naincorporaco de N Pelo milpho. Pesa. Agropec. Bras. 17, 1423–1432.

Hegazi N A, Monib M, Amer H A and Shokr E S 1983 Response of maize plants to inoculation with azospirilla and (or) straw amendment in Egypt. Can. J. Microbiol. 29, 888–894.

Hunter W J, Fahring C J, Olsen S R and Porter L K 1982 Location of nitrate reduction in different soybean cultivars. Crop Sci. 22, 944–948.

Hussain A, Arshad M, Hussain A and Hussain F 1987 Response of maize (*Zea mays*) to *Azotobacter* inoculation under fertilized and unfertilized conditions. Biol. Fertil. Soils 4, 73–78.

Jaworski E G 1971 Nitrate reductase assay in intact plant tissues. Biochem. Biophys. Res. Commun. 43, 1274–1279.

Kapulnik Y, Kigel J, Okon Y, Nur I and Henis Y 1981a Effect of *Azospirillum* inoculation on some growth parameters and N content of wheat, sorghum and panicum. Plant and Soil 61, 65–70.

Kapulnik Y, Sarig S, Nur I, Okon Y, Kigel Y and Henis Y 1981b Yield increases in summer cereal crops of Israel in fields inoculated with *Azospirillum*. Expt. Agric. 17, 179–187.

Kapulnik Y, Sarig S, Nur I and Okon Y 1983 Effect of *Azospirillum* inoculation on yield of field-grown wheat. Can. J. Microbiol. 29, 895–899.

Kapulnik Y, Gafny R and Okon Y 1985 Effect of *Azospirillum* spp. inoculation on root development and NO_3 uptake in wheat (*Triticum aestivum* cv. miriam) in hydroponic systems. Can. J. Bot. 63, 627–631.

Kishinevsky B and Bar-Joseph M 1978 *Rhizobium* strain identification in *Arachis hypogaea* nodules by enzyme-linked immunosorbent assay (ELISA). Can. J. Microbiol. 24, 1537–1543.

Lin W, Okon Y and Hardy R W F 1983 Enhanced mineral uptake by *Zea mays* and *Sorghum bicolor* roots inoculated with *Azospirillum brasilense*. Appl. Environ. Microbiol. 45, 1775–1779.

Millet E and Feldman M 1984 Yield response of a common spring wheat cultivar to inoculation with *Azospirillum brasilense* at various levels of nitrogen fertilization. Plant and Soil 80, 255–260.

Okon Y 1982 *Azospirillum*: Physiological properties, mode of association with roots and its application for the benefit of cereal and forage grass crops. Israel J. Bot. 31, 214–220.

Okon Y 1985 The physiology of *Azospirillum* in relation to its utilization as inoculum for promoting growth of plants. *In*

Nitrogen Fixation and CO_2 Metabolism. Eds. P W Ludden and J E Burris. pp. 165–174. Elsevier, New York.

Pacovsky R S, Paul E A and Bethlenfalvay G J 1985 Nutrition of sorghum plants fertilized with nitrogen or inoculated with *Azospirilum brasilense*. Plant and Soil 85, 145–148.

Rao A V and Venkateswarlu B 1986 Studies on the interactions between *Azospirillum* and pearl millet. *In* Cereal Nitrogen Fixation, pp. 37–42. ICRISAT (International Crops Research Institute for the Semi-Arid Tropics, India).

Reynders L and Vlassak K 1982 Use of *Azospirillum brasilense* as biofertilizer in intensive wheat cropping. Plant and Soil 66, 217–223.

Ruschel A P, Vose P B, Matsui E, Victoria R L and Tsai Saito S M 1982 Field evaluation of N_2 fixation and N-utilization by *Phaseolus* bean varieties determined by ^{15}N isotope dilution. Plant and Soil 65, 397–307.

Sarig S, Kapulnik Y, Nur I and Okon Y 1984 Response of non-irrigated *Sorghum bicolor* to *Azospirillum* inoculation. Expl. Agric. 20, 59–66.

Smith R L, Schank S C, Milam J R and Baltensperger A 1984 Response of *Sorghum* and *Pennisetum* species to the N_2 fixing bacterium *Azospirillum brasilense*. Appl. Environ. Microbiol. 47, 1331–1336.

Tien T M, Gaskins M H and Hubbell D H 1979 Plant growth substances produced by *Azospirillum brasilense* and their effect on the growth of pearl millet (*Pennisetum americanum* L.) Appl. Environ. Microbiol. 37, 1016–1024.

Tilak K V B R and Subba Rao N S 1987 Association of *Azospirillum brasilense* with pearl millet (*Pennisetum americanum* (L.) Leeke). Biol. Fertil. Soils 4, 97–102.

Umali-Garcia M, Hubbell D H, Gaskins M H and Dazzo F B 1980 Association of *Azospirillum* with grass roots. Appl. Environ. Microbiol. 39, 219–226.

Van Weeman B K and Schuurs A H W M 1971 Immunoassay using antigen-enzyme conjugates. FEBS Lett. 15, 232–236.

Villas Boas F C S and Döbereiner J 1981 Nitrogenase and nitrate reductase in rice plants inoculated with various *Azospirillum* strains. *In* Associative N_2 Fixation. Eds. P B Vose and A P Ruschel. pp. 2:231–239. CRC Press, Boca Raton, Florida, USA.

Vlassak K and Reynders L 1981 *Azospirillum* rhizocoenoses in agricultural practice. *In* Current Perspectives in Nitrogen Fixation. Eds. A H Gibson and W E Newton. Proceedings of the International Symposium on Nitrogen Fixation, 1–5 Dec. 1980. Canberra, Australia, Australian Academy of Science, Australia.

Wani S P 1986 Research on cereal nitrogen fixation at ICRISAT. *In* Cereal Nitrogen Fixation. pp 55–68. ICRISAT (International Crops Research Institute for the Semi-Arid Tropics, India).

Wani S P 1988 Nitrogen fixation potentials of sorghum and millets. *In* Biological Nitrogen Fixation: Recent Developments. Ed. N S Subba Rao. pp. 125–174. Oxford & IBH Publishing Co. Pvt. Ltd. New Delhi, India.

Wani S P, Dart P J and Upadhyaya M N 1983 Factors affecting nitrogenase activity (C_2H_2 reduction) associated with sorghum and millet estimated using the soil-core assay. Can. J. Microbiol. 29, 1063–1069.

Wani S P, Chandrapalaih S and Dart P J 1985 Response of pearl millet cultivars to inoculation with nitrogen-fixing bacteria. Expl. Agric. 21, 175–182.

Wani S P, Upadhyaya M N and Dart P J 1984 An intact plant assay for estimating nitrogenase activity (C_2H_2 reduction) of sorghum and millet plants grown in pots. Plant and Soil 82, 15–29.

F. A. Skinner et al. (Eds.), Nitrogen fixation with non-legumes, 263–272.

Effect of inorganic N and organic fertilizers on nitrogen-fixing (acetylene-reducing) activity associated with wetland rice plants

J. K. LADHA, A. TIROL-PADRE, G. C. PUNZALAN, M. GARCIA and I. WATANABE
Department of Soil Microbiology, International Rice Research Institute, P.O. Box 933, Manila, Philippines

Key words: acetylene-reducing activity, associative N_2 fixation, organic manure, rice (*Oryza sativa* L.), wetland rice

Abstract

Nitrogen-fixing (acetylene-reducing) activities associated with rice plants obtained from field experiments using different N sources were measured by a modified short-term acetylene reduction assay. Besides using urea as source of inorganic N fertilizer, the following different organic fertilizers were incorporated in the treatments: grain legume green manure (*Vigna radiata* and *Dolichos lablab*), non-grain legume green manure (*Sesbania* and *Crotolaria*), non-legume green manure (*Azolla*), and rice straw. Nitrogen fixation per plant (or per unit area) was either enhanced or not affected by the application of inorganic N and organic fertilizers. However, when expressed on a per unit plant dry weight basis, the N_2-fixing activities were generally lower, except in cases where some legume green manures such as *V. radiata* and *Crotolaria* were incorporated. Possible mechanisms of how rice plant associative N_2 fixation is affected by different N fertilizers are discussed.

Introduction

Moderate but sustainable yields of wetland rice can be obtained with no input of N fertilizer (Koyama and App, 1979). This has been attributed to plant-associative and free-living biological nitrogen fixation (BNF) (Watanabe, 1986). However, additional inorganic and/or organic N is required in order to obtain higher yields. Combined forms of N inhibit BNF under laboratory conditions. The degree of inhibition could, however, be different under field conditions due to biological and physical heterogeneity. Several studies have illustrated almost complete and long-lasting inhibitory effects of N fertilizer on the nitrogen-fixing (C_2H_2-reducing) activity of free-living cyanobacteria (see Roger and Kulasooriya, 1980). On the other hand, a critical and systematic study on the effects of combined N on rice plant-associative N_2 fixation is still lacking.

Based on long-term incubations of soil and/or excised roots (Balandreau *et al.*, 1975; Trolldenier, 1977; Matsuguchi, 1977; Gilmour *et al.*, 1978; Rao *et al.*, 1982, 1983) and intact plants (Watanabe *et al.*, 1981), N fertilizers have been shown to have both positive and negative effects on acetylene-reducing activity (ARA). Stimulatory effects of an organic fertilizer such as rice straw on plant ARA have also been reported (Ladha *et al.*, 1986).

In this study, ARA associated with rice plants from several wetland field experiments in which different N sources were applied were measured using a modified short-term acetylene reduction (AR) assay.

Materials and methods

Rice plants were obtained from the irrigated wetland rice fields of the International Rice Research Institute (IRRI) at Los Baños, Laguna, Philippines. The soil (Maahas clay) is an isothermic clayey mixed Aquic Tropudalf (pH 6.2; organic matter, 2.9%; total N, 0.15%; and CEC, 49 meq/100 g soil). The experimental details are shown in Table 1.

Table 1. Details of experiments conducted

Experiment No.	Treatment[a]	Crop No. in the sequence	Field layout[b]/ subplot size (m²)	No. of replicates	Season[c]	Rice variety	Time of ARA and plant biomass measurements (days after transplanting)
1	See Figs 1, 2	1 & 2	RCBD/15	3	WS 1985	IR50	22, 37, 50, 68 and 78 (heading)
2	See Table 2	3	SPD/30	3	DS 1985	IR54	74, 75, 76 (3 consecutive days at heading)
3	See Table 3	3 & 4	RCBD/49	4	DS 1986	IR54	43, 74 (heading)
4	See Table 4	3	SPD/32	4	WS 1986	IR54	74, 75, 76 (3 consecutive days at heading)
5	See Table 5	10	RCBD/16	4	DS 1987	IR64	32, 47, 70 (heading)
6	See Table 6	3	RCBD/16	4	DS 1986		33, 63 and 90 (heading)

[a] Experiments Nos. 3 and 6 received 20 kg P_2O_5/ha/crop as solophos before rice transplanting, and Experiment No. 5, 30 kg. Experiment No. 6 also received 30 kg K_2O ha/crop.
[b] RCBD, randomized complete block design, SPD, split plot design.
[c] DS, dry season, WS, wet season.

Basal treatment of inorganic N fertilizer involved application of urea 1 day before transplanting (DAT) and a top dressing at the panicle initiation (PI) stage. Green manures (GM) were grown in the same field before rice either in continuously flooded (Experiment No. 3) or in rain-fed conditions (Experiments 2 and 4). In Experiment No. 5, the green manures were from another field. *Sesbania* and *Crotolaria* were sown by broadcasting at seeding rates of 40–50 kg ha⁻¹ while other legumes were sown at a rate of 30 kg ha⁻¹. Rice seedlings about 20 days old were transplanted 1 to 7 days after incorporation of organic manure in 20- × 20-cm spacing. The field was then kept flooded. Pesticides were applied when needed.

Plant sampling, AR assay and determination of plant dry weight

Rhizospheric soil and two rice plants (hills) per block (total 6 plants per treatment) with similar tiller numbers and are representatives of the majority of the plants in the plot were sampled from the second or third rows along the field perimeter. AR assays were performed as described earlier (Ladha *et al.*, 1986). When ARAs were measured for three consecutive days, six plants per variety per day (total of 18 plants per variety) were assayed (Tirol-Padre *et al.*, 1987). The mean ARAs for each day and for the three-day periods were calculated.

Dry weights of root and shoot of plants used for AR assay were determined as described earlier (Ladha *et al.*, 1987).

Results

Effect of inorganic N fertilizer on ARA associated with rice plants (Experiment No. 1)

Plant ARA of IR50 and IR54 were measured several times during the crop cycle at different levels of inorganic N fertilizer. A combined two factor analysis of variance of ARA per plant and ARA per g plant dry weight showed significant differences due to N treatment (main plot) and growth stage (sub-plot). In IR54, only ARA expressed on a per plant basis was significantly different at different growth stages. On the other hand, total plant dry weights were significantly different in both varieties at all N levels (data not shown). At an early stage of plant growth (22–23 DAT), ARA of both varieties expressed on a per plant basis showed a slight inhibition specially in treatments with higher doses of N (Figs. 1–2). At maximum tillering and just before the second application of urea, ARAs of N-treated plants were always higher than in the control in both varieties. The responses of ARA to topdressing of N were different in both varieties; there were decreases in IR50 and increases in IR54. However, when only the fertilized treatments after the second application of N were compared, it was observed that the highest level of top-dressed N resulted in the lowest plant ARA in both varieties.

In this experiment, the inhibitory effect of N fertilizer on plant ARA was demonstrated by measurements made a few days after N application, but it was also shown that this inhibitory effect

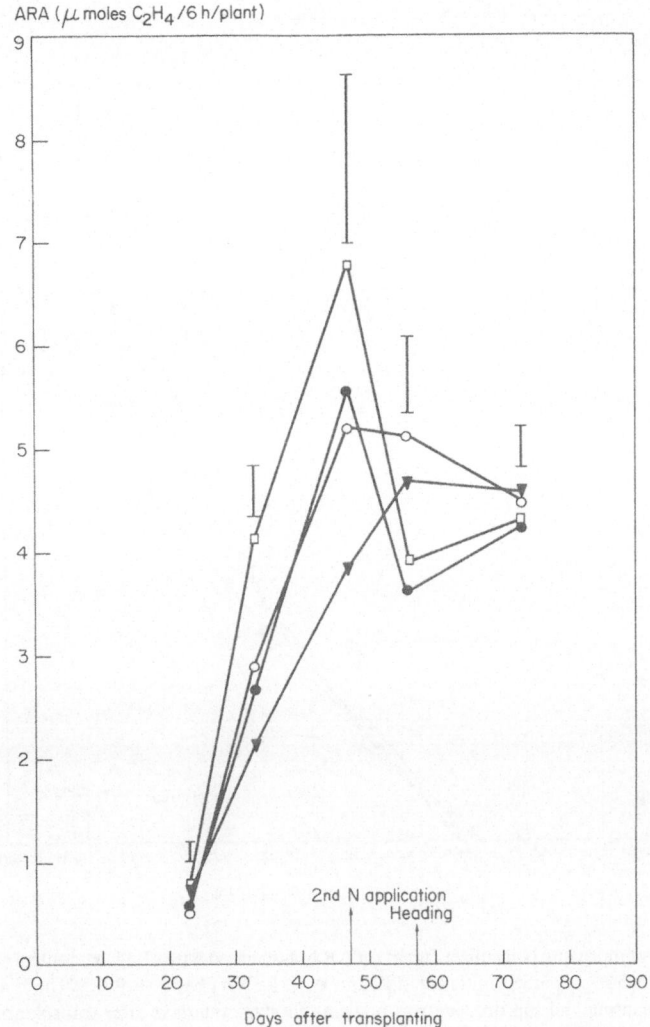

Fig. 1. Effect of different levels of inorganic N fertilizer (urea) on ARA associated with IR50. ▼, control — no N; ○, 60 kg N ha basal; □, 40 kg N ha⁻¹ basal + 20 kg N ha⁻¹ at panicle initiation [PI]; and ●, 70 kg N ha⁻¹ basal + 30 kg N ha⁻¹ at PI. Vertical bars represent standard error of the mean. Basal application of urea was made 1 day before trnsplanting and top dressing was made at PI initiation stage (47 days after transplanting). (Experiment 1).

could be overcome at later stages, and that the increased biomass due to N fertilizer could lead to a higher ARA/plant. The effect of inorganic N fertilizer on plant ARA is therefore dependent on the time interval between N application and ARA measurements and the response of the rice plant to fertilization in terms of dry weight.

Effect of incorporation of preceding leguminous and non-leguminous GM crops on ARA associated with rice plant (Experiments Nos. 2 and 3)

In Experiments 2 and 3, the effects of incorpor-

ation of different legume and non-leguminous green manure (GM) crops (grown and incorporated in the same field) on plant-associated ARA were compared with controls (no nitrogen) and with inorganic N fertilizer. In Experiment 2, the control treatments included weed-free and weedy fallow with and without inorganic N. Measurements of ARA and plant dry weights were made for 3 consecutive days at the heading stage in Experiment 2 and at both maximum tillering and heading stages in Experiment 3. The amount of N incorporated in different GM treatments and their effects on ARA are shown in Tables 2 and 3.

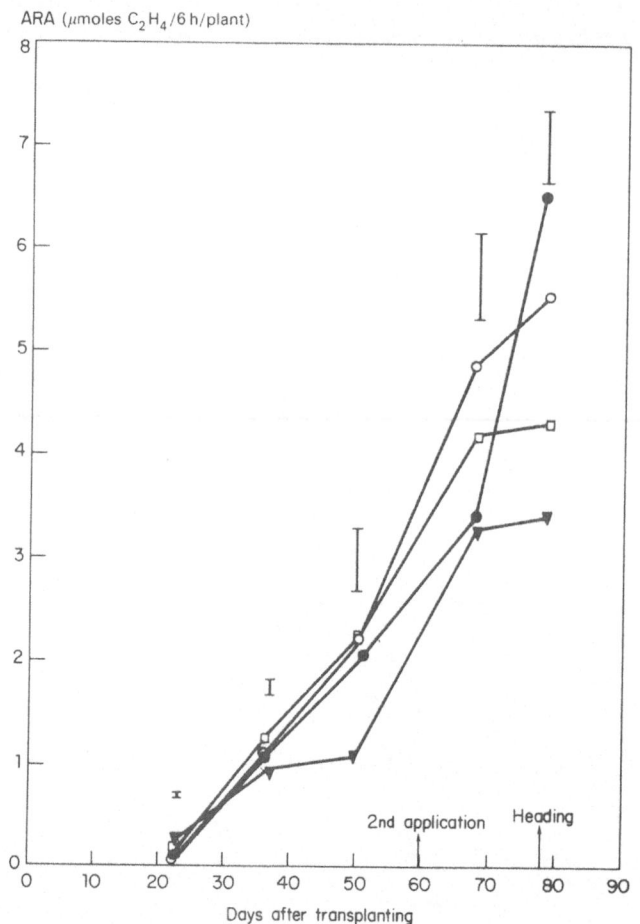

ARA (μmoles C_2H_4/6 h/plant)

Days after transplanting

Fig. 2. Effect of different levels of inorganic N fertilizer (urea) on ARA associated with IR54. ▼, control — no N; ○, 60 kg N ha^{-1} basal; □, 40 kg N ha^{-1} basal + 20 kg N ha^{-1} at panicle initiation [PI]; and ●, 70 kg N ha basal + 30 kg N ha^{-1} at PI. Basal application of urea was made 1 day before transplanting and top dressing was made at PI stage (60 days after transplanting). (Experiment No. 1).

Acetylene-reducing activity expressed per plant or per gram plant dry weight basis were not significantly different among control treatments — weed-free, weedy fallow, and weedy fallow plus 70 kg N ha^{-1}. However, ARAs per plant in all GM treatments were significantly higher than in all three controls. Acetylene-reducing activities expressed on a per gram plant dry weight basis were, on the other hand, significantly higher only in *Crotolaria juncea*- and mungbean (*Vigna radiata*)-incorporated treatments.

In Experiment No. 3, significant effects were found more often in the dry season (DS) than in the wet season (WS) (Table 3). Acetylene-reducing activities per plant were significantly higher at the heading stage in both (*Azolla* and *Sesbania*) GM treatments than in the control and urea treatments.

Acetylene-reducing activity per gram plant dry weight was only higher in the *Azolla*-incorporated treatment in the DS but the same treatment gave significantly lower activities in the WS. Acetylene-reducing activities at the maximum tillering stage, on the other hand, were unaffected in both seasons except that ARA per gram plant dry weight decreased significantly in *Azolla*- and *Sesbania*-incorporated treatments during the dry season.

Effect of inorganic N, organic (farmyard manure) and GM fertilizers on ARA associated with rice plants as affected by preceding upland or fallow crop (Experiment No. 4)

This experiment studied the effect of inorganic

Table 2. Effect of different leguminous green manure species on plant-associated acetylene-reducing activity (ARA) at heading stage of IR54 (IRRI, 1985 DS). Experiment No. 2

Treatment[a]	Amount of N ha^{-1} (as urea and/or green manure crop) applied	ARA (nmol $C_2H_4.6h^{-1}$)[b]	
		Per plant	Per gram plant dry weight
Control			
Weed free	0	2100d	70 c
Weedy fallow	21	2100d	75 bc
Weedy fallow + urea (70 kg N/ha)	21 + 70	2300cd	67 c
Incorporation of			
Crotolaria juncea	144	4700a	87 ab
Vigna radiata	93	4700a	97 a
Sesbania aculeata	173	4000ab	82 abc
S. aculeata + urea (35 kg N/ha)	173 + 35	4200a	82 abc
Dolichos lablab	76	3200b	83 abc
Standard error of mean		281	4.9

[a] Urea was applied in split: 1/2 basal and 1/2 at panicle initiation (55 days after trnsplanting). Green manure crop was grown for 60 days and incorporated. In weedy fallow treatments, weeds accumulated 21 kg N ha^{-1} in 60 days.
[b] In a column, means followed by a common letter are not significantly different at the 5% level by Duncan's Multiple Range Test.

N, organic (farmyard manure), and green manure on ARA as affected by preceding upland or fallow crop (Table 4). Acetylene-reducing activity per plant from treatments with inorganic N were significantly higher than with no N or with farmyard manure. On the other hand, ARA per gram plant dry weight was significantly higher only in the mungbean-incorporated treatment. Whether leaving the land fallow or growing a GM crop before rice affects ARA is not clear. However, as in the other experiments, it seems that ARAs associated with plants grown after upland or fallow crop were lower than ARAs associated with plants grown under continuous wetland cropping.

Table 3. Effect of inorganic and green manure N on plant-associated acetylene reducing activity (ARA) of IR54 (IRRI, 1986 DS and 1987 WS) Experiment No. 3

Treatment	Amount of N ha^{-1} (as urea or green manure applied)[a]	ARA (nmol $C_2H_4.6h^{-1}$)			
		Per plant		Per g plant dry weight	
		Maximum tillering	Heading	Maximum tillering	Heading
Dry season (1986)					
Control	0	2400	4500	121	86
Urea-applied	50	2300	5300	99	87
Azolla-incorporated	98	2000	6300**	68*	99*
Sesbania-incorporated	65	2200	6500**	89*	91
Wet season (1987)					
Control	0	1000	3200	83	107
Urea-applied	50	1200	3500	68	95
Azolla-incorporated	86	1100	3600	79	67*
Sesbania-incorporated	78	1200	4100	83	88

[a] Urea was applied in split: 1/2 basal and 1/2 at 50 days after transplanting; *Azolla microphylla* was grown for 30 days and incorporated three times before transplanting and one time, 1 month after transplanting. *Sesbania* was grown for 52 and 45 days in dry and wet seasons, respectively, and incorporated 1 day before transplanting.
*,**Significantly different from the control at 1 and 5% levels, respectively.

Table 4. Effect of preceding upland or fallow crop on rice plant-associated acetylene-reducing activity (ARA) at the heading stage of IR54 (IRRI, 1986 WS). Experiment No. 4

Cropping sequence[a]	Amount of N ha^{-1} (as urea, FYM, or green manure applied)[b]	ARA (nmol $C_2H_4.6h^{-1}$)[c]	
		Per plant	Per gram plant dry weight
F – R – M	20 + 0	1460 c	66 b
F – R + FYM – M	20 + 103	1400 c	62 b
F – R + urea – M	20 + 35	2000 a	69 b
F – R + urea – M	20 + 70	1800 ab	57 b
F – R + urea – M	20 + 105	2000 a	57 b
S – R – M	171	1800 a	59 b
V – R – M	100	1800 ab	82 a
Standard error of mean		105.4	3.8

[a] F, fallow; R, rice; M, maize; FYM, farmyard manure; S, *Sesbania aculeata*; V, *Vigna radiata*.

[b] Urea was applied in split: 1/2 basal and 1/2 at panicle initiation (55 days after transplanting). *Sesbania* and *Vigna* crops were grown for 60 days and incorporated. In fallow treatment, weeds accumulated 20 kg N ha^{-1}.

[c] In a column, means followed by a common letter are not significantly different at the 5% level by Duncan's Multiple Range Test.

Effect of long-term application of inorganic N, organic, and GM fertilizer (Experiment No. 5)

The effect of different sources of inorganic N fertilizer, organic, and green manures was studied while keeping the amount of N applied constant. Green manures incorporated in this experiment were obtained from another field. Acetylene-reducing activity was measured on plants of the 13th crop, grown continuously with the same treat-ments, at 32, 47 and 70 DAT (Table 5). Acetylene-reducing activities per plant were significantly different among treatments at 32 DAT. Prilled urea-treated plants had significantly lower ARAs than the control plants. Acetylene-reducing activities on a per gram plant dry weight basis among the treatments were, on the other hand, significantly different in all three samplings. Both inorganic N fertilizer treatments inhibited ARA per gram plant dry weight significantly compared to the control at 32 and 47 DAT.

Table 5. Effect of mineral N fertilizer, green manure, and organic manure on plant associated acetylene reducing activity (ARA) of IR64 (IRRI, 1987 DS). Experiment No. 5

Treatment[a]	ARA (nmol $C_2H_4.6h^{-1}$)[b]					
	32 DAT		47 DAT		70 DAT	
	ARA per plant	ARA per gram plant dry weight	ARA per plant	ARA per gram plant dry weight	ARA per plant	ARA per gram plant dry weight
Control	564 ab	142 a	1030 a	90 a	3280 a	102 ab
Prilled urea (best split)	317 c	53 b	830 a	41 bc	3817 a	85 ab
Urea supergranule	421 bc	82 b	603 a	32 c	3793 a	84 ab
Azolla incorporation	536 abc	129 a	1150 a	83 a	2872 a	75 ab
Sesbania rostrata incorporation	521 abc	124 a	860 a	78 a	3133 a	75 ab
Rice straw incorporation	625 ab	152 a	990 a	68 ab	2642 a	59 b
Rice straw compost incorporation	717 a	145 a	850 a	63 abc	3535 a	109 a
Standard error of mean	71	13	141	10	498	14

[a] Urea, green manure and organic manure were applied in amounts equivalent to 116 kg N ha^{-1}. Prilled urea was applied in split: 2/3 basal and 1/3 topdressed at 7 days before panicle initiation (40 days after transplanting). The total amount (dry weight t ha^{-1}) of *Azolla microphylla*, *Sesbania*, straw, and straw compost were 8.5, 8.8, 28.0 and 22.0, respectively.

[b] In a column, means followed by a common letter are not significantly different at the 5% level by Duncan's Multiple Range Test.

Table 6. Effect of two modes of straw application on plant-associated acetylene-reducing activity (ARA) of IR64 (IRRI, 1986 DS). Experiment No. 6

Treatment[a]	Amount of N ha^{-1} as urea or organic manure applied	ARA (nmol $C_2H_4.6h^{-1}$)					
		33 DAT		63 DAT		90 DAT	
		ARA per plant	ARA per gram plant dry weight	ARA per plant	ARA per gram plant dry weight	ARA per plant	ARA per gram plant dry weight
Control (0–30)	30	710	85	2700	80	5600	90
Urea (50–30)	80	860	79	3400*	95*	6900	104
Chopped straw applied before transplanting + urea (0–30)	60	910*	101*	4000*	124*	5900	92
Long straw applied after transplanting + urea (0–30)	60	830	94	2900	90	7500*	104
Standard error of mean		64	5	238	5	504	6

[a] Chopped straw (5 t ha^{-1}) was surface applied 21 days and incorporated 2 days before transplanting; long straw (5 t ha^{-1}) was surface applied 25 days after transplanting. All treatments received a top dressing of 30 kg N ha^{-1} at the panicle initiation stage. Urea treatment also received basal application of 50 kg N ha^{-1}.

Effect of integrated use of inorganic N and organic (straw) fertilizer on ARA associated with rice plants (Experiment No. 6)

The purpose of this experiment was to study the rice plant ARA as affected by different time and method of application of organic material (straw) with high C:N ratio in combination with inorganic N fertilizer. Although straw application has been found to stimulate non-rhizospheric N_2 fixation (Ladha *et al.,*.1986, 1987), N deficiency due to immobilization of N by straw has also been reported (Ponnamperuma, 1984). Therefore, both methods of straw application were combined with top-

dressings of N. For comparison, two controls (urea topdressing and urea basal with topdressing) were included. ARA was measured at 33 and 63 DAT and at the heading stage (90 DAT) (Table 6).

In general, rice plant ARA was higher in plots where straw and split urea were applied than in the control. Straw application before transplanting significantly increased ARA per plant and per gram plant dry weight at 33 and 63 DAT. Straw application after transplanting significantly increased only ARA per plant at heading.

The higher ARA at an early stage of plant growth due to straw application before rice transplanting might have coincided with the release of

Table 7. Summary of experimental results on the effect of inorganic N and organic fertilizers on rice plant-associated acetylene-reducing activity (ARA) at the heading stage[a]

Treatment category	No. of trials	N applied (kg ha)		% increase or decrease over control					
		Mean	STE	ARA/plant		ARA/g plant dry weight		Plant dry weight	
				Mean	STE	Mean	STE	Mean	STE
Grain legume GM	3	90	7.1	64	29.8	26.6	5.9	30.2	17.1
Non-grain legume GM	8	129.3	15.2	50.8	16.5	−2.6	7.6	45.7	8.7
Urea	17	77	8	28.0	6.8	−8.8	2.8	37.8	7.7
Non-legume GM	3	100	7	12.2	14.6	−16.4	16.1	41.6	16.6
Rice straw and farmyard manure	5	83	18	4.4	8.6	−4.6	10.0	15.4	8.1

[a] GM, green manure; STE, standard error of mean.

nutrients from straw decomposition as the half life of straw was reported to be about 43 days (Neue, 1985). This observation is further confirmed by higher ARA at the heading stage due to straw application after transplanting (25 DAT) (Table 6). Such results suggest the possibility of split application of straw to increase the associative N_2 fixation.

Table 7 summarizes the effects of inorganic and organic fertilizer on plant ARA and dry weight at the heading stage. Data for the heading stage alone were considered because the ARA was found to be highest at this stage (Ladha *et al.*, 1986, 1987). Based on the nature of fertilizer used, the treatments were divided into the following categories: (1) grain legume GM (*e.g., V. radiata* (mungbean) and *D. lablab* (lablab)); (2) non-grain legume GM (*e.g., Sesbania* and *Crotolaria*); (3) inorganic N fertilizer (urea); (4) non-legume GM (*Azolla*); (5) rice straw, and farmyard manure. The average amounts of N applied in the different fertilizer categories and the percentage increase or decrease of ARA and plant dry weight over their respective controls were determined. Acetylene-reducing activity per plant increased in all treatments except in the farmyard manure treatment which shows a slight decline (3.3%). Acetylene-reducing activity per gram plant dry weight was lower in all treatments except in the grain legume GM treatments which showed an average increase of 27%. On the other hand, the plant dry weight increased in all categories of fertilizer treatments.

Discussion

Inorganic and organic fertilizers may exert different effects on N_2 fixation associated with rice. One of these could be the inhibitory effects due to the combined form of N (as in the case of inorganic fertilizer) or to toxic compounds released by the decomposition of organic materials). The former may be more important than the latter. Such inhibitory effects, however, may not be very important and long lasting, especially in tropical rice fields, where there is intense biological activity. Furthermore, the inorganic N disappears in flooded soil systems after 30–50 days of application of fertilizer (Watanabe and Inubushi, 1986; Nagarajah, 1987).

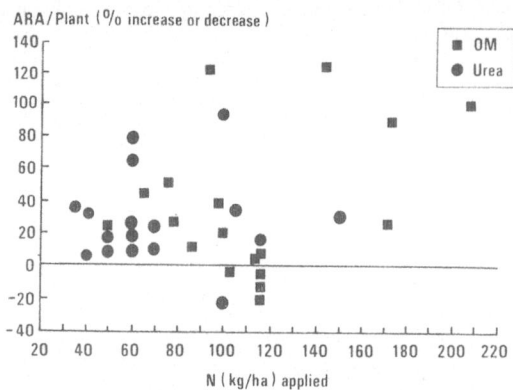

Fig. 3. Relationship between % increase or decrease in ARA per plant at heading and the amount of N (kg ha^{-1}) as urea and organic fertilizers applied.

The second type of effect may be a direct or indirect stimulatory effect. The direct effect could be due to the release of carbon substrates and other nutrients from the organic fertilizers, and the indirect effect could be increased plant growth and greater release of carbon compounds by the plant.

Infrequent and insignificant inhibition of ARA at the early growth stages, and frequent and significant stimulation of ARA at the later stages due to different fertilizers were found in the present study and in earlier studies (Trolldenier, 1977; Watanabe *et al.*, 1981). These findings support the assumptions made earlier. However, the stimulation in ARA per gram plant dry weight was not as much as that in ARA per plant; in fact, there was some decline in several cases, compared to the control.

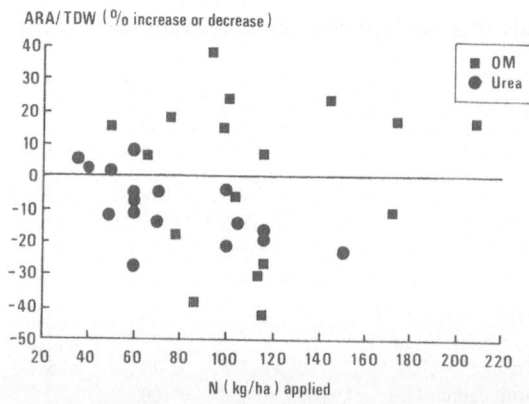

Fig. 4. Relationship between % increase or decrease in ARA per gram plant dry weight (ARA/TDW) at heading and the amount of N (kg ha^{-1}) as urea and organic fertilizers applied.

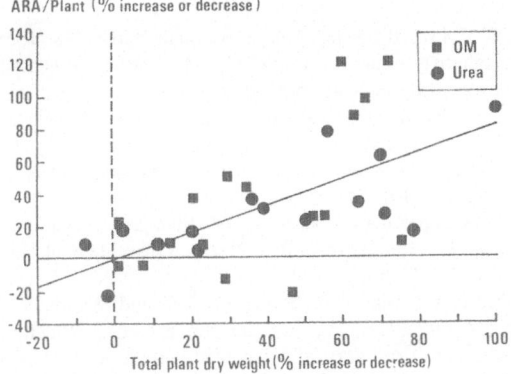

Fig. 5. Relationship between % increase or decrease in ARA per plant and % increase or decrease in total plant dry weight at the heading stage.

Interestingly, there were also no correlations between the increases or decreases in ARA per plant or ARA per gram plant dry weight and the amount of N applied (Figs. 3 and 4). This rules out the possibility that the inhibition or stimulation of ARA is a monotonic function of the amount of N. The relationship between percent increase or decrease of ARA per plant or ARA per gram plant dry weight and percent increase or decrease of total plant dry weight were also examined (Figs. 5 and 6). There was a correlation between ARA per plant and plant dry weight, but there was no correlation between ARA per gram plant dry weight and plant dry weight. The absence of correlation in the latter case is due to the fact that the increase in plant dry weight is much more than the increase in the specific ARA (ARA per gram dry weight). Possible reasons for this are not known.

The significant increases in specific N_2-fixing activity and the total plant dry weight at the heading stage in mungbean- and *Crotolaria*-incorporated treatments (Tables 2 and 4) are interesting and noteworthy. It may be worthwhile to investigate whether the rhizobia, which are being inoculated in large numbers through the incorporation of its host biomass, are making an active N_2-fixing association with the rice plant. Rhizobia of some legumes such as *Sesbania rostrata* used as GM in rice has been reported to be fixing N_2 *ex planta* (Dreyfus *et al.*, 1983).

It may be concluded that the application of inorganic N and different organic (including GM) fertilizers does not inhibit N_2 fixation associated with the rice plant. Nitrogen fixation on a per plant or per unit area basis is either enhanced or not affected but the effect is generally not proportional to the plant biomass. The enhancement of N_2 fixation in proportion to the plant biomass by incorporating some leguminous plants is interesting and worth investigating.

Acknowledgements

This work was supported by the United Nations Development Programme Fund. We thank Dr. S K De Datta and Dr. O P Meelu for allowing us to obtain rice plants from their field experiments, and Dr. D Patriquin for reviewing the paper.

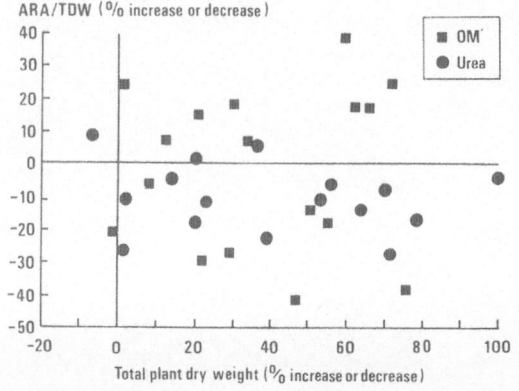

Fig. 6. Relationship between % increase or decrease in ARA per gram plant dry weight (ARA/TDW) and % increase or decrease in total plant dry weight at the heading stage.

References

Balandreau J, Rinaudo G, Fares-Hamad, J and Dommergues Y 1975 N_2 fixation in paddy soils. *In* Nitrogen Fixation by Free-living Microorganisms. IBP Series No. 6, 57–70.

Dreyfus B L, Elmerich C and Dommergues Y R 1983 Free-living *Rhizobium* strain able to grow on N_2 as the sole nitrogen source. Appl. Environ. Microbiol. 45, 711–713.

Gilmour J T, Gilmour C M, and Johnston T H 1978 Nitrogenase activity of rice plant root system. Soil Biol. Biochem. 10, 261–264.

Koyama T and App A A 1979 Nitrogen balance in flooded rice soils. *In* Nitrogen and Rice. pp. 95–104. International Rice Research Institute, P.O. Box 933, Manila, Philippines.

Ladha J K, Tirol-Padre A, Daroy M L, Punzalan G and Ventura W 1986 Plant-associated N_2 fixation (C_2H_2-reduction) by five rice varieties, and relationship with plant growth characters as affected by straw incorporation. Soil Sci. Plant Nutr. 32, 91–106.

Ladha J K, Tirol-Padre A, Daroy M L, Punzalan G and Ventura W 1986 Plant-associated N_2 fixation (C_2H_2-reduction) by five rice varieties, and relationship with plant growth characters as affected by straw incorporation. Soil Sci. Plant Nutr. 32, 91–106.

Ladha J K, Tirol-Padre A, Punzalan G C and Watanabe I 1987 Nitrogen-fixing (C_2H_2-reducing) activity and plant growth characters of 16 wetland rice varieties. Soil Sci. Plant Nutr. 33, 187–200.

Matsuguchi T 1977 Factors affecting heterotrophic nitrogen fixation in submerged rice soils. *In* Nitrogen and Rice. pp. 207–222. International Rice Research Institute, P.O. Box 933, Philippines.

Nagarajah S 1987 Transformation of green manure nitrogen in lowland rice soils. Proc. Symposium on Sustainable Agriculture—The Role of Green Manure Crops in Rice Farming Systems. CASAFA/IRRI, 25–29 May 1987. (In press).

Neue H 1985 Organic matter dynamics in wetland soils. *In* Wetland Soils: Characterization, Classification and Utilization. pp. 109–122. International Rice Research Institute, P.O. Box 933, Manila, Philippines.

Ponnamperuma F N 1984 Straw as source of nutrients for wetland rice. *In* Organic Matter and Rice. pp. 117–136. International Rice Research Institute, P.O. Box 933, Manila, Philippines.

Rao J L N, Prasad J S and Rajaramamohan Rao V 1982 Rice rhizosphere nitrogen fixation (C_2H_2 reduction) as influenced by nitrogen source. Curr. Sci. 50, 900–901.

Rao J L N, Reddy B B, Rajaramamohan Rao V 1983 Rhizosphere soil nitrogenase (C_2H_2 reduction) as influenced by the nitrogen management in intermediate deep water rice. J. Agric. Camb. 101, 547–551.

Roger P A and Kulasooriya S A 1980 Blue-green Algae and Rice. pp. 112. International Rice Research Institute, P.O. Box 933, Manila, Philippines.

Tirol-Padre A, Ladha J K, Punzalan G C and Watanabe I (1987) A plant sampling procedure for acetylene reduction assay to detect rice varietal differences in ability to stimulate N_2 fixation. Soil Biol. Biochem. (In press)

Trolldenier G 1977 Influence of some environmental factors on nitrogen fixation in the rhizosphere of rice. Plant and Soil 47, 203–217.

Watanabe I, De Guzman M R and Cabrera D A 1981 The effect of nitrogen fertilizer on N_2 fixation in the paddy field measured by *in situ* acetylene reduction assay. Plant and Soil 59, 135–139.

Watanabe I 1986 Nitrogen fixation by non-legumes in tropical agriculture with special reference to wetland rice. Plant and Soil 90, 343–357.

Watanabe I and Inubushi K 1986 Dynamics of available nitrogen in paddy soils. Soil Sci. Plant Nutr. 32, 37–50.

F. A. Skinner et al. (Eds.), Nitrogen fixation with non-legumes, 273–285.

Impact of soil environmental factors on rates of N_2-fixation associated with roots of intact maize and sorghum plants

D. B. ALEXANDER and D. A. ZUBERER

Texas Agricultural Experiment Station, Texas A&M University, Department of Soil and Crop Sciences, College Station, TX 77843, USA

Key words: ammonium, carbon, inoculation, nitrite, nitrate, oxygen

Abstract

We have evaluated the effects of oxygen partial pressure (pO_2), combined nitrogen, and the availability of organic substrates on nitrogen fixation (acetylene reduction) by bacteria associated with the roots of intact maize and sorghum plants. We also investigated the possibility of enhancing associative nitrogen-fixation by inoculating the soil in which the plants were grown with *Azospirillum*. Acetylene reduction (AR) activity was greatest when roots of intact plants were exposed to pO_2 between 1.3 and 2.1 kPa. Field-grown and greenhouse-grown plants supported similar levels of activity. Respiration inhibitors (2,4-dinitrophenol and sodium azide) eliminated AR activity at $2 \, kPa \, O_2$, whereas a fermentation inhibitor (sodium fluoride) only partially reduced the activity. Acetylene reduction activity was rapidly (1–3 h) inhibited by NH_4^+, NO_3^-, and NO_2^- at concentrations of 4–$20 \, mg \, N \, l^{-1}$. Rates of AR varied substantially among individual plants in each experiment and between experiments. Amendment with any of several organic substrates greatly increased AR activity when rates were low, suggesting that the lack of activity was caused by a shortage of available carbon in the rhizosphere. Inoculation with *Azospirillum* failed to increase rates of AR associated with maize plants. In several experiments the indigenous bacteria associated with uninoculated plants exhibited greater activity than the bacteria associated with inoculated plants.

Abbreviations. AR = acetylene reduction, ADS = anthrone-detectable sugars, DNP = dinitrophenol, DOC = dichromate-oxidizable carbon

Introduction

During the past decade much research has been conducted on those plant-microbe associations most commonly referred to as 'associative' symbioses, and which are endowed with the capacity to provide nitrogen for growth of non-leguminous plants (Burris, 1977; Neyra and Döbereiner, 1977; van Berkum and Bohlool, 1980). Such research has generated much new information regarding the types of N_2-fixing bacteria (diazotrophs) which inhabit the roots and rhizospheres of a vast array of grasses, and has led to the description of new species of these bacteria (Tarrand *et al.*, 1978; Magalhães *et al.*, 1983; Baldani *et al.*, 1986). The same research attracted interest in the use of such bacteria as inoculants for field-grown grasses and cereals (van Berkum and Bohlool, 1980; Okon, 1985) and considerable efforts have been made to test the efficacy of bacteria such as *Azospirillum* for their ability to enhance crop yields. Some investigators have reported significant yield increases due to inoculation (Albrecht *et al.*, 1981; O'Hara *et al.*, 1981; Okon, 1982). It is clear, however, that for every successful inoculation trial there are many which failed to indicate any benefit of inoculation (Barber *et al.*, 1979; Albrecht *et al.*, 1981; Smith *et al.*, 1984).

It has become increasingly clear that many factors, both biotic and abiotic, can influence the performance of these associations. Many of these factors were reviewed by Balandreau *et al.* (1978), who listed soil redox potential and water content among the factors limiting associative N_2-fixation. Clearly, soil environmental factors can exert stringent constraints on the performance of these associative symbioses and better understanding of the complex plant-microbe-soil system will be necessary if these systems are to be refined, perhaps engineered, and put to use in agriculture. For several years now we have investigated the impacts of several environmental factors on rates of nitrogen fixation (acetylene reduction; AR) associated with roots of maize and sorghum (Zuberer and Alexander, 1986; Alexander *et al.*, 1987). We demonstrated that AR was greatly increased when the pO_2 around intact root systems was reduced to *ca.* $2 kPa O_2$, and that the response to low pO_2 observed with roots exposed in a hydroponic system was identical to that of plants grown in a solid rooting medium (Alexander *et al.*, 1987). The objectives of the studies reported here were to define more clearly the effects of low pO_2, combined nitrogen (ammonium, nitrite, nitrate), exogenous carbon, and inoculation with diazotrophs on rates of nitrogen fixation (AR) associated with roots of healthy, intact, physiologically-active maize and sorghum plants under controlled conditions.

Materials and methods

Plant species and growth conditions

Zea mays L. (Mo17 × B73, Funk's G4522) was grown under ambient light in a greenhouse on the campus of Texas A&M Univ., College Station, TX. Air temperatures ranged from 20–24°C at night to 29–32°C during the day. During cool winter months (December to February), plants were grown on a light table in the laboratory and illuminated under ten very-high-output fluorescent tubes ($200 \mu mol$ photons $m^{-2} sec^{-1}$) or four 400-watt metal halide lamps ($1000 \mu mol$ photons $m^{-2} sec^{-1}$). The light period was 14 h and dark, 10 h.

The plants were grown in plastic pots (25 × 6.4 cm diam., 600 ml capacity) containing 450 g of a

1:1 mixture of Weswood silt loam (Fluventic Ustochrept, fine-silty, mixed, thermic) and sand (0.2–1 mm). Plants were fertilized weekly with 50 ml of Peter's liquid fertilizer (20N-20P-20K; $2.5 g l^{-1}$ supplemented with trace elements, W.R. Grace, Co.) and watered as needed with distilled water. Black, glazed pebbles were placed on the soil surface to retard evaporation and deter algal growth.

Two experiments were conducted with sorghum plants (*Sorghum bicolor* L. Moench; Funk's G522DR, Oro GXTRA) grown in the field at the Experimental farm of the Texas Agricultural Experiment Station (College Station, TX) during April and May of 1986. The soil was Weswood silt loam as above. Plants 30–33 cm tall were excavated with a spade, leaving the root-soil ball intact.

Exposure of intact plants to low pO_2 in the root zone

Greenhouse- or field-grown plants were brought to the laboratory where the soil was gently washed from the roots. The intact plants were transferred to polycarbonate cylinders (25 × 3.2 cm i.d.) fitted with threaded caps bearing stainless steel tubing for introduction of gas mixtures of controlled pO_2 as previously described by Zuberer and Alexander (1986). Gas mixtures (0 to $21 kPa O_2$) were prepared by blending oxygen or air with cylinder N_2 (99.995% purity). The oxygen content of the mixtures was verified by gas chromatography. Each plant was placed in a plastic cylinder through a slotted rubber stopper with the stem sealed to the stopper with a non-toxic, pourable, catalytically-cured silicone sealant (General Electric Co. RTV-11 silicone rubber compound). Gases were vented from the cylinders through a port just below the rubber stopper.

Once in the hydroponic apparatus, the plants were maintained in an N-free modification of Hoagland's solution 1 ((Hoagland and Arnon, 1950) containing ($mg l^{-1}$): KH_2PO_4, 136; K_2HPO_4, 174; $CaSO_4.2H_2O$, 172; $MgSO_4.7H_2O$, 494; $FeCl_3.6H_2O$, 7.0; Na_2H_2EDTA, 8.5; H_3BO_3, 1.43; $MnCl_2.4H_2O$, 0.90; $ZnSO_4.7H_2O$, 0.11; $CuSO_4.5-H_2O$, 0.04; $Na_2MoO_4.2H_2O$, 0.01. The solution was changed daily.

Acetylene reduction assays

To determine rates of acetylene reduction (AR) associated with roots of intact plants exposed to low pO$_2$, nutrient solution was withdrawn from the polycarbonate cylinders with a 160-ml syringe connected to the cylinders with Tygon tubing. The gas inlet and outlet ports were closed with pinch clamps, leaving the low pO$_2$ atmosphere in the cylinders. Acetylene was injected through a serum stopper in the side of the cylinder while the 160-ml syringe was used to withdraw simultaneously an equal volume of residual nutrient solution, leaving an atmosphere containing 11.0 kPa C$_2$H$_2$ at a pressure of 101 kPa (1 atm).

Ethylene was measured with a gas chromatograph equipped with a hydrogen-flame ionization detector and stainless steel columns (183 cm × 0.3 cm o.d) containing Porapak N (Waters Assoc., Framingham, MA.). A standard curve was prepared by injecting known concentrations of pure ethylene.

Oxygen consumption within sealed cylinders was monitored with a miniature Clark-type electrode (Diamond Electrotech Inc., type 730). The electrode, mounted in a serum stopper, was inserted through a 9-mm diam. hole in the cylinder, taking care not to make contact with a root. Only one cylinder could be monitored during each AR assay.

Following measurements of AR, the nutrient solution was returned to each cylinder and gas flow was resumed to purge acetylene and ethylene from the system.

Effects of respiratory inhibitors, combined nitrogen and exogenous organic substrates on root-associated acetylene reduction

To determine the effects of respiratory inhibitors, combined nitrogen (ammonium, nitrate, and nitrite), and exogenous organic substrates on root-associated AR, various inhibitors, N sources, or substrates were added to the nutrient solution immediately after establishing a baseline rate of AR for unamended plants exposed to low pO$_2$ for 18–22 h. Rates of AR in the presence of an inhibitor, N source, or substrate were then measured at various times after amendment of the solution. The inhi-

bitors included 2,4-dinitrophenol (DNP, 1 mM), sodium azide (1 mM), and sodium fluoride (NaF, 1 mM). Nitrogen sources included $(NH_4)_2SO_4$ (4.2–20 mg N l^{-1}), KNO$_2$ (5–20 mg N l^{-1}), and KNO$_3$ (5–20 mg N l^{-1}). The organic substrates used were glucose, lactose, DL-malate and succinate (0.5–1.0 g C.l^{-1}). All concentrations are given as the final concentration in the nutrient solution.

Oxygen consumption in the presence of respiratory inhibitors was measured with the O$_2$ electrode. The concentration of O$_2$ in one cylinder of each treatment was measured during consecutive AR assays. Ammonium, nitrite, and nitrate concentrations were determined colorimetrically. Acetylene reduction was determined as above.

Determination of soluble carbon and ammonium in roots of plants exposed to low pO$_2$

Soluble carbon in roots of plants exposed to low pO$_2$ was measured after extracting 1-gram samples of the roots with hot water. The samples were ground in a mortar and filtered extracts were measured for soluble carbohydrates using the anthrone method of Brink *et al.* (1960). Soluble carbon was determined using the dichromate method of Johnson (1949). Both are colorimetric methods. Absorbance values of extracts were compared to those of known concentrations of glucose, therefore the data are presented as glucose-equivalents. Ammonium in the extracts was measured using the catalyzed indophenol method of Chaney and Marbach (1962).

Effects of inoculating plants with nitrogen-fixing bacteria

Throughout a period of 6 years, numerous experiments were conducted in which greenhouse-grown maize plants were inoculated with various strains of *Azospirillum brasilense* (Cd, 13t), *Azospirillum lipoferum* (Br17, USA 5b) or enteric bacteria (*Klebsiella pneumoniae, Enterobacter cloacae*). Azospirilla were obtained from J. R. Milam, University of Florida, or J. Döbereiner, Km 47, Rio de Janeiro, Brazil.

Inocula consisted of cells grown in Tryptic Soy

broth (Difco), harvested by centrifugation, and re-suspended (twice) in sterile distilled water. Suspensions contained 10^6–10^8 cells ml^{-1}, and 1–5 ml of suspension was applied to the crown of plants growing in Weswood soil (autoclaved or nonautoclaved) or fritted clay (Alexander *et al.*, 1987).

Exposure of soil-grown, inoculated and uninoculated plants to reduced pO_2 and assays for AR were as described above. Plants grown in fritted clay were assayed *in situ* (roots undisturbed) as described by Alexander *et al.* (1987). Shoot and root dry weights were determined after drying to constant weight at 65°C.

Results

Effect of low pO_2 on root-associated AR

Acetylene reduction by roots of greenhouse-grown maize exposed to 0, 0.6, 1.3 and 2.1 kPa O_2 for 19 h is illustrated in Fig. 1a. Rates were greatest and not markedly different at 1.3 and 2.1 kPa after 1 h incubation with C_2H_2. Acetylene reduction was slower at 0.6 kPa O_2. In this experiment, AR at all O_2 levels commenced immediately and was linear for at least 1 h after which it declined at 1.3 and 2.1 kPa O_2. Oxygen consumption by a single plant exposed to 2.1 kPa O_2 is shown in Fig. 1b. Note that the decline in AR (Fig. 1a) coincides with the depletion of O_2 to near 0 after 1 h. Activity increased to the initial rate following adjustment of the pO_2 to 1.5 kPa and then declined as O_2 was depleted a second time.

Acetylene reduction by maize and sorghum plants exposed to 0, 1.5, 2.5, and 3.5 kPa O_2 is illustrated in Fig. 2. Both maize and sorghum showed greatest AR at 1.5 kPa O_2 followed by 2.5 kPa O_2. Greater AR was observed at 3.5 kPa O_2 than at 0 kPa O_2 with maize, while AR associated with sorghum was greater at 0 kPa than at 3.5 kPa O_2. In this particular pair of experiments, it was apparent that AR by sorghum roots was less sensitive to 0 and 3.5 kPa O_2 than that of maize roots. The sorghum plants were 13 days younger than the maize plants. The decrease in AR upon exposure to 3.5 kPa was striking.

Field-grown sorghum plants exhibited rates of AR nearly identical to those of greenhouse-grown

plants (Fig. 3). Acetylene reduction was immediate and linear for 2 h following 18 h at 2 kPa O_2. Cultivar Funk's G522DR showed greater rates of AR, however, it would require many more experiments to substantiate if this was truly a cultivar-specific trait. The results of this experiment confirmed those obtained with laboratory and greenhouse-grown plants.

Effects of inhibitors on root-associated AR

The effects of respiratory inhibitors on AR and O_2 uptake by the root-microbe association are illustrated in Fig. 4. Acetylene reduction was eliminated

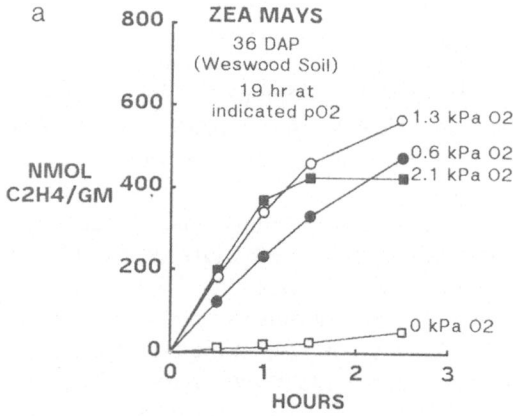

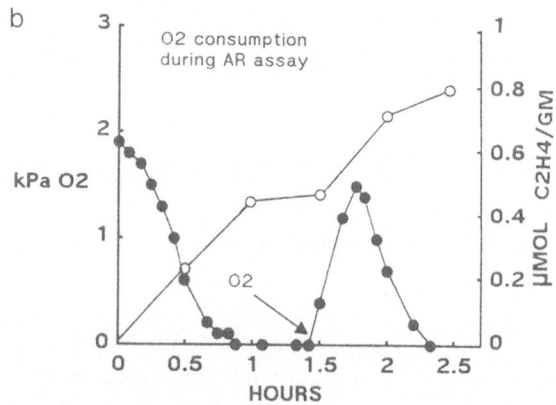

Fig. 1. AR activity and O_2 consumption associated with maize roots exposed to 0, 0.6, 1.3, or 2.1 kPa O_2 in the hydroponic apparatus. (**a**) AR activity after 19 h at the reduced pO_2. SE = 79.4, n = 3. (**b**) AR activity and O_2 consumption by a single plant exposed to 2.1 kPa O_2. (O) AR activity, (●) O_2 consumption. Plants were grown in the greenhouse and transferred to the hydroponic apparatus at 36 DAP.

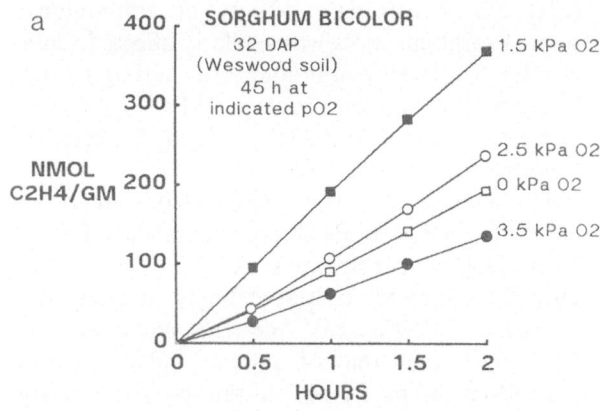

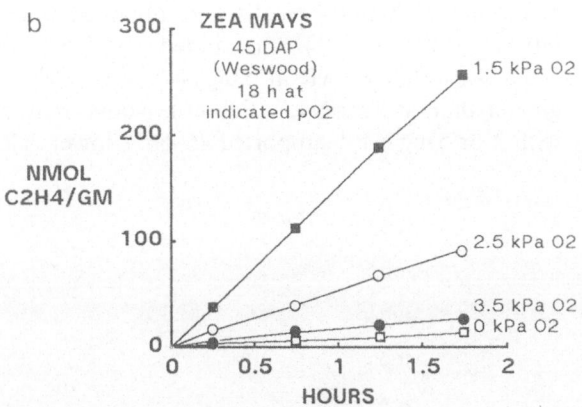

Fig. 2. AR activity associated with maize and sorghum roots exposed to 0, 1.5, 2.5, or 3.5 kPa O₂ in the hydroponic apparatus. (a) AR activity associated with sorghum roots after 45 h at the reduced pO₂. Plants were grown in the greenhouse and transferred to the hydroponic apparatus at 32 DAP. SE = 44.1, n = 3. (b) AR activity associated with maize roots after 18 h at the reduced pO₂. Plants were grown in the greenhouse and transferred to the hydroponic apparatus at 45 DAP. SE = 22.8, n = 3.

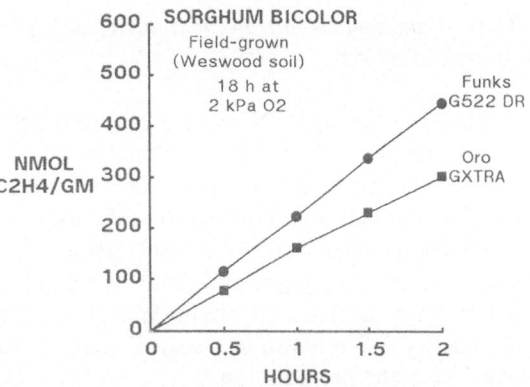

Fig. 3. AR activity associated with two varieties of sorghum grown in the field and exposed to 2 kPa O₂ in the hydroponic apparatus. Plants were grown at the Experimental Farm and transferred to the hydroponic apparatus at 28 DAP. SE = 49.5, n = 12.

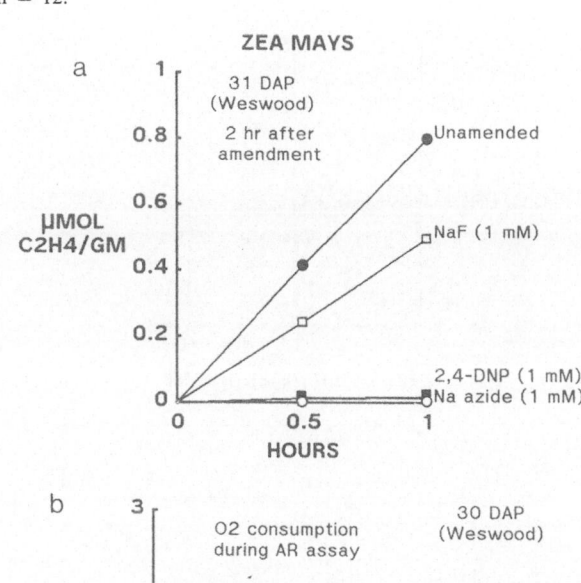

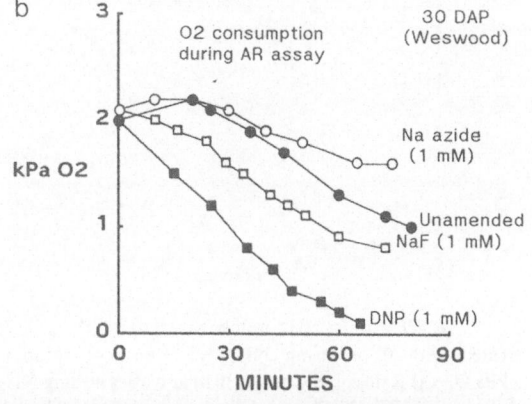

Fig. 4. AR activity and O₂ consumption associated with maize roots treated with sodium azide, DNP, or NaF. (a) AR activity 2 h after adding the inhibitors to the nutrient solution. SE = 0.077, n = 3. (b) O₂ consumption during an AR assay by roots of individual plants treated with the inhibitors. Plants were grown in the greenhouse and transferred to the hydroponic apparatus at 31 DAP.

by 1 m*M* DNP and sodium azide within 2 h of the addition of the inhibitors (Fig. 4a). Sodium fluoride only partially eliminated AR which was still linear in the presence of the inhibitor. DNP caused a substantial increase in O₂-consumption compared to untreated controls while sodium azide caused a decrease in O₂-consumption (Fig. 4b). These results were consistent with the known effects of the inhibitors on respiratory chain components. Rates of O₂-uptake by fluoride-treated and untreated plants were similar 20 min after initiation of treatment with the inhibitors.

Effects of ammonium, nitrate, and nitrite on root-associated AR

Addition of $16 \, mg \, l^{-1} \, N$ as ammonium quickly reduced AR (Fig. 5). Activity declined within 3 h after the addition of ammonium (Fig. 5b) and remained low until N was depleted from the solution, presumably by plant and microbial assimilation. Rates of AR were depressed until the level of ammonium-N declined to about $5 \, mg \, N \, l^{-1}$. The AR activity of ammonium-treated plants was greater than the pre-addition level once the NH_4^+ was depleted in the solution (Fig. 5b), however, it never reached rates equal to those of untreated plants. There was no significant difference in pre-treatment rates of AR for the two groups of plants

(Fig. 5a). Nitrogenase activity of ammonium-treated sorghum roots was similarly affected (data not shown). Twenty-four hours after adding 10 and $20 \, mg \, N \, l^{-1}$ to sorghum roots at $2 \, kPa \, O_2$, AR was depressed by 74% and 93%, respectively, relative to untreated controls.

The effects of nitrate and nitrite on root-associated AR at $2 \, kPa \, O_2$ are illustrated in Fig. 6. Increasing concentrations of nitrate-N (5–$20 \, mg \, l^{-1}$) caused correspondingly greater decreases in AR (Fig. 6a). Acetylene reduction was 62% lower than that of untreated plants within 1.5 h after adding $5 \, mg \, l^{-1}$ nitrate-N. The activity was immediate and linear before and after addition of the nitrate. Similar results were obtained when nitrite (2–$10 \, mg \, N \, l^{-1}$) was added to roots at $2 \, kPa \, O_2$, although AR at $5 \, mg \, N \, l^{-1}$ was slightly greater than at $2 \, mg \, N \, l^{-1}$ (Fig. 6b). Roots treated with 2 or $5 \, mg \, N \, l^{-1}$ supported 46–56% lower AR

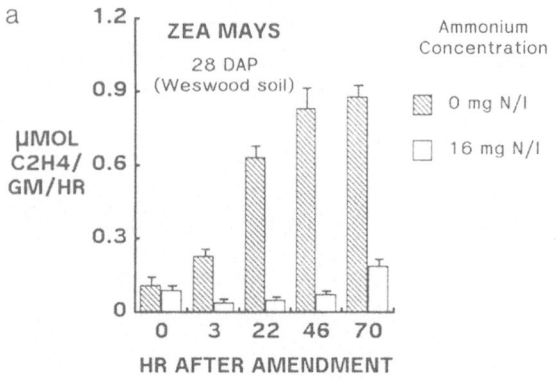

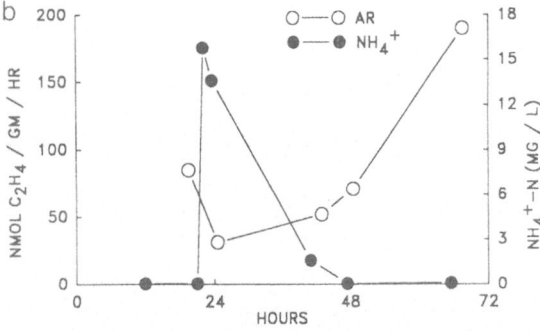

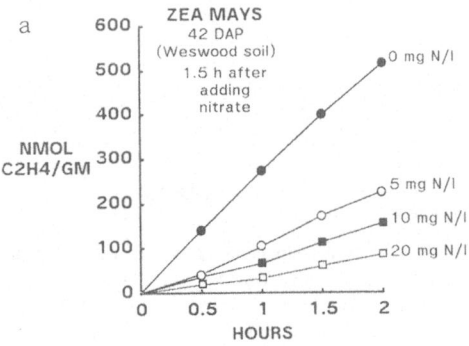

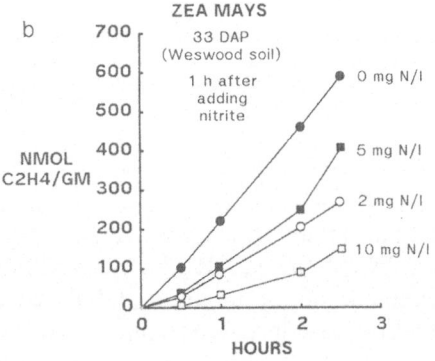

Fig. 5. AR activity and NH_4^+ uptake associated with maize roots treated with 0 or $16 \, mg \, NH_4^+ \text{-} N \, l^{-1}$ nutrient solution at $2 \, kPa \, O_2$. (a) Rates of AR before (0 h) and after adding NH_4^+ to the solution. The NH_4^+ was added after the roots had been exposed to $2 \, kPa \, O_2$ for 21 h. Data are means of six replications. Error bars are standard errors of the means. (b) AR activity (O) and NH_4^+ uptake (●) associated with the plants treated with $16 \, mg \, N \, l^{-1}$. SE = 0.103, n = 6 for NH_4^+ uptake data. SE = 61.6, n = 6 for AR data. Plants were grown in the greenhouse and transferred to the hydroponic apparatus at 28 DAP.

Fig. 6. AR activity associated with maize roots treated with nitrate or nitrite at $2 \, kPa \, O_2$. (a) Rates of AR 1.5 h after adding nitrate to the nutrient solution. SE = 26.7, n = 3. Plants were grown on the light table and transferred to hydroponic apparatus at 42 DAP. (b) AR activity 1 h after adding nitrite to the solution. SE = 94.8, n = 3. Plants were grown on the light table and transferred to hydroponic apparatus at 33 DAP.

than untreated roots 1 h after adding KNO_2 to the nutrient solution. The rapid decrease in AR upon the addition of nitrate or nitrite is noteworthy. As was observed with ammonium, nitrogenase activity of nitrate- and nitrite-treated roots increased as the concentration of N in solution decreased (data not shown).

The rates of removal of ammonium from solution differed according to the pO_2 of the system (Fig. 7). Removal (assimilation) of ammonium was significantly more rapid at 6 and $10\,kPa\,O_2$ than at 0 and $2\,kPa$. Disappearance of ammonium was slightly more rapid at $6\,kPa\,O_2$ than at $10\,kPa\,O_2$, though the difference was not statistically significant. Removal of ammonium was significantly less rapid at $0\,kPa\,O_2$ than at $2\,kPa\,O_2$, and ammonium was still present at levels sufficient to depress AR after 23 h with the anaerobic treatment (see Figs. 5 and 7). Ammonium was still detectable even after 71 h at $0\,kPa\,O_2$.

Effect of prolonged exposure to low pO_2 on root-associated AR and concentrations of carbohydrate and ammonium in roots

Over the course of several years of conducting AR assays on intact plants with roots at or near $2\,kPa\,O_2$, we have observed that AR usually increases during the period 24–48 h after initiating the low pO_2 treatment and often declines beyond this point (see Figs. 8 and 9 for example). We have

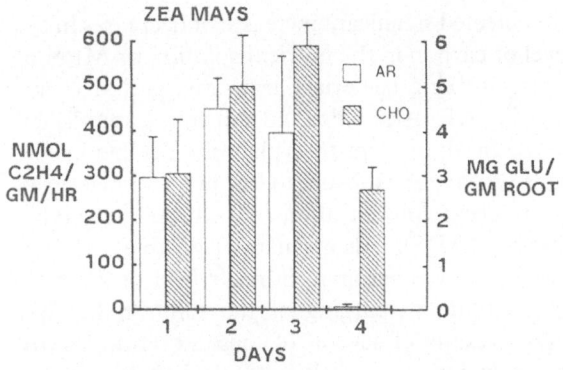

Fig. 8. Rates of AR and corresponding concentrations of ADS in maize roots exposed to $2\,kPa\,O_2$ for 4 days. Plants were grown in the greenhouse and transferred to the hydroponic apparatus at 45 DAP. Data are means of three replications. Error bars are standard errors of the means.

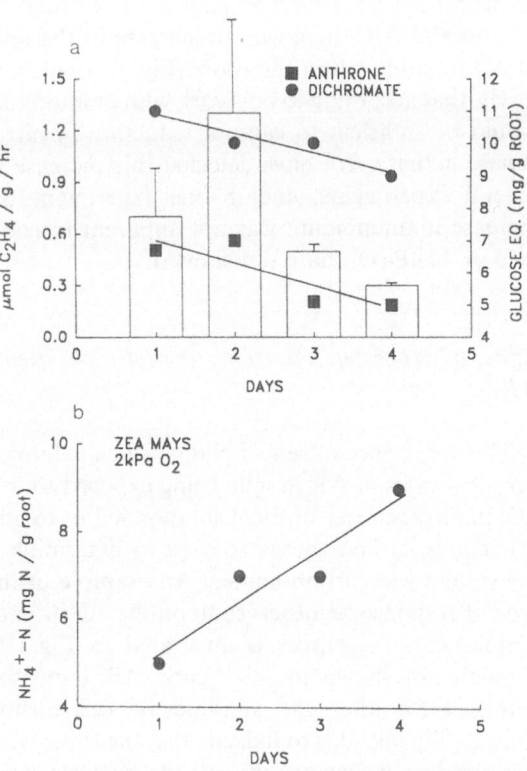

Fig. 9. Rates of AR and corresponding concentrations of (a) ADS and DOC, and (b) ammonium in maize roots exposed to $2\,kPa\,O_2$ for 4 days. Plants were grown in the greenhouse and transferred to the hydroponic apparatus at 21 DAP. Data are means of three replications. Error bars are standard errors of the means. SE = 0.43, n = 3 for ADS data. SE = 0.46, n = 3 for DOC data. SE = 0.99, n = 3 for ammonium data.

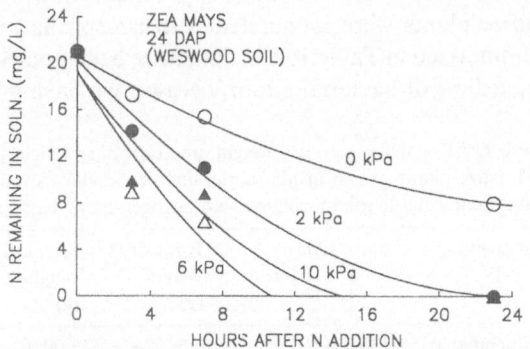

Fig. 7. Removal of ammonium from solution by maize roots exposed to 0, 2, 6, or $10\,kPa\,O_2$. Ammonium ($21\,mg\,N\,l^{-1}$) was added to the nutrient solution after the roots had been exposed to the reduced pO_2 for 88 h. Plants were grown in the greenhouse and transferred to the hydroponic apparatus at 24 DAP. Data are means of three replications.

not detected significant increases or decreases in the level of carbon in the nutrient solution or MPN of nitrogen-fixing bacteria during this period of increasing AR (Alexander and Zuberer, unpublished data). In an attempt to explain the decline in AR after prolonged exposure to low pO_2, we measured the internal concentration of anthrone-detectable sugars (ADS), dichromate-oxidizable carbon (DOC), and ammonium in roots held at $2\,kPa\,O_2$ by continuous sparging of the solution for 1–4 days. Results of several of these experiments are shown in Figs. 8 and 9. We observed a tendency for ADS to increase during the early phase of the exposure period and then to decline on or about day 4 (Fig. 8). Such an increase did not always occur, however. It is more apparent that the level of internal ADS and DOC tended to decline upon prolonged exposure to low pO_2 (Fig. 9a) and that AR also declined at or just before this point in time. Concomitant with the declining levels of internal carbon and AR, there was an increase in the level of ammonium within the roots (Fig. 9b) reaching levels, that according to our work with ammonium, would be sufficient to cause a reduction in nitrogenase activity. We have detected this increase in several experiments, and in one experiment the increase in ammonium was not apparent in roots held at $10\,kPa\,O_2$ (data not shown).

Effect of exogenous substrates on root-associated AR

The occurrence of sets of plants which exhibited very low rates of AR despite being exposed to low pO_2 in nitrogen-free nutrient solution led us to add exogenous carbon/energy sources to determine if the system was carbon-limited. An example of the type of response we observed upon the addition of various carbon sources is illustrated in Fig. 10. Though not shown in this figure, AR increased within 1–2 h after the addition of the carbon sources. We take this to indicate that the roots were colonized with diazotrophs but the bacteria were not able to reduce acetylene for lack of an ample supply of carbon. In this as well as in several other experiments, succinate proved most stimulatory to AR. It should be pointed out, however, that the lower rate of AR shown by the malate-amended plants was due to the use of the DL-malate. In

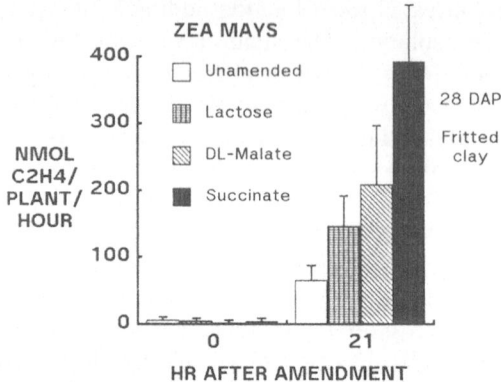

Fig. 10. Stimulation of AR activity associated with roots of intact maize plants by lactose, DL-malate, and succinate ($1.0\,g\,C\,l^{-1}$) at $2\,kPa\,O_2$. Activity at 0 h represents preamendment rates of AR. Plants were grown in fritted clay in the greenhouse and brought into the lab to measure AR at 28 DAP. Data are means of three replications. Error bars are standard errors of the means.

subsequent studies using L-malate as the carbon source, AR activity was equivalent to that of succinate-amended plants. The experiment illustrated in Fig. 10 was conducted at $2\,kPa\,O_2$. In similar experiments (data not shown) we observed increased AR levels at all pO_2 values up to and including $21\,kPa\,O_2$ (air). In general, organic acids and hexose sugars were most stimulatory to AR.

Effect of inoculation with N_2-fixing bacteria on root-associated AR

The results from at least 15 experiments in which maize plants were inoculated with azospirilla are summarized in Table 1. The data have been pooled, regardless of bacterial strain, because we have not

Table 1. AR activity, root dry weight, and shoot dry weight of 200 maize plants grown in nonsterile, uninoculated Weswood soil or in nonsterile soil inoculated with nitrogen-fixing bacteria

Treatment	nmol C_2H_4 $h^{-1}g^{-1}$ dry root	Root dry weight (g)	Shoot dry weight (g)
Uninoculated	133.7 a	1.63 a	2.00 a
Inoculated	93.6 b	1.80 a	2.06 a

Data are means of 100 replications from several experiments over a 3-year period.

Means in each column followed by the same letter are not significantly different at the alpha = 0.05 level (Duncan's multiple range test).

detected consistent strain-specific responses in AR or dry matter yield over a 3-year period in experiments with *A. brasilense* (13t, Cd) and *A. lipoferum* (USA 5b, Br17). It is evident in this data set that inoculation did not lead to greater rates of AR when plants were assayed at $2 kPa O_2$. In fact, AR was significantly greater among uninoculated plants colonized by indigenous diazotrophs. *Enterobacter cloacae*, *Klebsiella pneumoniae*, and *Azotobacter chroococcum* have been isolated from roots of maize plants grown in Weswood soil for these experiments. Acetylene reduction of inoculated and uninoculated plants was immediate and linear (data not shown).

We did not detect a significant increase in weights of shoots or roots resulting from inoculation. The dry weight of roots was 10% greater among inoculated plants, however, high plant-to-plant variation prevented statistical separation of the treatments. We have observed similar variation in rates of AR. Such high variation frequently confounds statistical detection of differences between treatments.

Discussion

Effect of pO₂ on root-associated AR

Oxygen partial pressure profoundly affected the expression of nitrogenase activity (AR) by bacteria associated with roots of intact maize and sorghum plants under controlled conditions. The highest rates of AR occurred within a narrow range of pO_2 between 1.3 and 2.1 kPa. At lower pO_2 AR activity decreased presumably because the bacteria associated with the roots were unable to generate enough energy through respiration to maintain a highly active nitrogenase system. At pO_2 higher than $2 kPa$ the toxic effects of O_2 limited nitrogenase activity. As little as $3.5 kPa O_2$ was sufficient to reduce AR by more than 50%. We have observed inhibition of root-associated AR by similar concentrations of O_2 in a solid rooting medium (Alexander *et al.*, 1987).

The response to pO_2 was quite sensitive, as evidenced by the decline in AR as O_2 was consumed during the AR assays and the immediate increase following the addition of a small amount of O_2 (see Fig. 1b). Our data indicate that the intact plant-root-microbe system responds to O_2 as would a microaerophilic organism. This is consistent with the microaerophilic nature of many diazotrophs stemming from the inherent O_2 sensitivity of the nitrogenase system. Our findings on the effect of O_2 on intact root-diazotroph associations have strong implications for the occurrence of associative N_2-fixation under field situations. If root-associated nitrogenase activity in the soil environment responds in a like manner, as our data with plants growing in fritted clay suggest it may (Alexander *et al.*, 1987), then it is difficult to imagine substantial activity occurring on roots in well-aerated soils. It seems more likely that under non-flooded conditions root-associated N_2-fixation is restricted to anoxic, hypoxic, or microaerobic microsites dispersed throughout the soil-root system. Data supporting this notion have been put forth by several investigators (Döbereiner *et al.*, 1973; Okon *et al.*, 1977; Nelson and Knowles, 1978; Haahtela *et al.*, 1983; Klugkist and Haaker, 1984) who have shown that lowered O_2 concentrations favour nitrogenase activity associated with washed excised roots or soil-grass cores.

Effects of inhibitors on root-associated AR

Acetylene reduction activity at $2 kPa O_2$ was closely linked to the ability of the N_2-fixing bacteria associated with the roots to respire O_2. Rates of AR decreased substantially when the O_2 in the cylinders was depleted during an AR assay or when a respiration inhibitor was added to the nutrient solution. The effects of the inhibitors are illustrated in the O_2-consumption data (Fig. 4b). Oxygen consumption ceased when the roots were treated with sodium azide because this compound blocks the transfer of electrons between cytochromes during oxidative phosphorylation. Dinitrophenol uncouples electron transport and ATP synthesis. Roots treated with DNP consumed O_2 very rapidly as reductant was channelled through the respiratory chain, but no ATP was produced due to the effect of the uncoupler. Both sodium azide and DNP eliminated AR activity associated with maize roots very rapidly ($< 2h$).

Sodium fluoride inhibits fermentation but has no effect on respiratory O_2-consumption (Saglio *et al.*, 1983). AR activity at $2 kPa O_2$ was partially inhibited by NaF, indicating that some of the activity was produced by bacteria growing fermentatively. Considering the mode of action of NaF (F^- inhibits enolase activity by forming an insoluble salt with Mg^+, a cofactor of the enzyme; Dawes and Large, 1982), however, it is difficult to conceive how respiration would be unaffected by this compound. The decrease in AR activity following treatment with NaF may also have been a result of impaired respiration.

Ammonium and nitrate strongly inhibited AR activity associated with roots of intact maize and sorghum plants at $2 kPa O_2$. Concentrations of ammonium as low as $4 mg N l^{-1}$ reduced AR activity by 93%, and concentrations of nitrate as low as $5 mg N l^{-1}$ reduced AR by 60%. Nitrate was somewhat less inhibitory probably because it affected nitrogenase activity only after being reduced to ammonium, though recently it has been suggested that nitrate may affect the enzyme directly (Hom *et al.*, 1980). Greater inhibition occurred with increasing concentrations of combined N.

Inhibition occurred rapidly (within 2–3 h), suggesting that a switch-off mechanism regulated AR activity in addition to repression of nitrogenase synthesis. This type of regulation has been described previously for ammonium (Drozd *et al.*, 1972; Laane *et al.*, 1980) and nitrate (Cejudo and Paneque, 1986) in pure cultures of *Azotobacter*, and for ammonium on excised roots of *Spartina alterniflora* (Yoch and Whiting, 1986). If repression alone were operating, AR activity would not decrease immediately after adding combined N to the nutrient solution. The activity would remain high until the existing pool of nitrogenase was diluted through attrition.

Similar levels of inorganic N have been shown to inhibit AR activity associated with excised roots and soil-grass cores (Day *et al.*, 1975; Neyra *et al.*, 1976; van Berkum *et al.*, 1976; Neyra and van Berkum, 1976; Ogan, 1983).

The inhibition of AR activity by ammonium and nitrate was transient. Rates of AR associated with the treated plants increased as the added N was removed from solution, presumably through assimilation by the roots and associated microorganisms. Acetylene reduction activity returned to the level exhibited by the untreated control plants within 24–48 h after adding 4–$5 mg N l^{-1}$ to the solution. Recovery was slower with higher concentrations of N.

Effects of prolonged exposure to low pO_2

Oxygen tension greatly affected the rate at which the roots took up ammonium. When $21 mg N l^{-1}$ was added to the nutrient solution, roots exposed to 6 or $10 kPa O_2$ removed more than half of the ammonium from solution within 3 h, while roots exposed to $0 kPa O_2$ removed only 18% of the ammonium. All of the added N was taken up within 23 h at the higher pO_2. Under anaerobic conditions some N remained in solution for more than 71 h. The rate of ammonium-assimilation was lower at $2 kPa O_2$ than at 6 or $10 kPa O_2$, but it was still sufficient to remove $21 mg N l^{-1}$ within 23 h.

Rates of AR varied considerably from one day to the next among plants exposed to reduced pO_2 for several days, suggesting that other environmental factors regulated AR when the pO_2 was maintained at favourable levels. Rates of AR consistently increased on the second day of exposure to $2 kPa O_2$, and decreased after 3 or more days. Acetylene reduction activity developed more slowly at $0 kPa O_2$, but rivalled the activity at $2 kPa O_2$ after 3 or more days of treatment. Oxygen deficiency causes roots to leak organic compounds and inorganic ions by disrupting the integrity of cell membranes (Grineva, 1961; Hiatt and Lowe, 1967). It is likely, then, that the initial increases in AR activity at low pO_2 were due at least in part to an increased supply of organic substrates on the roots. Nitrogen-fixing bacteria may have proliferated on the 'leaky' roots or the existing population may have increased its activity by generating more ATP and reductant with the increased supply of substrates.

Since we frequently observed a decline in AR during prolonged exposure of roots to low pO_2, we measured daily levels of anthrone-detectable sugars (ADS), dichromate-oxidizable carbon (DOC), and ammonium in the roots. We observed a general trend for levels of carbon to decrease during the period of low pO_2. Transport of fixed carbon from shoots to primary roots is inhibited by reduced pO_2 (Drew and Sisworo, 1979). This may explain the decrease in AR activity at $2 kPa O_2$ after 3 or more days of treatment. Once the initial supply of organic substrates leaked by the roots was exhausted, little more became available because the flow of fixed carbon into the roots decreased. We hypothesize that reduced pO_2 enhanced root-associated AR for short periods of time by enabling nitrogenase to function at a high rate, but prolonged exposure harmed the host plant such that the high activity could not be maintained for more than a few days.

The rapid response of 'AR-negative' plants to the addition of readily available carbon sources supports the findings above that the pool of available carbon is perhaps *the* principal limiting factor for high rates of associative N_2-fixation. Further, the observation that AR is markedly increased by the addition of exogenous carbon, even at $21\,kPa\,O_2$, suggests that while pO_2 strongly governs rates of AR it is the interaction of carbon level and pO_2 which is important in the expression of root-associated nitrogenase activity. Therefore, it is very important that we learn why some plants are considerably more active than others even though they are grown under seemingly identical conditions.

Along with the decline in ADS and DOC we observed increases in ammonium within the roots. The levels of ammonium which we detected would be sufficient to interfere with nitrogenase activity if encountered by the bacteria. Recall that AR was suppressed by as little as $4.2\,mg\,N\,l^{-1}$ supplied to roots at $2\,kPa\,O_2$. Declining carbon reserves and increasing levels of ammonium would almost certainly combine to restrict nitrogenase activity. van Berkum and Sloger (1983) reported similar findings with wild rice (*Zizania aquatica* L.).

There have been many reports in the literature of field and greenhouse studies in which inoculation with *Azospirillum* increased the dry matter yield of grasses. Some researchers have measured increases in shoot dry weight, total plant dry weight, and total N in maize, wheat, foxtail millet, and *Panicum maximum* (Okon, 1982). In some studies they have observed increases in the N content of the plants (Cohen *et al.*, 1980; Kapulnik *et al.*, 1981), but this response frequently does not occur (Okon *et al.*, 1983; Avivi and Feldman, 1982; Schank *et al.*, 1983). Okon (1985) reports a 75% success rate in obtaining yield increases following inoculation of summer crops with *Azospirillum*.

Other researchers, in addition to ourselves, have obtained conflicting results. While there have been some reports of positive responses to inoculation

(Albrecht *et al.*, 1981; O'Hara *et al.*, 1981), in many cases there was no response (Smith *et al.*, 1984; Albrecht *et al.*, 1981; Barber *et al.*, 1979).

In our research programme we have been unable to increase N_2-fixation by inoculating maize or sorghum with any of several diazotrophic bacteria, including *A. brasilense*, *A. lipoferum*, *Enterobacter*, and *Klebsiella*. We failed to detect consistent increases in AR associated with inoculated plants even when they were assayed at low pO_2. In fact, the pooled data from many individual experiments (100 inoculated and 100 uninoculated plants) over a period of several years showed statistically lower rates of AR in assays of inoculated plants conducted at $2\,kPa\,O_2$. Root growth was occasionally increased following inoculation with *Azospirillum*. In two experiments inoculated plants exhibited significantly greater root dry weights than uninoculated plants. Shoot dry weights were not significantly different in any of the experiments comprising the pooled data set. Thus it is clear that inoculating maize and sorghum plants with N_2-fixing bacteria did not lead to any consistent crop response in our experiments. The lack of response was most likely due to the failure of inoculants to establish and compete with indigenous bacteria or due to stringent limitations by soil environmental conditions.

Acknowledgements

We thank Dr. Sara Wright for the diazotrophic isolates of the Enterobacteriaceae. We are also grateful to the organizers of the Fourth International Symposium on Nitrogen Fixation with Nonlegumes for the opportunity to present this research and to the editors and reviewers for their helpful suggestions regarding the manuscript!

References

Albrecht S L, Okon Y, Lonnquist J and Burris R H 1981 Nitrogen fixation by corn-*Azospirillum* associations in a temperate climate. Crop Sci. 21, 301–306.

Alexander D B, Zuberer D A and Vietor D M 1987 Nitrogen fixation (acetylene reduction) associated with roots of intact *Zea mays* in fritted clay at reduced oxygen tensions. Soil Biol. Biochem. 19, 1–6.

Avivi Y and Feldman M 1982 The response of wheat to bacteria of the genus *Azospirillum*. Israel J. Bot. 31, 237–245.

Balandreau J, Ducerf P, Hamad-Fares I, Weinhard P, Rinaudo G, Millier C and Dommergues Y 1978 Limiting factors in grass nitrogen fixation. *In* Limitations and Potentials for Biological Nitrogen Fixation in the Tropics. Eds. J Döbereiner, R H Burris and A Hollaender. pp. 275–302. Plenum Press, New York.

Baldani J I, Baldani V L D, Seldin L and Döbereiner J 1986 Characterization of *Herbaspirillum seropedicae*, gen. nov., sp. nov., a root-associated nitrogen-fixing bacterium. Int. J. Syst. Bacteriol. 36, 86–93.

Barber L E, Russell S A and Evans H J 1979 Inoculation of millet with *Azospirillum*. Plant and Soil 52, 49–57.

Brink R H, Dubach P and Lynch D L 1960 Measurement of carbohydrates in soil hydrolysates with anthrone. Soil Sci. 89, 157–166.

Burris R H 1977 A synthesis paper on nonleguminous N_2-fixing systems. *In* Recent Developments in Nitrogen Fixation. Eds. W Newton, J R Postgate and C Rodriguez-Barrueco. pp. 487–511. Academic Press, London.

Cejudo F J and Paneque A 1986 Short-term nitrate (nitrite) inhibition of nitrogen fixation in *Azotobacter chroococcum*. J. Bacteriol. 165, 240–243.

Chaney A L and Marbach E P 1962 Modified reagents for determination of urea and ammonia. Clin. Chem. 8, 130–132.

Cohen E, Okon Y, Kigel J, Nur I and Henis Y 1980 Increase in dry weight and total nitrogen content in *Zea mays* and *Setaria italica* associated with nitrogen-fixing *Azospirillum* spp. Plant Physiol. 66, 746–749.

Dawes E A and Large P J 1982 Class I reactions: Supply of carbon skeletons. *In* Biochemistry of Bacterial Growth. Eds. J Mandelstam, K McQuillen and I Dawes. pp. 125–158. Blackwell Scientific Publications, Oxford.

Day J M, Neves C P and Döbereiner J 1975 Nitrogenase activity on the roots of tropical forage grasses. Soil Biol. Biochem. 7, 107–112.

Döbereiner J, Day J M and Dart P J 1973 Rhizosphere associations between grasses and nitrogen-fixing bacteria: Effect of O_2 on nitrogenase activity in the rhizosphere of *Paspalum notatum*. Soil Biol. Biochem. 5, 157–159.

Drew M C and Sisworo E J 1979 The development of water-logging damage in young barley plants in relation to plant nutrient status and changes in soil properties. New Phytol. 82, 301–314.

Drozd J W, Tubb R S and Postgate J R 1972 A chemostat study of the effect of fixed nitrogen sources on nitrogen fixation, membranes and free amino acids in *Azotobacter chroococcum*. J. Gen. Microbiol. 73, 221–232.

Grineva G M 1961 Excretion by plant roots during brief periods of anaerobiosis. Soviet Plant Physiol. 8, 549–552 (English translation).

Haahtela K, Kari K and Sundman V 1983 Nitrogenase activity (acetylene reduction) of root-associated, cold-climate *Azospirillum, Enterobacter, Klebsiella*, and *Pseudomonas* species during growth on various carbon sources and at various partial pressures of oxygen. Appl. Environ. Microbiol. 45, 563–570.

Hiatt A J and Lowe R H 1967 Loss of organic acids, amino acids, K and Cl from barley roots treated anaerobically and with metabolic inhibitors. Plant Physiol. 42, 1731–1736.

Hoagland D R and Arnon D I 1950 The water-culture method for growing plants without soil. Calif. Agric. Exp. Sta. Circ. No. 347.

Hom S S M, Hennecke H and Shanmugam K T 1980 Regulation of nitrogenase biosynthesis in *Klebsiella pneumoniae*: Effect of nitrate. J. Gen. Microbiol. 177, 169–179.

Johnson M J 1949 A rapid micromethod for estimation of non-volatile organic matter. J. Biol. Chem. 181, 707–711.

Kapulnik Y, Kigel J, Okon Y, Nur I and Henis Y 1981 Effect of *Azospirillum* inoculation on some growth parameters and N-content of wheat, sorghum and panicum. Plant and Soil 61, 65–70.

Klugkist J and Haaker H 1984 Inhibition of nitrogenase activity by ammonium chloride in *Azotobacter vinelandii*. J. Bacteriol. 157, 148–151.

Laane C, Krone W, Konings W N, Haaker H and Veeger C 1980 Short-term effect of ammonium chloride on nitrogen fixation by *Azotobacter vinelandii* and by bacteroids of *Rhizobium leguminosarum*. Eur. J. Biochem. 103, 39–46.

Magalhães F M, Baldani J I, Souto S M, Kuykendall J R and Döbereiner J 1983 A new acid-tolerant *Azospirillum* species. Anais Academia Brasiliera Ciência 55, 417–430.

Nelson L M and Knowles R 1978 Effect of oxygen and nitrate on nitrogen fixation and denitrification by *Azospirillum brasilense* grown in continuous culture. Can. J. Microbiol. 24, 1395–1403.

Neyra C A and Döbereiner J 1977 Nitrogen fixation in grasses. *In* Advances in Agronomy vol. 29. Ed. N C Brady. pp. 1–38. Academic Press, New York.

Neyra C A, Pereira P A, von Bülow J F W and Döbereiner J 1976 11th Reunião Brasileira Milho Sorgo, Piracicaba, São Paulo, Brazil. Cited by Neyra C A and Döbereiner J (1977) Nitrogen fixation in grasses. Advances in Agronomy vol. 29. Ed. N C Brady. pp. 1–38. Academic Press, New York.

Neyra C A and van Berkum P 1976 11th Reunião Brasileira Milho Sorgo, Piracicaba, São Paulo, Brazil. Cited by Döbereiner J (1977) Physiological aspects of N_2 fixation in grass-bacteria associations. *In* Recent Developments in Nitrogen Fixation. Eds. W Newton, J R Postgate and C Rodriguez-Barrueco. pp. 513–529. Academic Press, London.

O'Hara G W, Davey M R and Lucas J A 1981 Effect of inoculation of *Zea mays* with *Azospirillum brasilense* strains under temperate conditions. Can. J. Microbiol. 27, 871–877.

Ogan M T 1983 Factors affecting nitrogenase activity associated with marsh grasses and their soils from eutrophic lakes. Aquatic Bot. 17, 215–230.

Okon Y 1982 *Azospirillum* physiological properties, mode of association with roots and its application for the benefit of cereal and forage grass crops. Israel J. Bot. 31, 214–220.

Okon Y 1985 *Azospirillum* as a potential inoculant for agriculture. Trends Biotechnol. 3, 223–228.

Okon Y, Heytler P G and Hardy R W F 1983 N_2 fixation by *Azospirillum brasilense* and its incorporation into host *Setaria italica*. Appl. Environ. Microbiol. 46, 694–697.

Okon Y, Houchins J P, Albrecht S L and Burris R H 1977 Growth of *Spirillum lipoferum* at constant partial pressures of oxygen, and the properties of its nitrogenase in cell-free extracts. J. Gen. Microbiol. 98, 87–93.

Saglio P H, Raymond P and Pradet A 1983 Oxygen transport

and root respiration of maize seedlings. A quantitative approach using the correlation between ATP/ADP and the respiration rate controlled by oxygen tension. Plant Physiol. 72, 1035–1039.

Schank S C, Weier K L and MacRae I C 1981 Plant yield and nitrogen content of a digitgrass in response to *Azospirillum* inoculation. Appl. Environ. Microbiol. 41, 342–345.

Smith R L, Schank S C, Milam J R and Baltensperger A A 1984 Responses of *Sorghum* and *Pennisetum* species to the N₂-fixing bacterium *Azospirillum brasilense*. Appl. Environ. Microbiol. 47, 1331–1336.

Tarrand J J, Krieg N R and Döbereiner J 1978 A taxonomic study of the *Spirillum lipoferum* group, with descriptions of a new genus, *Azospirillum* gen. nov., and two species, *Azospirillum lipoferum* (Beijerinck) comb. nov. and *Azospirillum brasilense* sp. nov. Can. J. Microbiol. 24, 967–980.

van Berkum P and Bohlool B B 1980 Evaluation of nitrogen fixation by bacteria in association with roots of tropical grasses. Microbiol. Rev. 44, 491–517.

van Berkum P, Neyra C A and von Bülow J F W 1976 11th Reunião Brasileira Milho Sorgo, Piracicaba, São Paulo, Brazil. Cited by Döbereiner J (1977) Physiological aspects of N₂ fixation in grass-bacteria associations. *In* Recent developments in Nitrogen Fixation. Eds. W Newton, J R Postgate and C Rodriguez-Barrueco. pp. 513–529. Academic Press, London.

van Berkum P and Sloger C 1983 Interaction of combined nitrogen with the expression of root-associated nitrogenase activity in grasses and with the development of N₂ fixation in soybean (*Glycine max* L. Merr.). Plant Physiol. 72, 741–745.

Yoch D C and Whiting G J 1986 Evidence for NH₄⁺ switch-off regulation of nitrogenase activity by bacteria in salt marsh sediments and roots of the grass *Spartina alterniflora*. Appl. Environ. Microbiol. 51, 143–149.

Zuberer D A and Alexander D B 1986 Effects of oxygen partial pressure and combined nitrogen on N₂-fixation (C₂H₂) associated with *Zea mays* and other gramineous species. Plant and Soil 90, 47–58.

Session 10

Effects of plant genotype and
environmental factors on plant-
associated N_2-fixation

F. A. Skinner et al. (Eds.), Nitrogen fixation with non-legumes, 289–299.
© 1989 by Kluwer Academic Publishers.

Natural ^{15}N abundance as a method of estimating the contribution of biologically fixed nitrogen to N$_2$-fixing systems: Potential for non-legumes*

GEORGIA SHEARER and D.H. KOHL

Department of Biology, Washington University, St. Louis, MO 63130, USA

Key words: biological nitrogen fixation, ^{15}N, non-legumes

Abstract

The ^{15}N abundance of plants usually closely reflects the ^{15}N abundance of their major immediate N source(s); plant-available soil N in the case of non-N$_2$-fixing plants and atmospheric N$_2$ in the case of N$_2$ fixing plants. The ^{15}N abundance values of these sources are usually sufficiently different from each other that a significant and systematic difference in the ^{15}N abundance between the two kinds of plants can be detected. This difference provides the basis for the natural ^{15}N abundance method of estimating the relative contribution of atmospheric N$_2$ to N$_2$-fixing plants growing in natural and agricultural settings. The natural ^{15}N abundance method has certain advantages over more conventional methods, particularly in natural ecosystems, since disturbance of the system is not required and the measurements may be made on samples dried in the field. This method has been tested mainly with legumes in agricultural settings. The tests have demonstrated the validity of this method of arriving at semi-quantitative estimates of biological N$_2$-fixation in these settings. More limited tests and applications have been made for legumes in natural ecosystems. An understanding of the limits and utility of this method in these systems is beginning to emerge. Examples of systematic measurements of differences in ^{15}N abundance beween non-legume N$_2$-fixing systems and neighbouring non-fixing systems are more unusual. In principle, application of the method to estimate N$_2$-fixation by nodulated non-legumes, using the natural ^{15}N abundance method, is as feasible as estimating N$_2$-fixation by legumes. Most of the studies involving N$_2$-fixing non-legumes are with this type of system (*e.g.*, *Ceanothus, Chamabatia, Eleagnus, Alnus, Myrica,* and so forth). Results of these studies are described. Applicability for associative N$_2$-fixation is an empirical question, the answer to which probably depends upon the degree to which fixed N goes predominantly to the plant rather than to the soil N pool. The natural ^{15}N abundance method is probably not well suited to assessing the contribution of N$_2$-fixation by free-living microorganisms in their natural habitat, particularly soil microorganisms.

Introduction

Atmospheric N$_2$ is the ultimate source of N to the biosphere in most places in the world. This N undergoes numerous transformations, sometimes repeatedly over many years, before it is returned to the atmosphere or is incorporated into sediments. This cycling with its associated isotopic fractionation results in alteration of the natural abundance

of ^{15}N of the ultimate N source as it is transformed into non-atmospheric nitrogen to plants (*e.g.*, plant-available soil N). For this reason, the ^{15}N abundance of atmospheric N$_2$ is often distinct from that of other N sources. The natural ^{15}N abundance method for estimating the fractional contribution of biologically fixed N to nitrogen fixing systems is based on small, but easily measurable differences in ^{15}N abundance between atmospheric N$_2$ and other sources of N. The natural ^{15}N abundance method has been discussed in detail elsewhere (Kohl *et al.*, 1980; Mariotti *et al.*, 1983; Rennie and Rennie,

* This work was supported in part by subcontracts under grants from the US National Science Foundation (DEB79-21971 and BSR821618)

1983; Shearer and Kohl, 1986, 1987) and will be only briefly described here. Before doing so, it is useful to define the units used-per mil (‰) ^{15}N-in expressing ^{15}N abundance.

$$\delta^{15}N = 1000 \times \frac{R_{spl} - R_{std}}{R_{std}}‰$$

where the subscripts spl and std refer to experimental sample and standard sample (usually atmospheric N_2), and R is the quotient of the concentration of ^{15}N to the total N in the sample of interest. This may be written as R = (^{15}N/^{15}N + ^{14}N). (This definition of R is slightly different than that used by geochemists, who define R as equal to ^{15}N/^{14}N. But at natural ^{15}N abundance, the difference in the values of R by these two definitions is negligible.)

Brief description of the δ^{15}N method

Basis of the method

The method is exactly analogous to the more familiar isotope dilution method in which artificially enriched fertilizers are applied to the soil. In the natural abundance method, however, an endogenous rather than an added tracer is used. Figure 1 illustrates the principle. The horizontal line represents ^{15}N abundance. The ^{15}N abundance of biologically fixed N, N of nitrogen fixing plants A and B, and N in the N_2-fixing plant derived from sources other than atmospheric N_2 are shown along this axis. The ^{15}N abundance of plant A is much closer to that of bilogically fixed N than to that of N derived from other sources (primarily plant-available soil N in the case of terrestrial plants),

while the opposite is true of plant B. It seems reasonable to conclude that most of the N of plant A is derived from atmospheric N_2 and most of the N of plant B is derived from other sources. Notice that the direction of the difference between the two poles of the diagram is not specified. Soil N is usually more abundant in ^{15}N than is atmospheric N_2 (*e.g.,* Shearer *et al.,* 1978) but this is not universally the case (Shearer and Kohl, 1986; Binkley *et al.,* 1985).

In applying the method, the first question is how to establish the δ^{15}N value of the poles. It is inadequate to measure the δ^{15}N of atmospheric N_2 and soil N (in the case of terrestrial plants) because isotopic fractionation occurring during the processes and transformations preceding incorporation of N from these sources into plant tissues may cause isotopic alteration, and not all of the soil N is equally available to the plant. The δ^{15}N of biologically fixed N can be established by measuring the δ^{15}N of N_2-fixing plants which are forced to rely almost entirely on biological N_2-fixation by growing them hydroponically with N-free medium. (It may be necessary to correct the result for the input of seed N if the seeds are large or the plants small.) Figure 2 shows the frequency distribution of 35 measurements of δ^{15}N values for biologically fixed N. The histogram includes measurements of 15 legume species, four species of *Azotobacter* and one N_2-fixing non-legume species made in seven different laboratories. Almost all of the values are within 2‰ of atmospheric N_2. That is, isotopic alteration due to N_2 fixation is minimal. However, in many cases, the difference between the δ^{15}N value of fixed N and that of atmospheric N_2 is large enough that isotopic fractionation associated with N_2-fixation must be taken into account in arriving

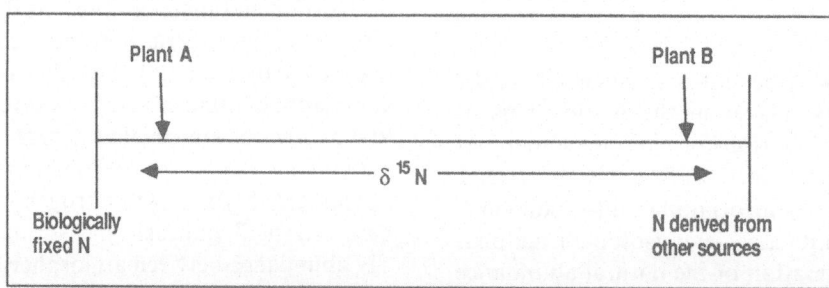

Fig. 1. Hypothetical diagram showing ^{15}N abundance of biologically fixed N, N of N_2-fixing plants A and B, and N derived from sources other than atmospheric N_2. In this illustration, about 90% of the N of plant A is derived from N_2-fixation and only about 10% of the N of plant B is derived from N_2-fixation.

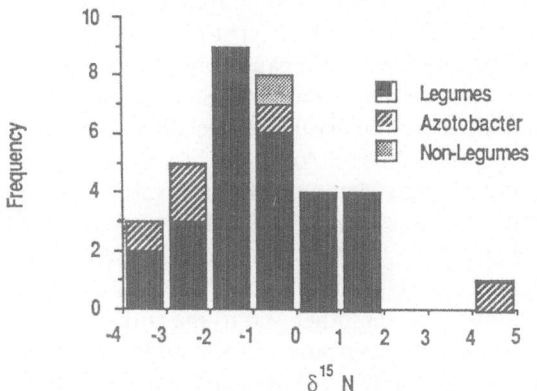

Isotopic Freactionation Associated
With Biological Nitrogen-Fixation

Fig. 2. Isotopic fractionation associated with N_2-fixation. In the absence of isotopic fractionation the $\delta^{15}N$ of the N_2-fixing organisms should be zero, the value for atmospheric N_2. The histogram includes 15 legume species, 4 species of *Azotobacter* and one N_2-fixing non-legume. Measurements were made in several different laboratories and reported by Hoering and Ford (1960), Delwiche and Steyn (1970), Kohl and Shearer, 1980, Mariotti *et al.* (1980), Bergersen and Turner (1983), Shearer *et al.* (1983), Steele *et al.* (1983), Domenach and Corman (1984), and Shearer and Kohl (1986).

at an estimate of the fractional contribution of fixed N, especially at sites in which the difference in ^{15}N abundance between atmospheric N_2 and N derived from other sources is small. There is a modest variation in the isotopic fractionation associated with N_2-fixation among plant species and cultivars (Amarger *et al.*,1979; Shearer and Kohl, 1986). It is therefore important to measure the $\delta^{15}N$ of fixed N in all of the cultivars of interest.

The question of the appropriate measure of the $\delta^{15}N$ of N derived from sources other than atmospheric N_2 is more complex. For terrestrial systems, some investigators use the $\delta^{15}N$ value of total soil N. This requires the assumption that the $\delta^{15}N$ of plant N that is derived from the soil is the same as that of total soil N. That this assumption is invalid, at least in some cases, is illustrated by Fig. 3 from a paper by Ledgard *et al.* (1984) which shows that the $\delta^{15}N$ of total soil N declined with depth in the profile, but at all depths was considerably higher than that of ryegrass (*Lolium rigidum*) grown on the soil. Inorganic N extracted from the soil at the beginning of the experiment was substantially lower in ^{15}N than were the plants, while inorganic

N mineralized over a period of 55 days (the same length of time that the plants were grown) had $\delta^{15}N$ values close to that of the ryegrass, regardless of depth. There was much smaller variation of $\delta^{15}N$ of plant N and mineralizable N than of toal N. The discrepancy between the $\delta^{15}N$ value of plants and the inorganic N extracted at the beginning of the experiment was most likely due to a change over time in the $\delta^{15}N$ of mineral N. Turner *et al.* (1987) measured the $\delta^{15}N$ and mineral N extracted from fallow and cropped soil as a function of time. The values for total soil N were rather constant and about the same in both soils ($\sim 8\permil$). Figure 4, taken from their paper, shows that the $\delta^{15}N$ values of mineral N extracted from fallow soil were also rather constant at about 5.5‰, but the $\delta^{15}N$ of mineral N extracted from soil under oats increased dramatically during the growing season (May through December in Australia where the experiment was done). The $\delta^{15}N$ of the increment of plant N [oats (*Avena sativa*) and wild mustard (*Brassica* sp.)] between emergence and the first harvest and between successive harvests matched closely that of the average $\delta^{15}N$ of mineral N between emergence and the first harvest or between successive harvests. Given the lack of correspondence between the $\delta^{15}N$ of total soil N and plant N, and given the time variation in the $\delta^{15}N$ of mineral

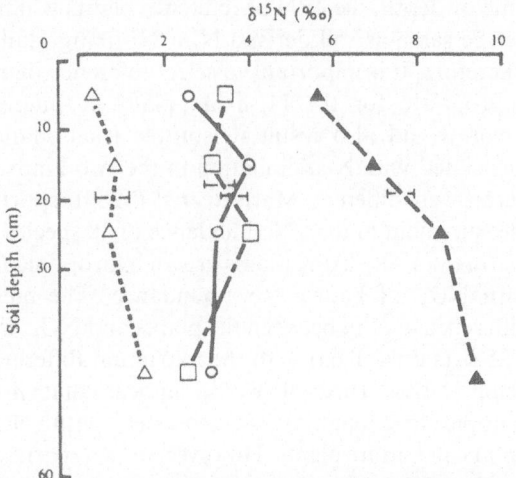

Fig. 3. Variation in the natural abundance of ^{15}N of plant and soil fractions: inorganic N at day 0 ($\triangle$), N mineralized during 55 days ($\square$), ryegrass (*Lolium rigidum*) N after 55 days growth on the same soil (O), and total soil N ($\blacktriangle$). Error bars denote one standard error on either side of the mean. From Ledgard *et al.* (1984)

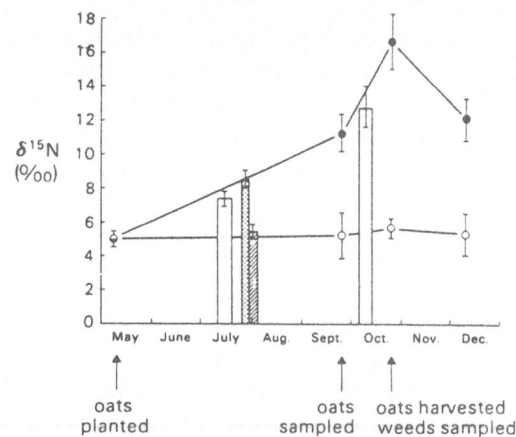

Fig. 4. ^{15}N abundance of extractable mineral N in continuously fallowed soil (O) and cropped soil (●). Histograms show values for the increments of plant N plotted at the midpoint between emergence and harvests and between successive harvests. Unmarked histograms, oats; hatched histogram, wild mustard in fallowed plots; dotted histograms, wild mustard in cropped soil. Error bars denote one standard error on either side of the mean. From Turner *et al.* (1987).

soil N, the most reasonable measure of the $\delta^{15}N$ value of soil-derived N within the N_2-fixing plant is the $\delta^{15}N$ value of a non-N_2 fixing reference plant. However, this measure is subject to error. For example, if the N_2-fixing plant takes N from the soil at different times or from different depths, and if the $\delta^{15}N$ of the plant-available soil N varies with time or depth, the $\delta^{15}N$ of reference plants will not be the same as soil derived N in N_2-fixing plants. Therefore, it is important to select reference plants appropriate for the N_2-fixing plant of interest. Error would also result if isotopic fractionation associated with N assimilation in the two kinds of plants were different. Mariotti *et al.* (1980) reported measurement of the ^{15}N abundance of 38 species of mature non-N_2-fixing plants grown hydroponically with NO_3^- of known ^{15}N abundance. The mean difference in $\delta^{15}N$ between the plants and NO_3^- was 0.25‰ (s.d. = 0.65) with the maximum difference being 2.2‰. Thus, it would appear that little isotopic fractionation is associated with NO_3^- uptake in mature plants. However, in two species of pearl millet (*Pennisetum americanum* and *P. mollissimum*), there was substantial fractionation in young seedlings. This fractionation diminished as the plants matured (Mariotti *et al.*, 1982). A strategy for dealing with the problem of selecting appropriate control plants is discussed later.

The next question concerns the most appropriate plant part for analysis. Ideally, the entire plant should be analyzed. This would eliminate any problem connected with isotopic fractionation during metabolic processes. Transport of isotorpically altered metabolites might cause variation in isotopic abundance among plant parts. Sampling the entire plant might not be feasible, especially in the case of large plants. The distribution of ^{15}N within the plant has been determined for several kinds of plants (Shearer *et al.*, 1980; Shearer *et al*, 1983; Mariotti *et al.*, 1980). The general result is that different above-ground tissues have ^{15}N abundances similar to each other and to the entire plant. In contrast to above-ground tissues, nodules of some, but not all, legumes are strikingly enriched in ^{15}N (Kohl *et al.*, 1982; Shearer *et al.*, 1982). Nodules of non-legumes which have been examined are not enriched in ^{15}N (Kohl *et al.*, 1982; Shearer *et al.*, 1982). Even when the ^{15}N abundance of nodules is elevated compared to the rest of the plant, there is little impact on the $\delta^{15}N$ of the rest of the plant since only a small fraction of the plant's N is in the nodules. The metabolic significance of the unique ^{15}N enrichment of N_2-fixing nodules is under study in our laboraory. Preliminary results of that study are reported elsewhere (Kohl and Shearer, 1988). Although the differences in $\delta^{15}N$ among above-ground tissues are small, some are consistent and significant. For example, stem tissue is lower than leaf tissue by 1 or 2‰ (Shearer *et al.*, 1983; Shearer and Kohl, 1986; Virginia *et al.*, 1987). This difference is large enough to affect estimates of the contribution of biological N_2-fixation, especially at sites where the difference in ^{15}N abundance between fixed N and soil-derived N is small. It is therefore desirable to determine the distribution of ^{15}N within the N_2-fixing and reference plants in order to select a part of the plant for analysis which has a ^{15}N abundance similar to the entire plant, or at a minimum, to select a part of the plant which has a ^{15}N abundance which deviates from the entire plant in the same direction and by the same magnitude in both N_2-fixing and reference plants.

Requirements of the method

The method is not applicable to all sites. It requires that the difference in $\delta^{15}N$ between N derived from atmospheric N and N derived from other sources be significant. The method is useless at sites in which variation in N derived from sources other than atmospheric N$_2$ is large compared to the mean difference between this N and biologically fixed N. A second requirement is that isotopic fractionation associated with N$_2$-fixation, assimilation of N from sources other than atmospheric N$_2$, and metabolism followed by transport, be taken into account. Finally, the method requires the assumption that the ^{15}N abundance of soil-derived N be the same in N$_2$-fixing and reference plants.

Strategies for dealing with requirements of the $\delta^{15}N$ method

When a new site is selected, it is advisable to evaluate the site for suitability for application of the δ^{15}N method by collecting as many different species of non-N$_2$-fixing reference plants as possible in order to determine the magnitude of the difference between the ^{15}N abundance of reference plants and atmospheric N$_2$ and to determine the variability of δ^{15}N at the site. The observed variation may be due to real spatial variability of plant-available soil N (horizontal or vertical) or to differences in the degree to which isotopic fractionation has altered the ^{15}N abundance of the source. In either case, if the observed variability is large compared to the difference between the mean δ^{15}N value and the δ^{15}N value of atmospheric N$_2$, the site may be regarded as unsuitable. If δ^{15}N values vary in a regular manner across the site, it is possible to compensate for this trend by collecting N$_2$-fixing and reference plants in pairs.

The ^{15}N abundance of biologically fixed N should be determined for all of the N$_2$-fixing plants to be assessed by growing inoculated plants hydroponically with N-free nutrient medium. The preferred inoculum is soil from the site, so as to increase the probability that the symbiont will be the same as under field conditions. The distribution of ^{15}N among plant parts of N$_2$-fixing and reference plants should be determined in order to select the appropriate plant part for analysis.

Advantages of the $\delta^{15}N$ method

This method provides a time-integrated estimate of the percentage of the total N of a N$_2$-fixing system that is derived from atmospheric N$_2$ (% Ndfa). The analyses need not be done *in situ*. Samples may be collected by an agronomist or ecologist for later analysis, perhaps in a different institution. The method need not disturb the system. By using the δ^{15}N method, the contribution of fixed N, even to a large deeply rooted plant, can be estimated without disturbing it.

Verification of the $\delta^{15}N$ method

A number of comparisons of estimates of the contribution of biologically fixed N based on the δ^{15}N method with estimates based on more conventional methods have been made, under both field and greenhouse conditions. The results of these tests hae been reviewed elsewhere (Shearer and Kohl, 1986; Virginia *et al.*, 1987). All of these tests were with legumes and most were with agricultural plants. The N$_2$-fixing plants, the method with which the δ^{15}N method was compared, and references are listed in Table 1. In all cases the δ^{15}N method compared favorably with other methods. In addition to these comparisons, certain other results bear on the validity of the δ^{15}N method, for example:

(i) The relative contribution of biologically fixed N to soybeans, as judged by the δ^{15}N method, varied as expected with soil amendment. It decreased with increasing rate of fertilization with NO$_3^-$, and increased when carbonaceous material was applied to the soil (Kohl *et al.*, 1980).

(ii) Feeder roots of a mesquite (*Prosopis glandulosa*) tree were removed, leaving major tap roots intact. Since root nodules are located on the deeper roots (Felker and Clark, 1982) and virtually all of the N in the soil profile is in the surface layer which was removed with the feeder roots, we anticipated that this treatment would result in a decrease in δ^{15}N in the mesquite tissues, reflecting an increase in the relative contribution of nitrogen fixation compared to uptake of soil N. A significant decrease in δ^{15}N was observed in this tree, but not in four neighbouring untrenched trees (Shearer *et al.*, 1983).

We conclude that the δ^{15}N method of estimating the relative contribution of N$_2$-fixation to N$_2$-fixing systems is an appropriate, although not highly pre-

Table 1. N$_2$-fixing plants used in comparisons between the δ^{15}N method and more conventional methods of estimating the fractional contribution of fixed N

Conventional method	N$_2$-fixing plant	Type of experiment	Reference
Acetylene reduction	Soybeans (*Glycine max*)	Field	Amarger *et al.* (1979)
	Lupins (*Lupinus luteus*)	Greenhouse	Amarger *et al.* (1977)
The N yield method	Soybeans	Greenhouse	Kohl *et al.* (1979)
		Field	Kohl *et al.* (1979)
		Field	Amarger *et al.* (1979)
		Greenhouse	Kohl *et al.* (1980)
		Field	Kohl *et al.* (1980)
	Clover (*Trifolium repens*)	Field	Steele (1983)
Isotope dilution with artificially ^{15}N enriched fertilizers	Soybeans	Field	Domenach and Chalamet (1979)
		Field	Domenach and Corman (1984)
	Clover (*Trifolium subterraneum*))	Field	Bergersen and Turner (1983)
		Greenhouse	Ledgard *et al.* (1985b)
		Field	Ledgard *et al.* (1985b)
		Field	Ledgard *et al.* (1985c)
	Psorothamnus spinosus	Field	Virginia *et al.* (1987)

cise, method in many settings, at least for the legumes which have been tested.

Application of the δ^{15}N method to N$_2$-fixing non-legumes

Almost all applications of and experience with the δ^{15}N method have been with legumes. We know of no instance in which estimates of % Ndfa have been made for non-legumes. This information requires growing the non-legume hydroponically with N-free medium. This is usually done only in large studies. With one exception, all studies done so far with non-legumes have been preliminary. These studies are described below.

Nodulated non-legumes

In principle, the δ^{15}N method for estimating the fractional contribution of biologically fixed N should be equally applicable to nodulated non-legumes and legumes. The limited number of studies described below would seem to bear this out.

An early survey. In our first survey of δ^{15}N values of N$_2$-fixing and non-fixing plants (Shearer and Kohl, 1978), we collected plants from agricultural fields and private gardens in central Illinois and from the

Missouri Botanical Garden. Included in this collection were *Alnus maritima* (2 specimens), *Myrica pensylvanica* (1 speciment) and *Eleagnus angustifolia* (3 specimens). Reference plants were collected near each specimen. Figure 5 shows the comparison of δ^{15}N values of N$_2$-fixing and reference plants. In all cases the δ^{15}N values of the N$_2$-fixing non-legumes were closer to that of atmospheric N$_2$ than were those of neighbouring reference plants. The

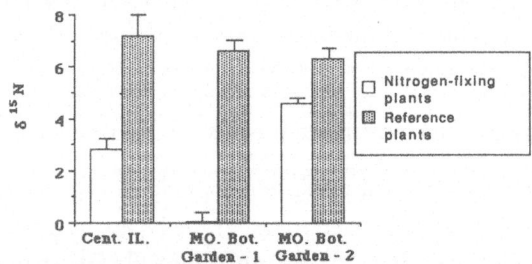

Fig. 5. The ^{15}N abundance of nodulated non-legumes and non-N$_2$-fixing reference plants in Missouri (Missouri Botanical Garden) and central Illinois. Plotted from data given in Shearer and Kohl (1978). The N$_2$-fixing plants growing in central Illinois were *Eleagnus angustifolia*. Neighbouring reference plants were *Taraxacum* sp., *Oxalis stricta, Chenopodium album, Polygonum hydropiper, Bromus arvensis, Viola missouriensis* Greene and *Physalis heterophylla* Nees. The N$_2$-fixing plants growing at site 1 at the Missouri Botanical Garden were *Alnus maritima*, with *Plantago* sp. and *Agropyron* sp. (reference plants) collected from the same bed. The nodulated non-legumes on site 2 of the Missouri Botanical GArden were *Myrica pensylvanica*. Reference plants for *Myrica* were *Oxalis stricta, Agropyron* sp. and *lamium amplexicaule*. Error bars denote one standard error.

results are consistent with the interpretation that most of the N of *Alnus maritima* was derived from biological N_2-fixation, and that in *Eleagnus angustifolia*, biologically fixed N made an intermediate contribution to the plant's N, while only a modest fraction of the N in *Myrica pensylvanica* was from fixation.

Chamabatia foliolosa. Virginia (1980) observed consistently low $\delta^{15}N$ values in this plant during a survey of the natural abundance of plant and soil samples in in selected California ecosystems. This observation led him to suspect that this species was capable of N_2-fixation. Although it is a member of a family (Rosaceae) which includes a number of nodulated, N_2-fixing plants, *Chamabatia foliolosa* was not, at that time, considered a N_2-fixer. Heisey *et al.* (1980), who later found nodules on these plants, confirmed N_2-fixation in *Chamabatia foliolosa* with the acetylene reduction assay and incorporation of $^{15}N_2$, thereby demonstrating the usefulness of the $\delta^{15}N$ method as a screen for N_2-fixation among species not known to have this capability.

N_2-fixation by Alnus sinuata *and* A. rubra *in mixed conifer/alder stands.* In an attempt to detect the transfer of biologically fixed N from alder to other N pools, Binkley *et al.* (1985) measured ^{15}N abundances of both alder and conifers, and soil N (total N, NH_4^+-N, and NO_3^--N) at four sites, and in alder grown hydroponically with N-free medium. At each site there was one stand of pure conifers and one or two mixed stands of conifers with alder. They found no consistent trend across all sites for $\delta^{15}N$ of plants *vs* soil N pools, and concluded that the ^{15}N abundance method does not provide a simple means for evaluating N dynamics at the forest sites studied. However, the data do show one consistent trend across the sites; namely, the $\delta^{15}N$ of alder fell between that of conifers and that of alder grown hydroponically with N-free nutrient medium at all sites, as shown in Fig. 6. The authors did not take this as evidence for N_2-fixation by alder, however, since they had no information about differences between the two types of plants in the degree of isotopic discrimination associated with "uptake, transport, and assimilation". We consider this to be unduly cautious for the following reasons. At four of the five mixed conifer/alder

stands the conifer was Douglas fir [*Pseudotsuga menziesii* (Mirb.) Franco]. In two of these stands the $\delta^{15}N$ of the Douglas fir was lower than that of alder and that of atmospheric N_2. At the other two stands the $\delta^{15}N$ abundance of Douglas fir was higher than that of alder and that of atmospheric N_2. If the difference between the two kinds of plants resulted from isotopic fractionation, this difference should at least be in the same direction, if not of the same magnitude, from site to site. Although isotope effects can vary with environmental conditions, it would be quite coincidental for this variation to result in conifers having a higher ^{15}N abundance than alder in stands in which the $\delta^{15}N$ of conifers was higher than atmospheric N_2, and a lower ^{15}N abundance than alder in stands in which the $\delta^{15}N$ of conifers was lower than that of atmospheric N_2. We consider it much more likely that the observed difference in $\delta^{15}N$ between conifers and alder was caused by the contribution of atmospheric N_2 to alder plants. While it was not possible to evaluate N transfer from N_2-fixing plants to non-fixing plants using the $\delta^{15}N$ method at these sites (a quite stringent demand on a natural abundance method), we interpret the results of Binkley *et al.* (1985) as an example of the ability of the method to detect N_2-fixation in this type of ecosystem.

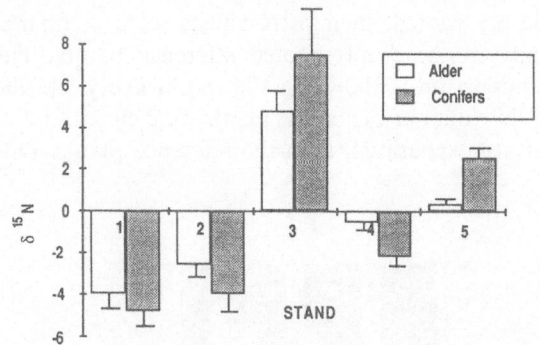

Fig. 6. Comparison of the ^{15}N abundance of alder with that of neighbouring conifers. Data from Binkley *et al.* (1985). Stand 1-was a mixed sitka alder (*Alnus sinuata*)/Douglas fir [*Pseudotsuga menziesii* (Mirb.) Franco] stand located at Mt. Benson, Vancouver Island. This was a young stand with infertile soil. Stand 2 was a mixed red alder/Douglas fir stand also located at Mr. Benson. Stand 3 was an old stand with infertile soils at Wind River, Washington; it was a mixed red alder/Douglas fir stand. Stand 4 was a young stand with fertile soil at Skykomish, Washington with red alder and Douglas fir. Stand 5 was an old stand with fertile soil in Cascade Head, Oregon with red alder and mixed conifers. Error bars represent one standard error.

N_2-*fixation in a chaparral ecosystem.* Mediterranean ecosystems are characterized by recurrent fire and rapid recovery by way of seed germination and resprouting. These systems contain many N_2-fixing species including nodulated non-legumes. In collaboration with P.W. Rundel (University of California, Los Angeles) we measured the ^{15}N abundance of N_2-fixing and reference plants collected from chamise chaparral sites in Sequoia National Park in California. Plants were collected at four sites in which the number of years since the last burn varied. Results were reported by Shearer and Kohl (1986) and Virginia *et al.* (1987). Figure 7 gives $\delta^{15}N$ values of reference plants growing at these sites. Reference plants included long-lived, deeply-rooted evergreen shrubs, *Adenostoma fasciculatum, Arctostaphylos viscida* and *Quercus dumosa,* and short-lived, drought-deciduous, shallow-rooted shrubs, *Malacothamus fremontii, Eriodictyon californicum,* and *Dendromecon rigida.* All reference plants had ^{15}N abundances lower than atmospheric N_2. At one of the sites, no shallow-rooted plants were collected. At the other three sites, deeply-rooted shrubs were significantly ($P < 0.001$) lower in ^{15}N abundances than shallow-rooted plants, indicating variation of $\delta^{15}N$ of plant-available soil N with depth at these site. *Ceanothus leucodermis* and *C. cuneatus* were collected from three of the four sites. Because these two species are deeply rooted, their $\delta^{15}N$ values were compared only to the deeply rooted reference plants. The comparison is shown in Fig. 8. At every site the $\delta^{15}N$ values of *Ceanothus* plants were closer to that of atmospheric N_2 than the reference plants. Dif-

ferences beween reference plants and *Ceanothus* (both species) at each site were significant ($P < 0.05$). The difference between *Ceanothus* (both species) and reference plants was larger on the 13 year post-burn site than on the 3 or 23 year post-burn sites, suggesting that a substantial period is required to achieve full capacity for N_2-fixation after the burn, and a decline of N_2-fixation in older shrubs. But the data suggest that *Ceanothus* derives a substantial fraction of its N from biological N_2-fixation at these sites, regardless of the length of time since a burn occurred.

Other N_2-fixing associations

Very little work has been done to asses the applicability of the $\delta^{15}N$ method to non-legumes other than nodulated non-legumes. The feasibility of this method for assessing N_2-fixation in associative N_2-fixation would seem to us to depend upon the

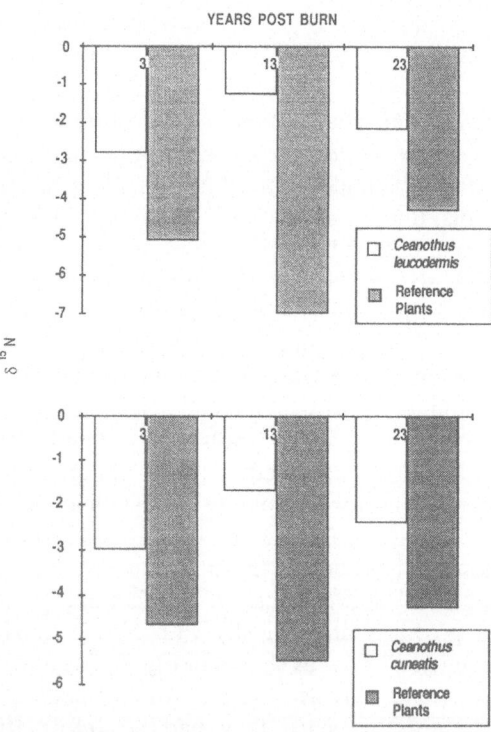

Fig. 8. Comparison of the ^{15}N abundance of Ceanothus with that of neighbouring deep-rooted reference plants in a chamise chaparral in Sequoia National Park in California. Data from a collaborative study with P W Rundel (University of California, Los Angeles).

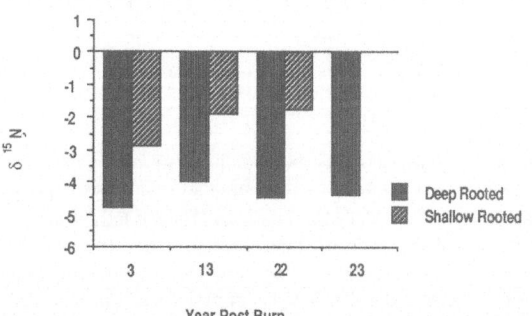

Fig. 7. The ^{15}N abundance of deep-rooted and shallow-rooted non-N_2-fixing plants in a chamise chaparral in Sequoia National Park in California. Data from a collaborative study with P W Rundel (Univeristy of california, Los Angeles).

closeness of the association. For example, the method should be useful for lichens and azolla. Indeed, as part of his broad survey of the ^{15}N abundance of plants and soils in California, Virginia (1980) collected lichens and fungi. He commented that lichens had much lower ^{15}N abundances than fungi, and wolf lichen (*Letharia vulpina*) had a lower ^{15}N abundance than any of the plants tested. On the other hand, N_2-fixation in the rhizosphere might be difficult to detect, particularly if the fixed N is substantially partitioned between the plant and the soil. In this case, it would be difficult to distinguish between the two poles represented in Fig 1 (N derived from biological N_2-fixation and N derived from other sources). The question of applicability of the $\delta^{15}N$ method to various types of associative N_2-fixing systems is empirical.

We know of only one example of the use of the $\delta^{15}N$ method for an associative N_2-fixing system, a rhizosheath grass, *Oryzopsis hymenoides*. Wullstein *et al.* (1979) and Harker and Wullstein (1981a, b) showed that the sheath was capable of N_2-fixation and that fixed ^{15}N was assimilated into tissues of these plants. In collaboration with L.H. Wullstein we measured the N abundance of *O. hymenoides* and reference plants collected from a sand dune (ridge and swale) and from a site with true soil (Fig. 9). These sand dune sites are the least favourable we have so far encountered for application of the $\delta^{15}N$ method, the mean $\delta^{15}N$ value of reference plants being only 1.5‰ different from atmospheric N_2. Despite this, we did observe a highly significant

difference between the rhizosheath grass and reference plants collected from the dune ridge ($P <$ 0.01), possibly because the N isotope effect for N_2-fixation is large in this system. (The mean $\delta^{15}N$ for *Oryzopsis* collected from the dune ridge was -2.7 ± 0.5, n = 6) The difference was not significant in plants collected from the dune swale, and the significance of the difference at the true soil site could not be evaluated since only one reference plant was available. It would be of interest to augment these results using a larger collection from more sites, and to compare the $\delta^{15}N$ method with acetylene reduction and $^{15}N_2$ assimilation by the plant, since input of biologically fixed N by rhizosheaths would be important in these nutrient poor sites.

Free-living N_2-fixing microorganisms

The application of the $\delta^{15}N$ method to N_2-fixing free-living microorganisms seems problematic, particularly in the case of soil microorganisms, where the complexity of the system precludes obtaining enough N from individual species under natural growing conditions for ^{15}N analysis. Moreover, to the degree that partitioning of the fixed N between the N_2-fixing microoganism and the surrounding medium is substantial, separation of the two poles shown in Figure 1 sould be narrowed. This point is illustrated by measurements on marine organisms. Wada and Hattori (1976) measured the $\delta^{15}N$ of a N_2-fixing cyanobacterium (*Trichodesmium* sp.), non-N_2-fixing microorganisms which grew in association with the cyanobacterium, and non-fixing species from deeper layers of a nutrient poor region of the Philippine Sea. The $\delta^{15}N$ values of *Trichodesmium* and non-fixing microorganisms growing in association with it were close to that of atmospheric N_2^-, and significantly lower than $\delta^{15}N$ values for non-fixing microorganisms growing in deeper layers. The similarity in $\delta^{15}N$ of the N_2-fixing organisms to both atmospheric N_2 and associated non-fixing organisms suggests both substantial inputs of fixed N and substantial transfer of fixed N to the non-fixing organisms in the surface water. However, this conclusion depends on the assumption that ^{15}N abundance of N sources other than atmospheric N_2 are the same in surface and deeper water layers.

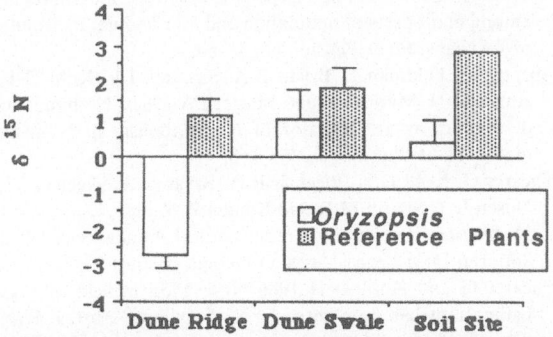

Fig. 9. Comparison of the ^{15}N abundance of a rhizosheath grass with that of non-N_2-fixing reference plants collected from a sand dune (ridge and swale) and a site with true soil. Data from a collaborative study with L H Wullstein. Error bars represent one standard error.

References

Amarger N, Mariotti A, Mariotti F, Durr J C, Bourguignon C and Lagacherie B 1979 Estimate of symbiotically fixed nitrogen in field grown soybeans using variations in ^{15}N natural abundance. Plant and Soil 52, 269–280

Amarger N, Mariotti A and Mariotti F 1977 Essai d'estimation du taux d'azoté fixé symbiotiquement chez le lupin par le traçage isotopique naturel (δ^{15}N). C.R. Acad. Sci. Paris 284, 2179–2182.

Bergersen F J and Turner G L 1983 An evaluation of ^{15}N methods for estimating nitrogen fixation in a subterranean clover-perennial ryegrass sward. Aust. J. Agric. Res. 34, 391–401.

Binkley D, Sollins P and McGill W B 1985 Natural abundance of nitrogen-15 as a tool for tracing alder-fixed nitrogen. Soil Sci. Soc. Am. J. 49, 463–471.

Delwiche C C and Steyn, P L 1970 Nitrogen isotope fractionation in soils and microbial reactions. Environ. Sci. Technol. 4, 929–935.

Domenach A M and Chalamet A 1979 Estimates d'azote par le soja a l'aide de deux methods l'analyses isotopiques naturels. C.R. Acad Sci. Paris Ser. D. 289, 291–294.

Domenach A M and Corman A 1984 Dinitrogen fixation by field grown soybeans: Statistical analysis of variations in δ^{15}N and proposed sampling procedures. Plant and Soil 78, 301–313.

Felker P and Clark P R 1982 Position of mesquite (*Prosopis* spp) nodulation and nitrogen fixation (acetylene reduction) in 3-m long phreatophytically simulated soil columns. Plant and Soil 64, 297–305.

Harker A R and Wullstein L H 1981a Transfer of ^{15}N$_2$ to closed systems in the field. Soil Biol. Biochem. 13, 535–536.

Harker A R and Wullstein L H 1981b Plant assimilation of nitrogen fixed in the rhizosheaths of *Oryzopsis hymenoides*. Abstracts, AAAS, Western Division, p. 17.

Heisey R M, Delwiche C C, Virginia R A, Wrona A F and Bryan B A 1980 A new nitrogen-fixing non-legume: *Chamaebatia foliolosa* (Rosaceae). Am. J. Bot. 67, 429–431.

Hoering T and Ford H T 1960 The isotope effect in the fixation of nitrogen by *Azotobacter*. J. Am. Chem. Soc. 82, 376–378.

Kohl D H, Shearer G and Harper J E 1979 The natural abundance of ^{15}N in nodulating and non-nodulating isolines of soybeans. *In* Proc. 3 rd Int. Conf. on Stable Isotopes. Eds. E R Klein and P Klein. pp 317–325. Academic Press, N.Y.

Kohl D H, Shearer G and Harper J E 1980 Estimates of N$_2$-fixation based on differences in the natural abundance of ^{15}N in nodulating and non-nodulating isolines of soybean. Plant Physiol. 66, 61–65.

Kohl D H and Shearer G 1980 Isotopic fractionation associated with symbiotic N$_2$ fixation and uptake of NO$_3^-$ by plants. Plant Physiol. 66, 51–56.

Kohl D H, Bryan B A, Feldman L, Brown P H and Shearer G 1982 Isotopic fractionation in soybean nodules. *In* Stable Isotopes. Proc. 4 th Int. Conf. Eds. H L Schmidt, H Förstel and K Heinzinger. pp 451–457. Elsevier, Amsterdam.

Kohl D H and Shearer G 1988 Ecosystem estimates of N$_2$-fixation and investigations of nodule metabolism using variations in the natural abundance of ^{15}N. *In* Advances in Legume Biology. Eds. C H Stirton and J L Zarucchi. In press. Monographs in Systematic Botany.

Ledgard S F, Freney J R and Simpson J R 1984 Variations in the natural enrichment of ^{15}N in the profiles of some Australian pasture soils. Aust. J. Soil Res. 22, 155–164.

Ledgard S F, Simpson J R, Freney J R and Bergersen F J 1985a Assessment of the relative uptake of added and indigenous soil nitrogen by nodulated legumes and reference plants in the ^{15}N dilution measurement of N$_2$-fixation: Glasshouse application of the method. Soil Biol. Biochem. 17, 233–238.

Ledgard S F, Simpson J R, Freney J R and Bergersen F J 1985b Field evaluation of ^{15}N techniques for estimating nitrogen fixation in legume-grass associations. Aust. J. Agric. Res. 36, 247–258.

Ledgard S F, Simpson J R, Freney J R and Bergersen F J 1985c Effect of reference plant on estimation of nitrogen fixation by subterranean clover using ^{15}N methods. Aust. J. Agric. Res. 36, 663–676.

Mariotti A, Mariotti F and Amarger N 1983 Utilization du traçage isotopique naturel par pour la mesure du taux d'azote fixé symbiotiquement par les légumineuses. Physiol. Veg. 21, 279–291.

Mariotti A, Mariotti F, Amarger N, Pizelle G, Ngambi J M, Champigny M L and Moyse A 1980 Fractionnements isotopiques de l'azote lors des processus d'absorption des nitrates et de fixation de l'azote atmosphérique par les plantes. Physiol. Veg. 18, 163–181.

Mariotti A, Mariotti F, Champigny M L, Amarger N and Moyse A 1982 Nitrogen isotope fractionation associated with nitrate reductase activity and uptake of NO$_3^-$ by pearl millet. Plant Physiol. 69, 880–884.

Rennie R J and Rennie D A 1983 Techniques for quantifying N$_2$-fixation in association with non legumes under field conditions. Can. J. Microbiol. 29, 1022–1035.

Shearer G, Kohl D H and Chien S H 1978 The nitrogen-15 abundance in a wide variety of soils. Soil Sci. Soc. Am. Proc. 38, 315–322.

Shearer G and Kohl D H 1978 ^{15}N abundance in N-fixing and non-N-fixing plants. *In* Recent Developments in Mass Spectrometry in Biochemistry and Medicine, Vol 1. Ed. A Frigerio. pp 605–622. Plenum Press, New York.

Shearer G, Kohl D H and Harper J E 1980 Distribution of ^{15}N among plant parts of nodulating and non-nodulating isolines of soybeans. Plant Physiol. 66, 57–60.

Shearer G, Feldman L, Bryan B A, Skeeters J L, Kohl D H, Amarger N, Mariotti F and Mariotti A 1982 ^{15}N abundance of nodules as an indicator of N metabolism in N$_2$-fixing plants. Plant Physiol. 70, 465–468.

Shearer G, Kohl D H, Virginia R A, Bryan B A, Skeeters J L, Nilsen E T, Sharifi M R and Rundel P W 1983. Estimates of N$_2$-fixation from variation in the natural abundance of ^{15}N in Sonoran Desert ecosystems. Oecologia (Berlin) 56, 365–373.

Shearer G and Kohl D H 1986 N$_2$-fixation in field settings: Estimations based on natural ^{15}N abundance. Aust. J. Plant Physiol. 13, 699–756.

Shearer G and Kohl D H 1987 Estimates of N$_2$-fixation in ecosystems: The need for and basis of the method. *In* Stable Isotopes in Ecological Research. Eds. P W Rundel, J R Ehleringer and K A Nagy. In press. Springer Verlag.

Steele K W 1983 Quantitative measurements of nitrogen turnover in pasture systems with particular reference to the role of ^{15}N. *In* Nuclear Techniques in Improving Pasture Management. pp 17–35. Int. Atomic Energy Agency, Vienna.

Steele K W, Bonish P M, Daniel R M and O'Hara, G W 1983 Effect of rhizobial strain and host plant on nitrogen isotopic fractionation in legumes. Plant Physiol. 72, 1001–1004.

Turner G L, Gault R R, Morthorpe L, Chase D L and Bergersen F J 1987 Differences in the natural abundance of ^{15}N in the extractable mineral nitrogen of cropped and fallowed surface soils. Aust. J. Agric. Res. 38, 15–25.

Virginia R A 1980 Natural Abundance of Nitrogen-15 of Presumed N_2-Fixing and Non N_2-Fixing plants from Selected Ecosystems. Ph.D. Thesis, University of California, Davis (Diss. Abstr. Int. 41).

Virginia R A, Jarrell W M, Rundel P W, Shearer G and Kohl D H 1987 The use of variation in the natural abundance of ^{15}N to assess symbiotic N_2-fixation by woody plants. *In* Stable Isotopes in Ecological Research. Eds. P W Rundel, J R Ehleringer and K A Nagy. In press. Springer-Verlag.

Wada E and Hattori A 1976 Natural abundance of ^{15}N in particulate organic matter in the North Pacific ocean. Geochim. Cosmochim. Acta 40, 249–251.

Wullstein L H, Bruening M L and Bollen W B 1979 Nitrogen fixation associated with sand grain root sheaths (rhizosheaths) of certain zeric grasses. Physiol. Plant. 46, 1–4.

F. A. Skinner et al. (Eds.), Nitrogen fixation with non-legumes, 301–310.
© 1989 by Kluwer Academic Publishers.

Survival and colonization of inoculated bacteria in kallar grass rhizosphere and quantification of N_2-fixation

KAUSER A. MALIK and RAKHSHANDA BILAL
Nuclear Institute for Agriculture & Biology, Faisalabad, Pakistan

Key words: colonization, inoculation, kallar grass, N_2-fixation, ^{15}N isotopic dilution

Abstract

Two experiments have been conducted, one in semi-solid Hoagland nutrient medium and the other in shallow pots containing saline soil. N_2-fixing bacteria belonging to *Azospirillum, Azotobacter, Klebsiella* and *Enterobacter* were inoculated separately on kallar grass grown in semi-solid nutrient medium. It was shown that inoculation affects root proliferation and also results in ^{15}N isotopic dilution. The % Ndfa ranged from 47–70 whereas no significant effect on the total nitrogen uptake was observed. The bacterial colonization of the root surface and the presence of enteric bacteria inside the root hair cells is reported. In a soil pot experiment, non-N_2-fixing *Polypogon monspeliensis* was used as a reference plant (control). A treatment receiving a high rate of nitrogen was also used as a non-N_2-fixing control. ^{15}N-labelled ammonium sulphate at $20\,kg\,N\,ha^{-1}$ and $90\,kg\,N\,ha^{-1}$ was used. The % Ndfa in the aerial parts of kallar grass was 12–15 when *P. monspeliensis* was used as reference plant whereas 37–39% Ndfa was estimated when the treatment receiving high nitrogen fertilizer was used as a non-N_2-fixing control. These investigations revealed some problems of methodology which are discussed.

Introduction

Leptochloa fusca (locally known as kallar grass), a highly salt-tolerant perennial plant, is used as a primary colonizer for salt-affected soils in Pakistan. It has a C4 photosynthetic pathway (Zafar and Malik, 1984) and grows luxuriantly in salt affected low fertility soils. Nitrogen fixation associated with the roots of this grass has previously been reported (Malik *et al.,* 1980, 1981). Quantification of the diazotrophic population and estimations of nitrogenase activity associated with its roots over a period of one year have also been reported previously (Zafar *et al.,* 1986) and several species of diazotrophic bacteria have been isolated from its roots (Reinhold *et al,* 1986, Bilal and Malik, Zafar *et al.,* 1987).

Microbiological studies, and assay of the nitrogenase activity (C_2H_2 reduction) of excised roots indicate the presence of associative nitrogen fixation in kallar grass. However, the role of such fixation in plant nutrition can only be ascertained when its contribution is quantified. The most reliable evidence of such inputs has been produced by ^{15}N isotopic dilution techniques (Chalk, 1985; Miranda and Boddey, 1987). Two recent studies at our Institute, using the isotope dilution technique (Malik and Zafar, 1985; Malik *et al.,* 1987) have indicated a high potential of nitrogen fixation associated with the roots of kallar grass grown in nutrient solution. However, under field conditions, selection of an appropriate non -N_2-fixing control plant, which could grow in highly salinized soils, has been a problem (Malik *et al.,* 1987). In order to overcome this, the treatment receiving high fertilizer N was taken as a control, assuming that the soil nitrogenase activity would be inhibited. The estimation of nitrogen fixation was based on the difference in 'A' values of different treatments. It was found that the rate of N fixation increased with the plant growth; maximum Ndfa (nitrogen derived from the atmosphere) was 54% or $32\,kg\,N\,ha^{-1}$ during its optimal growth between June and September (Malik *et al.,* in press).

Many workers have reported increase in biomass yield as a result of inoculation with N_2-fixing bacteria, especially azospirilla (Cohen *et al.*, 1980; Kapulnik *et al.*, 1983) However, responses may not entirely be due to N_2-fixation. The proliferation of roots due to bacterial inoculation may result in an increased uptake of nutrients from the soil (Patriquin *et al.*, 1983, Okon and Kapulnik, 1986) and lead to increased biomass. Therefore, to quantify the contribution of N_2-fixation, a distinction between the indirect effects, due to root proliferation and actual N_2-fixation will have to be made. As mentioned earlier, this can be best achieved by using the ^{15}N isotopic dilution technique.

In this communication we report the effect of inoculation on root morphology, root colonization and survival of N_2-fixing bacteria using the fluorescent antibody technique and quantification of nitrogen fixation using the ^{15}N isotopic dilution technique.

Materials and methods

Bacterial inoculum

A mixture of four N_2-fixing bacteria, namely *Klebsiella* sp. (NIAB1), *Beijerinckia* sp. (Iso2), *Enterobacter* sp. (AH6) and *E. agglomerans* (11), isolated from grass roots were used as inoculum for soil pot studies. For studies on the effect of inoculation on root morphology, five different diazotrophs were used. They were *Azospirillum brasilense* (Sp7), *A. brasilense* (K1), *Enterobacter* sp. (QH7), *Enterobacter* sp. (KCl-1), *Azotobacter* sp. (K9) and a non-fixing strain of *E. coli*. The bacteria were grown in nutrient broth with shaking. After 48 h of growth, bacterial cells were harvested by centrifugation at 10,000 rev. min^{-1} for 15 min. The pellet was washed thrice with phosphate buffer (0.1 M, pH 6.8) and resuspended in Hoagland nutrient solution. The optical density (O.D.) was adjusted to 0.5 at 600 nm.

Soil pot experiment

The soil was obtained from kallar grass fields at Biosaline Research Substation of NIAB near Lahore. It was a saline sodic soil of pH 9.5 and electrical conductivity of 8.5 $mScm^{-1}$. The soil was thoroughly mixed and amended with ^{15}N fertilizer in the form of ammonium sulphate. A 5 kg portion of this soil was added to plastic pots having 30 cm dia and 15 cm depth. The treatments were as follows:

T1 Ammonium sulphate (AS) 5% ^{15}N atom excess at 20 kg-N ha^{-1} (50 mg N/pot) inoculated with mixed inoculum.

T2 AS 5% ^{15}N atom excess at 20 kg-N ha^{-1} (50 mg N/pot) uninoculated.

T3 AS 1.0% ^{15}N atom excess at 90 kg-N ha^{-1} (225 mg N/pot) uninoculated.

T4 AS 5% ^{15}N atom excess at 20 kg-N ha^{-1} (50 mg N/pot) uninoculated and planted with *Polypogon*.

Treatments 1–3 were planted with kallar grass. *Polypogon monspeliensis*, an annual grass which invades the saline fields where kallar grass has been growing for few years, was used as non-N_2-fixing control. In an earlier survey of the plants growing in saline areas, no root-associated nitrogenase activity was detected in this grass.

Kallar grass cuttings (1.5 cm) having one node were washed and surface sterilized by dipping them in 5% NaOCl solution for 1 min and were then sown in acid-washed autoclaved moist sand. After rooting, the cuttings were washed with sterile distilled water and then dipped in inoculum mixture for 5 min. Five cuttings were planted in each pot and an additional 1 ml inoculum given around each plant root. Similarly, five seedlings of *P. monspeliensis* grown separately, were planted at the same time. Each treatment was replicated four times.

Semi-solid nutrient medium experiment

Seeds of kallar grass were surface sterilized by dipping in 5% NaOCl for 30 min followed by several washings with sterile water. Seeds were then transferred to water agar (1%) plates, and were kept in a controlled temperature growth room with a 16-h day and a 8-h night period. The temperature was maintained at 30°C $\pm$ 2°C and 25°C $\pm$ 2°C during the day and night, respectively. Light intensity during the day time was 20,000 lux. After 4 days the seedlings were transferred to long glass tubes (45 cm × 2.5 cm) containing 30 ml N-free

Hoagland nutrient solution plus 0.2% agar. Ammonium sulphate $500 \mu g$ ($^{15}NH_4)_2SO_4$ (ca. 5% ^{15}N atom excess) in solution form was added to each tube. After autoclaving the tubes, two seedlings were planted on the surface of semi-solid Hoagland nutrient medium with the help of a perforated autoclaveable perspex disc. The disc was supported by a glass rod so that only the roots were touching the semi-solid medium. After the establishment of seedlings in the culture tube, 1 ml of inoculum (10^8– 10^9 cells) of each bacterial strain was added separately. Each treatment was replicated six times. Inoculum of *E. coli* was also used as a non-N_2-fixing control with an uninoculated control treatment.

Plant analysis

The plants were harvested after 8 weeks in the case of the pot experiment and after 5 weeks in the case of the semi-solid nutrient solution experiment. The root morphology was examined visually and with phase contrast microscopy. The dry weight of shoots and roots were recorded. Total N was determined by the semimicro-Kjeldahl method (Bremner, 1965). Distillates were collected and concentrated for ^{15}N analysis. Samples were analysed by the Rittenburg method on a mass spectrometer fitted with a double inlet system (Varian Mat GD-150). Sodium hypobromite was used for releasing $^{15}N_2$.

Enumeration of diazotrophs by fluorescent antibody staining

Fluorescent antibody (FA) stain against *Klebsiella* sp NIAB-1 was used to enumerate them on the root surface of the plants grown in soil, at the end of the 8-week pot experiment. Five grams of roots from plants of each treatment were washed thoroughly, first with tap water and then with sterile saline (0.8% w/v) solution. The roots free of soil particles were then placed in a 250 ml conical flask containing 45 ml of saline and 30–40 sharp pebbles. The roots were shaken for 30 min to dislodge the bacteria and to bring them into suspension. A $30 \mu l$ sample from the supernatant solution was placed on a clean glass slide. The area of the

drop was calculated and found to be $12.56 \, mm^2$. The drop was air dried, heat-fixed and then stained with the fluorescent antibody. To stain, the smear was covered with two drops of FA and incubated for 20 min in a humid chamber, washed for 10 min with phosphate-buffered saline, pH 7.2 (PBS). The slides were rinsed with distilled water and after air-drying were mounted with 9:1 glycerol PBS mounting fluid. The flurorescent bacteria were counted with a Zeiss epifluorescence microscope fitted with an FITC filter system. A minimum of 25 fields were observed on each slide for counting the number of bacteria in each field. Average number of organisms per field was used to calculate the number of organisms present/g fresh weight of root.

Results

Pot experiment

Kallar grass and *P. monspeliensis* were harvested after 8 weeks of growth. The data for dry matter and nitrogen yield of the shoots and roots is presented in Table 1. Inoculation did not show a significant effect on the dry matter and nitrogen yield of kallar grass shoot and root. Maximum dry matter and nitrogen yield was obtained in the treatment where nitrogen was added at $90 \, kg \, ha^{-1}$. In the treatments receiving $20 \, kg \, N/ha$, shoots contained 0.81–0.84% N. However, there was a large variation in the root nitrogen of the various treatments.

In this study two treatments, namely T3 and T4, were regarded as non-N_2-fixing controls. The results of the ^{15}N enrichment, % Ndff, % FUE and the 'A' value are presented in Table 2. A significantly higher ^{15}N enrichment was observed in the case of the shoots of *P. monspeliensis,* indicating a dilution of ^{15}N in T1 and T2. However, there was no significant difference between these two treatments.

Two different methods were used for estimating the amount of N_2 fixed because of two types of non-fixing (T3 & T4) controls. In the case of T4 where *P. monspeliensis* was grown with the same rate of fertilizer application and ^{15}N enrichment as in T1 and T2, the following formula developed by Fried and Middleboe (1977) was used for quantification of nitrogen fixed:

Table 1. Dry matter yield and nitrogen accumulation by kallar grass and *Polypogon* grown at different N fertilizer rates

Treatment	DMY[1] (g/pot)			Nitrogen accumulation (mg/pot)		
	Shoot	Root	Total	Shoot	Root	Total N
T_1 Kallar grass, AS[2] 5% a.e.[3] 20 kg.ha^{-1} inoculated	7.61b	1.90a	9.51	61.87c	13.37ab	75.24
T_2 Kallar grass, AS 5% a.e. 20 kg.ha^{-1} uninoculated	7.01b	1.42a	8.43	57.27c	11.60b	68.87
T_3 Kallar grass, AS 1% a.e. 90 kg.ha^{-1}	10.06a	1.87a	11.93	94.38a	16.93a	111.31
T_4 *Polypogon*, AS 5% a.e. 20 kg.ha^{-1}	9.96a	1.89a	11.85	83.38b	9.54b	92.92

1. DMY, dry matter yield.
2. AS, ammonium sulphate.
3. a.e., ^{15}N atom excess.
Figures followed by the same letters are not significantly different at the 5% level as determined by Duncan's Multiple Ratio (DMR) test.
Each reading is the mean of four replicates.

$$\% \text{ Ndfa} = 1 - \frac{(^{15}\text{N \% atom excess) fs}}{(^{15}\text{N \% atom excess) nfs}} \cdot 100 \quad (1)$$

where fs is fixing system which is kallar grass and nfs is non-fixing system which, in this case, is *P. monspeliensis*.

In the case of T3 where kallar grass was grown with a high rate of fertilizer N (90 kg h^{-1}) at lower ^{15}N enrichment (1% a.e.) the following formula was used (Fried and Broeshart, 1975)

$$F = N_2\text{-fixed (mg/pot)} = (A_{fs} - A_{nfs}) . \% \text{ FUE}/100$$

$$\% \text{ Ndfa} = F/N \text{ yield. } 100 \quad (2) \qquad (2)$$

In this method of quantification, T3 has been taken as the non-fixing system (nfs) with the assumption that higher rates of nitrogenous fertilizer appli-

cation inhibit the nitrogenase activity of rhizospheric bacteria.

The estimation of nitrogen fixed using the above mentioned controls is presented in Table 3. The % Ndfa values of 12–15 were obtained for the shoots when *P. monspeliensis* was used as a control; by using T3 as a non-fixing control the values were 37–39%. Much lower values of fixation were obtained for roots with both methods. There was no significant difference between the inoculated and uninoculated treatments.

The survival of inoculated bacteria in the soil was studied by using fluorescent antibody, prepared against *Klebsiella* sp. (NIAB-I). The results of enumeration revealed that the plants receiving 20 kg N ha^{-1} had 3.5×10^5 *Klebsiella* spp. per g

Table 2. Atom % ^{15}N excess (% a.e.) and calculation of fertilizer nitrogen uptake and 'A' value for different treatments

Treatment	% a.e.		% Ndff		% FUE		'A' value (mg N/pot as AS equivalent)	
	Root	Shoot	Root	Shoot	Root	Shoot	Root	Shoot
T_1 Kallar grass, AS 5% a.e. 20 kg.ha^{-1} inoculated	0.908	0.892	18.060b	18.060c	4.253a	20.643c	222a	226a
T_2 Kallar grass, AS 5% a.e. 20 kg.ha^{-1} uninoculated	0.916a	0.925b	18.535b	18.735c	4.913c	23.177bc	220ab	217a
T_3 Kallar grass, AS 1% a.e. 90 kg.ha^{-1}	0.494b	0.503c	61.310a	62.443a	4.600a	26.170b	142c	135c
T_4 *Polypogon*, AS 5% a.e. 20 kg.ha^{-1}	0.977a	1.056a	19.765b	21.370b	3.753a	34.237a	203b	184b

Ndff, Nitrogen derived from fertilizer; FUE, fertilizer use efficiency; AS, ammonium sulphate; a.e., ^{15}N atom excess.
Figures followed by the same letters are not significantly different at the 5% level as determined by Duncan's Multiple Ratio (DMR) test.
Each reading is the mean of four replicates.

Table 3. Quantitative estimation of N_2-fixed by kallar grass growing in saline soil contained in pots.

Treatment	Estimation of Nitrogen fixed (%Ndfa)			
	T_3^*		T_4^{**}	
	Root	Shoot	Root	Shoot
T_1 Kallar grass, AS 5% a.e. 20 kg ha^{-1} inoculated	24.45	30.36	7.06	15.53
T_2 Kallar grass, AS 5% a.e. 20 kg ha^{-1} uninoculated	33.03	33.18	6.24	12.41

*, Estimated using T_3 (kallar grass AS 1% a.e. 90 kg/ha) as control by Formula -2 (see text). **, Estimated using T_4 (polypogon AS 5% a.e. 20 kg/ha) as control by Formula -1 (see text). AS, ammonium sulphate. a.e., ^{15}N atom excess.

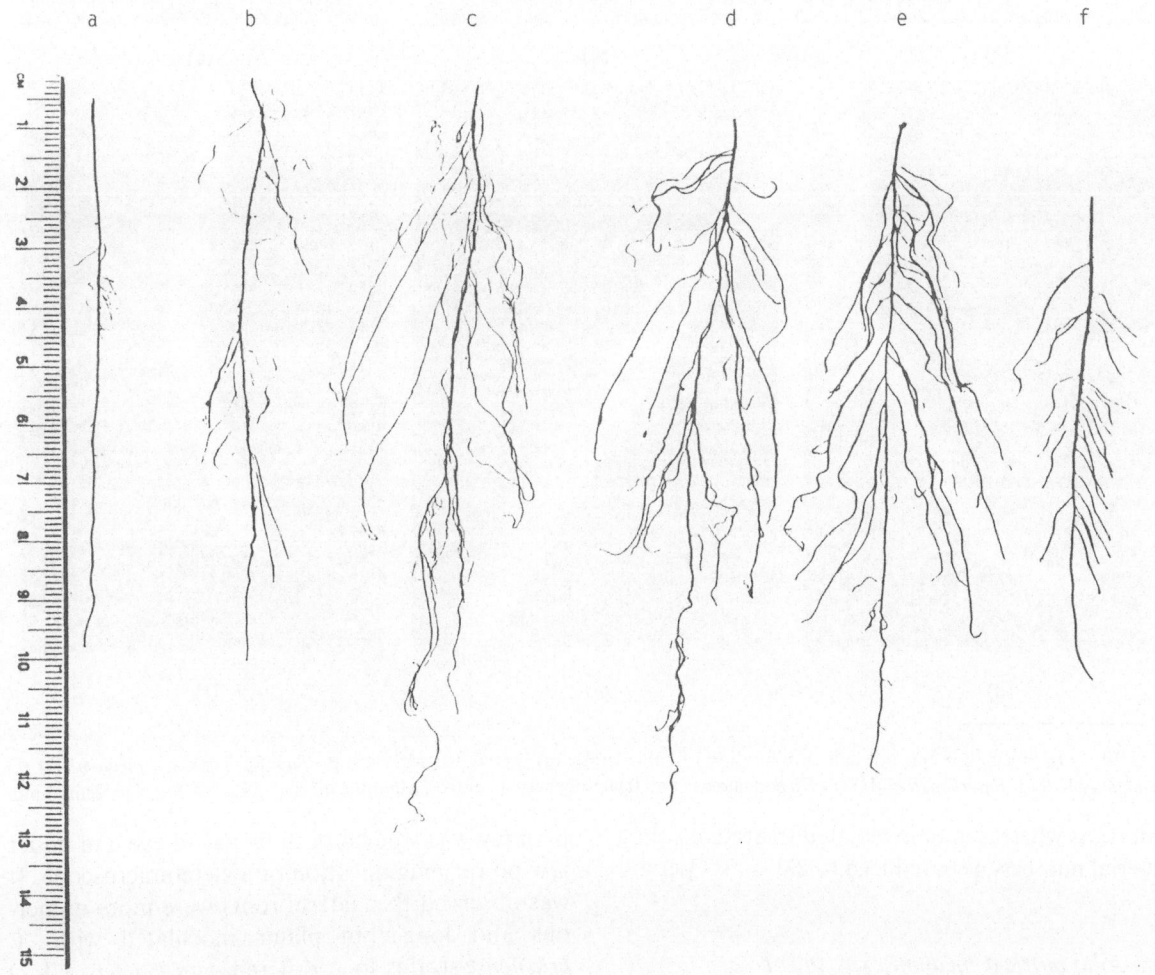

Fig. 1. Effect of inoculation with different diazotrophs on root morphology (a) Uninoculated (b) *Enterobacter* QH7 (c) *Azospirillum* sp-7 (d) *Azospirillum* K-1 (e) *Azotobacter* K-9 (f) *E. coli*.

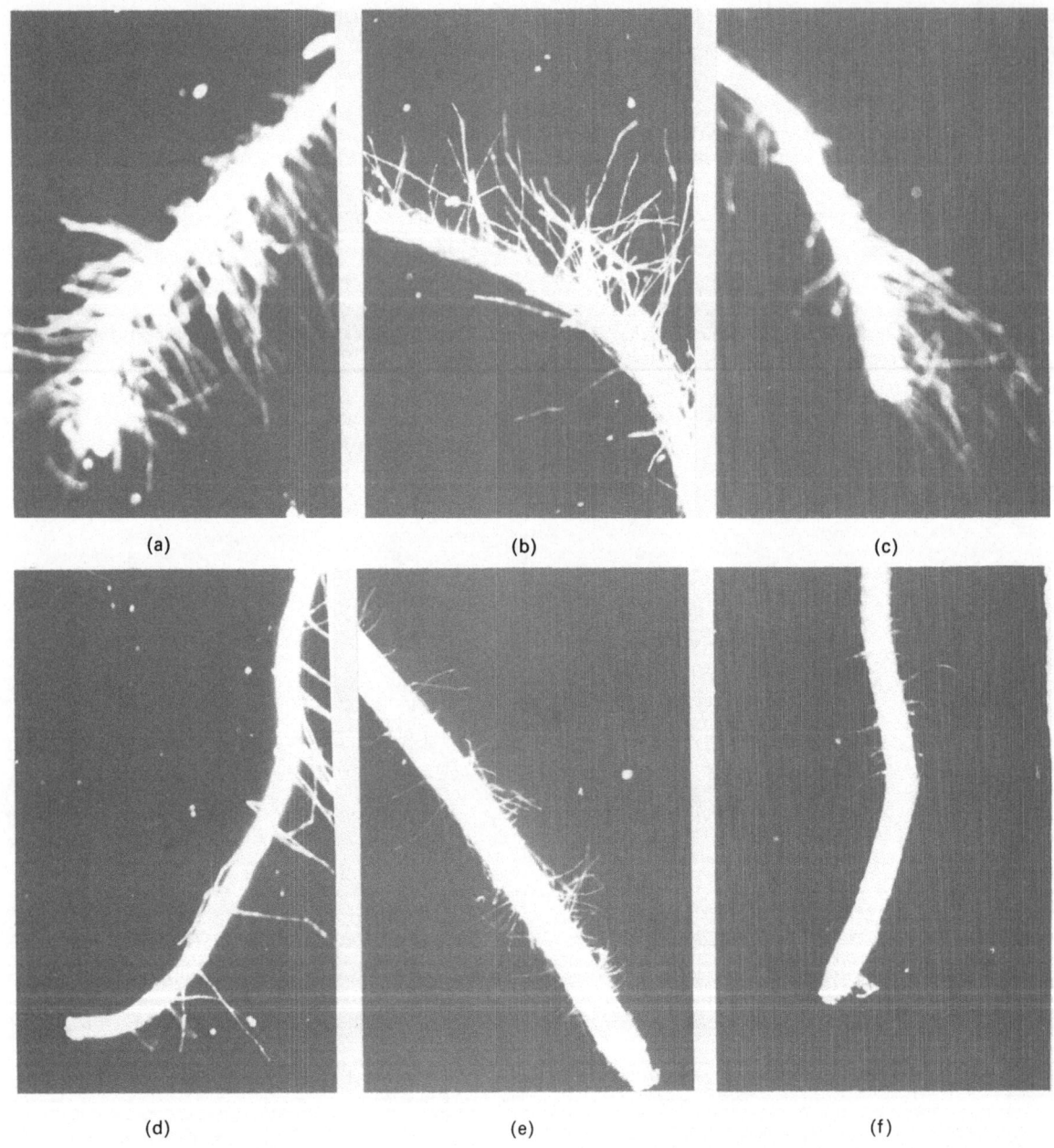

Fig. 2. Effect of inoculation on root hair proliferation of kallar grass. (a) Inoculated with *Azospirillum* Sp-7 (b) *Azospirillum* K-1 (c) *Azotobacter* K-9 (d) *Enterobacter* QH7 (e) *Escherichia coli* and (f) uninoculated control. Magnification × 27. —— Bar represents 1 mm

fresh roots whereas in uninoculated treatments, the bacterial numbers were reduced to 2.0 × 10⁵ per g.

Semi-solid nutrient medium experiment

After five weeks of growth, shoots with roots attached were carefully removed. Root mor-

phology was studied both by naked eye and under low power magnification of a stereomicroscope. It was observed that lateral roots were more numerous and longer in plants inoculated with *A. brasilense* strains sp-7, K-1 and *Azotobacter* sp. K-9 (Fig. 1). The inoculation with enteric bacteria *Klebsiella* sp. KCl-1 and *Enterobacter* sp. (QH7) showed a less pronounced difference over the control. Lateral roots tips, when observed under higher magnifications, with phase-contrast microscopy, revealed a difference in the root hair density and length (Fig. 2). The root hairs were longer and more numerous in all inoculated treatments as compared to the control. Such an effect was more pronounced with *A. brasilense* Sp-7 and K-1 and *Azotobacter* sp. (K-9) as compared with enteric bacteria KCl-1 and QH7 and the non-fixing control *E. coli*.

Upon termination of the experiment (after 5 weeks) 1 cm long pieces of lateral root tip region were washed and observed by phase-contrast microscopy to locate the bacteria. During the microscopic examination, only enteric bacteria strains namely KCl-1 & QH7 were found in abundance inside some of the root hair cells. The *A.*

brasilense strains Sp-7 and K-1 were observed attached mainly to the root hair surface (Fig. 3).

The results of dry weight, N uptake and isotopic dilution are presented in Table 4. There was no significant difference in the root dry weight except in the case of *A. brasilense* Sp-7 inoculation, where maximum root dry weight was recorded. Similarly, nitrogen uptake results also showed no significant difference. However the ^{15}N abundance results revealed nearly 3-fold dilution of ^{15}N enrichment in the kallar grass root + shoot when inoculated with N_2-fixing bacteria as compared to uninoculated and *E. coli* control treatment. Based on this dilution, % Ndfa has been calculated by equation (1). The Ndfa ranged from 47% to 70%. Maximum % Ndfa was obtained in the case of *Enterobacter* sp. (QH7) inoculation.

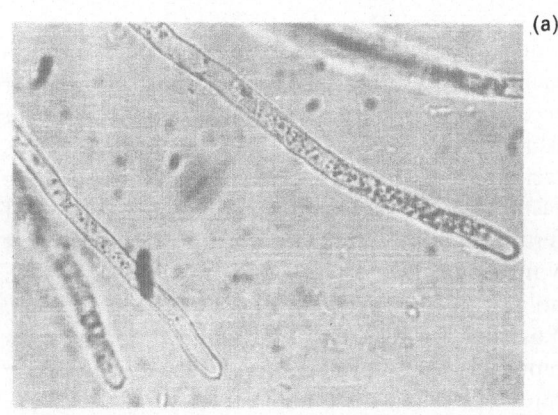

(a)

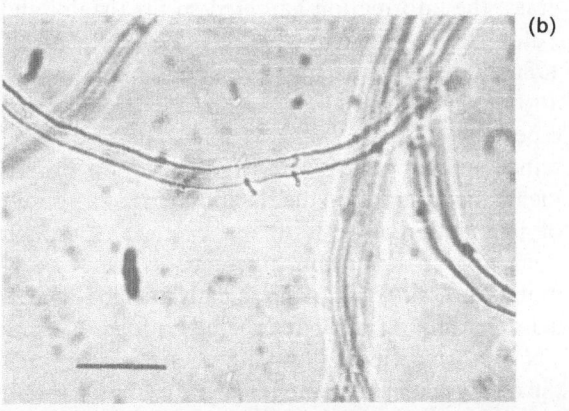

(b)

Fig. 3. Phase contrast microphotographs showing root hair colonization of Kallar grass (a) The *Enterobacter* sp QH7 are seen inside the root hair (b) and *Azospirillum* sp-7 attached to the root hair surface. Magnification × 600. Magnification bar = 20 μm.

Discussion

The results of the two experiments conducted in soil and semi-solid nutrient medium showed dilution of ^{15}N in the kallar grass as compared to the controls. The % Ndfa was much higher in the semi-solid nutrient solution than in the soil. The ^{15}N isotope dilution technique has been widely used for quantification of biological nitrogen fixation in legumes (Chalk, 1985) and more recently for the associative nitrogen fixation in grasses (Rennie, 1980; Boddey *et al.,* 1983; Giller *et al.,* 1986). The isotope dilution technique for quantification of N_2-fixation relies on the use of a non-fixing control against which the dilution of ^{15}N enrichment is assessed. The problems in selecting an appropriate non-fixing control have been discussed in some of the recent reviews and reports (Witty, 1983; Chalk 1985). In the case of perennial grasses growing in saline environments it is difficult to find an appropriate salt-tolerant non-fixing reference plant to be grown alongside kallar grass. Therefore in a previous study, kallar grass grown in an uninoculated fumigated soil was used as a non-fixing control (Malik *et al.,* 1987). In another field experiment, we were unable to grow salt-tolerant weeds such as *Desmostachya bipinnata* and *Typha* sp. as non-fixing controls in the ^{15}N microplots because of high salinity (Malik *et al.,* in press). In the present investigation *P. monspeliensis* was successfully grown in pots and was used as control. The treat-

Table 4. Effect of inoculation on the dry matter yield, total N uptake, ^{15}N enrichment and N_2-fixation in kallar grass grown in semi-solid nutrient solution

Inoculum	Dry matter (mg/tube)*			Total N accumulation μg/tube	[15]N (% a.e.)	% Ndfa[+]
	Root	Shoot	Total			
Enterobacter sp. (QH7)	2.85b	14.08b	16.93	446	0.8757	69.72
Azospirillum brasilense (K-1)	3.42b	16.50ab	19.92	446	1.2998	55.05
Klebsiella sp. (KCl-1)	3.12b	15.10b	18.22	455	1.5170	47.54
Azospirillum brasilense (sp-7)	4.72a	18.73a	23.45	462	1.0709	62.97
Azobacter sp (K-9)	3.48b	15.25b	18.73	452	1.4737	49.04
Escherichia coli	2.65b	15.65b	18.30	467	2.111	
Uninoculated	3.12b	15.15b	18.27	431	2.8919	

* Two plants per tube; [+] Nitrogen derived from atmosphere calculated using uninoculated treatment as control.
Figures followed by the same letters are not significantly different at the 5% level as determined by Duncan's Multiple Ratio (DMR) test.
Each reading is the mean of six replicates.

ment where kallar grass was grown with a high N fertilization rate was used as another non-fixing control. In this case 'A' value was determined for fixing and non-fixing system and from the difference, the amount of N_2-fixed was calculated according to Fried and Broeshart (1975).

Inhibition of soil nitrogenase activity by higher nitrogen fertilizer application has been used by some workers as a mean of obtaining a non-fixing control (Rennie *et al.,* 1978). Specific inhibitors of nitrogenase activity have also been reported (Vlassak *et al.,* 1976, Nohrstedt 1984). Their usefulness is, however, dependent on the extent of phytotoxicity and sustained bioactivity of the plants and microorganisms.

The estimates of nitrogen fixation determined from the 'A' value were relatively higher than those obtained by using *Polypogon* as a reference plant. *Polypogon* is an annual grass and has a profuse root system. As a requirement, the non-fixing reference plant should be closely related to the fixing plant so that their root systems have a similar pattern of N uptake *i.e.* both of them should explore a similar volume of soil (Boddey *et al.,* 1983). In our study, however, the reference plant and kallar grass are not closely related; the former is an annual plant and the latter a perennial. Since the plants were grown in shallow pots (30 cm × 15 cm) it was assumed that their root system would absorb available N in the same manner, but there is no evidence for or against the validity of this assumption.

The amount of N_2-fixed in the aerial parts of kallar grass as calculated from the difference in 'A'

value ranged between 21–24 mg/pot which is equivalent to 37–39% Ndfa. These estimates are similar to the ones obtained earlier in the field experiment using similar controls (Malik *et al.,* In press). This technique is based on the assumption that 'A' value of the soil N is constant regardless of the level of fertilizer addition to the non-fixing reference plant. Various workers have found this assumption to hold (Legg and Stanford, 1967; Aleksic *et al.,* 1968, Yoshida and Padre, 1977) while other workers reported that 'A' value decreased with increasing rates of fertilizer (Rennie, 1979; Boddey *et al.,* 1983).

In spite of the problems in methodology for quantification of the amount of N_2-fixed by kallar grass, the information gathered so far (Malik and Zafar, 1985; Malik *et al.,* 1987, Reinhold *et al.,* 1986) indicates a fairly high potential of N_2-fixation which can be of agronomic significance, especially in highly salinized soils. The results obtained in the semi-solid nutrient solution experiment showed dilution of ^{15}N abundance as a result of inoculation with N_2-fixing bacteria, although there was no significant difference in the N accumulation as a result of inoculation. Such a high isotopic dilution indicates the advantage of using ^{15}N methods which allow differentiation between the nitrogen coming from nutrient solution (or soil) and from the atmosphere. Moreover, it allows estimation of the biologically fixed nitrogen which has actually been taken up by the plant.

The enumeration of bacteria using FA against *Klebsiella* sp. (NIAB-1) indicated the ability of these bacteria to survive in association with roots as a result of inoculation. The indigenous population was also found to be fairly high ($2.0 \times 10^5 g^{-1}$ root). As a result of inoculation a 1.5-fold increase in the cell numbers was observed ($3.5 \times 10^5 g^{-1}$ fresh root). However, the number of bacteria is no indication of their being in a N_2-fixing condition.

During the present study, some further evidence has been reported regarding colonization of roots with inoculated bacteria. Some of the enteric bacteria have been observed to be inside the root hair cells. Maximum N_2-fixation has also been estimated in the case of inoculation with *Enterobacter* sp. (QH7). Localization of enteric bacteria on root hairs has also been reported by Haahtela *et al.* (1987) in the case of *Poa pratensis* and *Triticum aestivum*.

The results presented here further confirm earlier reports of nitrogen fixation associated with the roots of kallar grass growing in saline-sodic soils. It further indicates that such fixation makes a significant contribution to the nitrogen requirement of this plant. In this respect further studies on the mode of colonization of N_2-fixing bacteria and the method of exchange of fixed N and photosynthetic carbon between the bacteria and the roots will help in gaining a better understanding of the phenomenon.

Acknowledgement

Financial support for this research was partly provided by the U.S. National Academy of Sciences/National Research Council by means of a grant from the U.S. Agency for International Development. Thanks are due to Mr. Anwar-ul-Haq and Mr. Ghulam Rasool for their technical assistance. KAM would like to thank the Third World Academy of Sciences for a travel grant.

Reference

Aleksic Z, Broeshart H and Middleboe V 1968 The effect of nitrogen fertilization on the release of soil nitrogen. Plant and Soil 29, 474–478.

Bilal R and Malik K A 1987 Isolation and identification of a N_2-fixing zoogloea forming bacterium from Kallar grass histoplane. J. Appl. Bacteriol. 62, 289–294.

Boddey R M, Chalk P M, Victoria T L, Matsui E and Döbe-reiner J 1983 The use of the ^{15}N isotope dilution technique to estimate the contribution of associated biological nitrogen fixation to the nitrogen nutrition of *Paspalum notatum* cv batatais. Can. J. Microbiol. 29, 1036–1045.

Bremner J M 1965 Total nitrogen. *In* Methods of Soil Analysis. Eds. C A Black *et al.* pp. 1149–1178. American Society of agronomy, Madison WI USA.

Chalk P M 1985 Estimation of N_2-fixation by isotope dilution: An appraisal of techniques involving ^{15}N enrichment and their application. Soil Biol. Biochem. 17, 389–410.

Cohen E, Okon Y, Kiegel J, Nur I and Henis Y 1980 Increase in dry weight and total nitrogen content in *Zea mays* and *Setaria italica* associated with nitrogen fixing *Azospirillum* spp. Plant Physiol. 66, 246–249.

Fried M and Broeshart H 1975 An independent measurement of the amount of nitrogen fixed by a legume crop. Plant and Soil 43, 707–711.

Fried M and Middleboe V 1977 Measurement of amount of nitrogen fixed by a legume crop. Plant and Soil, 713–715.

Giller K E, Wani S P and Day J M 1986 Use of isotope dilution to measure nitrogen fixation associated with roots of sorghum and millet genotypes. Plant and Soil 90, 255–263.

Haahtela K, Laakso T, Nurmiaho Lassila E L and Korhonen T K 1987 Effect of inoculation of *Poa pratensis* and *Triticum aestivum* with root associated N_2-fixing *Klebsiella, Enterobacter* and *Azospirillum*. Paper presented at the 4 th International Symp. on N_2-fixation with non-legumes, Rio de Janeiro, Brazil Aug. 23–28, 1987.

Kapulnik Y, Sarig S, Nur I, Okon Y 1983 Effect of *Azospirillum* inoculation on yield of field grown wheat. Can. J. Microbiol. 29, 895–915.

Legg J O and Stanford G 1967 Utilization of soil and fertilizer N by oats in relation to the available N status of soils. Soil Sci. Soc. Am. Proc. 31, 215–219.

Malik K A and Zafar Y 1985 Quantification of root associated nitrogen fixation as estimated by ^{15}N isotopic dilution. *In* Nitrogen and the Environment. Eds. K A Malik, M I H Aleem and S H M Naqvi. pp. 161–171. Nuclear Institute for Agriculture & Biology, Faisalabad.

Malik K A, Zafar Y and Hussain A 1980 Nitrogenase acvivity in the rhizosphere of kallar grass. Biologia 26, 107–112.

Malik K A, Zafar Y and Hussain A 1981 Associative dinitrogen fixation in *Diplachne fusca* (Kallar grass). *In* BNF Technology for Tropical Agriculture. Eds. P H Graham and H C Harris. pp. 503–507.

Malik K A, Zafar Y, Bilal R and Azam F 1987 Use of ^{15}N isotope dilution for quantification of N_2-fixation associated with roots of Kallar grass (*Leptochloa fusca* (L.) Biology and Fertility of Soils. 4, 103–108.

Malik K A, Bilal R, Azam F and Sajjad M I 1988 Quantification of N_2-fixation and survival of inoculated diazotrophs associated with roots of Kallar grass. Plant and Soil (In press).

Miranda C H B and Boddey R M 1987 Estimation of biological nitrogen fixation associated with 11 ecotypes of *Panicum maximum* grown in N-15 labeled soil. Agron. J. 79, 558–563.

Nohrstedt H O 1984 Carbon monoxide as an inhibitor of N_2-ase activity (C_2H_2 reduction) in control measurements of endogenous formation of ethylene by forest soils. Soil Biol. Biochem. 16, 520–526.

Okon Y and Kapulnik Y 1986 Development and function of

Azospirillum-inoculated roots. Plant and Soil 90, 3–16.

Patriquin D G, Döbereiner J and Jain D K 1983 Sites and processes of association between diazotrophs and grasses. Can. J. Microbiol. 29, 900–915.

Reinhold B, Hurek T, Niemann E G and Fendrik I 1986 Close association of *Azospirillum* and diazotrophic rods with different root zones of Kallar grass. Appl. Environ. Microbiol. 52, 520–526.

Rennie R J 1979 Comparison of ^{15}N aided methods for determining symbiotic dinitrogen fixation. Revue D'Ecologie et de Biologie du Sol 16, 455–463.

Rennie R J 1980 ^{15}N-isotope dilution as a measure of dinitrogen fixation by *Azospirillum brasilense* associated with maize. Can. J. Bot. 58, 21–24.

Rennie R J, Rennie D A and Fried M 1978 Concepts of ^{15}N usage in dinitrogen fixation studies. *In* Isotopes in Biological Dinitrogen Fixation, pp. 107–130 International Atomic Energy Agency, Vienna.

Vlassak K, Herman K A H and van Rossen A R 1976 Dinoseb as a specific inhibitor of nitrogen fixation in soil. Soil Biol. Biochem. 8, 91–93.

Witty J F 1983 Estimating N fixation in the field using ^{15}N labelled fertilizer: Some problems and solutions. Soil Biol. Biochem. 15, 631–639.

Yoshida T and Padre B C I 1977 Transformation of soil and fertilizer nitrogen in paddy soil and their availability to rice plants. Plant and Soil 47, 113–123.

Zafar Y and Malik K A 1984 Photosynthetic system of *Leptochloa fusca* (L.) Kunth. Pak. J. Bot. 18, 109–116.

Zafar Y, Ashraf M and Malik, K A 1986 Nitrogen fixation associated with roots of Kallar grass (*Leptochloa fusca* L.) Kunth. Plant and Soil 90, 93–106.

Zafar Y, Malik K A and Niemann E G 1987 Studies on N_2-fixing bacteria associated with the salt tolerant grass, *Leptochloa fusca* (L.) Kunth. MIRCEN J. 3, 45–56.

F. A. Skinner et al. (Eds.), Nitrogen fixation with non-legumes, 311–319.

Selection of sugar cane cultivars for associated biological nitrogen fixation using ^{15}N-labelled soil

S. URQUIAGA, P.B.L. BOTTEON and R.M. BODDEY
EMBRAPA-UAPNPBS, Km 47, Seropédica 23851, Rio de Janeiro, Brazil

Key words: nitrogen fixation, ^{15}N dilution technique, quantification, sugar cane, varietal selection

Abstract

Ten varieties of sugar cane, and a non-N_2-fixing control crop (*Brachiaria radicans*), were planted in a concrete tank ($20 \times 6 \times 0.6$ m) filled with ^{15}N-labelled soil, with the objective of selecting those varieties of sugar cane which most benefited from associated biological nitrogen fixation (BNF). The tank contained *ca.* 80 tonnes of red-yellow podzolic soil (0.080 % N) uniformly mixed with ^{15}N-labelled compost made from sugar cane bagasse, filter cake and ^{15}N-labelled ammonium sulphate and potassium nitrate. The plant varieties were planted in 4 replicates (blocks) in 3.0 m rows. Temporal changes of ^{15}N enrichment of the plant available N in the soil were followed by analysing soil and leaf samples at 50 day intervals. At the final harvest (310 days after planting) all plant material was dried, weighed and analysed for total N and ^{15}N content. Several of the varieties yielded the equivalent of more than 200 t fresh cane per hectare, and all commercial varieties contained more than 250 kg N ha^{-1}. The ^{15}N enrichment of the soil mineral N decreased from *ca.* 1.0 to 0.16 atom % ^{15}N excess during the growth of the sugar cane. This suggests that the different patterns of N uptake by the plants could cause large differences in the enrichment of the plant N making it difficult to quantify BNF contributions accurately. This problem is discussed in detail, but it is apparent that for some of the varieties (Krakatau, SP 70-1143) a minimum of 25%, and perhaps as much as 55% of the plant N was derived from plant-associated BNF.

Introduction

Under Brazilian conditions, when sugar cane is harvested the trash is invariably burned off before cutting. Nitrogen fertilizer inputs are usually considerably less than the total nitrogen removed in the crop so that, in the long term, depletion of soil nitrogen reserves would be expected. The fact that this rarely seems to occur has led several authors to suggest that sugar cane may benefit significantly from contributions of biological nitrogen fixation (Döbereiner, 1961; Ruschel *et al.*, 1978; Purchase, 1980; Patriquin, 1982).

Recent work in Barbados, where cane is not subjected to preharvest burning, has suggested that biological nitrogen fixation (BNF) associated with cane trash may contribute up to 20 or 40 kg N ha^{-1} year^{-1} (Patriquin, 1982; Hill *et al.*, 1987). However,

the results of a recent N balance experiment performed in pots (64 kg soil per pot) at our institute in Rio de Janeiro (Lima *et al.*, 1987) suggest that plant-associated BNF could be a very important source of nitrogen for the sugar cane crop (perhaps as high as 200 kg N ha^{-1} year^{-1}). Furthermore, the results showed that there was a very strong effect of plant cultivar on this BNF contribution.

The total nitrogen balance technique suffers from two main disadvantages. firstly, for significant results to be obtained in reasonably short periods of time (years rather than decades) it is necessary to grow the plants in a limited quantity of soil (usually pots), which produces an abnormally compact rooting system which may affect plant-associated BNF. Secondly, such balances are exceedingly labour intensive, requiring large numbers of accurately performed nitrogen analyses. To avoid these problems the ^{15}N isotope dilution technique was employed to investigate the contribution of plant-associated BNF to 10 different varieties of sugar cane.

The principal assumption involved in the use of this technique is that in a soil labelled with ^{15}N the 'N$_2$-fixing' plants and the non-fixing control plants all obtain nitrogen from the soil with exactly the same ^{15}N enrichment. If ^{15}N is added to the soil as a single dose of soluble N fertilizer applied to the soil surface then there will be large temporal and spatial differences in ^{15}N enrichment in soil mineral N. Small differences between plants in the patterns of uptake of this nitrogen from the soil will cause the plants to accumulate N with different ^{15}N enrichments hence invalidating the use of the technique (Witty, 1983; Boddey, 1987). In an attempt to avoid, at least partially, these problems, the sugar cane varieties were grown in a concrete tank in soil mixed entirely with ^{15}N-labelled organic matter.

Materials and methods

Experimental layout

The experiment was planted in a reinforced concrete tank, $20 \times 6 \times 0.8$ m, provided with drains and filled to a depth of 60 cm with *ca.* 80 tonnes of soil (typic Hapludult – Table 1). In order to label the soil with ^{15}N, nine sacks of compost were prepared by mixing 80 g of ammonium sulphate (21.4 atom % ^{15}N) and 240 g of potassium nitrate (27.1 atom % ^{15}N) with 5 and 10 kg (fresh weight), respectively, of bagasse and filter cake from a sugar cane mill. The sacks of compost were maintained in a greenhouse (temperature 20–35°C) at *ca.* 70% moisture for 40 days, and mixed thoroughly at 7-day intervals to aid aeration. The compost initially heated up to 50–60°C and the colour darkened as decomposition progressed and the nitrogen was immobilized. This compost and the lime (600 g

m^{-2}) and fertilizers (35.3 g P, 32.8 g K and 12.0 g of fritted trace elements m^{-2}) were mixed into the soil as follows: The soil was distributed in a layer 10 cm thick and the amendments were spread evenly over marked areas (3×4 m), then incorporated using a hoe. This process was repeated in 10 cm layers until the tank was filled to a depth of 60 cm.

Filling of the tank with soil was completed on 18 th April 1986 and was planted on 29 th May. The tank was divided into 4 equal blocks each consisting of 11 rows (plots) of 3 m in length spaced 90 cm apart. For the sugar cane varieties (Table 2), whole stem pieces (*ca.* 1 m in length) were planted at a depth of 15 cm, and cuttings of the non-N$_2$-fixing control plant, *Brachiaria radicans* (cv. IRI 442) were planted at 5 cm depth. Several previous studies have demonstrated that this grass obtains negligible quantities of nitrogen from plant-associated BNF (Pereira *et al,* 1980, Boddey and Victoria, 1986; Miranda and Boddey, 1987).

Soil water monitored with ceramic cup tensiometers and the experiment was irrigated with tap water when soil suction exceeded 0.6 atmospheres.

At 100 days after emergence (DAE) of the sugar cane, the plants showed symptoms of copper deficiency and 2.5 g m^{-2} of CuSO$_4$.7H$_2$O were applied in solution to the soil surface.

Sampling and harvests

All the sugar cane plants emerged at approximately the same data, 10 th July 1986. By this time the *Brachiaria* had completely covered the plot area and was invading neighbouring sugar cane plots. For this reason all the *Brachiaria* was harvested (cut at 5 cm height) on 17 th July (6 days after emergence of the sugar cane).

Every 50 days soil samples were taken at 15 cm intervals to 60 cm depth for analysis of total available nitrogen and ^{15}N enrichment. Simultaneously, samples were taken of the third emergent leaf of

Table 1. Some physical and chemical properties of the Itaguaí series soil (Typic Hapludult)

Texture	pH(H$_2$O)	Total N (%)	Available P^a (g.g^{-1})	Exchangeable cations			
				Ca^{2+}	Mg^{2+}	K$^+$	Al^{3+}
						m-equiv. 100 g soil	
Loam sandy	4.7	0,08	1.2	1.3	0.9	0.06	0.2

^{a} Extracted with an aqueous solution of 0.05 *N* HCl and 0.025 *N* H$_2$SO$_4$

Table 2. Description and origins of the 10 varieties of sugar cane

Variety	Origin	Comments
CB 47–89	Campos (Rio de Janeiro) Brazil	Large BNF input estimated by Lima *et al.* (1987). [0.5%][a]
CB 45–3	Campos (Rio de Janeiro) Brazil	Variety most planted in NE Brazil [21.4%]
NA 56–79	Northern Argentina.	Variety most planted in all Brazil especially São Paulo state. [27.7%]
IAC 52–150	Inst. Agron. Campinas, São Paulo.	[3.3%]
SP 70–1143	Copersucar, São Paulo	Rust resistant variety recommended to replace NA 56–79. [8.9%]
SP 71–799	Copersucar, São Paulo.	Modern variety produces well on high fertility soils. [0.5%]
SP 79–2312	Copersucar, Sao Paulo.	Clone which produces cane of high sugar content (Brix). At present undergoing field tests.
Chunnee	(*Saccharum barberi*)	Not planted commercially but used in plant breeding.
Caiana	(*S. officinarum*)	Not planted commercially but used in plant breeding.
Krakatau	(*S. spontaneum*)	Not planted commercially but used in plant breeding.

[ ][a] Percentage of area planted to sugar cane in Brazil Source: Planalsucar Annual report 1986, Piracicaba, São Paulo.

each sugar cane variety and all the *Brachiaria* plots were harvested as before.

All aerial tissue from all plots was harvested from 10–13 th April 1987 (*ca.* 250 DAE) by separating stems, flag (green) leaves and senescent leaves of the cane plants and each fraction was weighed and sampled. For sampling, the cane stems were split longitudinally and cut into short lengths (*ca.* 15 cm), and this also aided subsequent drying.

Analyses

The samples of sugar cane stem were dried for a minimum of 72 h in a forced air oven at 65°C. All other plant material was oven dried for 48 h or more. All plant material was ground to 40 mesh (< 0.42 mm) using a Wiley mill. Total N analyses were performed in duplicate on 200 mg amounts of plant material using the procedure of Bremner and Mulvaney (1982). After titration against dilute sulphuric acid, 50–100 μmol of extra acid was added and the samples reduced in volume by evaporating at 65°C before being transferred to small (10 ml) vials and evaporated to dryness. These vials were connected to the mass spectrometer sample preparation unit similar to that described by Pruden *et al.* (1985). The mass spectrometer used for the ^{15}N analyses was a VG 900 (VG Isogas, Middlewich, Cheshire, UK) and N$_2$ gas was generated by injection of 1 ml of degassified lithium hypobromite solution via a silicone septum. Air contamination was corrected for by calculations based on readings of the 32 (0$_2$) or 40 (A) peaks.

The estimation of the concentration of mineral

nitrogen in the soil was performed by shaking 150 g of soil with 300 ml of 2M KCl solution for 30 min. After the soil had settled, 150 ml of the supernatant liquid were steam distilled with 1 g of magnesium oxide and 1 g of Devarda's alloy, and 150 ml of distillate were collected and titrated as described by Keeney and Nelson (1982). These distillates were acidified and dried as before for analysis of ^{15}N enrichment. Where samples contained less than 500 μg N (the minimum quantity measurable with the mass spectrometer) exactly 1 mg of unlabelled N (0.3666 atom % ^{15}N) was added to the sample before analysis.

Results and discussion

When plant material is incorporated into the soil the mineralization rate of this material is initially rapid and then subsequently decreases. If this material is labelled with ^{15}N and incorporated into an unlabelled soil, the contribution of the ^{15}N labelled material to net mineralization gradually declines with respect to the unlabelled N input from soil organic matter, and the ^{15}N label of the mineralized N also declines with time (Broadbent and Nakashima, 1974; Ladd *et al.*, 1981; Boddey *et al.*, 1984). Evidently the ^{15}N-labelled compost utilized in this experiment behaved similarly: as soil N availability decreased due to plant uptake and leaching losses etc. (Table 3), the ^{15}N enrichment of this mineral N also declined (Table 4).

The initially high availability of the soil N was reflected in the rapid growth of the *Brachiaria radicans* which completely covered the plots within 50

Table 3. Concentration of mineral nitrogen ($NH_4^+ + NO_3^-$)[a] in soil at four depths during the growth of sugar cane.

Depth (cm)	Concentration of mineral N (μg.g dry soil^{-1})				
	50 DAE[b]	100 DAE	150 DAE	200 DAE	250 DAE
0–15	11.8	7.7	3.9	4.4	3.0
15–30	14.8	7.3	4.0	4.3	2.9
30–45	12.4	7.2	3.6	4.1	2.1
45–60	13.1	6.2	3.0	3.9	2.6
Mean[c]	12.8	7.1	3.6	4.2	2.7

[a] Extracted with 2M KCl solution (Keeney and Nelson, 1982)
[b] DAE, Days after emergence of sugar cane.
[c] Coefficient of variation = 36.8%
Means of 4 replicates.

days after planting, and accumulated a high ^{15}N enrichment (Table 5). For the subsequent 93 days of growth (from 7–100 DAE of the sugar cane) the *Brachiaria* continued to grow vigorously although it was apparent from the decline in % N in the plants, that soil N availability was falling.

Sugar cane germinates slowly, first sending out supporting roots and then later lateral roots which absorb nutrients from the soil. The literature suggests that little or no nutrients are absorbed from the soil until shoot emergence (Bacchi, 1983). For this reason the rapid accumulation of nitrogen (and ^{15}N) by the *Brachiaria* up until the first harvest (7 DAE of the sugar cane) does not parallel the N accumulation of the sugar cane plants and thus the N and ^{15}N accumulation during the first 50 days of growth of the *Brachiaria* should be discounted if this plant is to be considered as a non-N_2-fixing control plant for the sugar cane (Table 5).

Sugar cane growth was vigorous throughout the experiment, the commercial varieties yielding the equivalent of between 175 and 230 t fresh cane ha^{-1} (65 to 70 t dry matter) within 250 days of emergence (Table 6). The variety of *Saccharum officinarum*, Caiana, was the only variety that germinated poorly, which accounts for the very low dry matter and cane yield. The *Saccharum barberi* variety, Chunnee, produced only approximately half of the dry matter and 40% of the cane yield of the commercial varieties, although the sugar content of the cane, as revealed by the brix reading (14.3%) was similar to the mean Brix of the commercial cultivars (14.4% —full data not shown). The *Saccharum spontaneum* variety, Krakatau, accumulated only slightly less dry matter than the commercial varieties but had a much lower cane yield and sugar content (Brix = 6.3%). However, this variety produced more green leaves with a higher N content, and hence more nitrogen, than any other variety (Table 6).

After 100 DAE of the sugar cane, growth of the *Brachiaria* was limited due to shading by the tall sugar cane plants, except in one plot (block 4) where the *Brachiaria* was in the end plot of the tank and hence only partially shaded. The data from this plot show that while the shading reduced nitrogen accumulation by the *B. radicans* the effect on overall N accumulation and ^{15}N enrichment was not great: In this plot the *Brachiaria* accumulated 4.95 and 2.80 g m^{-2} of nitrogen with an enrichment of 0.248 and 0.173 atom % ^{15}N excess at harvests at 150 and 250 DAE of the sugar cane, respectively. Using these data for the last two harvests of the *Brachiaria*, instead of the means of all 4 blocks (Table 5), the overall weighted mean ^{15}N enrichments of the *Brachiaria* are reduced slightly from

Table 4. ^{15}N enrichment of mineral nitrogen ($NH_4^+ + NO_3^-$)[a] at four depths during the growth of sugar cane.

Depth (cm)	^{15}N Enrichment of mineral N (atom % ^{15}N excess)					
	− 50 DAE[b]	50 DAE	100 DAE	150 DAE	200 DAE	250 DAE
0–15	1.434	0.464	0.336	0.155	0.143	0.102
15–30	1.643	0.778	0.374	0.192	0.174	0.134
30–45	1.149	0.916	0.391	0.198	0.190	0.167
45–60	0.571	0.485	0.394	0.190	0.185	0.137
Mean[c]	1,200	0,661	0,374	0,184	0,173	0,135

[a] Extracted with 2 *M* KCl solution (Keeney and Nelson, 1982)
[b] DAE, Days after emergence of sugar cane.
[c] Coefficient of variation = 42.6%
Means of 4 replicates.

Table 5. Accumulation of dry matter and labelled and unlabelled nitrogen by *Brachiaria radicans*

Harvest	DAE[a]	Dry matter (g.m^{-2})	% N	Total N (g.m^{-2})	^{15}N enrichment atm % ^{15}N excess	^{15}N recovery (mg ^{15}N excess. m^{-2})
1	7	242	3.28	7.96	1.338	106.7
2	50	338	2.43	8.19	0.840	68.9
3	100	362	1.67	6.04	0.381	23.1
4	150	201	1.86	3.70	0.253	9.3
Final	250	117	0.94	1.00	0.160	1.7
Total-all harvests		1259	2.15[b]	26.89	0.780[c]	566.0
Total-last 4 harvests		1017	1.89[b]	18.93	0.546[c]	278.0

[a]DAE, Days after emergence of sugar cane

[b]Weighted mean % N = $\dfrac{\text{Total N}}{\text{Total dry matter}} \times 100$

[c]Weighted mean atom % ^{15}N excess = $\dfrac{\text{Total recovery of excess }^{15}\text{N (mg)}}{\text{Total nitrogen (g)} \times 10}$

0.780 to 0.721 atom % excess for all harvests, and from 0.546 to 0.497 atom % excess for the mean of the last 4 harvests.

While it was not possible to take sequential harvests of the sugar cane without destroying the plants, leaf samples were taken for analysis of nitrogen (Table 7) and ^{15}N enrichment (Table 8). These latter data represent the cumulative ^{15}N enrichment over the whole growth period and cannot be compared with the ^{15}N enrichment of the soil mineral N (Table 4) which is a point measurement. Again there can be no strict comparison of the ^{15}N enrichment of the *Brachiaria* with that of the 3rd leaf sample of the sugar cane varieties as the former

is a mean label of the material from the whole plot while the latter is only a leaf sample which will not necessarily represent the mean enrichment of the whole plant. However, the ^{15}N enrichments of the various parts of each sugar cane variety (stem, green leaves and senescent leaves) at the final harvest were similar and not very different from the enrichment of the 3rd leaf sample at this harvest (Table 8 and 9). For this reason the fact that the ^{15}N enrichment of the N accumulated by the *Brachiaria* from 7 to 50 DAE of the sugar cane is similar to the enrichments of the 3rd leaf samples of most of the sugar cane varieties at the first sampling (50 DAE), suggests that there was little or no contribution of

Table 6. Fresh cane yield, dry matter accumulation and N content of 10 sugar cane varieties during 250 days of plant growth

Variety	Fresh cane yield (t.ha^{-1})	Dry matter (g m^{-2})				%N		
		Stem	Green leaves	Senescent leaves	Total	Stem	Green leaves	Senescent leaves
CB 47–89	208ab	4505a	961a	1856a	7323a	0.19b	1.14ab	0.36b
CB 45–3	219a	4995a	981ab	1754a	7730a	0.22b	1.10ab	0.37ab
NA 56–79	229a	5387a	767ab	1573a	7727a	0.19b	1.12ab	0.37ab
IAC 52–150	199abc	4173ab	1131ab	2195a	7499a	0.18b	1.14ab	0.33b
SP 70–1143	175abc	4119ab	875ab	1570a	6564a	0.19b	1.22a	0.41ab
SP 71–799	193abc	4272ab	757ab	1886a	6916a	0.21b	1.15ab	0.38ab
SP 79-2312	214ab	5465a	591ab	1362a	7418a	0.29ab	1.04bc	0.40ab
Chunee	89cd	2017ab	445ab	1194ab	3656ab	0.36a	0.90c	0.35b
Caiana	37d	635b	224b	246b	1107b	0.38a	1.14ab	0.44ab
Krakatau	100bcd	3144ab	1218a	1540a	5902ab	0.22b	1.23a	0.50a
Coefficient of variation (%)	28.3	39.7	44.0	29.5	33.8	19.0	5.5	14.6

(means of 4 replicates).

Table 7. Nitrogen content (%) of samples taken from third emergent leaf during the growth of ten sugar cane varieties

Variety	Concentration of total nitrogen (%)				
	50 DAE	100 DAE	150 DAE	200 DAE	250 DAE
CB 47–89	2.34	1.94	1.66	1.76	1.69
CB 45–3	2.34	1.85	1.72	1.77	1.69
NA 56–79	2.34	1.91	1.82	1.88	1.80
IAC 52–150	2.36	2.02	1.64	1.65	1.67
SP 70–1143	2.21	1.94	1.81	1.87	1.93
SP 71–799	2.43	2.18	1.76	1.72	1.94
SP 79–2312	2.31	1.90	1.82	1.84	1.69
Chunnee	2.16	1.98	1.59	1.49	1.50
Caiana	2.40	2.20	1.98	1.93	1.99
Krakatau	2.61	2.33	2.02	1.99	1.98
Means	2.35A	2.03B	1.78C	1.80C	1.79C

CV = 14.7%
(means of 4 replicates)

BNF to the sugar cane varieties during this initial growth period. This would be expected as at this time there was still considerable available soil nitrogen as revealed by the analyses of both soil mineral N (Table 3) and the N content of the leaf samples at 50 DAE (Table 7). Experience to date suggests that BNF associated with grasses is only significant when soil nitrogen is deficient (Boddey *et al.*, 1983; Miranda and Boddey, 1987).

As it can be assumed that until shoot emergence the sugar cane cultivars absorbed negligible quantities of N from the soil, from sugar cane emergence

Table 8. ^{15}N enrichment of nitrogen in samples taken from third emergent leaf during the growth of ten sugar cane varieties

Variety/ Species	^{15}N enrichment (atm % ^{15}N excess)				
	DAE[a]				
	50	100	150	200	250
CB 47–89	0.946	0.544	0.467	0.399	0.344
CB 45–3	0.897	0.538	0.495	0.402	0.352
NA 56–79	0.800	0.515	0.457	0.357	0.322
IAC 52–150	0.855	0.486	0.449	0.382	0.330
SP 70–1143	0.828	0.533	0.466	0.392	0.300
SP 71–799	0.896	0.528	0.483	0.410	0.336
SP 79–2312	0.854	0.583	0.508	0.460	0.370
Chunnee	0.950	0.588	0.511	0.568	0.334
Caiana	0.740	0.518	0.441	0.378	0.324
Krakatau	1.014	0.607	0.513	0.316	0.246
Mean[b]	0.878a[c]	0.544b	0.479c	0.396d	0.326e

[a] DAE, Days after emergence of sugar cane
[b] Coefficient of variation = 14.7%
[c] Means followed by the same letter are not significantly different at $P = 0.05$ (Tukey test).
(means of 4 replicates).

to 250 DAE the commercial varieties of sugar cane, and the *S. spontaneum* variety, Krakatau, all accumulated more nitrogen than the *Brachiaria radicans* (Table 9). Hence, either the sugar cane varieties removed more N from the soil than the *Brachiaria,* or there was an input of N from BNF. Our previous studies have shown that this cultivar of *Brachiaria radicans* is very efficient at removing nitrogen from N deficient soils (Boddey and Victoria, 1986; Miranda and Boddey, 1987). In comparison with 11 ecotypes of *Panicum maximum,* only one was able to remove as much N from an N deficient soil (the same soil type as used in this experiment – Miranda and Boddey, 1987).

The much lower ^{15}N enrichment of the sugar cane varieties at the final harvest compared to that of the *Brachiaria,* again suggests that there was a input of unlabelled nitrogen from plant-associated BNF. However, an alternative explanation for these observations must be considered: Both the *Brachiaria* and the sugar cane varieties (except Caiana) grew vigorously during the first 50 days of growth of the sugar cane and accumulated nitrogen at similar ^{15}N enrichments. If the sugar cane was able to continue to absorb large quantities of nitrogen after 150 DAE when the ^{15}N enrichment of the available soil N was very low (Table 4) then this could have caused the sugar cane varieties to accumulate (overall) a much lower ^{15}N enrichment than the *Brachiaria*. However, the analysis of soil mineral N (Table 3) and the N analysis of *Brachiaria* (Table 5), showed that soil nitrogen availability during this last 100 days of plant growth was very low and it seems inpossible that the sugar cane

Table 9. Total N accumulation and ^{15}N enrichment of 10 sugar cane varieties and *Brachiaria radicans* after 250 days of plant growth.

Variety	Total N whole plant (g.m^{-2})	Atom % ^{15}N excess				Total ^{15}N recovery (mg ^{15}N exc. m^{-2})
		Stem	Green leaves	Senescent leaves	Weighted mean whole plant	
CB 47–89	26.5	0.309	0.315	0.325	0.316	84.4
CB 45–3	25.7	0.291	0.308	0.315	0.305	85.2
NA 56–79	24.6	0.314	0.300	0.319	0.309	75.2
IAC 52–150	27.1	0.308	0.281	0.298	0.292	80.7
SP 70–1143	24.5	0.258	0.249	0.275	0.259	63.7
SP 71–799	24.4	0.308	0.293	0.278	0.292	70.4
SP 79–2312	20.2	0.339	0.325	0.319	0.345	58.5
Chunnee	15.2	0.384	0.321	0.292	0.341	51.9
Caiana	4.9	0.306	0.309	0.300	0.308	15.9
Krakatau	29.4	0.256	0.213	0.274	0.241	70.7
B. radicans						
All harvests	26.9	–	–	–	0.780	209.6
Last 4 harvests	18.9	–	–	–	0.546	103.0

(means of 4 replicates).

varieties removed much soil N during this period. It is to be concluded therefore, that while the decline in the enrichment of available soil N makes it impossible to quantify accurately the BNF input to the sugar cane varieties, the data show that these plants obtained very significant contributions of nitrogen from plant associated BNF. Using the weighted mean ^{15}N enrichment of the *Brachiaria* at the last four harvests these estimates range from 37% to 56% of plant nitrogen derived from BNF (Table 10). A more conservative estimate is obtained if the ^{15}N enrichment of the variety SP 79-

Table 10. Estimates of the contribution of biological nitrogen fixation to 10 sugar cane varieties

Variety	Total N accumulation (kg ha^{-1})	^{15}N Enrichment (atm % ^{15}N exc.)	Estimates of the contribution of BNF			
			A[a]		B[b]	
			%[c]	kgN.ha^{-1}	%[c]	kgN.ha^{-1}
CB 47–89	265ab	0.316bc	41.9	111	9	24
CB 45–3	257ab	0.305bcd	44.1	110	18	46
NA 56–79	246ab	0.313bc	42.9	106	11	27
IAC 52–150	271ab	0.300bc	46.2	125	16	43
SP 70–1143	245abc	0.261bc	52.3	128	26	64
SP71–799	244abc	0.298bc	46.1	112	1-6	39
SP 79–2312	202abc	0.344bc	41.7	84	cont.	0
Chunnee	152c	0.349b	37.2	56	2	3
Caiana	49d	0.309bc	42.2	21	11	5
Krakatau	294a	0.255c	55.6	163	31	91
B. radicans						
Last 4 harvests	189bc	0.546a				
Significance	**	**				
Coefficient of variation	27%	11.4%				

[a]Values calculated using the isotope dilution technique using *B. radicans* (last 4 harvests) as control.
[b]Values calculated using the isotope dilution technique using sugar cane variety SP 79–2312 as control.

[c]% Nitrogen derived from BNF $= (1 - \dfrac{\text{atom \% }^{15}\text{N excess in cane variety}}{\text{atom \% }^{15}\text{N excess in control plant}}).\ 100$

2312 is used as a control, but this still yields estimates of BNF contributions of up to $91\,kg\,ha^{-1}$.

Conclusions

The data show that the contribution of BNF to the N-nutrition of sugar cane can be highly significant under environmental conditions which closely resemble those existing in the field. While no microbiological investigations were performed in this study, our colleagues at here at EMBRAPA have recently isolated a novel N_2-fixing bacteria, *Acetobacter diazotrophicus*, from sugar cane roots and stems from various sites in Brazil (Cavalcante and Döbereiner, 1988), which is a strong candidate for one of the diazotrophs responsible for the observed plant-associated BNF.

There is good evidence of differences in N_2-fixing capability between the sugar cane varieties. The variety SP 70-1143 is known to grow well in conditions of low soil fertility and the data suggest that of the commercial varieties this one benefited most from BNF. Finally, the *S. spontaneum* variety, Krakatau, which showed the highest BNF contribution of all, while useless for commercial sugar cane production owing to its low sugar content, may become a useful parent in any program to breed for high N_2-fixing sugar cane varieties.

Acknowledgements

The authors would like to thank Dr Johanna Döbereiner for her guidance and encouragement and Drs Caio Cardoso and James Irving of Copersucar, Piracicaba, for their helpful advice and provision of planting material. The authors also acknowledge the dedicated technical assistance of Roberto G de Souza, Jorge C de Souza, Fernando V Cunha, Claudio F Augusto and Silmo Soares Perrut.

Financial support for this research was provided by the U.S. National Academy of Sciences, National Research Council, through a grant from the U.S. Agency for International Development.

References

Bacchi O O S 1983 Botánica da cana de açúcar. *In* Nutrição e Adubação da Cana de Açúcar no Brasil. Ed. J Orlando Filho.

pp. 25–37. Instituto do Açúcar e Alcool (Planalsucar), Piracicaba, São Paulo, Brazil.

Boddey R M 1987 Methods for the quantification of nitrogen fixation associated with gramineae. CRC Critical Rev. Plant Sci. 6. 209–266.

Boddey R M and Victoria R L 1986 Estimation of biological nitrogen fixation associated with *Brachiaria* and *Paspalum* grasses using ^{15}N labelled organic matter and fertilizer. Plant and Soil 90, 265–292.

Boddey R M, Chalk P M, Victoria R L and Matsui E 1983 The ^{15}N isotope dilution technique applied to the estimation of biological nitrogen fixation associated with *Paspalum notatum* cv. batatais in the field. Soil Biol. Biochem. 15, 25–32.

Boddey R M, Chalk P M, Victoria R L and Matsui E 1984 Nitrogen fixation by nodulated soybean under tropical field conditions estimated by the ^{15}N isotope dilution technique. Soil Biol. Biochem. 16, 583–588.

Bremner J M and Mulvaney C S 1982 Nitrogen – total. *In* Methods of Soil Analysis Part II. Chemical and Microbiological properties. Monograph 9(2). Eds. A L Page, R H Miller and D R Keeney. pp. 595–624. Am. Soc. Agron., Madison, Wisconsin, USA.

Broadbent F E and Nakashima T 1974 Mineralization of carbon and nitrogen in soil amended with carbon-13 and nitrogen-15 labeled plant material. Soil Sci. Soc. Am. Proc. 38, 313–315.

Cavalcante V A and Döbereiner J 1988 A new acid-tolerant nitrogen-fixing bacterium associated with sugar cane. Plant and Soil 108. 23–31.

Döbereiner J 1961 Nitrogen fixing bacteria of the genus *Beijerinckia* Derx. in the rhizosphere of sugar cane. Plant and Soil 15, 211–217.

Hill N M, Patriquin D G and Sircom K 1987 Oxygen inhibition-temperature hypothesis to explain high levels of aerobic nitrogen fixation in plant litters in warm climates. *In* Symposium on The Contribution of Biological Nitrogen Fixation to Plant Production, Cisarua, Indonesia, August 3–7.

Keeney D R and Nelson D W 1982 Nitrogen – inorganic forms. *In* Methods of Soil Analysis. Part II. Chemical and Microbiological Properties. Monograph 9(2). Eds. A L Page, R H Miller and D R Keeney. pp. 643–698. Amer. Soc. Agron., Madison, Wisconsin, USA.

Ladd J N, Oades J M and Amato M 1981 Distribution and recovery of nitrogen from legume residues decomposing in soils sown to wheat in the field. Soil Biol. Biochem. 13, 251–256.

Lima E, Boddey R M and Döbereiner J 1987 Quantification of biological nitrogen fixation associated with sugar cane using a ^{15}N aided nitrogen balance. Soil Biol. Biochem. 19, 165–170.

Miranda C H B and Boddey R M 1987 Estimation of biological nitrogen fixation associated with 11 ecotypes of *Panicum maximum* grown in ^{15}N labelled soil. Agron. J. 79, 558–563.

Patriquin D G 1982 Nitrogen fixation in sugar cane litter. Biol. Agric. Hort. 1, 39–64.

Pereira P A A, Döbereiner J and Neyra C A 1980 Nitrogen assimilation and dissimilation in five genotypes of *Brachiaria* spp. Can. J. Bot. 59, 1475–1479.

Pruden G, Powlson D S and Jenkinson D S 1985 The measurement of ^{15}N in soil and plant material. Fertilizer Res. 6, 205–218.

Purchase B S 1980 Nitrogen fixation associated with sugar cane. Proc. South African Sugar Cane Technologists Association, June. pp. 173–176.

Ruschel A P, Victoria R L, Salati E and Henis Y 1978 Nitrogen fixation in sugar cane (*Saccharum officinarum*.). *In* Environmental Role of Nitrogen-fixing Blue-green Algae and Asymbiotic Bacteria. Ed. U Granhall, Bull. Ecol. (Stockholm) 26. 297–303.

Witty J F 1983 Estimating N_2-fixation in the field using ^{15}N-labelled fertilizer: Some problems and solutions. Soil Biol. Biochem. 15, 631–639.

F. A. Skinner et al. (Eds.), Nitrogen fixation with non-legumes, 321–330.
© 1989 by Kluwer Academic Publishers.

Occurrence of effective nitrogen-scavenging bacteria in the rhizosphere of kallar grass

T. HUREK[1], BARBARA REINHOLD[1], B. GRIMM[2], I. FENDRIK[1] and E.-G. NIEMANN[1]
[1]Institute of Biophysics, [2]Institute of Botany, University of Hannover, Herrenhäuser Str. 2, D-3000 Hannover 21, FRG

Key words: kallar grass, N_2-fixation, N_2O, nitrogen scavengers

Abstract

Bacteria occurring in high numbers on the rhizoplane of kallar grass grown at a natural site in Pakistan were effective scavengers of traces of combined nitrogen from the atmosphere. Bacteria grew under appropriate conditions in nitrogen-free semi-solid malate medium in the form of a typical subsurface pellicle which resulted in a significant nitrogen gain in the medium within 3 to 4 days of incubation; this could be also measured by ^{15}N-dilution. Bacteria grew and incorporated nitrogen under an atmosphere containing NH_3 and N_2O. A rapid and strong binding of strain W1 to roots of kallar grass grown in hydroponic culture was found by using a ^{32}P-tracer technique. We obtained no evidence for diazotrophy of our strains because they failed to grow on nitrogen-free media when gases of high purity were used. No $^{15}N_2$ was incorporated when bacteria were grown on $^{15}N_2$ although a nitrogen gain was found, no acetylene reduction was observed and no homology with DNA containing sequences of nifHDK structural genes for the nitrogenase components from *Klebsiella pneumoniae* were detected. Owing to close contact of these bacteria with roots of kallar grass, utilization of scavenged nitrogen by the plant may have to be taken into account.

Introduction

During estimation of populations of non-diazotrophic and diazotrophic bacteria on the rhizoplane and in the endorhizosphere of kallar grass (*Leptochloa fusca* (L.) Kunth), high numbers of bacteria were counted (about $10^8 g^{-1}$ root dry weight); these formed a subsurface pellicle in nitrogen-free semi-solid medium but failed to reduce acetylene (Reinhold *et al.*, 1986). They made up 20% of the total aerobic microflora on the rhizoplane but did not occur in high numbers in the root interior (Reinhold *et al.*, 1986, 1987a). From immunocological and physiological evidence we concluded that these bacteria are closely related to *Pseudomonas* sp. strain W1. This bacterium showed the same growth in semi-solid medium, failed to reduce acetylene and did not incorporate $^{15}N_2$ when incubated under an atmosphere enriched with ^{15}N.

Occurrence of growth in semi-solid medium, which is not accompanied by acetylene reduction, has also been reported by others during counting of diazotrophic microflora (Graciolli and Ruschel, 1981; Patriquin and McClung, 1978; Thomas-Bauzon *et al.*, 1982; Zafar *et al.*, 1986). Impressive simulation of nitrogen-fixation in cultural tests has been attributed to the ability of some putative N_2-fixing bacteria to be very efficient scavengers of traces of fixed nitrogen (Hill and Postgate, 1969).

In order to assess the ability of our strains with the designation W to grow on nitrogen-free media, several tests including N-balance studies, chemostat cultures, $^{15}N_2$ incorporation, and hybridization with a nifHDK-containing probe were carried out. For correlation of the colonization pattern found in Pakistan with adsorption behaviour, a binding experiment was included, using a ^{32}P-tracer technique. We found no evidence for diazotrophy, but strains proved to be effective nitrogen scavengers.

Materials and methods

Bacterial cultures

Bacterial strains were isolated by B. Reinhold from the following sources: diazotrophic spirillum ER8 and *Pseudomonas* sp. W1, W4, and W5 from roots of kallar grass, grown in Punjab, Pakistan, from samples treated after transport to Germany; strain W7 from the highest most probable number (MPN) dilution of the rhizoplane fraction from samples treated immediately after collection (Reinhold *et al.*, 1986); *Azospirillum brasilense* 68 from sedge roots originating from a soil with access to sea water near Rio de Janeiro (Reinhold *et al.*, 1988). Maintenance of stock cultures and long term preservation were performed as described previously (Hurek *et al.*, 1987; Reinhold *et al.*, 1985). Precultures were grown on combined nitrogen in a synthetic malate medium (SM) containing (g.l^{-1} distilled water): D,L-malate, 5; NH_4Cl, 0.5; yeast extract, 0.1 (Reinhold *et al.*, 1985). Twenty ml of this medium were inoculated with a loopful of bacteria from stock cultures and incubated at 35°C with reciprocal shaking (100 rev. min^{-1}).

Growth in nitrogen-free, semi-solid medium

Five ml of a mid-log-phase pre-culture were centrifuged for 10 min at 2,300 g and washed three times with nitrogen-free liquid SM medium. Samples corresponding to 0.1 μg of N, as estimated by total nitrogen analysis were usually used as inocula for growth experiments in nitrogen-free medium. Autoclaved semi-solid medium with 2 g agar l^{-1} was dispensed in 15 ml vials which were sealed either with sterile stoppers or sterile screw-caps not tightly screwed. A 5 ml portion of this medium contained 18 $\pm$ 1 μg N as estimated by total nitrogen analysis. The atmosphere consisted of normal air, industrial grade gases (98–99% pure) or gases of high purity (minimum 99.995% pure). When normal air was used, the cap was usually not tightly screwed to allow gas exchange. In experiments where the headspace consisted of high purity gases, tubes were sealed with teflon-coated plastic stoppers; where the headspace consisted of industrial grade gases, vials were sealed with rubber stoppers (Thomas Scientific, no. 1780J22, USA). If

not indicated otherwise, strains were grown for 96 h at 35°C. Wet digestion of the cultures were carried out according to Bergersen, (1980), and ammonia was measured enzymatically by a test combination (Boehringer, Mannheim, FRG). Ethylene was detected gaschromatographically (Reinhold *et al.*, 1986).

For ^{15}N-dilution experiments, strains were grown in liquid, nitrogen-free SM-medium to which 100 mg of $^{15}NH_4Cl$ was added (25.4% ^{15}N; Isocommerz, Berlin, GDR). A washed cell suspension containing ^{15}N-labelled bacteria corresponding to 8.9 $\pm$ 0.6 μg N was used as inoculum for 5 ml of the semi-solid, nitrogen-free SM-medium in a 15 ml vial, sealed with a rubber stopper. This medium contained only 1.5 $\pm$ 0.2 μg N because agar was replaced by agarose. Cultures were grown for 96 hours at 35°C under an atmosphere of air in our laboratory. ^{15}N enrichment was estimated by emission spectroscopy (Reinhold *et al.*, 1986).

Growth on $^{15}N_2$

$^{15}N_2$ experiments were done essentially as outlined by Reinhold *et al.* (1986). Fifteen-ml vials were sealed with rubber stoppers.

Growth in chemostat

Studies were done essentially as described previously (Hurek *et al.*, 1987). The pH was maintained constant at 7.0 and growth temperature was 35°C. N-free SM-medium containing sodium-L-malate was used.

DNA hybridization

Clone pBRSA30 containing the *nif*HDK structural genes for the nitrogenase components from *Klebsiella pneumoniae* was kindly provided by C. Elmerich, Institute Pasteur, Paris. DNA (0.1 μg) was labelled by nick translation with [^{32}P] dCTP. DNA of strians W1 and *Azospirillum brasilense* 68 were isolated by a modification of the method of Marmur (Reinhold *et al.*, 1987b). DNA was digested to completion with EcoRI (Boehringer, Mannheim, FRG), electrophoresed on a 0.6% agarose gel and transferred to nylon filters. The filter was prehybridized for 4 h and hybridized for 40 h in 50% formamide, 5 × Denhardt's solution, 100 μg/ml denatured salmon sperm DNA, 0.1% SDS and 5 × SSC (1 × SSC = 0.15 M NaCl/0.015 M Na-citrate) at 42°C, after which it was washed twice for 5 min each time in 2 × SSC and 0.1% SDS at room temperature and subsequently once for 15 min in 2 × SSC and 1% SDS at 60°C. The filter was exposed to Kodak XAR-5 film for autoradiography with an intensifying screen for 5 h at −80°C.

Aseptic growth of Kallar grass

Kallar grass seeds were surface sterilized according to Nilsson, (1957). The procedure was modified by shaking seeds for 45 min in 2% streptomycin sulphate and further for 45 min in a sterile solution of 0.2% $HgCl_2$ after removal of the antibiotic. Subsequently the seeds were washed 5 times for 30 min each with sterile demineralized water. Seeds were germinated on water agar containing 15 g of agar in 1 litre tap water and also on nutrient agar plates (Merck, no. 5450, FRG) to test for sterility. For germination, agar plates were incubated in a growth chamber with an 18-h day 6-h night cycle at 28°C and 24°C, respectively and 20 klx with Osram HQIL lamps. Three 5-day-old seedlings were transplanted (1 per tube) into tubes (Kloss *et al.,* 1984) each containing 40 ml of sterile Fåhraeus medium (FM) (Fåhraeus, 1957) supplemented with 0.3 mM KNO_3, and incubated in the growth chamber under the same conditions as described above. After 5 days, seedlings were transferred to tubes each containing 9 ml FM-medium supplemented with 100 mM NH_4Cl and 30 mM Na_2SO_4 (FMS-medium). After incubation for 2 days in the growth chamber, plants were ready for binding experiments.

Labelling of strain W1 with ^{32}P-orthophosphate

Bacteria were grown in tryptic soy broth (TSB) containing (g.l^{-1} of distilled water): pancreatic digest of casein USP, 17; papaic digest of soy meal, 3; NaCl, 5; glucose, 2.5, and supplemented with ^{32}P-orthophosphate (10 μCi ml^{-1}, Amersham Buchler, FRG). Cultures were incubated for 3 h at 37°C with reciprocal shaking (100 rev. min^{-1}). Cells were washed three times with TSB-medium and suspended in 6.5 ml of FMS-medium.

Conditions and determination of binding

Plants were inoculated with one ml of ^{32}P-labelled cell suspension (complete binding mixture contained 10 ml FMS-medium) and incubated in the growth chamber at 24°C and 5 klx illumination. Roots of the plants were totally immersed in this suspension. Subsequently roots were dipped for 3 min in 100 ml FMS-medium with slow stirring and were then carefully rinsed with 10 ml of this medium. Roots were separated from the shoot and dried at 105°C to constant weight (average root dry weight was 141 μg with a minimum value of 80μg and a maximum value of 230 μg). Dry roots were sealed in plastic foil and inserted into the borehole of a plexiglass solid with the shape of a normal 20 ml scintillation vial. The appearing Cerenkov-radiation was registered and monitored by a liquid scintillation counter connected to a multi-channel analyzer (Bunnenberg *et al.,* 1984; Bunnenberg *et al.,* 1987) Total cell counts were made with a Thoma counting chamber, and colony counts were done on nutrient agar plates.

Results

Binding of strain W1 to roots of kallar grass

In order to evaluate the potential of organisms with the designation W to adsorb to roots of kallar grass, the isolate W1 was incubated with intact plants grown in hydroponic culture in a minimal medium without carbon source.

Figure 1 shows that binding of W1 cells to roots of kallar grass was rapid and reached a maximum binding after about 2 h, after which binding remained constant for at least 12 h. Total cell and colony counts of unattached cells after 14.25 h were 2.3 × 10^8 cells ml^{-1} and 2.1 × 10^8 cells ml^{-1}, respectively. Thus, no growth of our bacteria in the binding mixture occurred and all cells remained alive. For evaluation of radioactive isotope leakage from labelled bacteria, 4 × 10^9 cells sealed in a dialysis tube were incubated with plants for 14.25 h. Specific radioactivity was 7.0 × 10^{-4} counts.

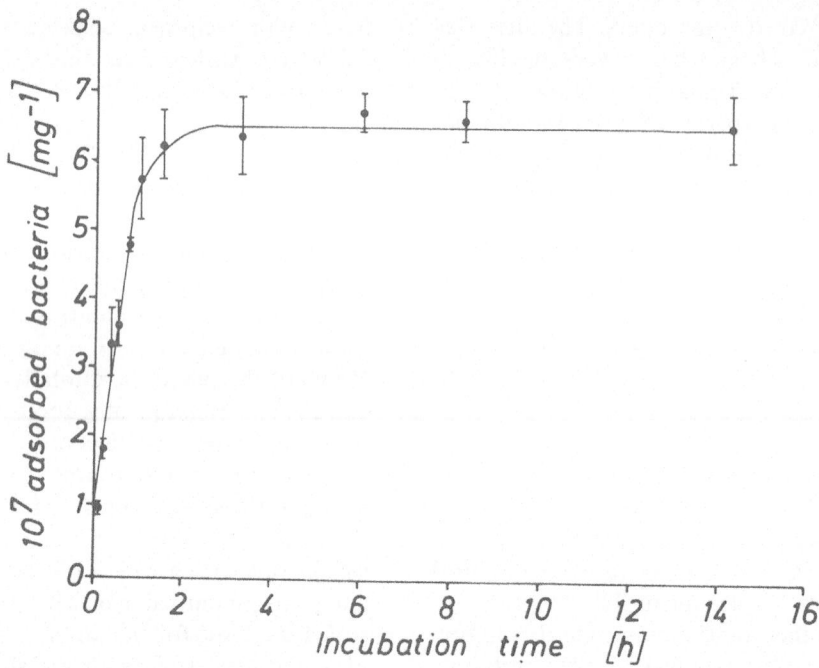

Fig. 1. Kinetics of ^{32}P-labelled strain W1 binding to roots of kallar grass. Bacterial numbers are averages of three replicates in two experiments and expressed as number of bacteria bound per mg root dry weight. Specific radioactivity was 7.9×10^{-4} counts. min^{-1} per cell and 7.0×10^{-4} counts. min^{-1} per cell, respectively, and concentration of cells in the medium was 2.0×10^8 ml^{-1} in each experiment.

min^{-1} per cell. The activity was 19.2 counts. min^{-1} compared to the background count of uninoculated plants of 20.1 counts. min^{-1}. Thus, radioactivity associated with roots of plants after inoculation was due to the adsorbed bacteria.

The number of bacteria bound depended on the initial concentration of unattached cells (Fig. 2). The double logarithmic plot revealed a linear binding response of up to 2×10^8 cells ml^{-1} in the binding mixture. The curve suggests that at a higher cell density (1.1×10^9 bacteria ml^{-1}) binding was approaching saturation. Incubation of plants with 1.1×10^9 bacteria ml^{-1} for 8 h revealed $1.3 \times 10^8 \pm 0.2 \times 10^8$ attached cells mg^{-1} root dry weight, indicating that even at this high inoculum density binding was already at an maximum after 3 h incubation.

Hybridization to sequences known to be related to N$_2$-fixation

Among N$_2$-fixing organisms the structural genes for nitrogenase are highly conserved (Ruvkun and

Ausubel, 1980) and there is evidence for homology with DNA fragments involved in an alternative N$_2$-fixation system (Bishop et al., 1986; Robson, 1986) first proposed in *Azotobacter* (Bishop et al., 1980). In order to find DNA-fragments of strain W1 which are homologous to structural genes for nitrogenase (*nif*), Southern hybridization between labelled pBRSA30 containing the *nif*HDK structural genes from *Klebsiella pneumoniae* and DNA from W1, and from *Azospirillum brasilense* 68 as a positive control, were carried out. Under hybridization conditions of low stringency the hybridization patterns resulted in no band with DNA from W1 and the one band of apparently 4.5 kb for DNA for the *Azospirillum* strain (Fig. 3).

Growth in nitrogen-free, semi-solid medium

When various strains with the designation W were grown in a N-free, semi-solid malate medium, we found a significant increase of nitrogen under a headspace of air (Table 1), all bacteria forming a subsurface pellicle within 12 h (Fig. 4). A similar

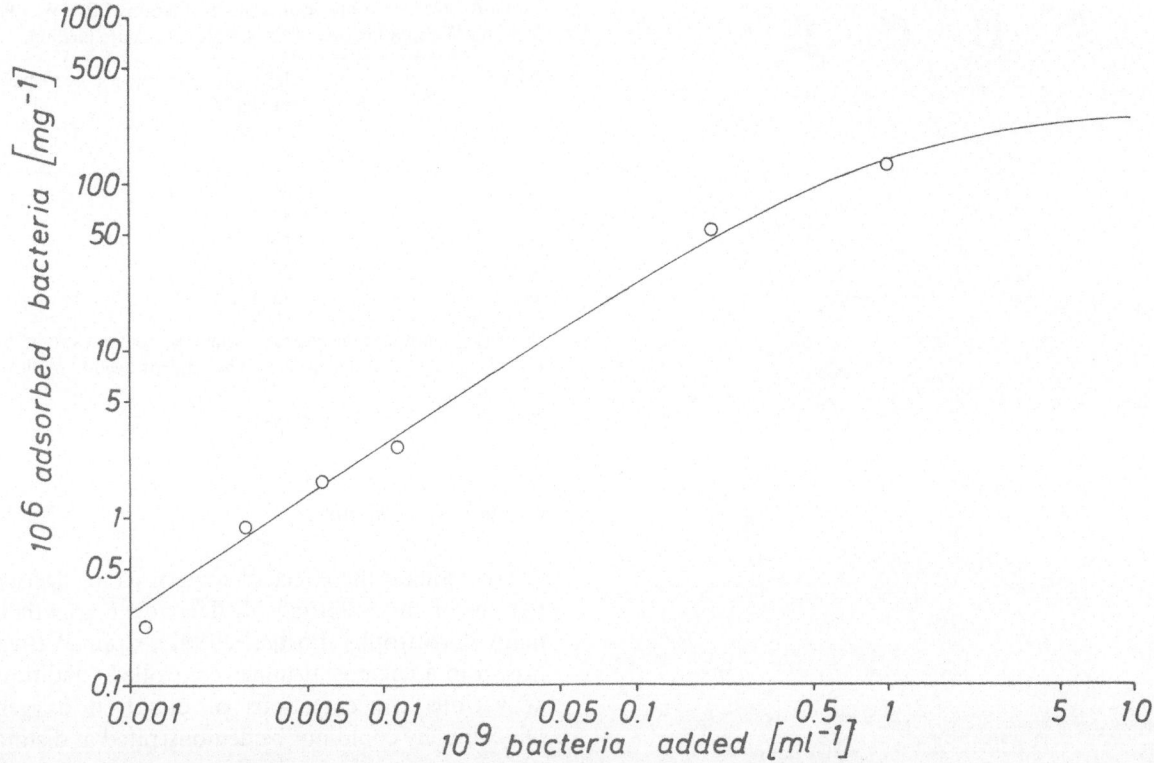

Fig. 2. Relationship between numbers of strain W1 added and numbers of bacteria bound to roots of kallar grass. Inoculation time was 3 h. Bacterial numbers are averages of three replicates and expressed as number of bacteria bound per mg root dry weight (maximum standard deviation 25%). Specific radioactivity was 7.2×10^{-4} counts. min^{-1} per cell.

nitrogen gain was detected, when the diazotrophic strain ER8 was grown under headspaces of air and also of synthetic air, which did not support growth of the other strains. Thus strains with the designation W had the ability to scavenge effectively traces of combined nitrogen occurring as impurities in the atmosphere, when they had access to laboratory air.

For confirmation, ^{15}N-labelled strains W1 and W7 were grown under comparable conditions with the growth tube sealed by a rubber stopper. Both strains diluted ^{15}N significantly by incorporation of nitrogen (Table 2). However, nitrogen gain was too high to be explained by initial impurities of combined nitrogen in the air alone.

In order to evaluate the possible involvement of the rubber stopper in the release of combined nitrogen, strain W1 was grown in an atmosphere of 20% high purity O_2 and 80% high purity N_2 enriched with 25% ^{15}N. After 96 h incubation at 37°C we found a nitrogen gain of $25 \pm 4 \mu g\,N$ but no sig-

nificant ^{15}N-enrichment $(0.38 \pm 0.03\%\quad ^{15}N)$. Hence our rubber stoppers released a significant amount of combined nitrogen, which could be incorporated.

To confirm previous results that these bacteria do not reduce acetylene in N-free media although a typical pellicle can be formed (Reinhold *et al.,* 1986), strain W1, and the diazotrophic strain ER8 as a N_2-fixing positive control, were incubated under various concentrations of acetylene, and the ethylene formed, as well as the nitrogen gain, were measured. Both strains formed a subsurface pellicle and incorporated nitrogen significantly although only strain ER8 formed ethylene, the unpurified acetylene possibly acting to some extent as a nitrogen source for W1 (Fig. 5). Thus no evidence for the involvement of nitrogenase in the increase of nitrogen incorporation for strain W1 was found.

The ability of our strains to make use of combined nitrogen applied as atmospheric NH_3 and N_2O is shown in Table 3. No growth and no nitro-

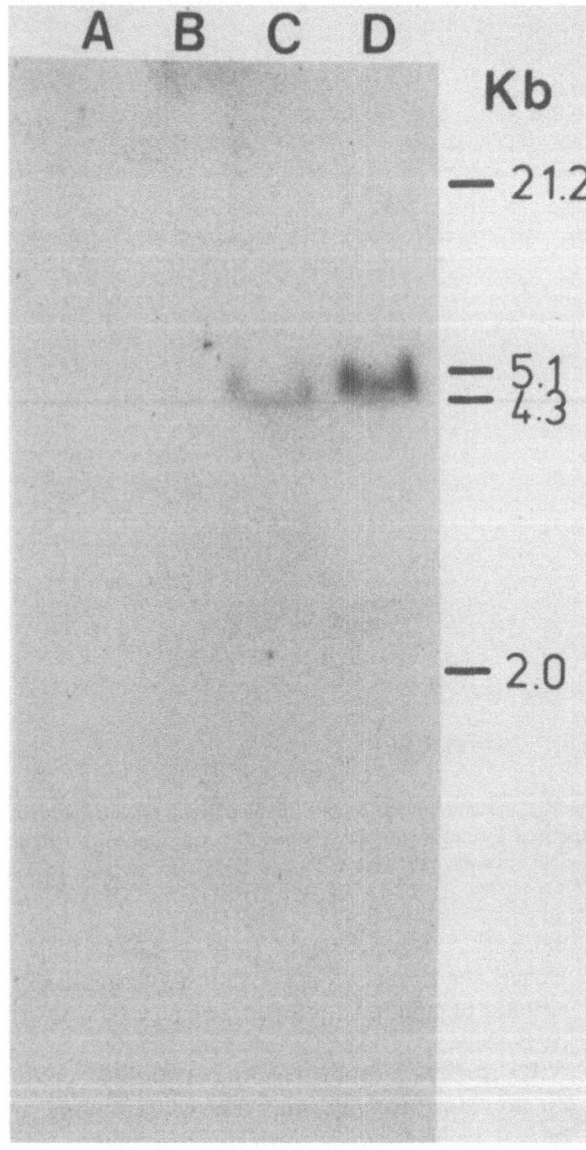

Fig. 3. EcoRI restriction digests of DNA from strains W1 (4 μg A, 12 μg B) and *Azospirillum brasilense* 68 (4 μg C, 12 μg D) probed with [³²P]dCTP-labelled pBRSA30 containing the *nif*HDK structural genes for nitrogenase components from *Klebsiella pneumoniae*.

gen gain were detected when strains W1 and W7 were incubated under a headspace of high purity N₂ and O₂, the vial being sealed with a teflon-coated cap. In contrast to this result a significant nitrogen incorporation was found for both strains when the atmosphere was supplemented with different amounts of NH₃ and N₂O.

Table 1. Increase in nitrogen content of strains ER8, W1, W4, W5 and W7 in a nitrogen free, semi-solid malate medium

Bacterial strain	Nitrogen gain under a headspace[a] of (μg N)	
	air	synthetic air[b]
ER8	35 ± 4	38 ± 3
W1	43 ± 10	≤ 1
W4	50 ± 5	≤ 1
W5	44 ± 4	≤ 1
W7	42 ± 6	≤ 1

[a]Means ± standard deviation. Data represent means of five replicates each. For strains W1 an average of twendy replicates is given.
[b]Gasses of high purity were used.

Growth in a chemostat

To evaluate the effect of oxygen as an decisive parameter in exhibiting N₂-dependent growth in many diazotrophs (Postgate, 1982), strain W1 was grown in a chemostat under controlled conditions at various concentrations of dissolved oxygen. Diazotrophy could not be demonstrated at distinct concentrations of low O₂. When concentrations of dissolved O₂ of 10 μM O₂, 1 μM O₂ and ≤ 5 nM O₂ were applied, all cultures were diluted out and no steady state was established.

Discussion

In accordance with our previous studies (Reinhold *et al.*, 1986) we found no evidence for diazotrophy of strains with the designation W: (i) there was not growth on N-free, semi-solid medium, when gases of high purity were used; (ii) no ¹⁵N₂ was incorporated when bacteria were incubated under an atmosphere enriched with ¹⁵N; (iii) cultures were washed out when they were grown in a chemostat at a dilution rate of 0.08 h⁻¹ in a N-free medium at low concentrations of dissolved O₂; (iv) no acetylene was reduced although growth in form of a subsurface pellicle was accompanied by a significant nitrogen gain; (v) no homology with DNA containing sequences for the polypeptide of the Fe-protein component of nitrogenase (*nif*H) and the alpha (*nif*D) and beta subunits (*nif*K) of the MoFe-protein component of the enzyme could be detected.

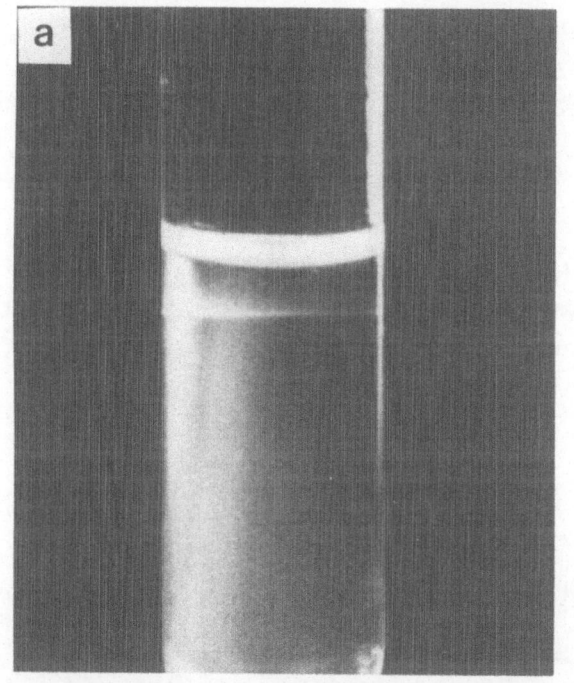
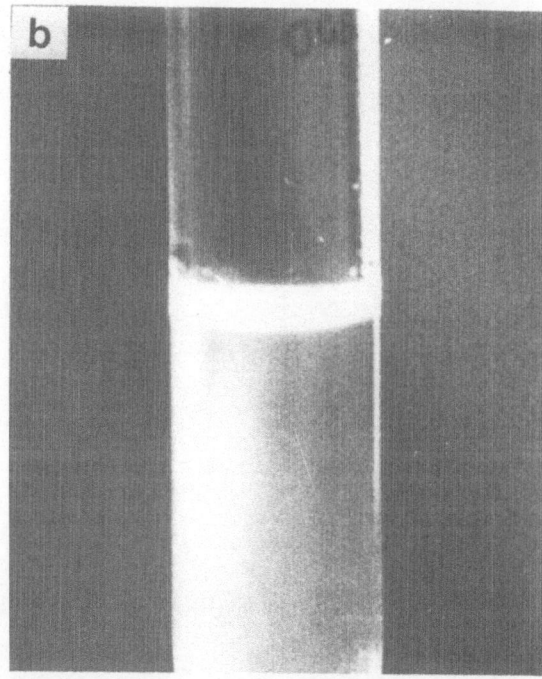

Fig. 4. Growth of strain W1 in form of a subsurface pellicle in N-free, semi-solid malate medium under normal air. Appearance after an incubation time of 10 h (a) and 24 h (b).

However, impressive growth in a N-free, semi-solid malate medium occurred when high purity gases were not used and the 15-ml vial was sealed with a rubber stopper. A similar phenomenon has been described by Hill and Postgate, 1969 for several strains from different places and attributed to the feature of these organisms to simulate N_2-fixation by scavenging traces of combined nitrogen, probably ammonia and oxides of nitrogen from the air. Also, Zolg and Ottow, (1975) described a nitrogen-scavenging *Pseudomonas* species which can grow in 5 days on agar plates prepared with re-distilled water and incubated under air washed with $1 N$ H_2SO_4. In our case, possibly the acetylene, which is known to contain NH_3 as an impurity, (Turner and Gibson, 1980) and clearly our rubber stoppers, harboured additional nitrogenous nutrients for our bacteria. Hence we regard our strains with the designation W to be effective nitrogen scavengers.

The extensive colonization of the rhizosphere of kallar grass in Pakistan by nitrogen scavenging strains in even higher numbers than those of *Azospirillum* spp. (Reinhold *et al.*, 1986, 1987b), results in a close contact of these bacteria with kallar grass roots. A basis for this phenomenon may be the rapid and strong adsorption of strain W1 to roots of this plant. Maximum attachment of *A. brasilense* Cd to corn roots occurred after 4.5 h of incubation (Okon and Kapulnik, 1986) and of *Rhizobium leguminosarum* J357 to pea roots after 8 h (Kato *et al.*, 1980), whereas we found maximum attachment after only 2–3 h. Furthermore, the high numbers of cells bound to the root at high concentrations of

Table 2. ^{15}N-dilution by strains W1 and W7 grown in a nitrogen-free, semi-solid malate medium under a headspace of air

Bacterial strain	[15]N % of the inoculated[a] medium at the beginning	Nitrogen gain[a] (μg N)	[15]N % of the medium at the[a] end of inoculation
W1	19 ± 0.5	47 ± 4	5.1 ± 0.5
W7	19 ± 2	43 ± 5	4.8 ± 0.6

[a]Means ± standard deviation. Data represent means of three replicates each.

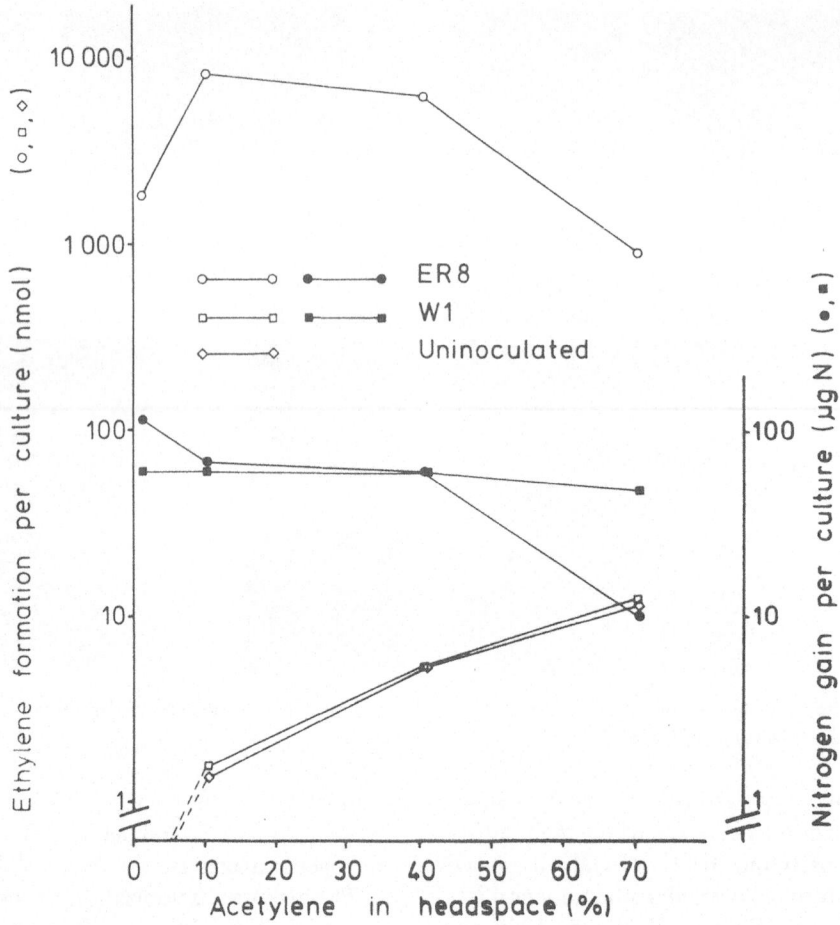

Fig. 5. Ethylene formation and increase in nitrogen content of strains W1 and ER8 at different concentrations of acetylene in a N-free, semi-solid malate medium. Headspace consisted of 20% industrial grade O_2 and 80% industrial grade N_2. Nitrogen was replaced by acetylene. Data represent means of three replicates each (maximum standard deviation 25%).

Table 3. Growth of strains W1- and W7 at various concentrations of N_2O and NH_3 in headspace in a semi-solid, nitrogen-free malate medium within 72 hours

Bacterial strain	Supplement to[a] headspace	Nitrogen gain[b] (μg N)
W1	–	$\leqslant 1$
	0.5% N_2O	8 ± 2
	2.5% N_2O	10 ± 3
	0.5% NH_3	10 ± 1
	5.0% NH_3	224 ± 29
W7	–	$\leqslant 1$
	0.5% N_2O	11 ± 2
	2.5% N_2O	34 ± 4
	0.5% NH_3	21 ± 4
	5.0% NH_3	44 ± 6

[a]Headspace consisted of 10% high purity O_2 and 90% high purity N_2; gas supplements replaced nitrogen gas.
[b]Means $\pm$ standard deviation. Data represent means of three replicates each.

cells in the binding mixture, although roots were washed intensively, indicated that adsorption was not as weak as suggested for the binding of *A. brasilense* Cd to wheat roots (Bashan *et al.,* 1986).

Nitrous oxide (N_2O) is the most prevalent oxide of nitrogen in the atmosphere and occurs there in concentrations of about $450 \,\mu$g m^{-3} (Robinson and Robbins, 1972). N_2O concentrations in the soil air of up to 0.65% (v/v) have been reported (Dowdell and Smith, 1974), the concentration of N_2O increasing with moisture (Flühler *et al.,* 1976a,b) and being greater in soils with zero-tillage than in tilled soil (Burford *et al.,* 1981). It can be formed by denitrification, nitrification (Bremner and Blackmer, 1978, 1979) and chemical decomposition of nitrite in soils (Smith and Chalk, 1980; Van Cleemput *et al.,* 1976). The site in Pakistan, from which the nitrogen-scavenging bacteria were isolated, was often flooded, not tilled and had a pH of 9.7 which also favours volatilization of ammonia.

Growth of strains W1 and W7 under an atmosphere enriched with NH_3 and N_2O may indicate that they can make use of these nitrogen sources. Since N_2O-utilization has been demonstrated for diazotrophs by nitrogenase alone (Hardy and Knight, 1966) or by a dissimilation nitrous oxide reductase together with nitrogenase (Stephan *et al.*, 1984), the preliminary evidence for N_2O-utilization of probably non-diazotrophic bacteria presented here has to be further confirmed by $^{15}N_2O$ studies.

The ability of these bacteria to scavenge nitrogen in pure culture makes it likely that they also scavenge nitrogen at the natural site in Pakistan. Although it appears unlikely that plants can make use of this nitrogen immediately, we think one has to consider the possibility that they can make use of it at some time because of close contact with the bacteria. Hence, occurrence of high numbers of nitrogen scavenging bacteria in the rhizosphere of plants may result in a nitrogen gain, simulating N_2-fixation in ^{15}N-dilution and N-balance studies in the field.

Acknowledgements

We are endebted to G Erbfeige and Dr C Bunnenberg, Niedersächsisches Institut für Radioökologie, Hannover, FRG for carrying out the Cerenkovspectroscopy measurements for us.

References

Bashan Y, Levanony H and Klein E 1986 Evidence for a weak active external adsorption of *Azospirillum brasilense* Cd to wheat roots. J Gen. Microbiol. 132, 3069–3073.

Bergersen F J 1980 Measurement of nitrogen fixation by direct means. *In* Methods for Evaluating Biological Nitrogen Fixation. Ed. F J Bergersen. pp. 65–110. John Wiley and Sons, Chichester.

Bishop P E, Jarlenski D M L and Hetherington D R 1980 Evidence for an alternative nitrogen fixation system in *Azotobacter vinelandii*. Proc. Natl. Ac. Sci. USA 77, 7342–7346.

Bishop P E, Premakumar R, Dean D R, Jacobsen M R, Chisnell J R, Rizzo T M and Kopczynski J 1986 Nitrogen fixation by *Azotobacter vinelandii* strains having deletions in structural genes for nitrogenase. Science 232, 92–94.

Bremner J M and Blackmer A M 1978 Nitrous oxide: Emission from soils during nitrification of fertilizer nitrogen. Science 199, 295–296.

Bremner J M and Blackmer A M 1979 Effects of acetylene and soil water content on emission of nitrous oxide from soils. Nature (London) 280, 380–381.

Bunnenberg C, Kraul K and Kühn W 1987 Analysis of betanuclides by Cherenkov-spectrometry. Nucl. Instr. Meth. A255, 346–350.

Bunnenberg C, Kühn W and Glubrecht H 1984 Non-destructive analyses of aerosol filters by spectroscopy of Cerenkov radiation. Atomkernenergie Kerntech. 44, 315–316.

Burford J R, Dowdell R J and Crees R 1981 Emission of nitrous oxide to the atmosphere from direct drilled and ploughed clay soils. J. Sci. Food Agric. 32, 219–224.

Dowdell R J and Smith K A 1974 Field studies of the soil atmosphere. II. Occurrence of nitrous oxide. J. Soil. Sci, 25. 231–238.

Fåhraeus G 1957 The infection of clover root hairs by nodule bacteria studied by a simple glass slide technique. J. Gen. Microbiol. 16, 374–381.

Flühler H, Ardakani M S, Szuszkiewicz T E and Stolzy L H 1976a Field measured nitrous oxide concentrations, redox potentials, oxygen diffusion rates, and oxygen partial pressures in relation to denitrification. Soil Sci. 122, 107–114.

Flühler H, Stolzy L H and Ardakani M S 1976b A statistical approach to define soil aeration in respect to denitrification. Soil. Sci. 122, 115–123.

Graciolli L A and Ruschel A P 1981 Microorganisms in the phyllosphere and rhizosphere of sugar cane. *In* Associative N_2-fixation, Vol II. Ed. P B Vose and A P Ruschel. pp. 91–101. CRC-Press, Inc, Boca Raton.

Hardy R W F and Knight E 1966 Reduction of N_2O by biological N_2 fixing systems. Biochem. Biophys. Res. Commun. 23, 409–414.

Hill S and Postgate J R 1969 Failure of putative nitrogen-fixing bacteria to fix nitrogen. J. Gen. Microbiol. 58, 277–285.

Hurek T, Reinhold B, Fendrik I and Niemann E-G 1987 Rootzone-specific oxygen tolerance of *Azospirillum* spp. and diazotrophic rods closely associated with Kallar grass. Appl. Environ. Microbiol. 53, 163–169.

Kato G, Maruyama Y and Nakamura M 1980 Role of bacterial polysaccharides in the adsorption process of the *Rhizobium*-pea symbiosis. Agric. Biol. Chem. 44, 2843–2855.

Kloss M, Iwannek K-H, Fendrik I and Niemann E-G 1984 Organic acids in the root exudates of *Diplachne fusca* (Linn.) Beauv. Environ. Exp. Bot. 24, 179–188.

Nilsson P E 1957 Aseptic cultivation of higher plants. Arch. Mikrobiol. 26, 285–301.

Okon Y and Kapulnik Y 1986 Development and function of *Azospirillum*-inoculated roots. Plant and Soil 90, 3–16.

Patriquin D G and McClung C R 1978 Nitrogen accretion and the nature and possible significance of N_2 fixation (acetylene reduction) in a Nova Scotian *Spartina alterniflora* stand. Marine Biol. 47, 227–242.

Postgate J R 1982 The Fundamentals of Nitrogen Fixation. Cambridge University Press, Cambridge.

Reinhold B, Hurek, T, Baldani I and Döbereiner J 1988 Temperature and salt tolerance of *Azospirillum* spp. from salt-affected soils in Brazil. *In Azospirillum* IV: Genetics, Physiology, Ecology. Ed. W Klingmüller. pp 234–241. Springer Verlag, Berlin.

Reinhold B, Hurek T and Fendrik I 1985 Strain-specific chemotaxis of *Azospirillum* spp. J. Bacteriol. 162, 190–195.

Reinhold B, Hurek T and Fendrik I 1987a Cross reaction of predominant nitrogen-fixing bacteria with enveloped, round bodies in the root interior of Kallar grass. Appl. Environ. Microbiol. 53, 889–891.

Reinhold B, Hurek T, Fendrik I, Pot B, Gillis M, Kersters K, Thielemans S and De Ley J 1987b *Azospirillum halopraeferens* sp. nov., a nitrogen-fixing organism associated with roots of Kallar grass *Leptochloa fusca* (L.) Kunth. Int. J. Syst. Bacteriol. 37, 43–51.

Reinhold B, Hurek T, Niemann E-G and Fendrik I 1986 Close association of *Azospirillum* and diazotrophic rods with different root zones of Kallar grass. Appl. Environ. Microbiol. 52, 520–526.

Robinson E and Robbins R C 1972 Emissions, concentrations and fate of gaseous atmospheric pollutants. *In* Air Pollution Control, part II. Ed. W Strauss. pp. 1–93. Wiley Interscience, New York.

Robson R L 1986 Nitrogen fixation in strains of *Azotobacter chroococcum* bearing deletions of a cluster of genes coding for nitrogenase. Arch. Microbiol. 146, 74–79.

Ruvkun G B and Ausubel F M 1980 Interspecies homology of nitrogenase genes. Proc. Natl. Ac. Sci. USA 77, 191–195.

Smith G J and Chalk P M 1980 Gaseous nitrogen evolution during nitrification of ammonia fertilizer and nitrite transformations in soils. Soil Sci. Soc. Am. J. 44, 277–282.

Stephan M P, Zimmer W and Bothe H 1984 Denitrification by *Azospirillum brasilense* Sp7 II. Growth with nitrous oxide as respiratory electron acceptor. Arch. Microbiol. 138, 212–216.

Thomas-Bauzon D, Weinhard P, Villecourt P and Balandreau J 1982 The spermosphere model. I. Its use in growing, counting, and isolating N_2-fixing bacteria from the rhizosphere of rice. Can. J. Microbiol. 28, 922–928.

Turner G L and Gibson A H 1980 Measurement of nitrogen fixation by indirect means. *In* Methods for Evaluating Biological Nitrogen Fixation. Ed. F Bergersen. pp. 111–138. John Wiley and Sons, Chichester.

Van Cleemput O, Patrick W H and McIlhenny R C 1976 Nitrate decomposition in flooded soil under different pH and redox potential conditions. Soil Sci. Soc. Am. J. 40, 55–60.

Zafar Y, Ashraf M and Malik K A 1986 Nitrogen fixation associated with roots of Kallar grass (*Leptochloa* (L.) Kunth). Plant Soil 90, 93–105.

Zolg W and Ottow J C G 1975 *Pseudomonas glathei* sp. nov., a new nitrogen-scavenging rod isolated from acid lateric relicts in Germany. Z. Allg. Mikrobiol. 15, 287–299.

GENERAL SUMMARY AND CONCLUSIONS

Various aspects of N_2-fixation associated with non-legumes, many of which were considered controversial at the 3rd International Symposium in Finland in 1984, became much clearer at this meeting and many possibilities of practical usefulness of these systems began to emerge. It certainly became clear that a better understanding of the mechanisms of plant-diazotroph interactions is essential for their better exploitation.

The infection mechanisms of the legume-*Rhizobium* symbiosis were discussed in comparison with other less perfect associations and the advantage of interdisciplinary approaches to such research became clear. It was shown that in legumes bacteria can penetrate live plant cells only if there is no secondary cell wall, and rhizobia do not have to be released from the infection thread in order to fix N_2.

Primitive legume species, including all of the Caesalpinioideae studied so far and a few of the Papilionoideae, form nodules in which, in common with *Parasponia*, the bacteria are not released from the infection threads but still fix N_2 efficiently. *Parasponia* was reported to fix as much as $280 \, \mathrm{kg}$ $N \cdot ha^{-1} \cdot year^{-1}$, but this plant is still the only known member of a non-leguminous genus able to nodulate with *Rhizobium*.

Other non-legumes which nodulate with *Frankia* represent another link between the legume symbiosis and the looser associations. Considerable progress has been made in cross-inoculation studies, which provide a basis for inoculation of such species. Within the Rhamnaceae all members have been shown to nodulate except *Adolfii* spp.

Major achievements have been made in the *Azolla* symbiosis. Using the decapitated megaspore technique it is now possible to construct heterologous *Azolla/Anabaena azollae* symbioses which permit the study of the relative contributions of the two partners. An *Anabaena* isolate from *Azolla* was shown to have an unusually high heterocyst frequency and nitrogenase activity. More applied aspects of the *Azolla* symbiosis, particularly the spore germination technique, are also of great interest. Another finding was that the *Azolla-Anabaena azollae* symbiosis may involve a third partner. Bacteria, identified as *Arthrobacter* species which are consistently observed in the fern apex, live in the fern leaf cavities.

Further studies of heterotrophic microorganisms have shown that the nitrogen-fixing trait is more widespread taxonomically than previously supposed, and new N_2-fixing species and even genera have been isolated and classified. The screening for new diazotrophic taxa deserves to be continued. However, the recognition of a new N_2-fixing species or genus must be made very cautiously on the basis of several isolates. The general rules, as defined for example in Bergey's Manual, have to be applied when the creation of a new taxon is suggested.

The cosmopolitan character of *Azospirillum* spp. has been confirmed. Azospirilla exist in various climatic conditions, and survive relatively well even during the winter in temperate regions. However, the existing differences in the densities of *Azospirillum* populations between tropical and temperate soils indicate the preference of azospirilla for tropical conditions.

Investigations of various physiological aspects of growth and nitrogen fixation of root-associated diazotrophs were reported. A new diazotroph from sugar cane, *Acetobacter diazotrophicus*, possessed the following very remarkable properties: (a) diazotrophic growth at very low pH; (b) osmotolerance to high sucrose concentrations (up to 30%); (c) high O_2 tolerance for N_2-fixation and (d) lack of nitrate reductase.

The osmotolerant growth of a *Klebsiella* strain NIAB 1 from the rhizosphere of kallar grass was shown to be mediated by a plasmid which could express this trait in other members of the Enterobacteriaceae. The metabolosm of L-arabinose by *Herbaspirillum seropedicae* was investigated and shown to be similar to that of *Azospirillum brasilense*. The metabolism of indoleacetic acid by *A. brasilense* and by *Zea mays* infected with the same species was studied using gas chromatography/mass spectrometry and the roots were found to be the predominant site of auxin metabolism. The production of nitrite by *A. brasilense* apparently mimics auxin activity under certain conditions. This suprising result shows that the nitrogen metabolism of root-associated bacteria can be associated with morphological effects on plant roots.

Several prototrophic mutants, capable of excreting ammonium during diazotrophic growth, were isolated from *A. brasilense*. Characterization studies suggested a defect in the de-adenylation system for glutamine synthesis.

Azospirillum spp. have very different abilities to mobilize and assimilate ferric iron for growth. Some species can use siderophores from other microbes and plants to support growth under iron-deficient conditions. Efficient iron uptake may be very important for the competitive ability of the strains, and genetic improvement of this trait appears to be highly desirable.

Research on the regulation of nitrogen metabolism and nitrogen fixation in root-associated diazotrophs should be particularly rewarding in isolates from the interior tissue of roots, such as the 'straight rods' isolated from kallar grass. The conditions of the microhabitat in the roots and the occurrence of plant-derived substances such as amino acids or their derivatives, phenolics, oligosaccharides or even plant hormones, may be involved in the eventual differentiation from a free-living to a root endophyte state characterized by associative N_2-fixation.

Genetic studies on these recently discovered nitrogen-fixing bacteria are only just beginning. However, the techniques for performing genetic and molecular analysis of gram-negative and gram-positive bacteria can now easily be transposed to newly discovered organisms. In addition, the molecular basis of nitrogen fixation and of its regulation, established for *Klebsiella pneumoniae* M5ai, provides a useful model for the genetic analysis of new bacteria.

This model was indeed illustrated during the Symposium when information on the organization of *nif* DNA was given for four bacterial species: *Azospirillum brasilense, Herbaspirillum seropedicae, Enterobacter* sp. and *Bacillus azotofixans*. In *Enterobacter* the *nif* genes were found to be contiguous and organized in the same order as in *K. pneumoniae*, except for *nifJ*. An unusual feature of the system is the plasmid location of the *nif* genes in rhizosphere bacteria.

Attempts have been made to identify genetic determinants involved in the bacteria-plant interaction, taking as a model system, genes involved in

the interaction of *Rhizobium* and *Agrobacterium* with plants. Gene banks of *Azospirillum* total DNA were screened by hybridization with heterologous probes or by genetic complementation. At present it is not known if the various plasmid clones isolated carry genes involved in the bacteria-plant interaction. However, it is possible by transposon site-directed mutagenesis to mutate the cloned regions in the original host genome. It remains to correlate the mutations obtained with the modification of the plant phenotype in order to identify the function of the cloned DNA. *Azospirillum brasilense* Sp7 is now the best known strain genetically although it may not have a particularly great ability to interact with plants. As the number of known nitrogen-fixing species has increased tremendously in the last few years, it is difficult for the geneticist to choose a model strain for each particular species. For progress to be made it is important that geneticists concentrate their efforts on one and the same strain, as occurred in the case of strain M5a1 of *K. pneumoniae*.

Evidence was presented to show that there are significant differences in terms of chemotactic behaviour and O_2 tolerance between the populations of N_2-fixing bacteria isolated from the rhizoplane and rhizosphere of grasses and cereals. It was shown that certain enteric N_2-fixing bacteria were capable of colonizing various temperate grasses, and that fimbriae were always expressed during this colonization. Such plants exhibited associated nitrogenase (acetylene reduction) activity but only occasionally was firm evidence obtained (using ^{15}N techniques) that fixed N was transferred to the host plant.

More information concerning the establishment of antibiotic-labelled *Azospirillum* (and for the first time, *Herbaspirillum*) strains in cereal roots was obtained and this was found to correlate with observed growth responses. It was apparent however, that other rapid techniques for identifying bacterial strains in roots are needed, and some progress in this direction was reported with various serological techniques, although none yet developed has sufficient specificity to differentiate individual strains.

Results obtained using the ^{15}N natural abundance technique for quantification of BNF were presented, and showed that with careful use this technique could be most useful for such quantification in undisturbed ecosystems. Only two papers were presented on quantification of BNF associated with grasses (sugar cane and kallar grass) and it was suggested that for this area of research to be based on a firm footing many more studies were required.

It was shown that even the apparently simple task of identifying bacteria isolated from kallar grass roots as N_2-fixers can be fraught with difficulties. These organisms gave all signs of being N_2-fixers; they formed a pellicle in N-free semi-solid media and gained nitrogen, and when grown in ^{15}N-labelled substrates under air they presented an isotope dilution. They did not reduce acetylene, however, but were efficient scavengers of combined nitrogen and were eventually shown not to be able to fix atmospheric N_2.

Field studies on lowland rice showed considerable differences in nitrogenase activity between cultivars and the effect of combined nitrogen on this activity. Although combined nitrogen can inhibit plant-associated nitrogenase activity in model greenhouse systems (solution culture etc) in rice, sorghum and maize, the data from field studies on wetland rice show that the depression in activity may be temporary. As the N fertilizer stimulates growth of the plant the subsequent nitrogenase activity may even be greater on a per plant or area basis when N-deficient soil conditions return.

Yields of various cultivars of pearl millet were shown to respond to inoculation with *Azotobacter* and *Azospirillum* spp. in many experimental sites in India. Data from studies performed in model systems show that maize and sorghum can support relatively high associated nitrogenase activity (0.5–$1.5\,\mu$ mole C_2H_4 g root^{-1} h^{-1}) under microaerobic conditions similar to those sometimes experienced in the field and that there are considerable differences in activity between different plant genotypes and bacterial strains. However, there are many indications that nitrogen fixation is often not the most important factor responsible for the yield responses observed in the field.

It is apparent that there is a strong need to continue to investigate the mechanisms through which indigenous bacteria, or those added as inoculants, exert a beneficial or detrimental effect on plant growth. It is only through such an understanding that the ability to manipulate these systems for agronomic benefit may eventually be achieved.

Titles of posters presented at the meeting

Session 1. Mechanisms of infection of plants by N_2-fixing microorganisms

1. S. M. de Faria, and J. I. Sprent, Epidermial infection of *Mimosa scabrella* by rhizobia
2. S. M. de Faria, and J. I. Sprent, A comparison of infected cells in some woody legumes with those of *Parasponia*

Session 2. *Frankia* and *Parasponia* symbiosis

3. L. Longeri and M. Abarzua, Ultrastructure of *Frankia* isolated from *Discaria trinervis, D. serratifolia* and *Retanilla ephedra*
4. B. Sougoufara, E. Duhoux, M. Corbasson and Y. Dommergues, Improvement of nitrogen fixation by *Casuarina equisetifolia* through clonal selection

Session 3. *Azolla* and cyanobacteria

5. J. R. Milam, S. L. Albrecht, C. Latorre and K. T. Shanmugam, Effect of inoculation of ammonia-excreting cyanobacteria on rice
6. W. Balloni, F. Favilli and M. C. Margheri, Field experiments of bacterization with N_2-fixing photosynthetic bacteria of tomato plants
7. L. Giovanetti, L. Tomaselli, F. Favilli and S. Biagiolini, Soil colonization by N_2-fixing cyanobacteria
8. F. Favilli, S. Taruntoli and L. Giovannetti, *Azolla* biomass as biofertilizer for the growth of wheat, maize and sunflower
9. M. R. Tredici, L. Vagnoli, S. Biagiolini and M. Vincenzini, Salinity tolerance in *Azolla* spp.
10. L. Tomaselli, M. C. Margheri, L. Giovanetti, C. Sili and F. Bocci, Consideration on the taxonomy of *Azolla* spp. cyanosymbionts
11. N. Boonkerd, S. Choonluchanon and P. Swatdee, Preservation and germination of *Azolla* spores
12. J. S. Yuncs, M. T. Suzuki, H. Z. Souza and M. C. Schreiber, Preliminary assays on nitrogenase activity in the cyanobacterium *Anabaena* sp. strain RST 87 OI isolated from flooded soils of Taim, RS, Brazil

Session 4. Isolation and identification of plant-associated N_2-fixing bacteria

13. J. M. Harris, M. R. Davey, J. A. Lucas and K. A. Powell, The survival of *Azospirillum brasilense* in temperate soil
14. E. Paiva, M. E. Carvalho and M. J. V. V. D. Peixoto, Immunoelectrophoresis characterization of species and strains of *Azospirillum*

Session 5. Physiology of *Azospirillum* and other diazotrophs

15. J. A. Qureshi, P. Haq and K. A. Malik, Osmoregulation in *Klebsiella* spp.
16. J. M. C. S. Dias, M. C. R. Facciotti, W. Schmidell, G. P. A. Prado and I. Balleroni, Activation energy determination for the growth of *Azospirillum brasilense* sp245 on fructose or glycerol
17. A. L. Mathias, L. U. Rigo, S. Funayama and F. O. Pedrosa, L-Arabinose metabolism by *Herbaspirillum seropedicae*
18. K. R. S. Teixeira, M. P. Stephan and J. Döbereiner, Physiological studies of *Saccharobacter nitrocaptans*, a new acid-tolerant N_2 fixing bacterium
19. D. M. S. Mano, W. de Souza and T. Langenbach, Morphological modifications of *Azospirillum lipoferum* resulting from exposure to the organochloride dicofol
20. J. M. Jasmin, P. Arruda, A. M. Monteiro, A. Crozier and G. Sandberg, Endogenous indoles and metabolism of indole-3-acetic acid in cultures of *Azospirillum brasilense*
21. M. R. Alonso and M. C. Marzocca, Cultures of *Azospirillum* spp. with ethanol as the sole carbon source
22. P. Gamard, G. Laguerre and R. Bardin, Use of the fluorescent antibody and the Elisa techniques for *Azospirillum lipoferum* population studies
23. K. R. S. Teixeira, M. P. Stephan and J. Döbereiner, Effect of oxygen on the respiratory rate and nitrogen fixation of a new bacterium belonging to the group of acetic acid bacteria

Session 6. Genetics of *Azospirillum* and other diazotrophs

24. H. Bueno Machado, S. Funayama, L. U. Rigo and F. O. Pedrosa, Isolation and genetic characterization of *Azospirillum brasilense* mutants derepressed for nitrogenase activity
25. E. M. de Souza, S. Funayama, L. U. Rigo and F. O Pedrosa, Cloning of *nif*-A-like gene of *Herbaspirillum seropedicae*

336

Session 7. Mechanisms of association of N_2-fixing bacteria with grasses and cereals

26. C. H. Bellone and M. A. Monzon de Asconegui, Electron microscopy observations of *Azospirillum* spp. cells within sugar cane (*Saccharum* sp. L.) roots
27. S. L. Albrecht, D. H. Hubbell and M .H. Gaskins, Effect of energy sources on nitrogen fixation by free-living bacteria associated with the roots of grass plants
28. B. D. de Lima, E. G. C. Penido and J. Döbereiner, Inoculation *in vitro* of wheat straw cultures with *Bacillus azotofixans*
29. J. O. Mandel, J. I. Baldani and J. Döbereiner, Oil, a new possibility to viabilize the inoculation of cereals with *Azospirillum* sp.
30. P. Gamard, T. Dorji, C. Steinberg and R. Lensi, Denitrifying activity and population dynamics of an *Azospirillum* strain inoculated into a sterilized soil
31. R. Bilal, M. Arshad and K. A. Malik, Root colonization studies of kallar grass by fluorescent antibody technique

Session 8. Mechanisms of response of cereal crops to *Azospirillum* inoculation

32. E. Fallik, Y. Okon and M. Fischer, Effect of *Azospirillum brasilense* on maize root seedlings function
33. M. Fulcieri and L. Frioni, *Azospirillum* inoculation in corn
34. A. Pidello and G. Chapo, Changes in some rhizosphere soils parameters in soil-plant systems inoculated with *Azospirillum* sp7
35. E. Perotti, L. Menendez, G. Rivera and A. Pidello, *In vitro* survival of *Azospirillum* sp7 in relation to different bioedaphic conditions in a vertic molisol of humid pampa
36. V. L. Baldani, J. I. Baldani, J. Mandel, R. Rocha and J. Döbereiner, Inoculation of field grown wheat with *Azospirillum brasilense* inoculants in peat and oil

Session 9. Quantification of plant-associated biological nitrogen fixation

37. A. C. de Oliveira and I. E. Marriel, The effect of sampling methods on nitrogenase activity in excised roots of maize
38. H. A. Burity, M. A. Faris, B. C. Coulman, T. C. Ta and U. M. T. Cavalcanti, Dinitrogen fixation and nitrogen transference in alfalfa, red clover and birdsfoot trefoil associated with grass pastures
39. C. B. Miranda, J. C. G. Costa, S. S. Urquiaga and R. M. Boddey, Quantification of biological nitrogen fixation associated with 25 genotypes of *Panicum maximum*

Session 10. Effects of plant genotype and environmental factors on plant-associated N_2-fixation

40. G. P. Ardizzi, H. Giorgetti, R. Martinez, H. Iglesias, F. Margiotta and O. Montenegro, Effect of inoculation with *Azospirillum* spp. on field-grown wheat in arid soils (preliminary results)

Developments in Plant and Soil Sciences

31. N. J. Barrow, Reactions with variable-charge soils. 1987. ISBN 90-247-3589-0
32. D. P. Beck and L. A. Materon, Eds., Nitrogen fixation by legumes in Mediterranean agriculture. 1988.
ISBN 90-247-3624-2
33. R. D. Graham, R. J. Hannam and N. C. Uren, Eds., Manganese in soils and plants. 1988.
ISBN 90-247-3758-3
34. J. G. Torrey and J. L. Winship, Eds., Applications of continuous and steady-state methods to root biology.
1989. ISBN 0-7923-0024-6
35. F. A. Skinner, R. M. Boddey and I. Fendrik, Eds., Nitrogen fixation with non-legumes. 1989.
ISBN 0–7923–0059–9
36. B. C. Loughman, O. Gašparíková and J. Kolek, Eds.,
Structural and functional aspects of transport in roots. 1989. ISBN 0–7923–0060–2